Energy and Transport Processes

AA003090

Core Programming Topic at the 2011 AIChE Annual Meeting

Minneapolis, Minnesota, USA
16 – 21 October 2011

ISBN: 978-1-61839-728-7

Printed from e-media with permission by:

Curran Associates, Inc.
57 Morehouse Lane
Red Hook, NY 12571

Some format issues inherent in the e-media version may also appear in this print version.

Printed by Curran Associates, Inc. (2012)

AIChE
3 Park Avenue
New York, NY 10016-5991

Phone: (203) 702-7660
Fax: (203) 775-5177

www.aiche.org

Additional copies of this publication are available from:

Curran Associates, Inc.
57 Morehouse Lane
Red Hook, NY 12571 USA
Phone: 845-758-0400
Fax: 845-758-2634
Email: curran@proceedings.com
Web: www.proceedings.com

TABLE OF CONTENTS

(10a) Fermentation Cost Improvement Through Propagation Advances for a Commercially Relevant Biomass-to-Ethanol Process

Kate Brandon Sutton, B. Emme, L. Jones, S. McHatton, C. Kang, C. Corbett, L. Putnam and M. Torry-Smith, Novozymes North America Inc., Franklinton, NC

Compared to their starch-to-ethanol counterparts, pentose-utilizing organisms for biomass-to-ethanol processes are less robust while having to ferment mixed sugars in a significantly more inhibitory substrate. This inhibition can generally be overcome with a high yeast pitch to maximize xylose uptake rates in fermentation. The cost driven trade-off of a high yeast pitch during fermentation is the necessity of dedicated onsite propagation, as opposed to purchase of all yeast. Novozymes' Advance Biofuel cost model suggest the majority of the fermentation cost when using high yeast pitches can be attributed to the propagation variable costs, most significantly from the carbohydrate source. Therefore, maximizing the cost performance of C5 yeast propagation is important to ensure an economically feasible biomass-to-ethanol fermentation step.

Novozymes has identified two approaches to reduce the fermentation cost: first, increase propagation cell yields from sugars to reduce the realized organism cost, preferably on cheap carbohydrate sources, or second, produce a more robust organism from propagation with a higher realized xylose uptake rate in fermentation which would enable fermentation at a lower yeast pitch in a reasonable processing time. This presentation will discuss cost model results in conjunction with reactor data to pinpoint potential savings and opportunities to lower overall fermentation cost for second generation ethanol. Sensitivities on cost from propagation and fermentation process variables targets will be presented. Propagation optimization with respect to higher realized xylose uptake rate in fermentation will be addressed as well. Lastly, the holistic minimum ethanol selling price implications for process optimizations will be given.

See more of this Session: Biological Conversions and Processes for Renewable Feedstocks I
See more of this Group/Topical: Sustainable Engineering Forum

(10b) Combined Computational and Experimental Investigations of Processive Cellulases for Engineering Activity Improvements

Gregg T. Beckham[1], **Larry Taylor**[2], Michael F. Crowley[3], William S. Adney[3] and Michael E. Himmel[3], (1)National Bioenergy Center, National Renewable Energy Laboratory, Golden, CO, (2) Biosciences Center, NREL, Golden, CO, (3)Biosciences Center, National Renewable Energy Laboratory, Golden, CO

The Family 7 cellobiohydrolase (Cel7A) from *Hypocrea jecorina* (*Trichoderma reesei*) is one of the most well-characterized, active cellulases known for deconstruction of cellulose to soluble sugars. With the aim to engineer enhanced cellulase specific activity, we have investigated the Cel7A mechanism on cellulose with molecular dynamics simulations and biochemical experiments. We have elucidated new functions for the carbohydrate-binding module, determined the effect of glycosylation on the stability of the linker and carbohydrate-binding module, examined the role of aromatic and polar residues in the catalytic domain tunnel, and examined the structure of the catalytically-active complex of Cel7A on crystalline cellulose up to the microsecond time scale. These new insights combined with experimental activity data have led to a large-scale experimental screening of new enzymes to examine both natural diversity and engineered enzymes for activity and stability improvements. Our combined computational and experimental approach has demonstrated that unprecedented improvements in specific activities of processive cellulases are possible, which when combined in industrial cocktails, will result in lower enzyme loadings for cell wall deconstruction and lower overall production costs for renewable biofuels.

See more of this Session: Biological Conversions and Processes for Renewable Feedstocks I
See more of this Group/Topical: Sustainable Engineering Forum

(10c) Exploration of Significant Enzyme Activities Required for the Deconstruction of Ionic-Liquid Pretreated Biomass

Christopher J. Barr[1], Jeffery Mertens[2] and Constance Schall[1], (1)Chemical and Environmental Engineering, University of Toledo, Toledo, OH, (2)NCAUR, USDA-ARS, Peoria, IL

Terrestrial biomass is a second generation renewable feedstock, comprised of polysaccharides derived from cellulose and hemicellulose, which can be used to produce carbon-based chemicals and fuels. The recalcitrance to deconstruction of biomass is one of the main impediments in the conversion of biomass to monomeric sugars and the production of sugar platform fuels and chemicals. Ionic Liquid (IL) pretreatment creates unique biomass substrates composed of amorphous cellulose with largely intact hemicellulose and lignin. IL pretreated substrates differ from many other pretreatments which remove hemicellulose and retain cellulose crystallinity. This substrate poses an interesting challenge for enzymatic deconstruction using traditional commercial mixtures.

Mono-component cellulase and hemicellulase enzymes from *Aspergillus nidulans* have been expressed, purified, and characterized. Key enzyme activities and/or interactions that significantly contribute to the conversion of a variety of lignocellulosic biomass substrates to monomeric sugars were examined by enzyme combinations in series of hydrolysis experiments. Auxiliary hemicellulases seem to be essential to the mono-component enzyme mixtures. Investigation of these accessory enzymes also yields insight into an understanding of the differences in hemicellulose structure between feedstocks. Some auxiliary hemicellulases include acetyl xylan esterases (AXE), arabinofuranosidases, and α-glucuronidases. In addition to mono-component enzyme mixtures, next generation commercially-available enzyme mixtures were characterized and formulated using key activities to optimize hydrolysis at low enzyme loadings. These simplified mixtures will be dependent on the substrate. As part of these studies, an analysis of variance (ANOVA) was used to statistically model the main and interaction effects of key enzyme activities for each substrate.

See more of this Session: Biological Conversions and Processes for Renewable Feedstocks I
See more of this Group/Topical: Sustainable Engineering Forum

(10d) Inhibition of ENZYMATIC Hydrolysis by Soluble Sugars ON MODEL CELLULOSE Thin Film USING QUARTZ CRYSTAL Microbalance

Hsin-Fen Li[1], Ravinder Garlapalli[1], Michael D. Flythe[2], Sue E. Nokes[3], Stephen E. Rankin[1] and Barbara L. Knutson[1], (1)Chemical and Materials Engineering, University of Kentucky, Lexington, KY, (2)Department of Animal and Food Sciences, University of Kentucky, Lexington, KY, (3)Department of Biosystems and Agricultural Engineering, University of Kentucky, Lexington, KY

INHIBITION OF ENZYMATIC HYDROLYSIS BY SOLUBLE SUGARS ON MODEL CELLULOSE THIN FILM USING QUARTZ CRYSTAL MICROBALANCE

Hsin-Fen Li [1], Ravinder Garlapalli[1]. Michael D. Flythe [2,3], Sue E. Nokes [4], Stephen E. Rankin[1], Barbara L. Knutson[1]

[1]University of Kentucky, Department of Chemical and Materials Engineering, Lexington, Kentucky 40506

[2]USDA, Agricultural Research Service, Forage-Animal Production Research Unit; Lexington, Kentucky 40546

[3]University of Kentucky, Department of Animal and Food Sciences, Lexington, Kentucky 40546

[4]University of Kentucky, Department of Biosystems and Agricultural Engineering, Lexington, Kentucky 40506

Soluble sugars produced from the hydrolysis of cellulose are one of the most promising feedstocks for production of fuels and commodity chemicals from renewable plant-based resources. The recalcitrance of cellulose to degradation is well documented and attributed to multiple sources, including high cellulose crystallinity and the hydrogen binding framework that is embedded in the matrix substances (lignin and hemicellulose). Cellulase is a cellulytic enzyme that possess a catalytic domain capable of hydrolyzing the beta-1, 4-glycosidic bonds in cellulose. However, during biomass conversion to soluble sugars, its hydrolytic efficiency is limited by the presence of degradation products including cellobiose. To study the inhibitory effect of cellobiose and the interaction of the inhibited enzyme and the cellulose substrate, the change in mass and surface properties of cellulose thin films were measured using Quartz Crystal Microbalance with Dissipation (QCM-D). The cellulose thin film was synthesized from thermochemically dissolved cellulose and was coated on a quartz crystal resonator. The mass of the spin-coated cellulose film was detected by the resonance frequency, and the cellulose surface coverage was observed by atomic force microscopy (AFM). Cellulase adsorption, activity, and inhibition on model cellulose thin films as were measured directly as changes to the piezoelectric properties of the cellulose thin films by QCM-D. At 5 g/L, cellobiose completely inhibits cellulase (0.5 %v/v) activity. The inhibited cellulases bind reversibly to the cellulose and reduce subsequent enzymatic hydrolysis rates by more than 50%. Significantly higher activity is observed when cellulase concentration is decreased from 1 %v/v to 0.1 %v/v, suggesting that cellulase acts progressively on the cellulose chain, and the hydrolysis is restricted by saturated enzyme conditions.

See more of this Session: Biological Conversions and Processes for Renewable Feedstocks I
See more of this Group/Topical: Sustainable Engineering Forum

(10e) Selection and Characterization of Acid-Stable Cellulases Against Pretreated Oil Palm Empty Fruit Bunch From a Colombian Extreme Environment

Luis Miguel Medina[1], Luisa Cabezas[2], Isabella Bahamon[1], Laura Palma[1], Andrs Gonzlez[1] and Silvia Restrepo[2], (1)Chemical Engineering, Centro de Diseo de Productos y Procesos, Universidad de los Andes, Bogot, Colombia, (2)Biological Sciences, Laboratorio de Micologa y Fitopatologa (LAMFU), Universidad de los Andes, Bogot, Colombia

The rapid growth of the oil palm industry around the world generates environmental concerns, with a production of more than 28.0 million tons per year of oil palm around the world, generating a great amount of by-products and wastes, such as liquid effluent, gaseous emission, and solid waste materials. Nowadays, solid waste is used commonly as manure, nevertheless the composting process is reported to be complex[1]. Therefore this agro-industry should take into consideration the use of sustainable technology. A material balance of an oil palm plant yields, in terms of mass, for every Ton of fresh fruit processed 220 kg of oil palm empty fruit bunch (OPEFB), 125-130 kg of fiber, around 50 kg of shell and 195-200 kg of palm oil [2]. Studies on the lignocellulosic content of the OPEFB have shown that display a high cellulose and hemicellulose and low lignin content [2]. This feature allows to use this waste as a substrate to obtain simple sugars, which can be used as carbon source in different fermentation processes to enable the production of alcohols that can be used as fuels. The biodegradation of cellulose is carried out by three classes of enzymes: endoglucanase (carboxymethylcellulase), exogluconase (cellobiohydrolase) and β-glucosidase, which together with a preliminary treatment of cellulose, manage to break the glycosidic bonds to produce simple sugars [3]. However, these enzymes hydrolyze in different parts of the polymer so the relation between the substrate and the enzyme should not be extrapolated due to different lignocellulosic structure [4]. Taking all this into account the objective of this study was to explore the cellulolytic potential of endophytes from a Colombian extreme environment for the saccharification of oil palm empty fruit bunch. It was determined the cellulolytic capacity for the saccharification *in vitro* of the pretreated oil palm empty fruit bunch of one hundred Colombian native endophytes isolated from young leaves attacked by insects (with herbivory) as well with no symptom of attack (without herbivory) from *Espelieta grandiflora* and *E. corymbosa* plants from the Páramo Crúz Verde. The microorganisms were screened to asses which of them have cellulolytic capacity on solid agar with CMC as substrate followed by Congo red staining method [5]. The microorganisms that showed cellulolytic activity were grown on cellulase inducting medium, using pretreated OPEFB as substrate, the activity of the three main components of cellulase (CMCase, β-glucosidase and Cellobiohydrolase) was measured, and total cellulolytic activity (Filter paper activity. FPA), using the procedures described by Zhang *et. al (2009)*. Candidates obtained, (two entophytes), were partially purified through partial precipitation with ammonium sulfate (90%) [6] and chromatography using DAE-shepadex A-50 column previously equilibrated with sodium acetate buffer (pH 5, 0.05 M) and the fractions were withdrawn with a linear NaCl gradient, the fractions with cellulolytic activity were pooled and then concentrated by ultrafiltration with a centriprep (Amicon) having a 10 kDa molecular mass cut-off. The screenings allowed us to identify 28 endophytes with cellulolytic activity, and the best obtained were a *Penicilium sp.* and an *Aspergillus sp.* with maximum enzyme activities, CMCase, FPase, exogluconase and β-glucosidase, of 0.50-0.16, 0.035-0.032, 0.87-0.99 and 0.068-0.01 U/ml respectively. CMCase and the FPA activities of the partially purified cellulase was assessed at pHs ranging from 1 to 10, and temperatures ranging from 30 to 80 °C in order to find the optimum conditions for the cellulases. The cellulases from *Penicillium sp. and Aspergillus sp.* were stable at low pH which could be a valuable tool for hydrolysis of cellulose under acidic conditions.

References

1. Yusoff, S., *Renewable energy from palm oil - innovation on effective utilization of waste.* Journal of Cleaner Production, 2006. **14**(1): p. 87-93.

2. Umikalsom, M.S., et al., *The treatment of oil palm empty fruit bunch fibre for subsequent use as substrate for cellulase production by Chaetomium globosum Kunze.* Bioresource Technology, 1997. **62**(1-2): p. 1-9.

3. Sun, Y. and J. Cheng, *Hydrolysis of lignocellulosic materials for ethanol production, a review.* Bioresource Technology, 2002. **83**(1): p. 1-11.

4. Zhang, Y.H.P., M.E. Himmel, and J.R. Mielenz, *Outlook for cellulase improvement: Screening and selection strategies.* Biotechnology Advances, 2006. **24**: p. 452-481.

5. Kasana, R., et al., *A Rapid and Easy Method for the Detection of Microbial Cellulases on Agar Plates Using Gram's Iodine.* Current Microbiology, 2008. **57**(5): p. 503-507.

6. Scopes, R.K., *Protein purification principles and practice* Second ed. Spinger Advanced texts in chemistry, ed. C.R. Cantor. 1982: Springer-Verlag.

See more of this Session: Biological Conversions and Processes for Renewable Feedstocks I
See more of this Group/Topical: Sustainable Engineering Forum

(10f) Effect of AFEX Pretreatment Degradation Products On Enzymatic Hydrolysis and Microbial Fermentation by Saccharomyces Cerevisiae 424A (LNH-ST)

Xiaoyu Tang[1], Leonardo da Costa Sousa[1], Shishir Chundawat[1], Mingjie Jin[1], Lau Ming[1], James Humpula[1], Nirmal Uppugundla[1], Kevin Chambliss[2], Ramin Vismeh[3], A. Daniel Jones[3], Zeyi Xiao[4], Bruce Dale[1] and **Venkatesh Balan**[1], (1)Chemical Engineering and Material Science, Michigan State Univeristy, Lansing, MI, (2)Chemistry and Biochemistry, Baylor University, Waco, TX, (3) Biochemistry and Molecular Biology, Michigan State University, East Lansing, MI, (4)School of Chemical Engineering, Sichuan University, Chengdu, CA, China

Several degradation products are formed or released during AFEX pretreatment process, which can be divided into five groups: 1) carbohydrates, 2) aliphatic acids, 3) nitrogenous compounds, 4) furans, 5) aromatic compounds. Some of these degradation products are expected to be inhibitory to *Saccharomyces cerevisiae* 424A(LNH-ST), thus affecting the xylose fermentation, as well as various saccharolytic enzymes. In this work, the effect of various AFEX degradation products on co-fermentation by *Saccharomyces cerevisiae* 424A(LNH-ST) were tested individually and in combinations. A synthetic media composed of similar nutrient composition as AFEX corn stover hydrolyzate was designed and successfully used as a control fermentation media in this study. Nitrogenous compounds showed the inhibitory effect on xylose fermentation, followed by carbohydrates, aliphatic acid and aromatic compounds. The inhibitory effect caused by nitrogenous compounds and carbohydrates were attributed to the presence of amides and oligomeric xylan, respectively. To our knowledge, this is the first report of inhibitory effect of oligomeric xylan and amides on xylose fermentation by *Saccharomyces cerevisiae*. When testing the amides (feruloyl amide, coumaroyl amide and acetamide) and their corresponding acids (ferulic acid, coumaric acid and acetic acid) individually, the amides were less inhibitory than their acids with respect to cell growth and xylose fermentation. From this point, the shift of degradation products from organic acids to amides may be responsible for the high fermentability of AFEX pretreated lignocellulosic biomass. We also evaluated the extent of enzymatic inhibition by various molecular fractions of AFEX treated corn stover degradation products. The pretreated biomass extracted products were fractionated using ultracentrifugation, gel filtration, and C18 resin solid phase extraction and further analyzed by mass spectrometry. The results indicate that both large and small molecule compounds play a role in enzyme/ microbial inhibition.

See more of this Session: Biological Conversions and Processes for Renewable Feedstocks I
See more of this Group/Topical: Sustainable Engineering Forum

MODELLING SIMULTANEOUS SACCHARIFICATION AND FERMENTATION OF MICROCRYSTALLINE CELLULOSE USING COMPUTATIONAL FLUID DYNAMICS

Josebus Maree van Zyl, Stellenbosch University, Stellenbosch, RSA
Thomas Michael Harms, Stellenbosch University, Stellenbosch, RSA
Eugéne van Rensburg, Stellenbosch University, Stellenbosch, RSA
Willem Heber van Zyl, Stellenbosch University, Stellenbosch, RSA
Lee Rybeck Lynd, Dartmouth College, Hanover, NH

Presentation ID#: 233146

Biography

J.M. van Zyl
email: jmvz@sun.ac.za
Tel: +27 (0)21 808 4272

Extended Abstract

Cellulose is a polysaccharide and one of the most abundant organic compounds found on the Earth surface [1]. It holds the potential to reduce human reliance on fossil derived fuels by producing ethanol through the hydrolysis and fermentation of cellulose. Ethanol is a highly combustible liquid fuel which has been shown to be a viable alternative to petrol. Producing ethanol from cellulose however has proved most challenging, particularly overcoming the recalcitrance of cellulose to enzymatic hydrolysis.

Cellulose to ethanol conversion has successfully been demonstrated at laboratory scale for a variety of feedstock and production methods and in a few cases at demonstration scale. Commercialization of cellulosic ethanol is however limited by the lack of large capital funds required to construct such facilities. A compounding effect is that investors still consider the technology as immature and high risk and as such are hesitant to commit. Obtaining the necessary capital funding requires de-risking the technology as well as policy frameworks and incentives from government.

Engineers have recently begun the development of numerical models for implementation with computational fluid dynamic (CFD) software that will allow investigation of secondary effects of cellulosic ethanol production scale-up. The advantage of CFD is the modelling of the flow patterns and conditions that could exist in a commercial scale fermenter. This includes shear-rates, power requirements, degree of particle suspension and through user-coded modules the reaction-rates of biological or chemical processes. This methodology could allow a significant reduction in the risk associated for investors at a reasonably low financial expense.

Presented in this paper are results of the reaction kinetic modelling (Figure 1) of microcrystalline cellulose to ethanol conversion using cellulases from the fungus *Trichoderma reesei* and an industrial grade yeast *Saccharomyces cerevisiae*, commonly used in the food industry [2]. Observations from the results indicated that the number of available bonding sites on the particle surface appeared to control the rate at which hydrolysis occurred. Enzymes appear to catalyse a reaction at a constant rate thus more sites available to attach should result in higher reaction rates. However due to the fibril structure of microcrystalline cellulose a large number of bonding sites are available resulting in a large number of exoglucanase and endoglucanase enzymes initially attaching to sites on the cellulose chains. Once the initial available sites diminish, a large percentage of the exoglucanase enzymes detach from the particle surface and enters the fermentation broth, effectively reducing the rate of hydrolysis. It thus suggests that if the work rate of the endoglucanase enzyme could be significantly increased, high hydrolysis rates could be achieved as more reduction ends become available for the exoglucanase enzymes.

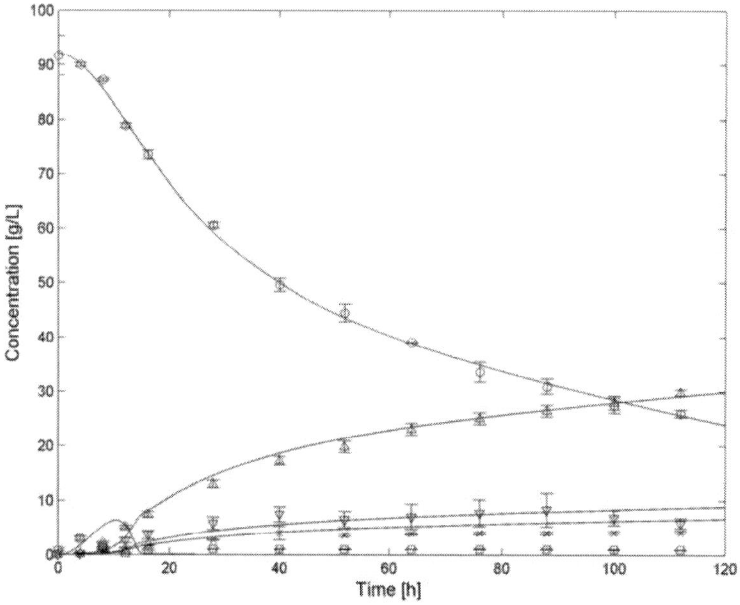

Figure 1: Kinetics model for hydrolysis and fermentation of Avicel PH-101
SSF of Avicel (○) forming glucose (□) fermented to ethanol (Δ) and glycerol (x). Additional yeast cells grown (∇). Simulated results are presented with solid lines.

Microcrystalline cellulose is an insoluble compound which needs to be modelled as particles. It is further required to determine the effects of each substance in the fermentation broth on the viscosity of the system. Results indicated microcrystalline particles occur in various shapes and sizes and exhibit significantly higher drag resistance when the average particle sized is in the order of 50 µm and smaller. This effect is possibly due to the combination of particle topology, structured water [3] and near surface effects [4]. Furthermore the presence of particles has the greatest influence on the apparent dynamic viscosity of the fermentation broth (Figure 2). A numerical model was developed for use in CFD codes which could predict the effects of the particles along with ethanol and glycerol on the apparent dynamic viscosity of the fermentation broth with an average error of 11.1 %.

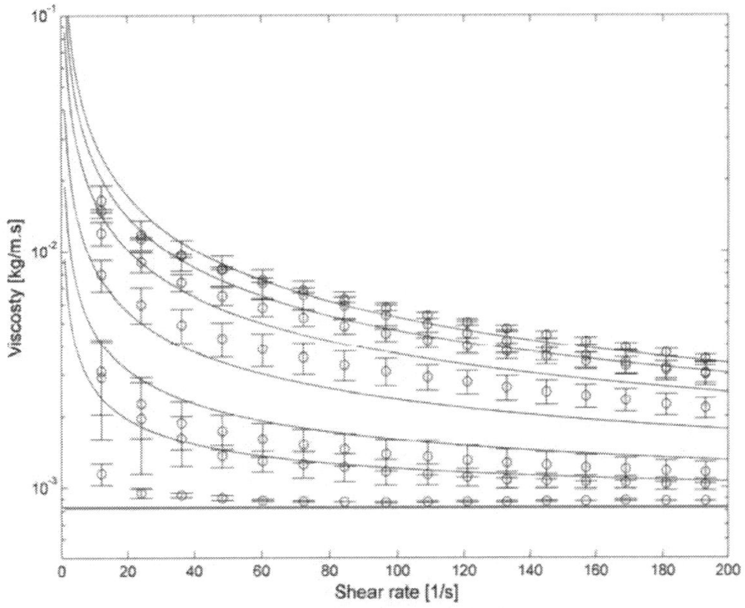

Figure 2: Dynamic viscosity influence of Avicel PH-101 particles

Experimental results (○) with error-bars indicating the standard deviation compared to the numerical predictions (–)

The above mentioned models were incorporated into the commercial CFD package STAR-CCM+ [5] to evaluate the dominant flow patterns and particle distribution within a typical batch mixing vessel. The kinetic models were used to calculate the efficiency of the mixing vessel in-terms of ethanol production rate based on the substrate and reactants concentrations. Results indicated that as cellulose conversion occurs the dynamic viscosity of the fermentation medium increases while the particle sizes decrease. This leads to improved particle suspension and homogeneity which in-turn approaches the optimal conditions for biological hydrolysis and thus ethanol production.

References

1. Lynd LR, Weimer PJ, van Zyl WH, and Pretorius IS, 2002, Microbial cellulose utilization: Fundamentals and Biotechnology, Microbiology and Molecular Biology Reviews, 66 (3): 506-577.
2. van Zyl JM, van Rensburg E, van Zyl WH, Lynd LR, and Harms TM, 2010, A kinetic model for simultaneous saccharification and fermentation of avicel with saccharomyces cerevisiae, Biotechnology and Bioengineering. 108:924-933.
3. J.F. Matthews, C.E. Skopec, P.E. Mason, P. Zuccato, R.W. Torget, J. Sugiyama, M.E. Himmel, J.W. Brady, Computer simulation studies of microcrystalline cellulose Iβ, Carbohyd. Res. 341 (2006) 138-152.
4. J. Zheng, G.H. Pollack, Long-range forces extending from polymer-gel surfaces. Phys. Rev. E, 68 (2003) 031408.
5. STAR-CCM+, version 6.02.007, CD-Adapco, London, UK

(13a) Mechanistic Modeling of Fast Pyrolysis of Cellulose to Predict Bio-Oil Composition

Vinu Ravikrishnan and Linda J. Broadbelt, Chemical and Biological Engineering, Northwestern University, Evanston, IL

Fast pyrolysis, a potential strategy for the production of transportation fuels from biomass, involves a complex network of competing reactions, which result in the formation of bio-oil, non-condensable gaseous species, and solid char. Bio-oil is a mixture of anhydro sugars, furan derivatives, and oxygenated aromatic and low molecular weight (LMW) compounds. Currently, the successful modeling of fast pyrolysis reactors for biomass conversion is hampered by lumped kinetic models, which fail to predict the bio-oil composition. Hence, a fundamental understanding of the chemistry and kinetics of biomass pyrolysis is important to evaluate the effects of process parameters like temperature, residence time and pressure on the composition of bio-oil.

We have recently developed a comprehensive mechanistic model to characterize the primary products of fast pyrolysis of pure cellulose. The model incorporates the following condensed phase reactions with ionic and electrocyclic fragmentation mechanisms for the formation of various LMW products in bio-oil: glucosidic bond scission, retro-aldol, retro-Diels-Alder, 1,2-dehydration, 1,3-dehydration, hydrolysis, cyclic Grob fragmentation and enol-keto transformation reactions. The kinetic rate coefficients for most of the above reaction steps were obtained from the literature, based on either experimentally determined Arrhenius parameters for cellulose and model compounds, or theoretically computed values of activation energies using quantum chemical calculations. A computational framework based on continuous distribution kinetics was constructed to solve the kinetic model. The model tracks 96 distinct polymer species and 25 LMW products, and incorporates over 350 reactions. The model predictions compare well with the experimental data (Patwardhan *et al.*, *J. Anal. Appl. Pyrolysis* **2009**, 86, 323-330) for the formation of important products like levoglucosan (60 wt.%), glycolaldehyde (6 wt.%), 5-hydroxymethyl furfural (3 wt.%), furfural (1 wt.%) and formic acid (6 wt. %). Levoglucosan was found to be formed by the initiation of ionic chain ends by mid-chain glucosidic bond scission, followed by end-chain ionic unzipping. Some of our efforts in extending the kinetic model of pyrolysis of pure cellulose to cellulose with mineral matter and hemicellulose will also be discussed in this presentation.

See more of this Session: Catalytic Conversion of Renewable Resources to Synthesis Gases and Pyrolysis Oils
See more of this Group/Topical: Sustainable Engineering Forum

(13b) Mixed Rare Earth Oxides (REOs) for Desulfurization and Tar Removal From Gasifier Effluents

Rui Li, Joseph Bridges, Weishi Kong and Kerry.M. Dooley, Chemical Engineering, Louisiana State University, Baton Rouge, LA

The removal of H_2S, NH_3 and tars from syngas generated in biomass or coal/biomass gasification plays an important role in further processing of syngas. We have developed materials capable of both H_2S adsorption and tar reforming at high temperatures (~600-650°C). These are mixed rare earth oxides (REOs) such as CeO_2/La_2O_3, prepared by templated sol-gel methods, and then impregnated with transition metals (Mn, Fe, etc.). Desulfurization tests have been carried out using model syngas mixtures. Successive desulfurization-regeneration cycles some of these materials (e.g., Mn/Ce/LaO_x) demonstrated reasonable sulfur capacities in the presence of water and CO_2, along with regenerability. Naphthalene and propane have been used as models for tar reforming/cracking, and the effects of H_2S, CO and CO_2 on the model reactions were examined. H_2S and CO_2 are strong inhibitors, while CO is not. Fresh, used and regenerated sorbents and catalysts have been characterized by BET, TPR, XRD, XANES and EXAFS.

See more of this Session: Catalytic Conversion of Renewable Resources to Synthesis Gases and Pyrolysis Oils
See more of this Group/Topical: Sustainable Engineering Forum

(13c) Multi-Scale Study On the Pyrolysis of Sustainable Biomass Feedstock

Jessica D. Murillo, Chemistry, Environmental Sciences, Tennessee Technological University, Cookeville, TN, Joseph J. Biernacki, Chemical Engineering, Tennessee Technological University, Cookeville, TN and C. Pat Bagley, College of Agricultural and Human Sciences, Tennessee Technological University, Cookeville, TN

Multi-Scale Study on the Pyrolysis of Sustainable Biomass Feedstock

Jessica D. Murillo

Department of Chemistry, Environmental Sciences, Tennessee Technological University, Cookeville, TN 38505

Joseph J. Biernacki

Department of Chemical Engineering, Tennessee Technological University, Cookeville, TN 38505

C. Pat Bagley

College of Agricultural and Human Sciences, Tennessee Technological University, Cookeville, TN 38505

Pyrolysis of lignocellulosic material for the production of biofuels and value-added products offers a short- to long-term renewable energy option to help curb fossil fuel use, gain energy independence, and mitigate global climate change. The main components of lignocellulsic material are cellulose, hemicellulose and lignin; however, there are many issues with using such a complex chemical feedstock for the production of renewable fuels. During pyrolysis the main components degrade via primary and secondary reactions, producing a spectrum of products, including volatiles, char, and bio-oil. To gain a more complete understanding of the pyrolytic behavior of lignocellulosic biomass, a deeper knowledge of the morphology and chemical structure is required.

Preliminary meso-scale experimental results from the pyrolysis of tall fescue hay, sorghum-sudangrass, Bermuda grass, switchgrass, and alfalfa hay, have paved the way for study at microscopic lengths. The physical and chemical characteristics of each grass hay was studied using X-ray diffraction, mid-infrared spectroscopy (MIR) and electron scanning microscopy to determine the microstructure and to quantify the content of lignin, cellulose, and hemicelluloses. Comparison studies on the reaction kinetics of grass samples and isolated pseudo-components were performed using thermogravimetric analysis (TGA).

(13d) Catalytic Cracking of Oak Pyrolytic Vapors Via Bench-Scale In-Situ Fixed-Bed Catalysis

David J. Mihalcik, Akwasi A. Boateng, Charles A. Mullen and Neil Goldberg, Eastern Regional Research Center, Agricultural Research Service, USDA, Wyndmoor, PA

Pyrolysis oils derived from lignocellulosic feedstocks by way of fast pyrolysis offer the possibility of a distinct alternative to the use of petroleum-based fuels. However, an inherent drawback associated with pyrolysis oil its high oxygen content which is due to the nature of the lignocellulosic biomass from which it derives. Extensive efforts are being put forth to treat pyrolytic vapors in situ using zeolite catalysts. For our experiment, the catalytic conversion of oak pyrolysis vapors was performed at 500°C using a 2.5 kg/hr fluidized-bed fast pyrolysis reactor that was fitted with a re-heated (425°C) split-stream fixed-bed catalytic cracker containing zeolite-type catalyst. The placement of the split-stream fixed-bed catalytic reactor was varied within the condenser train to determine the effect of the removal of increasing amounts of water and oxygenated components from within the pyrolytic vapor stream. The cracked pyrolytic vapors resulting from the fixed-bed catalyst were collected by way of immediate contact with a dry-ice acetone bath followed by secondary collection using a methanol spray condenser. Karl Fisher titration and elemental and quantitative GC/MS analyses were performed to determine relevant trends within the experiment. The recovered liquid products that passed through the fixed-bed catalytic reactor displayed a remarkable decrease in the amount of oxygen containing species and a significant increase in the presence of aromatic hydrocarbons.

See more of this Session: Catalytic Conversion of Renewable Resources to Synthesis Gases and Pyrolysis Oils
See more of this Group/Topical: Sustainable Engineering Forum

(13e) A Dual-Stage Laminar Entrained Flow Reactor (LEFR) and Plug Flow Reactor (PFR) System for Studying Biomass Pyrolysis and Gasification

Mark W. Jarvis, Calvin Mukarakate, David Robichaud and Mark Nimlos, National Renewable Energy Laboratory, Golden, CO

An understanding of pyrolysis vapor cracking kinetics and mechanisms are important for developing approaches for reducing tars during gasification and ultimately reducing the cost of producing transportation fuels from gasification and fuel synthesis. In order to reduce tars using predictive design approaches such as Computation Fluid Dynamics (CFD), it is necessary to have robust chemical reaction models containing mechanism that lead to the formation of these species. Existing reaction models only contain limited, highly generalized mechanisms for the formation of tars and are not robust enough to be predictive. In particular, the initial steps of gasification are inadequately characterized. These steps are primarily cracking reactions that break down the pyrolysis vapor species into smaller molecules ultimately leading to syngas and tars. In this study, we have undertaken an experimental investigation of these types of cracking reactions. The solid phase vaporization reactions were separated from the gas phase cracking reactions by conducting experiments in a two-stage reactor. In the first stage, particles of biomass were pyrolyzed in a Laminar Entrained Flow Reactor (LEFR) at a low temperature (450°C) and in the second phase the vapors were cracked at variable residence times (0.2 to 1.5 s) and temperatures (430° to 680°C) in a Plug Flow Reactor (PFR). The tar species and other product gases were measured using a Molecular Beam Mass Spectrometer (MBMS). The data from the MBMS showed trends in the tar species concentrations as a function of residence time and temperature, which will allow the development of more accurate and robust chemical kinetics models of gasification. Eventually, these kinetic models will be used in CFD models of gasification in order to optimize it for reduced tar formation and increased syngas yield.

See more of this Session: Catalytic Conversion of Renewable Resources to Synthesis Gases and Pyrolysis Oils
See more of this Group/Topical: Sustainable Engineering Forum

(13f) Reforming of Residual Tars and Oils From Biomass Gasification

Lyman Frost[1], Elango Elangovan[1] and J. Hartvigsen[2], (1)Ceramatec, Inc, Salt Lake City, UT, (2) Electrochemical, Ceramatec, Inc, Salt Lake City, UT

One of the barriers to wider use of many kinds of biomass is the fact that they generate refractory tars and oils when gasified. These compounds cause difficulty when they are passed to downstream components in the system by causing plugging and drop in performance. The compounds can be condensed and removed but this involves extensive heat exchange, increased cost, and removal to a hazardous waste landfill.

Working as a subcontractor to Emery Energy, a Salt Lake City company with expertise in biomass gasification, Ceramatec will be applying its non-thermal plasma reformer technology to eliminate any refractory tars and oils that exit the Emery gasifier. Preliminary laboratory work indicates that it is possible for the plasma reformer to operate on the gas exiting the gasifier and convert these tars and oils to additional hydrogen and carbon monoxide. This work is being done at Western Research Institute in Laramie, Wyoming under a US Department of Energy award to Emery Energy.

This paper will present an overall process structure of the gasification process and the non-thermal plasma reformer in particular. Operational characteristics and laboratory results related to the reforming of the refractory tars and oils will be presented. The larger pilot sized facility at Laramie will also be described. Data is expected to be available from the larger pilot sized facility in Laramie by the time for the conference. The resultant gas stream is of a quality that could be used by a high temperature fuel cell.

See more of this Session: Catalytic Conversion of Renewable Resources to Synthesis Gases and Pyrolysis Oils
See more of this Group/Topical: Sustainable Engineering Forum

(13g) Low Temperature Gasification of Impregnated Biomass Feedstock

Foster Agblevor, Biological Engineering, Utah State University, Logan, UT, Ofei Mante, Biological Systsems Engineering, Virginia Polytechnic Institute and State University, Blacksburg, VA, Allen Aradi, Afton Chemical Inc, Richmond, VA and Tze-Chi Jao, Afton Chemical, Richmond, VA

Hybrid poplar biomass feedstocks were impregnated with catalysts and gasified at three different temperatures in a bubbling fluidized bed reactor. The impregnated biomass were gasified in a nitrogen atmosphere at 400 °C, 600° C, and 800 °C. The gas yield increased with increasing temperature as expected and the char and liquid yields decreased with increasing temperature. Te catalytic effect was observed at 400 C when the impregnated gasification was compared with the control that did not contain any catalyst. At 400 C, the impregnated sample had a gas yield of 51wt% compared to 41 wt% for the non-catalytic gasification at 450 C. The char yield was 11 wt% compared to 14 wt% for the non-catalytic run at 450 C. The char yield decreased from 11 wt% to 6.5 wt% when the gasification temperature was increased from 400 to 800 C. Although there was 88 wt% gas yield at 800 C, the thermal effect was very strong and the catalyst was not effective in cracking the lignin degradation products. Thus, this catalyst needs considerable improvement to effectively crack the lignin fraction of the biomass pyrolysis products.

See more of this Session: Catalytic Conversion of Renewable Resources to Synthesis Gases and Pyrolysis Oils
See more of this Group/Topical: Sustainable Engineering Forum

(38a) Designing Nanostructures for High Performance Batteries and Supercapacitors

Yi Cui, Materials Science and Engineering, Stanford University, Stanford, CA

Storing energy electrochemically involves electronic and ionic processes and chemical transformation inside and at the interface of materials. The ability to engineer nanostructures and their interfaces afford the great opportunities for controlling these fundamental processes, which can ultimately lead to high performance devices such as batteries and supercapacitors. Here I will present several examples on designing nanostructures and their interfaces to realize high performance energy storage devices. One example is on designing nanowires and heterostructured nanowires for ultrahigh capacity storage of lithium ions in silicon battery anodes. The challenges of large volume expansion and solid-electrolyte-interphase (SEI) have been overcome. In another example, we have designed nanoporous carbon to encapsulate sulfur for lithium ion battery cathodes, which can addresses many issues related to sulfur. In the third example, paper and textile batteries and supercapacitors have been developed.

See more of this Session: Nanomaterials for Energy Storage I
See more of this Group/Topical: Topical 5: Nanomaterials for Energy Applications

(38b) Nanoscale Organic Hybrid Materials (NOHMs): Opportunities In Electrical Energy Storage

Lynden A. Archer, Chemical and Biomolecular Engineering, Cornell University, Ithaca, NY

This talk focuses on two new classes of nanoparticle organic hybrid materials (NOHMs), which manifest liquid-like ion transport properties at high particle loadings. In one group (Nanoparticle Ionic Materials (NIMs)), electrostatic and steric forces dominate the interparticle potential and appear to be responsible for the system's ability to relax to an equilibrium state. In the other group, only steric forces are present, yet the system is able to reach equilibrium in the absence of a suspending solvent. I will discuss equilibrium structure, rheology, and ionic transport properties of these hybrids. I will also study them for applications as anodes, cathodes, and electrolytes for advanced secondary lithium batteries.

See more of this Session: Nanomaterials for Energy Storage I
See more of this Group/Topical: Topical 5: Nanomaterials for Energy Applications

(38c) Novel Quaternary Ammonium Ionic Liquid Electrolyte for Stable, High Capacity Si Nanowire Based Li-Ion Battery

Vidhya Chakrapani, Florencia Rusli, Michael A. Filler and Paul A. Kohl, School of Chemical & Biomolecular Engineering, Georgia Institute of Technology, Atlanta, GA

*Email: paul.kohl@chbe.gatech.edu

In recent years, there has been interest in silicon nanowires as an anode for Li-ion battery because of its large gravimetric capacity of 4200 mAh/g. 1D nanostructure are essential because they permit radial strain relaxation of mechanical stress which occurs due to the volumetric expansion (>300%) as result of Li insertion. However, commercialization of silicon nanowire-based material has been impeded by number of key issues such as severe capacity fade, limited cycle life and mechanical degradation as result of structural changes. Overcoming these challenges requires optimization of anode structure and choice of better electrolyte than those commonly used. The electrolyte should enable the formation of a stable solid-electrolyte-interface (SEI). Furthermore, the use of organic electrolyte in lithium batteries poses a flammability issue and limits its operating temperature range.

In this work we show a novel quaternary ammonium based ionic liquid as an electrolyte for high capacity and high stability silicon nanowire based anode. Battery testing in these electrolytes showed steady reversible capacity of 2000 mAh/g and columbic efficiency ~97% even after 50 cycles. The efficiency for the first cycle was 91% with very little loss in the discharge capacity indicating good SEI formation in these ionic liquids. Results are discussed in the terms of the nature of this SEI layer and the role of additives that promotes its formation. New strategies were developed to minimize the capacity fade and obtain battery life > 600 cycles.

See more of this Session: Nanomaterials for Energy Storage I
See more of this Group/Topical: Topical 5: Nanomaterials for Energy Applications

Examination of Tropical Cyclogenesis using the High Temporal and Spatial Resolution JRA-25 Dataset

Masato Sugi
Forecast Research Department, Meteorological Research Institute, Japan
Correspondence: msugi@mri-jma.go.jp

INTRODUCTION

Imperfect models alone cannot draw a reliable picture of the real world, while temporally and spatially insufficient observation data alone cannot draw a very detailed picture of the real world. Reanalysis may be regarded as an optimum combination of an imperfect model and insufficient observation data, which can produce the most reliable and detailed dataset to describe the real world.

Regarding the future tropical cyclone (TC) activity, a recent very high resolution GCM experiment indicates a reduction in the TC frequency as a whole, with an increase in the frequency of very intense TCs (Oouchi et al. 2006). On the other hand, in the recent IPCC report, it is summarized as "Based on a range of models, it is likely that future tropical cyclones (typhoons and hurricanes) will become more intense, with larger peak wind speeds and more heavy precipitation associated with ongoing increases of tropical SSTs. There is less confidence in projections of a global decrease in numbers of tropical cyclones." (IPCC, 2007). There is no doubt that for a more reliable projection of the future tropical cyclone activity, it is necessary to better understand the physical processes associated with the genesis and development of tropical cyclones, and to assess the capability of GCMs to simulate these processes.

In the present study, in order to understand the physical processes of tropical cyclogenesis, we examine the formation stage of typhoons using the high temporal and spatial resolution JRA-25 dataset. Because of the high quality of the tropical precipitation in the JRA-25 dataset (Onogi et al. 2006), we can expect tropical disturbances and their environmental circulations are analyzed reasonably well in JRA-25.

DATA

The main data used in the present study is the 6 hourly, 1.25 degree latitude-longitude grid, standard pressure level JRA-25 data. For the JRA-25 precipitation data, 1.125 degree latitude-longitude grid, 6 hour forecast data is used. The CDA file is used to check if the TCR data (tropical cyclone wind profile retrieval data) is used or not in each time. To examine the convective activity around the tropical cyclones, 1 hourly, 0.05 degree latitude-longitude grid GMS satellite black body temperature (TBB) data is used. In addition, data from a very high resolution (horizontal grid size 20km) GCM experiment is used for the purpose of comparison. The period of analysis is the three months from August to October in 2004. During this period 14 typhoons are formed, 8 in August, 3 in September and 3 in October. Formation stages of 12 typhoons out of the 14 typhoons are examined, and three of them, T0416 (CHABA), T0418 (SONGDA) and T0422 (MA-ON), are analyzed in some detail.

DEFINITON OF TC GENESIS

In the present study, we define TC genesis time as the time when TCR data is first used for the TC in JRA-25. TCR data is included in JRA-25 when maximum wind $\geq$ 30 kts in the Neumann's best track data (Hatsushika et al. 2006). On the other hand, the JMA TC genesis time is defined as the time when maximum wind $\geq$ 34 kts, when tropical depression (TD) becomes named typhoon (TS). Because of the difference in the threshold value for TC genesis, the TCR genesis time is usually 6 to 18 hours earlier than the JMA genesis time. In addition, we should note that the maximum winds estimated from satellite images are considerably different in different best track dataset. The maximum wind in JTWC best track data is usually stronger than that in JMA best track data, particularly in the mature stage of TCs. For example, the maximum wind speed of T0416 (CHABA) at mature stage is 110kts in JMA best track data, while it is 155kts in JTWC best track data.

INTENSIFICATION OF VORTEX

We made maps of vorticity at 925hPa every 6 hours from August to October in 2004 over the tropical western North Pacific (Figures not shown). In these maps we can see many weak vortices, some of which develop into typhoons. The central question regarding TC genesis is why some of these vortices develop into TCs but many others do not. We first examine the development processes of the vortices that eventually developed into typhoons. Intensification of vortices during the TC genesis period, from 2 days before the genesis time to 2 days after the genesis time, is shown in Figure 1 for the 12 typhoons formed in August to October in 2004. There was a concern that the inclusion of TCR data might cause a discontinuous intensification of vortex, but the Figure 1 shows that there is no such gaps in the intensity before and after the genesis time. In some case there is a rapid intensification around the genesis time. For example, a rapid intensification can be seen around genesis time of T0419 or T0422, but the intensification continues sometime after the genesis time, and it may be real intensification rather than the artificial one caused by the inclusion of TCR data at genesis time. All the intensification curves of vortices of the 12 TCs are plotted together in Figure 2, with their average. The average curve shows slow intensification before the genesis time and more rapid intensification after the genesis time. This suggests that a self intensification process starts to take place around the genesis time.

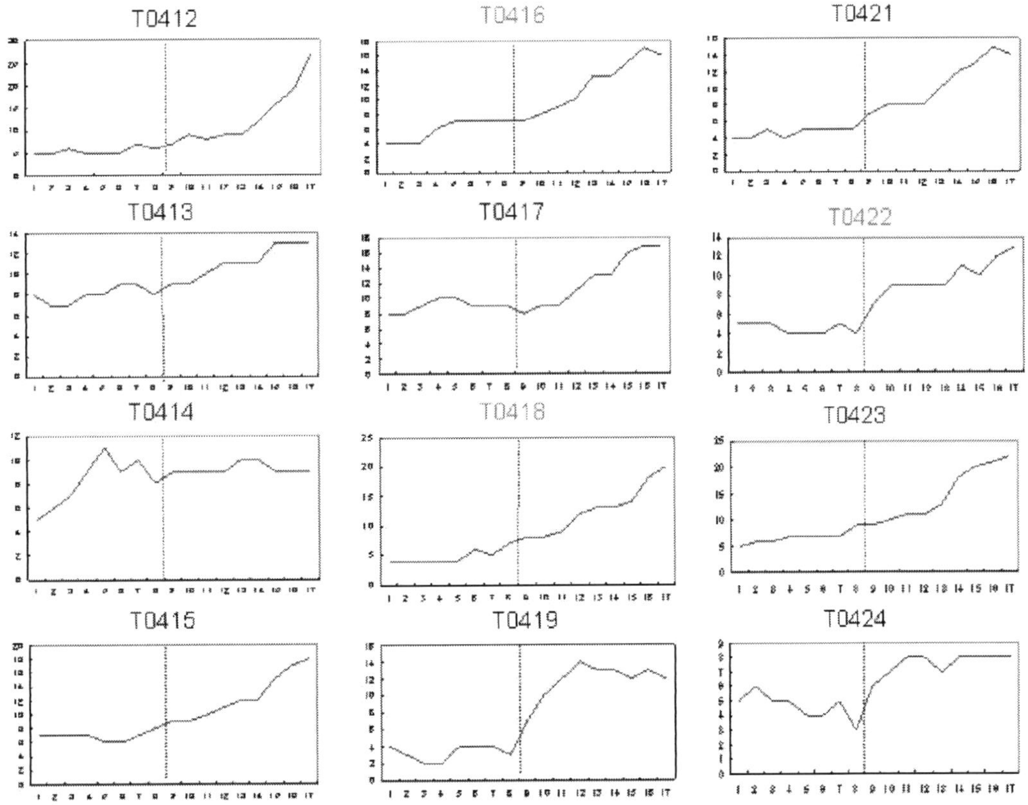

Figure 1 Variation of the intensity of vortex at 925hPa during the 4 day period centered at the genesis time for 12 typhoons formed in August to October of 2004.

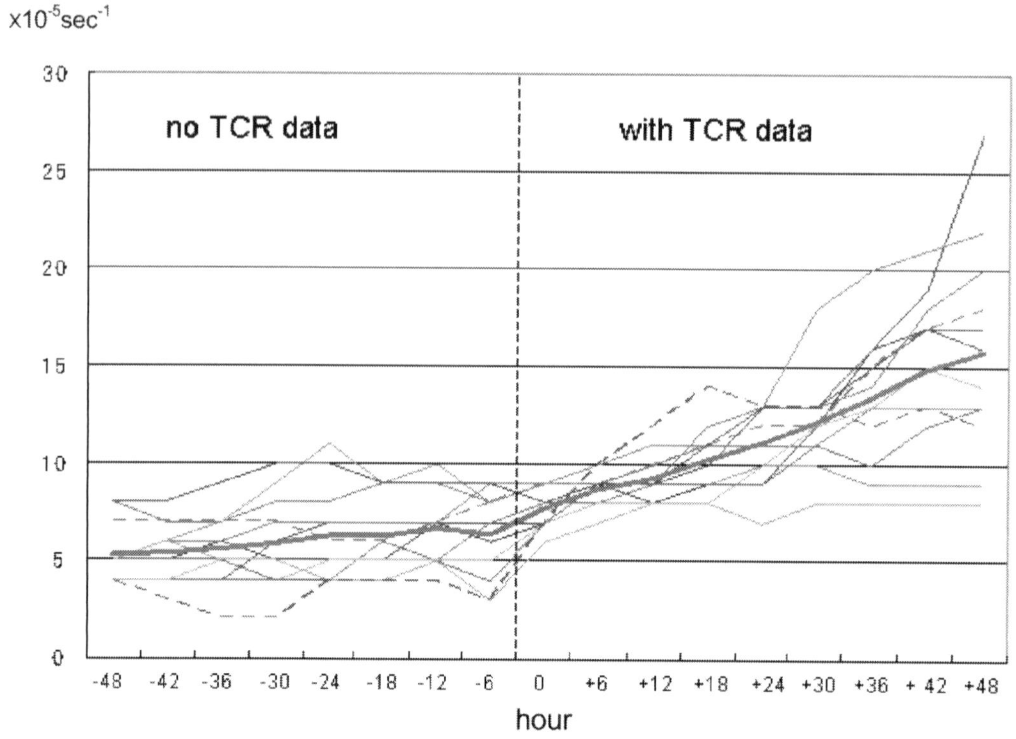

Figure 2 The same as Figure 1 but all curves for 12 typhoons are plotted together, with their average (thick red curve). 0 hour is the genesis time.

CASE STUDIES

We examine genesis processes of three typhoons in some detail. Figure 3 shows the evolution of vorticity, sea level pressure, convective clouds (GMS TBB), precipitation and vorticity generation term $-(\zeta+f)D$ for the TC0416 (CHABA) during the 4 days period centered at the genesis time (18UTC of 18 August 2004). The vertical cross sections of vorticity show the upward development of the vortex from bottom to top, suggesting that the vorticity is generated in the boundary layer and advected upward to free atmosphere by convections. Before the genesis time when the TCR data is first included, a development of vortex together with the surface low pressure area can be seen, indicating that the tropical cyclogenesis process is well captured by the JRA-25 data. Concentrated active convective clouds can be seen near the vortex center after the genesis time, while some scattered active convective clouds can be seen before the genesis time. The role of these scattered convective clouds during the genesis stage seems to be producing a large scale convergence in the boundary layer, leading to an intensification of the vortex through the vorticity generation term. Although the 6-hour forecast precipitations in the JRA-25 are not well concentrated near the vortex center, we can see vorticity generation in the boundary layer near the vortex center. This indicates that the large scale convergence associated with the ensemble of scattered convections around the vortex play an important role in the development of vortex.

The evolutions of the other two typhoons (T0418 and T0422) are almost similar to T0416 (Figures not shown), although there are some minor differences. For both typhoons, we can see upward development of vortex similar to T0416 during the TC genesis period. A development of surface low pressure area and an intensification of vorticity

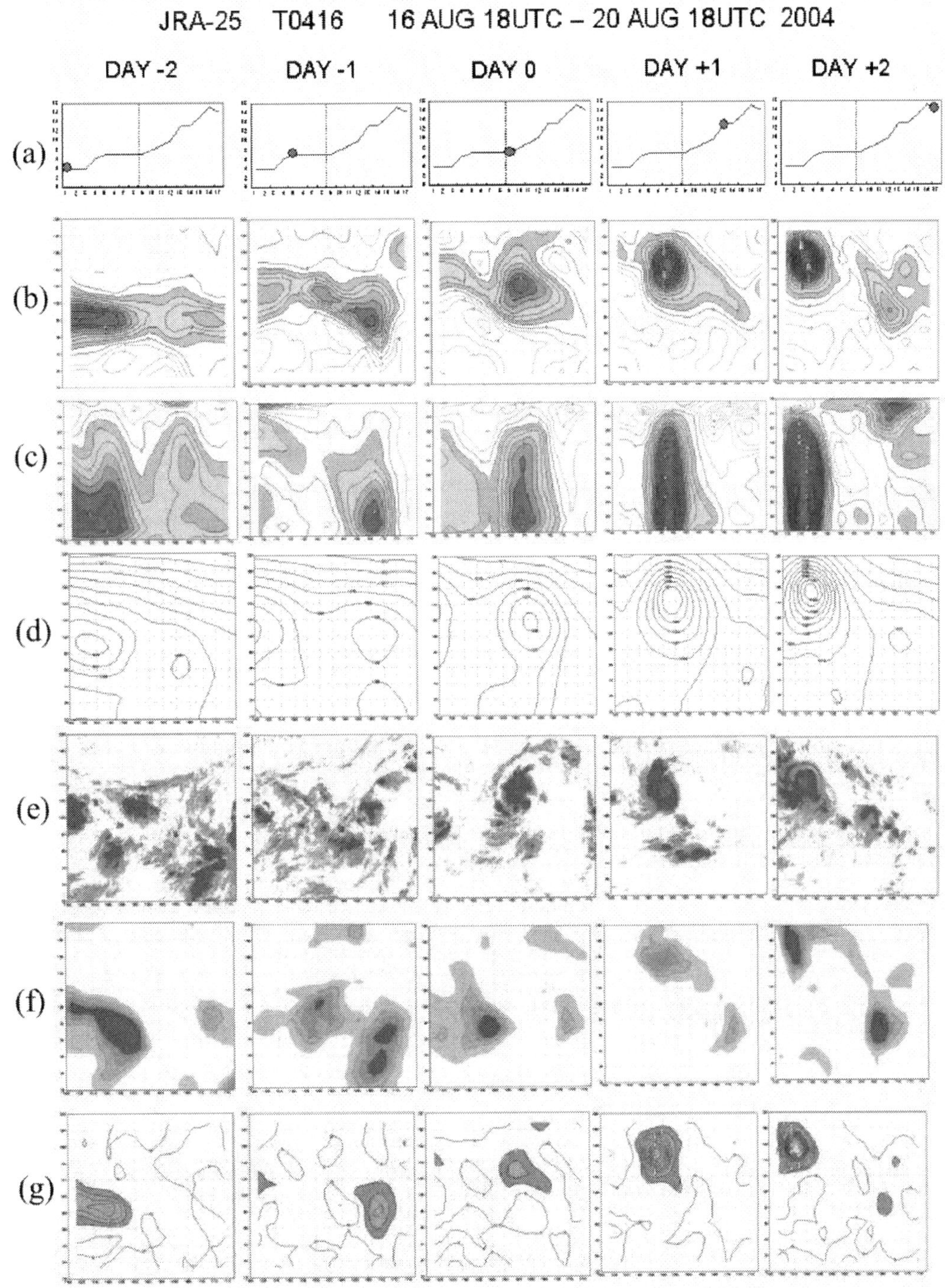

Figure 3 Evolution of T0416 during 4 days centered at the genesis time (18UTC 18 August 2004). (a) Variation of intensity of vortex. (b) Vorticity at 925 hPa. (c) Vertical cross section of vorticity. (d) Sea level pressure. (e) Convective clouds. (f) 6hr forecast precipitation. (g) vorticity generation term ($-(\zeta+f)D$)

generation term around the vortex center can be seen in both cases. We can also see some scattered convective clouds around the vortex center before the genesis time, and concentrated convective clouds near the vortex center as in the T0416 case. Furthermore, a similar evolution of vortex during the TC genesis period is also seen in the very high resolution GCM simulation.2.

CONCLUSIONS

The main conclusions of this study are:

1) High resolution JRA-25 dataset is very useful for TC genesis study. TC genesis process is reasonably well captured by the JRA-25. TCR data makes no significant gaps in the intensity of TC before and after TC genesis.

2) There are many weak vortices ($\leq 5 \times 10^{-5}$ sec^{-1}) in the boundary layer over the tropical western North Pacific. Some of them develop into typhoons but many of them do not develop.

3) It seems that the vorticity generation in the boundary layer through convergence of absolute vorticity ($-(\zeta+f)D$) and its upward advection to free atmosphere by convection are the most important processes for the development of the vortex.. Further quantitative analysis of these processes is an important future work.

4) During the genesis stage of TC, scattered active convective clouds can be seen around the vortex center. The role of these scattered convective clouds during the genesis stage seems to be producing a large scale convergence in the boundary layer, leading to an intensification of the vortex through the vorticity generation term.

5) To assess the capability of GCMs to simulate TC genesis processes, a comparison of GCM simulation with cloud resolving model (CRM) simulation is also an important future work.

ACKNOWLEDGMENT

The author would like to express sincere thanks to the members of JRA-25 project, who helped the author to use the high resolution JRA-25 dataset and CDA files. Thanks are extended to Dr. Mike Fiorino who prepared the high quality TCR data for the JRA-25.

REFERENCES

Hatsushika, H., J. Tsutsui, M. Fiorino, and K. Onogi, 2006: Impact of Wind Profile Retrievals on the Analysis of Tropical Cyclones in the JRA-25 Reanalysis. *J. Meteor. Soc. Japan*, **84**, 891-905.

IPCC 2007: Climate Change 2007: The Physical Science Basis. Cambridge University Press, 996pp

Onogi, K., J. Tsutsui, H. Koide, M. Sakamoto, S. Kobayashi, H. Hatsushika, T. Matsumoto, N. Yamazaki, H. Kamahori, K. Takahashi, S. Kadokura, K. Wada, K. Kato, R. Oyama, T. Ose, N. Mannoji and R. Taira 2007: The JRA-25 Reanalysis. *J. Meteor. Soc. Japan*, **85**, 369-432.

Oouchi, K., J. Yoshimura, H. Yoshimura, R. Mizuta, S. Kusunoki, A. Noda, 2006: Tropical Cyclone Climatology in a Global Warming Climate as Simulated in a 20km-Mesh Global Atmospheric Model: Frequency and Wind Intensity Analysis. *J. Meteor. Soc. Japan*, **84**, 259-276.

(38e) Mechanism of Li Adsorption On Carbon Nanotube-Fullerene Hybrid System: A First-Principles Study

Wonsang Koh[1], Ji Il Choi[2], Kevin Donaher[3], Seung Geol Lee[2] and **Seung Soon Jang**[2], (1)School of Physics, Georgia Institute of Technology, Atlanta, GA, (2)School of Materials Science and Engineering, Georgia Institute of Technology, Atlanta, GA, (3)Department of Applied Physics and Applied Mathematics, Columbia University, New York, NY

The lithium (Li) adsorption mechanism on the metallic (5,5) single wall carbon nanotube (SWCNT)-fullerene (C_{60}) hybrid material system is investigated using first-principles method. It is found that the Li adsorption energy (-2.649 eV) on the CNT-C_{60} hybrid system is lower than that on the peapod system (-1.837 eV) and the bare CNT (-1.720 eV), indicating that the Li adsorption on the CNT-C_{60} hybrid system is more stable than on the peapod or bare CNT system. This is due to the C_{60} of high electron affinity and the charge redistribution after mixing CNT with C_{60}. In order to estimate how efficiently Li can utilize the vast surface area of the hybrid system for increasing energy density, the Li adsorption energy is calculated as a function of the adsorption positions around the CNT-C_{60} hybrid system. It turns out that Li preferably occupies the mid-space between C_{60} and CNT and then wraps up the C_{60} side and subsequently the CNT side. It is also found that the electronic properties of the CNT-C_{60} system, such as band structure, molecular orbital and charge distribution, are influenced by the Li adsorption as a function of the number of Li atoms. From the results, it is expected that the CNT-C_{60} hybrid system has enhanced the charge transport properties in addition to the Li adsorption, compared to both CNT and C_{60}.

See more of this Session: Nanomaterials for Energy Storage I
See more of this Group/Topical: Topical 5: Nanomaterials for Energy Applications

(38f) Vanadium Oxide Nanowire - Carbon Nanotube Binder Free Flexible Electrodes for Supercapacitors

Kenneth J. Balkus Jr. and Sanjaya D. Perera, Chemistry, University of Texas at Dallas, Richardson, TX

Vanadium pentoxide (V_2O_5) layered nanostructures are known to have very stable crystal structures and high faradaic activity. The low electronic conductivity of V_2O_5 greatly limits the application of vanadium oxide as electrode materials and requires combining with conducting materials using binders. It is well known that the organic binders can degrade the overall performance of electrode materials and need carefully controlled compositions. In this study, we develop a simple method for preparing freestanding carbon nanotube (CNT)-V_2O_5 nanowire (VNW) composite paper electrodes without using binders. Coin cell type (CR2032) supercapacitors are assembled for using the nanocomposite paper electrode as the anode and high surface area carbon fiber electrode (Spectracarb 2225) as the cathode. The supercapacitor with CNT-VNW composite paper electrode exhibits a power density of 5.26 kW Kg^{-1} and an energy density of 46.3 Wh Kg^{-1}. Li-VNWs and CNT composite paper electrodes can be fabricated in similar manner and show improved overall performance with a power density of 8.32 kW Kg^{-1} and an energy density of 65.9 Wh Kg^{-1}. The power and energy density values suggest that such flexible hybrid nanocomposite paper electrodes may be useful for high performance electrochemical supercapacitors.

See more of this Session: Nanomaterials for Energy Storage I
See more of this Group/Topical: Topical 5: Nanomaterials for Energy Applications

(39a) Towards Quantum-Dot Solar Cells From Abundant Nontoxic Materials

Ankur Khare[1], Andrew W. Wills[2], Boris D. Chernomordik[1], David J. Norris[3] and **Eray S. Aydil**[1], (1) Chemical Engineering and Materials Science, University of Minnesota, Minneapolis, MN, (2) Department of Chemistry, University of Minnesota Twin Cities, Minneapolis, MN, (3)Optical Materials Engineering Laboratory, ETH Zurich, Zurich, Switzerland

Quantum confinement of electrons and holes in nanometer size crystals (quantum dots or QDs), endows them with properties that may be advantageous for efficient solar-to-electric energy conversion. First, electronic energy levels and optical absorption in QDs can be manipulated by changing their size. This allows the optimization of their optical absorption for maximum overlap with the solar spectrum. Second, the ability to manipulate energy levels through size raises the possibility to make inexpensive multijunction solar cells by judiciously layering different size QDs. Third, it has been suggested that quantum confinement may slow energy dissipative electron and hole relaxation rates such that two new physical processes, multiple exciton generation and hot electron extraction may now compete with relaxation and lead to higher photocurrents or higher photovoltages, respectively. Finally, QDs can be prepared in large quantities as stable colloidal solutions under mild conditions and deposited on surfaces of various planar or nanostructured substrates as thin films through inexpensive high-throughput coating processes to form photovoltaic devices. For these reasons, solar cells based on QDs may have the potential to achieve high power conversion efficiencies at low cost and are promising candidates for third generation photovoltaic devices.

Most quantum dots made to date are based on PbSe and PbS QDs because we know the most about their photophysical properties and synthesis. Lead causes irreversible neurological, reproductive, renal and cardiovascular damage to humans and its use has been banned from many products. Quantum dots based on nontoxic and abundant materials are needed. Copper zinc tin sulfide (Cu_2ZnSnS_4 or CZTS) is emerging as a promising new sustainable semiconductor for photovoltaics. CZTS has a high absorption coefficient in the visible range of the electromagnetic spectrum and an electronic bandgap (~1.5 eV) that is ideal for photovoltaics. We have developed a novel and facile synthesis method for making CZTS nanocrystals and achieved quantum confinement in nanocrystals with diameters less than 3 nm. Our synthesis method is based on decomposition of metal dithiocarbamates in an organic solvent such as dodecene. Presence of oleylamine in the reaction mixture reduces the decomposition temperature of copper, zinc and tin dithiocarbamates to a narrow temperature range (170-220 °C). Rapid injection of oleylamine and tin diethyldithiocarbamate into a hot mixture of copper and zinc dithiocarbamate in dodecene initiates a burst of nanocrystal nucleation followed by growth. We show that using this approach avoids the formation of binary and ternary sulfide impurities such as zinc sulfide or copper tin sulfide. Moreover, we have been able to control the sizes of CZTS nanocrystals from 2-7 nm by varying the experimental conditions such as the synthesis temperature and time. In this talk we will present our latest results towards making CZTS quantum dot solar cells.

See more of this Session: Nanomaterials for Photovoltaics I
See more of this Group/Topical: Topical 5: Nanomaterials for Energy Applications

(39b) Colloidal Nanomaterials As the Building Blocks for Next-Generation Solar Cells

Jeffrey M. Pietryga, Chemistry Division, Los Alamos National Laboratory, Los Alamos, NM

The use of colloidally synthesized nanomaterials in photovoltaic devices is attractive because of the low-cost and scalability of solution-based fabrication methods. Moreover, the facile control over optoelectronic properties, via the fine control over structure these methods afford, can potentially be exploited in highly efficient next-generation solar cells. Design, synthesis and characterization of novel nanomaterials to enable advanced-concept solar cells are an important part of the ongoing effort within the Center for Advanced Solar Photophysics, a DOE Energy Frontier Research Center. I will discuss a number of specific examples from this work that explore the range of complexities achievable through inexpensive wet-chemical methods. Themes to be examined include the straightforward rendering of established bulk materials into quantum-confined nanocrystal form, band-gap engineering in semiconductor nanoheterostructures, and finally the controlled coupling of plasmonic metal nanoparticles with semiconductor nanocrystals in a diverse series of hybrid superstructures.

See more of this Session: Nanomaterials for Photovoltaics I
See more of this Group/Topical: Topical 5: Nanomaterials for Energy Applications

(39c) Band Gap Tuning with CZTS-Based Nanocrystal-Ink Solar Cells

Grayson Ford[1], Rakesh Agrawal[1] and **Hugh W. Hillhouse**[2], (1)School of Chemical Engineering, Purdue University, West Lafayette, IN, (2)Chemical Engineering, University of Washington, Seattle, WA

$Cu_2ZnSn(S,Se)_4$ is of particular interest due to the achievement of 9.6% efficient laboratory scale devices using a non-vacuum, hydrazine-based deposition process. While hydrazine poses safety challenges from a manufacturing perspective, an alternative non-vacuum deposition approach has achieved 7.2% efficiency via the selenization of Cu_2ZnSnS_4 (CZTS) nanocrystals deposited by roll coating. The resulting $Cu_2ZnSn(S,Se)_4$ (CZTSSe) film has a band gap of approximately 1.1 eV, similar to $CuInSe_2$. $CuInSe_2$ device performance is improved by incorporation of Ga, which widens the band gap. We hypothesized that similar bandgap tuning and improvements to CZTS devices may be possible by substituting a lower atomic number group IV element for some of the Sn. Substitution with germanium is an attractive route, as germanium's crustal abudance of 1.5 mg/kg compares favorably with the 2.3 mg/kg crustal abundance of Sn, the rarest element in CZTS. Here we report a synthesis of $Cu_2Zn(Sn_{1-x}Ge_x)S_4$ (CZTGS) nanocrystals and the performance of $Cu_2Zn(Sn_{1-x}Ge_x)(S_ySe_{1-y})_4$ (CZTGSSe) solar cells fabricated by sintering the nanocrystal films with elemental selenium vapor. We find that the band gap of the CZTGS nanocrystals and the CZTGSSe solar cell can be rationally controlled by adjusting the Ge/(Sn+Ge) ratio. Solar cells fabricated from Cu_2ZnGeS_4 nanocrystal films yielded a power conversion efficiency of 0.51%. However, $Cu_2Zn(Sn_xGe_{1-x})S_4$ nanocrystals with a Ge/(Ge+Sn) ratio 0.7 yielded devices with an efficiency of 6.8% when synthesized to be Cu-poor and Zn-rich. This result opens the possibility of forming Ge gradients to direct minority carriers away from high recombination interfaces and significantly improve the device efficiency of CZTSSe-based solar cells.

See more of this Session: Nanomaterials for Photovoltaics I
See more of this Group/Topical: Topical 5: Nanomaterials for Energy Applications

(39d) Electrostatic Stabilization of Silicon Nanocrystal Colloids for Solution-Processed Photovoltaics

Lance M. Wheeler and Uwe R. Kortshagen, Mechanical Engineering, University of Minnesota, Minneapolis, MN

The tunable optoelectronicproperties of semiconductor nanocrystals (NCs) have garnered significant attention for photovoltaicdevice integration. Additionally,solution processing of semiconductor nanocrystals (NCs) provides a low-cost,scalable technique for the fabrication of third-generation solar cells. Group II-VI and IV-VI compoundsemiconductors have dominated the field of NC photovoltaics, as solution-basedsynthesis processes are well established. The process results in a stericallystabilized colloid from which ligand-capped NCs will naturally self-assemblewhen cast into a film. The ligandscan then be removed or exchanged to form a dense, electronically-coupled NCfilm.

Solution synthesis of Silicon NCshas proven difficult due to the high temperatures needed. Nonthermal flow-through plasmasynthesis offers an effective alternative. Typical plasma processes using silane as the siliconprecursor lead to free-standing, hydrogen-terminated Si NCs that will not,however, stabilize in solution. A hydrosylilation reaction will attach alkylligands to the surface to achieve a stable colloid similar to group II-VI andIV-VI NCs, but post-processing of cast films to remove or exchange the ligandsis problematic.

Here we report nonthermal plasmasynthesis of Si NCs with an alternative silicon precursor, silicontetrachloride. The presence of Clduring synthesis allows for tuning of the surface chemistry of the Si NCs. FTIR analysis reveals Cl present on thesurface, which facilitates a novel colloidal stabilization in aprotic polarsolvents. The Si NCs are readilydispersed and stabilized in aprotic dipolar solvents. The solutions are optically transparent and have remainedstable for over 8 months. Webelieve the electronegativity of the chlorine surface leads to solvation of theNCs by electrostatic interaction with solvent molecules.

Solutions of Cl-terminated Si NCsare spin- or drop-cast into self-assembled films that do not require a post-treatmentto remove ligands. Electronmicroscopy and Scanning Probe Microscopy (SPM) reveal the smooth and continuousfilm morphology needed for photovoltaic device structures. Electrical characterization reveals photoconductiveSi NC films with differential dark conductivities of 1.09×10^{-7}S/cm. This is competitive with nano-and micro-crystalline Si filmsdeposited using high-vacuum techniques such as PECVD.

This work was supported in partsby the National Science Foundation under grant CBET-0756326 and by DOE underthe EFRC Center for Advanced Solar Photophysics.

(39e) Atomic Layer Deposition of Nanoscale Materials for Energy Conversion

Marja N. Mullings, Jonathan R. Bakke, Carl Hgglund and Stacey F. Bent, Chemical Engineering, Stanford University, Stanford, CA

Fora number of technologies, including microelectronics and energy conversiondevices, feature sizes are becoming increasingly small, and the need forprecise control over costly materials continues to grow. In addition to decreasinglength scales, it is important that limited natural resources are carefully used.In fact, many photovoltaic and fuel cell components are comprised of a varietyof materials that may be expensive, toxic and/or rare. Placing such materialsonly where needed to derive the most desirable properties will be essential inreducing the costs of energy conversion devices. Atomic layer deposition (ALD)is uniquely poised to accomplish these goals.

Atomiclayer deposition is a growth technique consisting of self-limiting surfacereactions, which allow for sub-nanometer control of growth. The ability tofinely tune growth allows for careful control over material properties. Furthermore,ALD provides excellent capabilities for depositing a variety of materials onhigh surface area substrates including those that are highly structured.

Thiswork focuses on the ALD and materials characterization of transparentconducting oxides (TCOs) and surface passivation techniques that direct thegrowth of functional materials. Transparent conducting oxides (TCOs) are highlyconductive and transparent materials used in thin film solar cells, serving aselectrical contacts, chemical barriers and antireflection coatings. One of themost common and robust TCO materials, indium-doped tin oxide (ITO), has manyattractive properties; however, alternatives must be found to address thediminishing supplies, increasing demands and subsequently escalating costs ofindium. Replacing indium with zinc to create zinc tin oxide (ZTO) provides acost-effective and earth-abundant alternative. Various material properties ofZTOs are not well understood and there is little control over stoichiometryacross growth systems. At present, it is known that ZTO takes on an unstabletrigonal ilmenite structure ($ZnSO_3$) and stabilizes into the cubic spinel(Zn_2SnO_4) and tin dioxide (SnO_2) under hightemperature conditions.

Bydeveloping and carefully characterizing a ZTO ALD system, these structures willbe better understood and eventually optimized for use as TCOs. Atomic layerdeposition reaction conditions for tin oxide (SnO_x) and zinc oxide(ZnO), which are alternative TCOs, and ZTO will be described, including growthrates and temperature windows in our custom built reactors. Material propertiesof the ZTO films will be characterized with x-ray photoelectron spectroscopy(XPS), scanning Auger electron spectroscopy (AES), spectroscopic ellipsometry,scanning electron microscopy (SEM), x-ray diffractometry (XRD) and four-pointprobe measurements providing insight into surface, structural and electronicproperties.

(39f) Development of a Scalable Bottom-up Nanomanufacturing Platform for Highly Efficient Photovoltaics

Blayne Phillips and **Peng Jiang**, Chemical Engineering, University of Florida, Gainesville, FL

Solar energy is clean, abundant, and renewable, yet it is vastly under-utilized in the current world. Although crystalline silicon solar cells are the premier candidate for renewable and environmentally friendly solar energy conversion, high manufacturing cost has prevented such cells from economically competing with the combustion of fossil fuels as a source of electricity on a large scale. We have developed a scalable bottom-up nanomanufacturing platform for mass-producing highly efficient photovoltaics by exploring effective photon management using surface-plasmon-enabled transparent conducting electrodes and subwavelength antireflective coatings. In this methodology, self-assembled colloidal crystals with remarkably large domain sizes are used as templates to create periodic metallic nanohole arrays and subwavelength moth-eye antireflection coatings. The resulting metallic nanohole arrays exhibit high and tunable optical transmission and high electrical conductivity, promising as transparent conducting electrodes for flexible photovoltaics. The subwavelength-structured moth-eye antireflection coatings show much improved broadband antireflective properties than traditional quarter-wavelength coatings. This novel nanomanufacturing platform combines the simplicity and cost benefits of bottom-up self-assembly with the scalability and compatibility of top-down microfabrication.

See more of this Session: Nanomaterials for Photovoltaics I
See more of this Group/Topical: Topical 5: Nanomaterials for Energy Applications

(43a) Modelling Acid Gas Reactions In Air Products Sour Compression Process

Frank Petrocelli[1], Vince White[2] and Jeremy Immer[1], (1)Tonnage Gases, Equipment and Energy - Technology, Air Products and Chemicals, Inc., Allentown, PA, (2)Tonnage Gases, Equipment and Energy - Technology, Air Products PLC, Walton-on-Thames, United Kingdom

Oxyfuel combustion in a coal-fired power station produces a raw CO_2 product containing contaminants such as water vapor plus oxygen, nitrogen and argon derived from the excess oxygen for combustion, impurities in the oxygen used, and any air leakage into the system. There are also acid gases present, such as SO_3, SO_2, HCl and NO_X produced as byproducts of combustion. The overall process for purifying the CO_2 takes the impure CO_2 from the power boiler and cools it to condense water vapor, remove traces of ash and dissolve soluble gases such as SO_3 and HCl. The cooled, raw CO_2 is then compressed to about 30 bar and the CO_2 is purified by partial liquefaction and phase separation.

In 2006, Air Products first presented reactions that gave a pathway for SO_2 to be removed as H_2SO_4 and NO and NO_2 to be removed as HNO_3 during the compression of the raw CO_2 from the Oxyfuel combustion process. This proprietary process is now widely known as Air Products' sour compression process. Then, in 2008, we presented initial results from the Oxycoal-UK project in which these reactions were studied experimentally to provide the important reaction kinetic information at elevated pressure that was so far missing from the literature. This experimental work was carried out at lab scale in Imperial College London using synthetic flue gas, and then using actual flue gas via a sidestream from Doosan Babcock's 160kW coal-fired oxyfuel rig. From these experimental results, it was clear that the main reaction pathways are viable and the rates of reaction sufficient to produce the desired results: SO_x and NO_x removal by compression and contact with water. And, in 2010, we presented further results on experimental work on improving our understanding of the sour compression process.

Simultaneous with the above efforts, Air Products has pursued two opportunities to demonstrate the sour compression process at larger scale: 1) a U.S. Department of Energy (DOE) project to design and construct a proprietary CO_2 purification system for slip stream pilot-scale tests at the site of a 15 MW_{th} tangentially-fired (T-fired) oxycombustion test facility; and 2) a slip stream pilot plant to demonstrate proprietary Oxyfuel purification technology at Vattenfall's 30 MWth wall-fired Oxyfuel boiler research and development facility in Schwarze Pumpe, Germany. Both projects allow us to better understand the sour compression reactions, using real Oxyfuel-derived raw CO_2, with a vapor-liquid contacting device equivalent to that forseen for full scale plants.

At present, our pursuit of larger demonstration plant projects for the Oxyfuel CO_2 purification technology relies, in part, on further enhancing our understanding of the sour compression process. To that end, in late 2010 Air Products commissioned a purpose built lab that is being used for studying the sour compression reactions in more detail than previous experimental work allowed. To date, the experimental program has explored the effect of key process variables, including SO_2 and NO_x concentrations, pressure, residence time, and humidity; through experiments run under several different gas-liquid contacting configurations. Analysis of these data has allowed us to better elucidate the fundamental reaction network and to develop a robust kinetic model that is being utilized for our design work on several commercial scale demo plant projects.

In this paper, we will present improvements we have made in our understanding of the reaction network and in our process model, as a result of the recent lab work.

See more of this Session: Oxycombustion of Coal and Other Fuels I
See more of this Group/Topical: Energy and Transport Processes

(43b) The Effect of Compression and Intercooling On the Composition of a Pulverized Coal Oxycombution Flue Gas

Stephen J. Gerdemann, Casey Carney, John Clark, Sivaram Harendra, Thomas Ochs and Danylo Oryshchyn, National Energy Technology Laboratory, U.S. Department of Energy, Albany, OR

Oxy-fuel combustion uses conventional technology to combust pulverized coal in denitrified air resulting in a flue gas which is primarily CO_2 and water. Integrated Pollutant Removal (IPR®) is an NETL patented process that compresses the oxycombustion flue gas to pipeline pressure. The IPR process as it is realized at the Jupiter Oxygen Burner Test Facility is a spray tower (direct-contact heat exchanger) followed by 4 stages of compression with intercooling. The spray tower can be run either with water or with water plus a reagent used to control the pH and precipitate the reactive components of the gas. Latent and sensible heat given up by the processed gas at the cooling steps are partially recovered and in a commercial process could be used to preheat the boiler feed water. A FTIR and a gas chromatograph were used to measure the entering gas composition, the exiting composition as well as the composition at several intermediate points within the IPR system. The experimental results are compared to an ASPEN model of the system. The entering flue gas is approximately 30% water vapor with the remainder mostly CO_2. The gas also contains SO_2, various oxides of nitrogen (mostly from nitrogen in the coal), O_2, CO, HCl and small amounts of other gasses derived from these. The initial spray tower removes much of the water and soluble gasses. More water and soluble gases are removed in each intercooling step until the exit gas is primarily CO_2. The increasing pressure also changes the relative amounts of the different oxides of nitrogen.

See more of this Session: Oxycombustion of Coal and Other Fuels I
See more of this Group/Topical: Energy and Transport Processes

2011 AIChE Annual Meeting; Abstract # 225838; Energy and Transport Processes

A novel high pressure oxyfuel process with staged combustion

Author(s): **Krish R. Krishnamurthy**, Clean Energy and Innovation Management, Linde LLC, Murray Hill, NJ and Hanno Tautz, Research and Development, Linde AG, Linde Engineering Division, Pullach, Germany

Introduction

The application of pure oxygen for the combustion of coal provides the advantage to reduce the flue gas flow significantly and to operate the combustion and steam production at higher pressure. This results in higher electrical net efficiency and a reduction of investment costs.

High pressure oxyfuel process concept

There are already several oxyfuel processes in development. One approach is to keep the technology as close as possible to the existing atmospheric boiler concept. In this case, the combustion temperature of coal with oxygen has to be limited by dilution of the pure oxygen with a recycled CO_2 flow.

Another approach is the application of elevated pressure to the combustion process. In this case as well, a recycle flow of CO_2 or condensed water or a mixture of this is necessary to limit the combustion temperature.

To overcome the problem of intensive cooling of coal combustion with oxygen in the new process covered by this presentation, the oxidation is performed in two steps. In the first step, the coal is gasified with pure oxygen and, in the second step, the syngas produced by the gasifier is burnt in several stages in a unique heat exchanger equipment and process for generating steam. Between the two steps above, a cyclone and hot gas filtration system are employed for gas cleaning from solids.

The above process avoids the need for recycle CO_2, reduces cooling water requirement and enables increased preheating of boiler feed water. As the combustion is mainly in reducing atmosphere, the NOx formation is strongly reduced. At the pressure of 80 bars, the water can be condensed at about 50°C. With a further reduction of the temperature to 30°C CO_2 is condensed in a relatively pure condition and needs none or minimum purification requirement prior to pumping to 120 bars. The desulphurization can be done in the water phase by neutralization of H_2SO_4 with caustic, see figure 1.

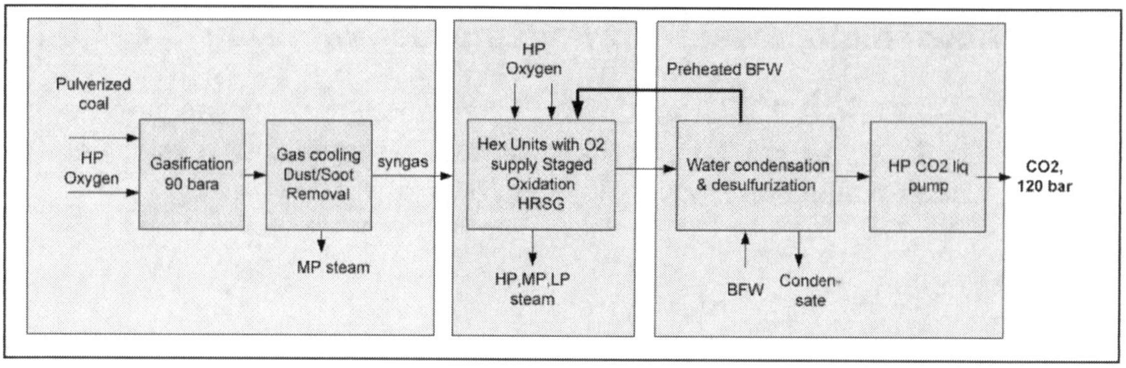

Figure 1: Process for high pressure oxyfuel process with staged combustion

Thermodynamic consideration

(i) Increase of thermal energy input by condensation energy

From the thermodynamic viewpoint, the combustion of coal at elevated pressure has the advantage in that the thermal energy of water condensation can be used at reasonable temperatures for the preheating of boiler feed water. The difference of this thermal energy is shown in a QT diagram in figure 2. The values are calculated based on a power plant with 2200 MW thermal energy input, for the combustion of pre-dried lignite with an approximately 22,000 kJ/kg ash free lower heating value. At 80 bars, the condensation starts at a temperature of about 145°C compared with about 80°C at atmospheric pressure. With this higher condensation temperature the boiler feed water can be preheated to about 160°C instead of 100°C and about 110 MW more thermal power can be used for the process.

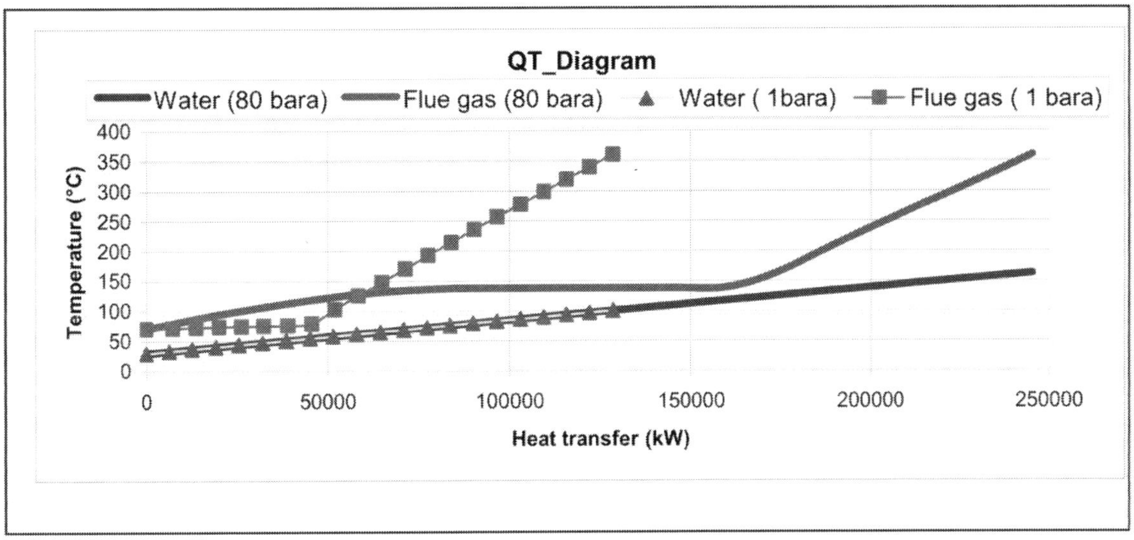

Figure 2: QT-diagram for atmospheric and high pressure flue gas case.

(ii) Increase of turbine efficiency with higher steam temperature

There are several groups investigating high temperature materials development, which can be applied in steam boilers for steam generation up to 700°C at 350 bars. With the currently available materials, the introduction of these operating conditions is not expected within the next several years in atmospheric boiler plants.

The high pressure oxyfuel process gives the opportunity to apply these process parameters already with existing materials. The reasons allowing this possibility of higher temperature and higher pressure steam generation are highlighted below:

- Reduction of differential pressure between steam and flue gas by 80 bars
- Increased heat transfer because of elevated pressure
- Reduction of heat and fouling resistance because of gas cleaning in front of heat exchanger section

Therefore the flue gas and material temperature and the tube material wall thickness in the heat exchangers can be lower than in atmospheric boilers for the same heat transfer power. The expected increase of electrical efficiency with the 700°C technology and equipment optimization is 4% points.

Key equipment for steam production

The production of high pressure steam at temperatures above 500°C needs specially constructed heat exchanger equipment. Typically used tube sheets in a shell and tube arrangement would become too thick at the high temperatures and pressures. A suitable construction for high pressure flue gas and staged combustion can be a spiral wound heat exchanger, see figure 3.

The steam can be overheated in tube bundles, which are connected over different tube baffles to the high pressure pipes. With this, the diameter and the mechanical load of the tube sheets can be reduced. Additional burning or oxidizing gases can be introduced in the heat exchanger.

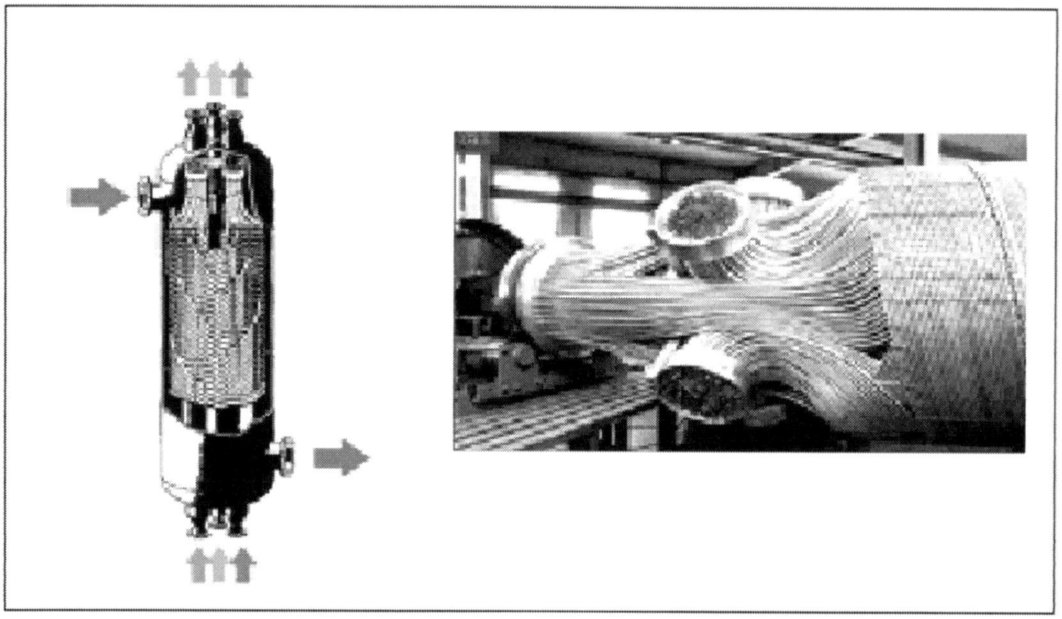

Figure 3: Spiral wound heat exchanger for high temperature steam production.

Results of process simulation

A detailed process simulation was conducted for the calculation of the electrical efficiency of the high pressure oxyfuel process and is compared with published data for the atmospheric oxyfuel process and an IGCC process with CO_2 capture. The results can be seen in table 1:

Table 1: Efficiencies of oxyfuel processes with CCS.

	Atmospheric Oxyfuel	IGCC & CCS	HP Oxyfuel 600 °C steam	HP Oxyfuel 700 °C steam
Net efficiency (%)	40 /3/	35 /2/	42	42+4=46
CO_2 capture (%)	90	90	99	99

Process concept and construction

The high pressure oxyfuel process results in a significant reduction of effective gas flow. This leads to a compact plant layout, which can be seen in figs 4 and 5.

A preliminary estimation of equipment weight shows a strong reduction of material for the boiler construction. The main equipment can be pre fabricated in a workshop, transported and directly installed at the construction site. With this the erection time and costs can be reduced significantly.

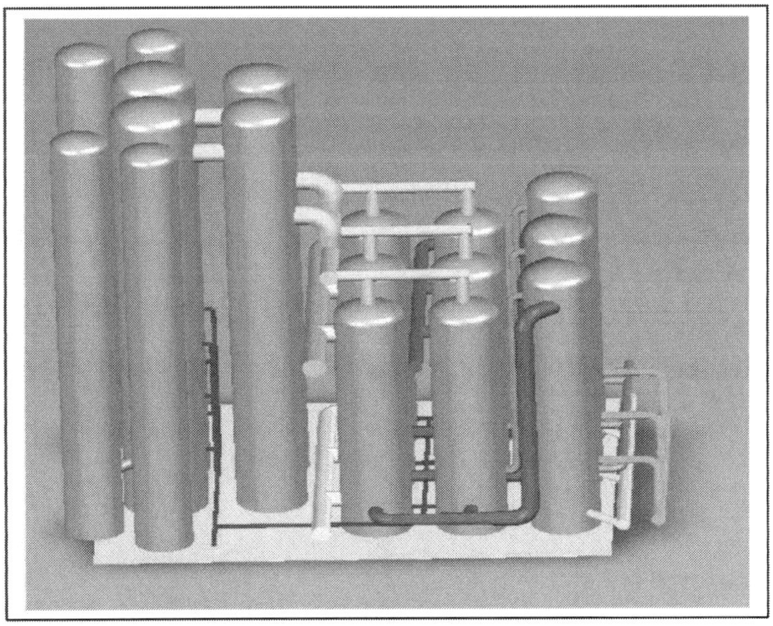

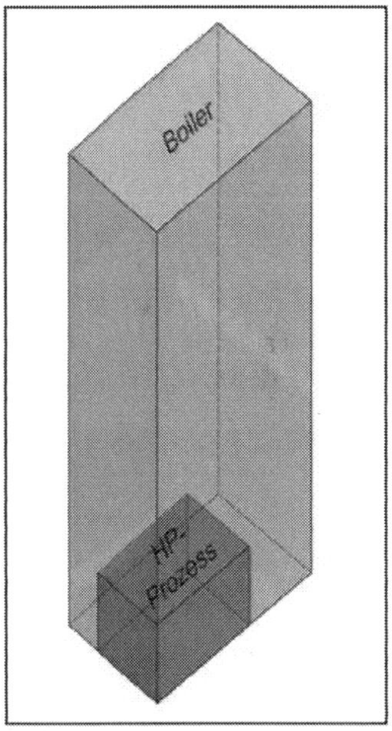

Figure 4: Sketch of high pressure oxyfuel process
Figure 5: Size proportion Boiler and HP oxyfuel plant

Summary

The high pressure oxyfuel process involving staged combustion of gasified syngas to generate steam offers a number of advantages summarized below:

- Higher electrical efficiency by > 2-6 % points compared to atmospheric oxyfuel process.
- Strongly reduced cost by unit standardisation, prefabrication and material reduction.
- Strong reduction of boiler size and material demand.
- Simplified gas cleaning and CO_2 capture
- Option for 700°C steam technology possible
- The production of syngas gives the possibility to use the gasification for poly-generation processes in times of lower electricity demand.

(43d) High Temperature Oxygen Production

Michael K. Opoku, Chemical and Biological Engineering, South Dakota School of Mines and Technology, Rapid City, SD and Jan A. Puszynski, Chemical and Biological Engineering Department, South Dakota School of Mines and Technology, Rapid City, SD

A potentialalternative to produce clean coal technologies would be to combust fossil fuelsin pure oxygen instead of air, which contains approximately 78% nitrogen byvolume. If nitrogen was removed from the process, flue gas streams would have amuch higher concentration of CO_2, reducing or eliminating the needfor costly CO_2 capture. Moreover, NO_X emissions and thesubsequent need for processing would be reduced significantly. The main problemwith this method is separating oxygen from the air. This is usually completedcryogenically which requires a lot of energy (for a typical 500MW coal-firedpower station supplying pure oxygen requires at least 15% of the electricitythe plant generates annually). Cryogenic air separation technology forproducing high purity oxygen is energy intensive and requires air to be cooledto about - 300 degree Fahrenheit for separation. The cold oxygen must be heatedto high temperatures in order to couple this air separation process with coalpower plants. The high cost of producing oxygen, however, has made oxy-combustiona cost-prohibitive option for commercial use in most power plants. Developingtechnologies in oxygen and ion transport membranes have the potential to reducethe cost of oxygen production and increase oxy-combustion's cost-effectiveness.

In this research, multicomponentoxides containing Mn in different oxidation states were explored as reactantsfor two-step high temperature high purity oxygen production. High temperaturereduction and oxidation of these compounds were carried out in an inert gas andair atmosphere, respectively, using TGA-DSC. For synthesized manganese-based oxidematerials to separate air, they should be able to incorporate oxygen atoms into their crystal structure and release oxygen at the same or elevatedtemperatures while varying the oxygen partial pressure of the system.

$$Air + MO_{red(s)} \rightarrow MO_{oxid(s)} + gases(mostly\ N_2)$$

$$MO_{oxid(s)} \rightarrow MO_{red(s)} + O_{2(g)}$$

The results obtainedfrom TGA-DSC showed all investigated compounds can produce high purity oxygenat elevated temperatures. Details of multicomponent oxides preparation, characterization,and high temperature conversion of the oxides under reduction and oxidationconditions will be discussed.

See more of this Session: Oxycombustion of Coal and Other Fuels I
See more of this Group/Topical: Energy and Transport Processes

(43e) Design, Construction and Operation of a Pilot Scale Integrated Pollutant Removal (IPR) Process for Oxy-Combustion

Stephen J. Gerdemann and Danylo Oryshchyn, National Energy Technology Laboratory, U.S. Department of Energy, Albany, OR

NETL has built an Integrated Pollutant Removal (IPR®) system at Jupiter Oxygen's 15MWth oxy-fuel test facility in Hammond, Indiana. An ASPENPLUS model of this installation is also in use to inform the development of the next stage of the IPR system-development. The installed system is designed to process a slip stream of flue gas from an oxy-coal boiler. This flue is composed primarily of CO_2 (~55%) and water (~30%) with smaller amounts O_2, N_2, Ar and other combustion products such as NO_X and SO_X. The IPR process is designed to output a relatively pure (>90%) CO_2 stream at pipeline pressure while recovering as much energy as possible and integrating that recovery into the overall combustion process. Water condensed from the flue-gas is also recovered. Water-treatment process design is being refined through experimentation.

IPR consists of a spray tower (direct-contact heat exchanger) followed by 4 stages of compression (3, 14, 40 & 140 atm) with intercooling between each stage. The process has been completely instrumented to flow rate, temperature and pressure at each stage. Material samples include gas and liquid throughout IPR, and solids (where they fall out in the first IPR operation – a direct contact heat exchange spray tower). In the field, IPR has shown that the achieved the desired output and has further defined challenges for IPR design and the design of its instrumentation, for example accurate flow measurement of a corrosive/erosive fluid stream, minimizing corrosion in components, and in-stream condensation which also efficiently removes soluble gasses. This paper discusses these challenges, presents results from IPR operation, and identifies opportunities for optimization of an oxy-combustion IPR system, also visiting the design of the system model in ASPEN plus.

See more of this Session: Oxycombustion of Coal and Other Fuels I
See more of this Group/Topical: Energy and Transport Processes

(66a) Ionic Liquids: A Potential Role In the Sulfur Family of Thermochemical Cycles

Nick AuYeung, Victoria Johnson, Malachi D. Bunn and Alexandre F.T. Yokochi, School of Chemical, Biological and Environmental Engineering, Oregon State University, Corvallis, OR

The Sulfur-family of thermochemical water splitting cycles has already been identified as an ideal application of ionic liquid solvents for SO_2 absorption.[1] Here, ionic liquids are proposed as suitable reaction media in both the Sulfur-Sulfur cycle (a new cycle based upon the Sulfur-Iodine Cycle) and also the Hybrid Sulfur (Westinghouse) Cycle.

Using ionic liquids as a reaction medium for the Bunsen reaction of the Sulfur-Iodine Cycle, we identified conditions where a stream of H_2S is produced based on a normally undesired side-reaction. This led to a new "Sulfur-Sulfur" thermochemical cycle based upon the production and steam reformation of H_2S:[2]

Step 1A, liquid phase: $4I_2+4SO_2+8H_2O \leftrightarrow 4H_2SO4 + 8HI$

Step 1B, liquid phase: $8HI+H_2SO_4 \leftrightarrow H_2S+4H_2O+4I_2$

Step 2, gas phase: $3H_2SO_4 \leftrightarrow 3H_2O+3SO_2+1.5O_2$

Step 3, gas phase: $H_2S+2H_2O \leftrightarrow SO_2+3H_2$

Instead of the initial Bunsen reaction (Step 1A), the Hybrid Sulfur Cycle begins with an electrochemical reaction between H_2O and SO_2:

$$SO_{2(aq)}+2H_2O \leftrightarrow H_2SO_{4(aq)}+H_{2(g)}$$

As product acid is formed, solubility of SO_2 decreases, requiring an overpotential to drive the reaction forward.[3] Possessing a wide electrochemical window, high conductivity, and high SO_2 solubility, an ionic liquid electrolyte ([BMIM.BF4]) was used for the electrochemical reaction between H_2O and SO_2 in the Hybrid Sulfur Cycle. Cyclic voltammetry studies will be presented showing the electrochemical reaction at SO_2 concentrations much higher than those achievable in aqueous systems.

1. Lee, K. Y.; Taek Gonga, G.; Songb, K. H.; Kima, H.; Junga, K.-D.; Kima, C. S., Use of ionic liquids as absorbents to separate SO2 in SO2/O2 in thermochemical processes to produce hydrogen. *International Journal of Hydrogen Energy* **2008**.

2. Yokochi, A., AuYeung, N. In *A Proposed New Sulfur-Sulfur Thermochemical Cycle*, AIChE National Meeting, Nashville, Tennessee, Nashville, Tennessee, 2009.

3. O'Brien, J. A.; Hinkley, J. T.; Donne, S. W.; Lindquist, S. E., The electrochemical oxidation of aqueous sulfur dioxide: A critical review of work with respect to the hybrid sulfur cycle. *Electrochimica Acta* **2010,** *55* (3), 573-591.

See more of this Session: Advances In Thermochemical Hydrogen Production
See more of this Group/Topical: International Congress on Energy 2011

(66b) Liquid-Liquid Equilibrium Compositions for the Iodine-HI-Water System of the Sulfur-Iodine Cycle At Elevated Temperatures and Pressures

Adam S. Klett, Mark C. Thies and David A. Bruce, Chemical and Biomolecular Engineering, Clemson University, Clemson, SC

Liquid-liquid equilibrium (LLE) phase behavior was investigated for the ternary system iodine-hydroiodic acid-water (I_2-HI-H_2O) at the elevated temperatures and pressures of interest for the reactive distillation column of the Sulfur-Iodine Cycle. A continuous-flow apparatus, with wetted parts fabricated from tantalum-tungsten alloys, was designed and constructed for the highly corrosive conditions of this work. A central feature of the apparatus is the equilibrium view cell, which allows for the observation and discernment of vapor-liquid, liquid-liquid, and liquid-liquid-vapor equilibria for HIx systems.

LLE compositions for all components present were measured for both the lighter, water-rich phase as well as for the heavier, iodine-rich phase at temperatures to 250 °C at pressures approximately 10 bar above the region of VLLE. HI feed compositions from 2-25 wt % were investigated, with the latter compositions nearing the system's plait point.

Implications of the results for the desired separation of the components of interest will also be discussed.

See more of this Session: Advances In Thermochemical Hydrogen Production
See more of this Group/Topical: International Congress on Energy 2011

(66c) Multiple Thermochemical Water-Splitting Cycles for H2 Generation Using Sol-Gel Derived Ferrites In A Packed Bed Reactor

Rahul Bhosale[1], Rajesh Shende[1] and Jan Puszynski[2], (1)Chemical and Biological Engineering, South Dakota School of Mines and Technology, Rapid City, SD, (2)Chemical and Biolgocial Engineering Department, South Dakota School of Mines and Technology, Rapid City, SD

Multiple Thermochemical Water-Splitting Cycles for H_2Generation Using Sol-Gel Derived Ferrites In A Packed Bed Reactor.

Rahul R. Bhosale,Rajesh V. Shende, Jan A. Puszynski

Department of Chemical & Biological Engineering,

South Dakota School of Mines & Technology, Rapid City,SD 57701

Two-stepthermochemical water-splitting process which utilizes mixed valence metal oxidei.e. ferrite based redox reactions is one of the green ways of producing H_2.In this process, the first step belongs to endothermic reduction of ferrite atelevated temperatures by releasing O_2. Whereas,the second step corresponds to the oxidation of the reduced ferrite by taking $O_2$from water and producing H_2 via water-splitting reaction at lowertemperatures. As the current research trends in H_2 generation viathermochemical water-splitting process are focused towards higher levels of H_2production at lower operating temperatures, it is believed that ferrite nanoparticlessynthesized by using sol-gel technique will significantly improve the H_2yield[1]. Sol-gel derived ferrite nanoparticles with higher specificsurface area are believed to be potential candidates to overcome thehydrodynamic limitations associated with this process and might provide highernumber of active sites for water-splitting reactions resulting into higheryields of H_2. In this investigation, phase pure ferrites thatinclude Ni-/Zn-/Sn-/Mn-/Co-ferrite and doped ferrites such as Ni-Zn-/Ni-Mn-/Ni-Co-/Co-Zn-/Co-Mn-ferritewere synthesized using sol-gel technique. As-synthesizedgels were aged for 24 h and heated rapidly in a muffle furnace at 600°Cin air environment and quenched. The calcined powder thus obtained wascharacterized using powder X-ray diffraction, BET surface area analyzer, scanningelectron microscopy (SEM) and transmission electron microscopy (TEM). Theferrite powder was placed in a packed bed reactor and water-splitting reaction wasperformed at different temperatures. After water-splitting reaction, thematerial was regenerated in the temperature range of 900°C 1100°Cand reused for multiple thermochemical cycles. An attempt was made to perform tensof thermochemical cycles at lower water-splitting and regeneration temperaturesof 800 1100°C. Synthesis and characterization of ferrites and transientH_2 profiles obtained using online H_2 sensor duringmultiple thermochemical cycles at various experimental conditions will bepresented. Furthermore, efforts are underway to investigate the water-splittingreaction kinetics and transport modeling of the thermochemical water-splitting ina packed bed reactor.

1. Rahul R. Bhosale, Rajesh V. Shende, Jan A. Puszynski, Thermochemicalwater-splitting for H_2 generation using sol-gel derived Mn-ferritein a packed bed reactor, *International Journal of Hydrogen Energy*, 2011(In press).

See more of this Session: Advances In Thermochemical Hydrogen Production
See more of this Group/Topical: International Congress on Energy 2011

(66d) Nickel Ferrites In the Hercynite Water Splitting Cycle

Christopher Lawrence Muhich[1], Alan W Weimber[2] and Charles B. Musgrave[2], (1)Department of Chemical and Biological Engineering, University of Colorado, Boulder, Bouler, CO, (2)Chemical and Biological Engineering, University of Colorado, Boulder, Bouler, CO

We have deposited nickel ferrites onto alumina supports using atomic layer deposition (ALD) in order to compare the nickel ferrite hercynite solar thermal water splitting (STWS) cycle to the cobalt ferrite hercynite STWS cycle. STWS occurs by cycling metal oxides through a high temperature reduction step and a lower temperature oxidation step where H_2O oxidizes the oxide and produces hydrogen gas. We have carried out redox cycling in a stagnation flow reactor with in situ mass spectrometery to study oxygen and hydrogen formation rates as well as thermodynamic simulations to understand the entropic/ enthapic effects at work. In addition, ALD of ferrites on zirconia supports was carried out to compare the nickel ferrite hercynite cycle with the traditional nickel ferrite cycle.

See more of this Session: Advances In Thermochemical Hydrogen Production
See more of this Group/Topical: International Congress on Energy 2011

THERMOCHEMICAL HYDROGEN PRODUCTION BY ALKALI METAL OXIDES

Hiroki Miyaoka[1], *Takayuki Ichikawa*[2] *and Yoshitsugu Kojima*[2]
[1]*Institute for Sustainable Sciences and Development, Hiroshima University, Higashi-Hiroshima, Japan*
[2]*Institute for Advanced Materials Research, Hiroshima University, Higashi-Hiroshima, Japan*

Introduction

Hydrogen is one of attractive energy storage materials to establish a sustainable energy cycle based on natural primary energy such as solar, wind, and water. Hydrogen has high gravimetric energy density compared with the batteries as electricity storage materials, suggesting that it is suitable media for storage and transportation of the large amount of energy. As the hydrogen production methods from the natural energy, the water-electrolysis by using hydro and wind power generation has already established. For the solar power, the utilization of photovoltaic cell, photo catalysts, and solar– heat power– generation are researching and developing as the electric hydrogen production. In addition to such methods, the hydrogen production techniques by a water– splitting via thermochemical reactions are also attractive because a high efficiency can be expected and they can be easily scaled up as a chemical plant for mass production although it is difficult to enlarge the system in the case of the electrochemical methods.

So far, the attractive thermochemical water-splitting reactions for the hydrogen production have sorted by using database. As a result, the researching thermochemical reactions such as I-S process requires more than 800 °C to generate hydrogen [1-4], suggesting that the heat source to run the systems is limited. If the operating temperature of thermochemical reactions could be lowered, various kinds of heat sources can be utilized for a wide use. In order to realize the lower reaction temperature, we focus on the hydrogen production by water-splitting via thermochemical reactions based on alkali metal oxides as follows,

$$2MOH+2M \rightarrow 2M_2O+H_2 \quad (H_2 \text{ generation}), \tag{1}$$
$$2M_2O \rightarrow M_2O_2+2M \quad (\text{metal separation}), \tag{2}$$
$$M_2O_2+H_2O \rightarrow 2MOH+1/2O_2 \quad (\text{hydrolysis}), \tag{3}$$

where M is alkali metals, lithium Li, sodium Na, and potassium K. The only K-system was reported before, and more than 800 °C was necessary to produce hydrogen [1]. However, it is possible to reduce the operating temperature by controlling the thermodynamics of the reactions.

In this work, we tried to control the reaction temperature of above thermochemical reactions based on alkali metal oxides, Na and K-system, by using nonequilibrium process from material science and chemical engineering points of view. In this paper, the results of Na-system is mainly shown and discussed.

Experiments

Commercial sodium Na (99.9 %, Aldrich), potassium K (99.95 %, Aldrich), sodium hydroxide NaOH (99.998 %, Aldrich), and potassium hydroxide KOH (99 % up, Kojundo Chemical Laboratory) were used as the starting materials. Sodium oxide Na_2O was synthesized by the heat treatment at 300 °C under dynamic vacuum condition for 20 h of NaH (95 %, Aldrich) and NaOH mixture, which was prepared by ball-milling (P7, Fritsch) under 1 MPa of H_2 for 2 h. Sodium peroxide Na_2O_2 was synthesized by heating Na_2O under O_2 atmosphere to 100 °C. Potassium oxide K_2O was synthesized by the reaction between K and O_2 by heating to 100 °C, and potassium peroxide K_2O_2 was prepared by reacting K_2O with O_2 at room temperature.

In this study, it is a key-point that the nonequilibrium condition is somehow realized to reduce the reaction temperature. The thermodynamics of chemical reaction is generally expressed by following Gibbs free energy ΔG,

$$\Delta G = \Delta H - T\Delta S, \quad (4)$$

$$\Delta S = \Delta S^0 + R\ln\left(\frac{p_0}{p_{pro}}\right), \quad (5)$$

where ΔH is the enthalpy change, ΔS^0 is the standard entropy change, R is gas constant, p_0 is standard pressure (constant), and p_{pro} is a partial pressure of the gaseous product. The value of last entropy term in equation (5) is increased by reducing partial pressure of product p_{pro}, indicating that the equilibrium condition of the reaction is changed. As a result, it is expected that the reaction proceeds at lower temperature. For H_2 generation reaction (1), the experimental system was especially designed and assembled by ourselves to realize the nonequilibrium condition. To remove the generated H_2 from the reaction field, the Nb_2O_5-catalyzed magnesium Mg was used, where it was reported by Hanada *et al.* and can absorb H_2 at room temperature [5]. The reaction (1) was performed with 1:1 molar ratio of M:MOH in the temperature range from 350 to 500 °C under Ar atmosphere for 20 h in a closed system. Here, the molar ratio of Mg was chosen to be 1:1 to M, suggesting that the half of Mg is hydrogenated when the reaction proceeds completely. For the metal separation reaction (2), the cooling part was equipped into the reactor to separate the metal vapor from the reaction field by the condensation at there. The heat treatment of M_2O was carried out below 500 °C for 20 h under vacuum condition in a closed system. To investigate the hydrolysis reaction (3), M_2O_2 was heated to 100 or 200 °C together with excess mole amount of liquid H_2O in 0.1 MPa of Ar in a closed system. All the samples were handled in a glove box (MP-P60W, Miwa MFG) filled with purified Ar gas (>99.9999%) to avoid the influence of air.

The solid samples before and after each reaction was identified by X-ray diffraction XRD measurement (RINT-2100, Cu Kα radiation, Rigaku), where all the samples were covered by a polyimide sheet (Kapton®, Du Pont-Toray) in the glove box to avoid the influence of air during the XRD measurement. To quantitatively evaluate the H_2 generated by reaction (1), the Nb_2O_5-catalyzed Mg was analyzed by a thermal desorption mass spectroscopy MS (M-QA200TS, Anelva) connected to thermogravimetry-differential thermal analysis TG-DTA (TG8120, Rigaku). In this thermal analysis, high purity helium (He) gas (>99.9999%) was flowed as a carrier gas and the heating rate was fixed at 5 °C min^{-1}. The gaseous products after the hydrolysis reaction (3) were analyzed by gas chromatography GC (GC-14B, Shimazu), where the reactor was soaked into an iced water during the measurement to freeze the remaining H_2O in it. In the GC measurement, the temperature of the column was set to be 100 °C, and high purity Ar gas was flowed as a carrier gas.

Results and Discussion

Reaction (1): 2MOH+2M → 2M$_2$O+H$_2$

Figure 1 shows XRD pattern of the product on H_2 generation reaction (1). The diffraction peaks corresponding to the starting materials of NaOH and Na were still observed, suggesting that the reaction was not completed. The other peaks generated by the reaction were assigned to Na_2O and NaH. The formation of Na_2O indicates the progress of the expected reaction (1). NaH phase was impurity and would be formed by the reaction between the remaining Na and H_2, which was generated due to the reaction (1), during the cooling process after the heat treatment. Here, NaH should disappear when the reaction proceeds completely because NaH decomposes to Na and H_2 above 300 °C.

In the experiment for the H_2 generation reaction (1) by endothermic reaction, the Nb_2O_5-catalyzed Mg was used as the absorbent of the generated H_2 to realize the nonequilibrium condition. Figure 2 shows the MS and TG-DTA results of the Nb_2O_5-catalyzed Mg. The H_2 desorption was clearly observed as shown in Fig. 2, indicating that the H_2 was generated by the reaction between NaOH and Na. The weight loss by the H_2 desorption was about 2.8 mass%. From the result, the yield of the reaction (1) at 350 °C for 20 h can be estimated to be about 70 % because the half amount of Mg should be hydrogenated in the case of the complete progress on the reaction (1). Furthermore, it was confirmed that the reaction (1) can also proceed at 300 °C. In this case, the reaction yield was about 10 %.

Reaction (2): 2M$_2$O → M$_2$O$_2$+2M

For the reaction (2) by endothermic reaction, the water cooling system was especially attached on the upper part of the reactor to condense the metal vapor at there, resulting that the nonequilibrium condition would be realized. In Figure 3, XRD patterns of the as-synthesized Na_2O, the product at the cooling part, and at bottom part of the reactor after the reaction (2) at 500 °C. From the XRD pattern of the as-synthesized Na_2O, it was confirmed that Na_2O as a main phase included a small amount of NaOH as impurity. After the heat treatment, the metallic and ductile material was found at the cooling part, and it was identified as mainly Na by XRD measurement as shown of middle part in Fig. 3. This result indicates that the expected reaction proceeds. The XRD pattern of the product at bottom part showed many diffraction peaks. The peaks with strong intensity corresponded to Na_2O and NaOH although Na_2O_2 is the expected product of the reaction (2). It was difficult to assign the other weak peaks even though the position of some peaks might be close to Na_2O_2, suggesting that unexpected reactions, e.g. a reaction between Na_2O_2 phase and the reactor, might be occurred to form unidentified phases. The reaction yield was roughly estimated by comparing the ratio of diffraction peaks intensity on Na_2O before and after the reaction, and the yield was estimated to be about 70 %. From the results, the possibility of progress on the reaction (2) at 500 °C was demonstrated because the Na generation was clearly clarified. However, the further research by using the material to prevent above corrosion is necessary to identify the product of the reaction (2).

Reaction (3): M$_2$O$_2$+H$_2$O → 2MOH+1/2O$_2$

The hydrolysis reaction (3) is exothermic reaction, thereby it can proceed even at room temperature thermodynamically. In this work, the reaction was activated by the heating to 100 °C. Figure 4 shows the XRD patterns of the as-synthesized Na_2O_2 and the product after the reaction with H_2O. As shown in upper XRD pattern, the as-synthesized Na_2O_2 contained a small amount of NaOH as impurity. After the hydrolysis reaction, it was confirmed from the XRD results that Na_2O_2 was completely changed to NaOH hydrate, indicating that the hydrolysis reaction proceeded. Here, the formation of NaOH hydrate was caused by a reaction between NaOH generated by the reaction (3) and the remaining H_2O because the excess H_2O was used in this experiment.

In order to identify the gaseous product, GC was performed. The gas chromatograms are shown in Figure 5. By comparison with the results of H_2 and O_2 mixture as reference, it was clearly found that O_2 and a small amount of H_2 existed in the reactor. This result indicates that O_2 was generated by the hydrolysis reaction. The H_2 would be caused by the hydrolysis reaction of the small amount of impurity including the as-synthesized Na_2O_2.ommercial sodium Na (99.9 %, Aldrich), potassium K (99.95 %, Aldrich), sodium hydroxide NaOH (99.998 %, Aldrich), and potassium hydroxide KOH (99 % up, Kojundo Chemical Laboratory)

Conclusion

In this work, the reaction conditions of the water-splitting reaction by alkali metal (Na and K) oxides were investigated as the possible thermochemical hydrogen production system. Especially, the nonequilibrium process was introduced to lower the reaction temperature below 500 °C.

For the reaction (1) $2MOH+2M \rightarrow 2M_2O+H_2$, it was clarified that H_2 can be generated from the NaOH and Na mixture at 350 °C with the formation of Na_2O by removing the H_2 from the reaction field, and then about 70 % of reaction yield was obtained for 20 h. In the case of the reaction (2) $2M_2O \rightarrow M_2O_2+2M$, we succeeded in condensing the generated M by using the cooling system and realizing the nonequilibrium condition. As a result, it is possible that about 70 % of the reaction proceeds at 500 °C for 20 h. However, the side-reaction would be occurred by the reaction between Na_2O_2 and the reactor, suggesting that the corrosion should be prevented to understand the reaction details. It was confirmed that the hydrolysis reaction (3) $M_2O_2+H_2O \rightarrow 2MOH+1/2O_2$ easily proceeded. By the heating to 100 °C, Na_2O_2 was completely changed to NaOH hydrate and O_2 as the expected products.

The K-system was also investigated by using the same experimental systems to utilize the nonequilibrium process and analyses. Only summary for the K-system is mentioned below although the details were omitted in this paper because the obtained results were similar to the Na-system. By the reaction (1), it was confirmed that H_2 was generated at 500 °C however the reaction yield was lower than the Na-system. The reaction (2) proceeded at around 500 °C, and then K was clearly condensed at the cooling part. In this case, the corrosion of the reactor was also found like the Na-system. For the hydrolysis reaction (3), it was clarified that KOH hydrate and O_2 were formed as the reaction products by heating to 200 °C, suggesting that the reaction can proceed easily. However, the K-system required higher activation temperature compared with the Na-system.

From above results, it was demonstrated that the possibility of the H_2 generation below 500 °C by the water-splitting reaction using alkali metal oxide systems. Therefore, it is expected that these systems is developed as one of the possible thermochemical hydrogen production methods for a wide use.

References

1. Yalçn, S. (1989), "A review of nuclear hydrogen production", *Int. J. Hydrogen Energy* 14, pp. 551-561.

2. Sakurai, M. Nakajima, H. Amir, R. Onuki, K. and Shimizu, S. (2000), "Experimental study on side-reaction occurrence condition in the iodine-sulfur thermochemical hydrogen production process", *Int. J. Hydrogen Energy* 25, pp. 613-619.

3. Kodama, T. Kondoh, Y. Yamamoto, R. Andou, H. and Satou, N. (2005), "Thermochemical hydrogen production by a redox system of ZrO_2-supported Co(II)-ferrite", *Solar Energy* 78, pp. 623-631.

4. Gokon, N. Murayama, H. Umeda, J. Hatamachi, T. and Kodama, T. (2009), "Monoclinic zirconia-supported Fe_3O_4 for the two-step water-splitting thermochemical cycle at high thermal reduction temperatures of 1400-1600°C", *Int. J. Hydrogen Energy* 34, pp. 1208-1217.

5. Hanada, N. Ichikawa, T. Hino, S. and Fujii, H. (2006), "Remarkable improvement of hydrogen sorption kinetics in magnesium catalyzed with Nb_2O_5", *J. Alloys Compd.* 420, Aug 31, pp. 46-49.

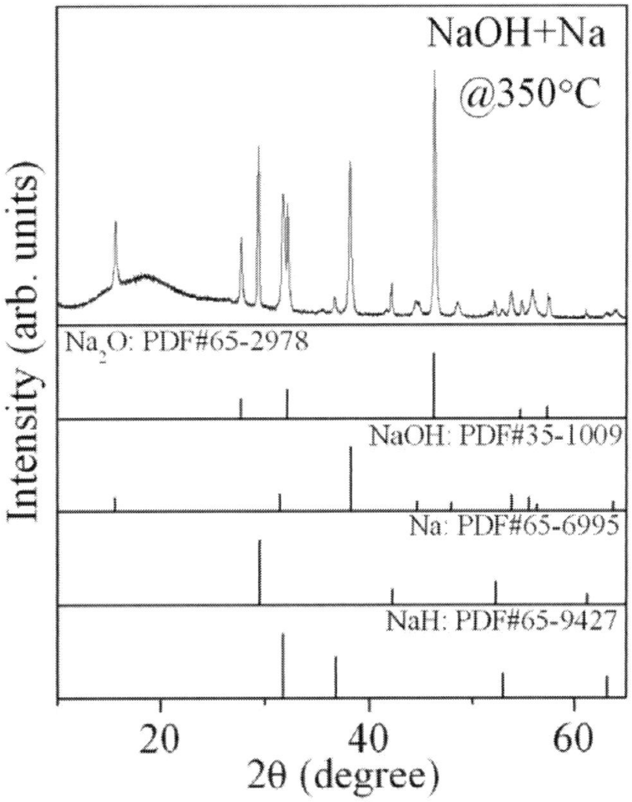

Figure 1 XRD pattern of the product on H_2 generation reaction (1). XRD patterns of Na_2O (PDF #65−2978), NaOH (PDF #35−1009), Na (PDF #65−6995), and NaH (PDF #65−9427) are referred from database.

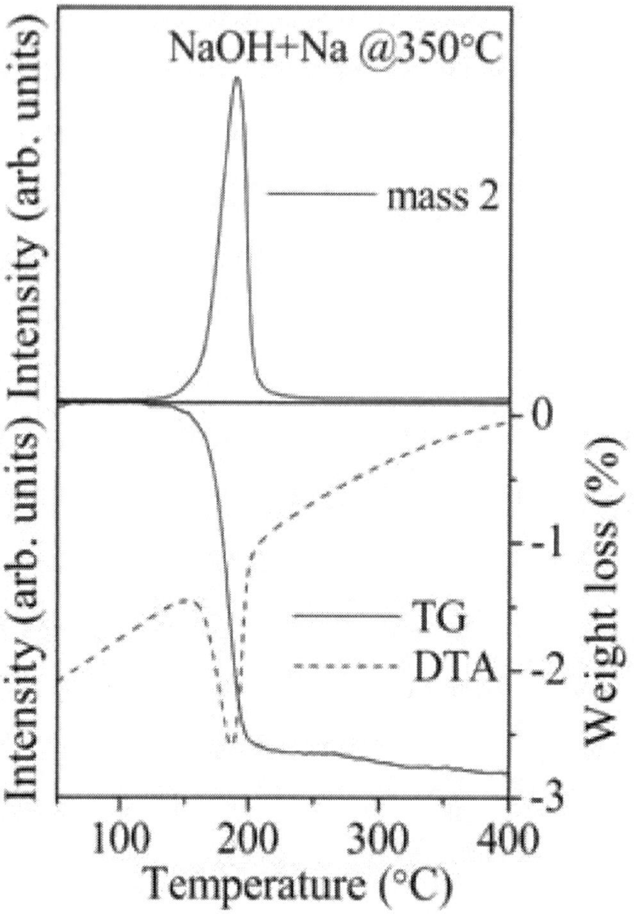

Figure 2 MS and TG−DTA profile of the Nb_2O_5−catalyzed Mg using as the absorbent of H_2 generated by reaction (1).

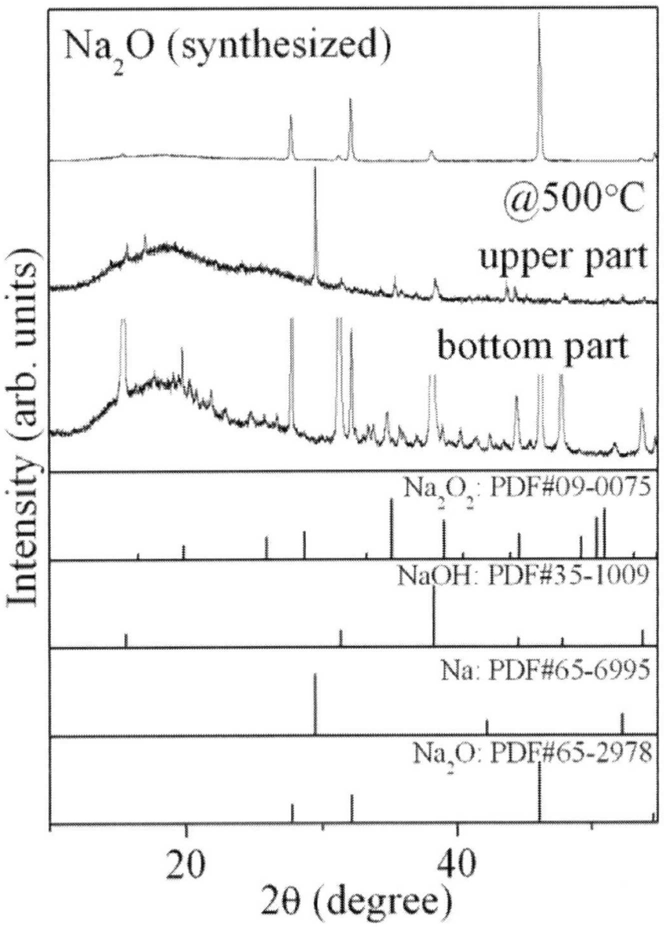

Figure 3 XRD patterns of the as−synthesized Na_2O, the product condensed at the cooling part (upper part) and at bottom part of the reactor by the reaction (2) at 500 °C. XRD patterns of Na_2O_2 (PDF #09−0075), NaOH (PDF #35−1009), Na (PDF #65−6995), and Na_2O (PDF #65−2978) are referred from database.

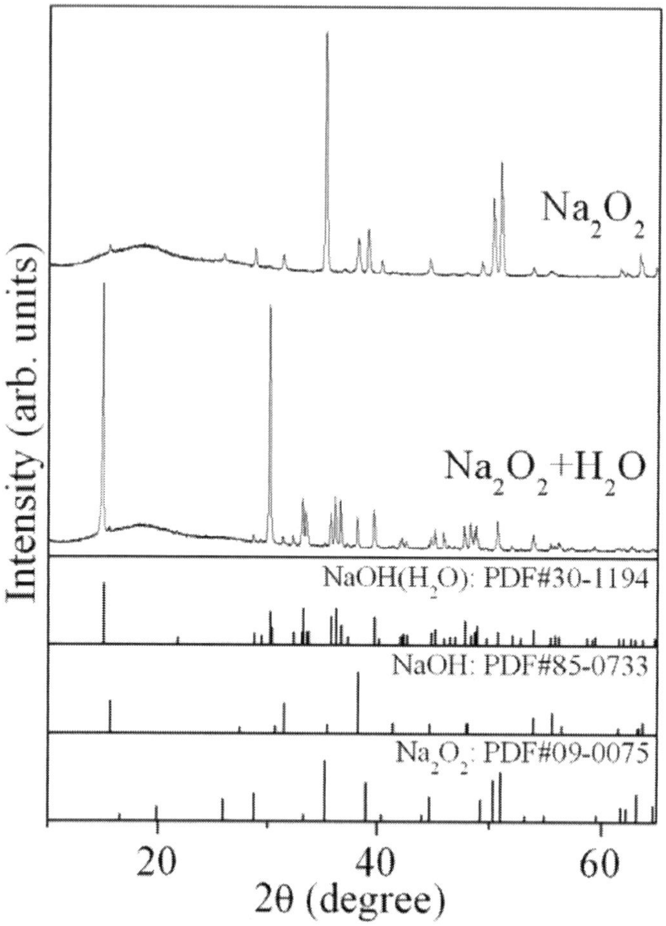

Figure 4 XRD patterns of as−synthesized Na_2O_2 and the product by the hydrolysis reaction (2). XRD patterns of NaOH hydrate (PDF #30−1194), NaOH (PDF #35−1009), and Na_2O_2 (PDF #09−0075) are referred from database.

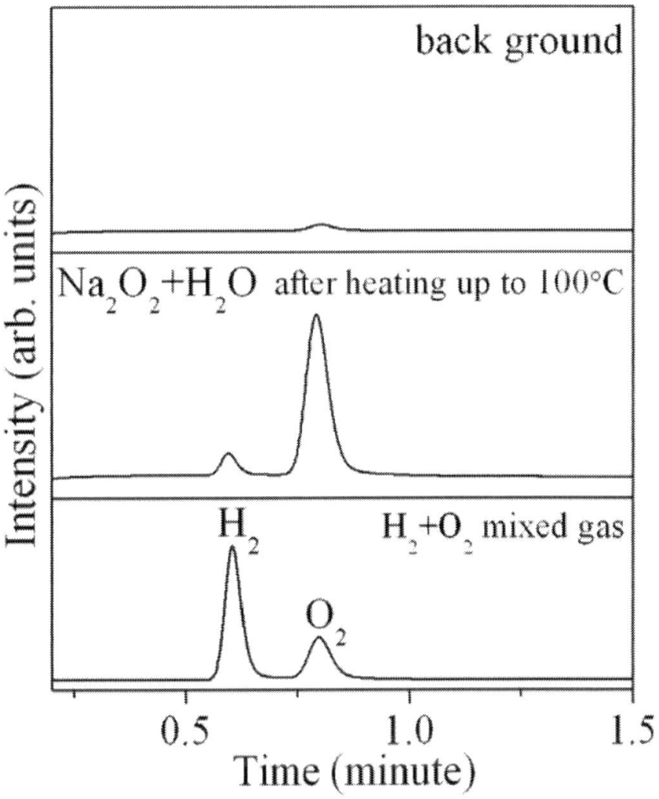

Figure 5 Chromatograms of background, the gaseous products by the hydrolysis reaction (3), and H_2 and O_2 mixture as reference.

Preliminary Cost Estimation of Nuclear Hydrogen Production Plant coupled with Sulfur-Iodine Cycle

Jong-Ho Kim[1*], Tae-Heun Lee[1], Ki-Young Lee[1], Yong-Wan Kim[1] and Kyung-Jin Yang[2]

(1)Korea Atomic Energy Research Institute,Daejeon-si, South Korea
(2)KEPCO-ENC,Yongin-si, South Korea

[1] corresponding author : (T)82-42-868-8846, (F) 82-868-8086, jhkim5@kaeri.re.kr*

Abstract

A preliminary study of cost estimates on the VHTR with SI cycle plant for the nuclear hydrogen production was performed from the perspective of Levelized Unit Hydrogen Production Cost(LUHC).
G4-ECONS developed by EMWG of GIF was appropriately modified to calculate the LUHC for the VHTR coupled with SI cycle. The result indicated estimated hydrogen production cost could be competitive with current techniques of hydrogen production from fossil fuels if CO2 capture and sequestration is required.
The parametric study was done with a modified G4-ECONS for the major factors that would affect the hydrogen production cost. The result indicated that the fixed charge rate and electricity costs have a more significant influence on hydrogen production cost than the construction time.

1. Introduction

Hydrogen is an environmentally attractive fuel that has the potential to replace fossil fuels, but contemporary hydrogen production primarily based on fossil fuels is not practically economical. To achieve the hydrogen economy, it is very important to produce a massive amount of hydrogen in a clean, safe and efficient way. Nuclear production of hydrogen would allow massive production of hydrogen at economic prices while avoiding the release of carbon dioxide. Nuclear production of hydrogen could thus become the enabling technology for the hydrogen economy.
A Very High Temperature Reactor (VHTR) with an outlet coolant temperature of 950 °C is considered as an efficient reactor to couple with the thermo-chemical Sulfur Iodine (SI) cycle to achieve the hydrogen economy[1-2].
KAERI is developing thermo-chemical hydrogen production process called SI cycle utilizing the heat from the VHTR with outlet coolant temperature up to 950 °C.

KAERI's research focuses on the development of the key technologies required for the realization of a nuclear hydrogen production economy. The key technologies comprise development of computational tools and methodologies to analyze the nuclear reactor systems, development of materials and components, development of TRISO coated particle fuel fabrication technology, and development of the SI cycle.

The economic assessment was performed for an nth-of-a-kind plant consisting of PMR coupled with SI plant. The PMR equipment costs were based on benchmarking H2-MHR[2-5].

As a preliminary study of cost estimates for the VHTR coupled with SI cycle, G4-ECONS developed by EMWG of GIF was appropriately modified to calculate the LUHC, assuming 36 months of plant construction time, 5 % of annual interest rate and 12.6 % of fixed charge rate.

2. Description of the Nuclear Hydrogen Production Plant

The massive hydrogen production system using nuclear energy consists of a VHTR, an intermediate loop and SI cycle. The high temperature heat from the VHTR is fully dedicated to the SI (Sulfur-Iodine) hydrogen production as shown in Figure 1.

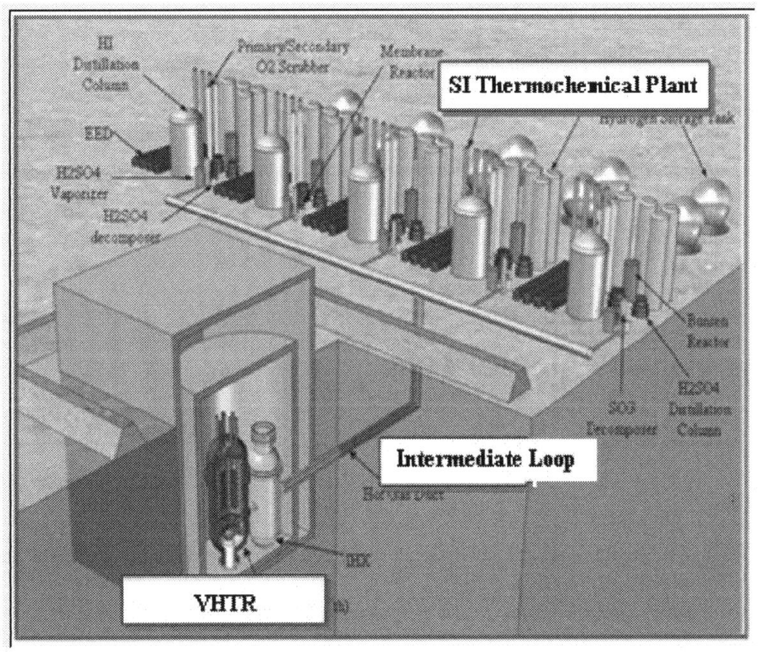

Figure 1. Schematic Diagram of the Nuclear Hydrogen Production Plant

The VHTR is an advanced helium gas-cooled, graphite moderated, thermal neutron spectrum reactor with a core outlet temperature greater than 950°C specified to support production of hydrogen by SI cycle. The reactor thermal power is 200 MWth. It is assumed that no electricity is generated by the VHTR. The VHTR uses block type fuels, graphite moderated with passive safety RCCS system. The VHTR assumes an once-through, LEU (<20% 235U) fuel cycle.

The nuclear heat is used directly to drive a SI cycle. The pressurized helium gas is used as a working fluid to transfer heat from VHTR to SI cycle through the intermediate heat transport loop.

The SI cycle produces hydrogen through a series of chemical reactions. It consists of 3 chemical reactions. The net reaction of the SI cycle is the decomposition of water into hydrogen and oxygen, leaving all process chemicals fully recycled.

The schematic diagram of the SI cycle is shown in Figure 2.

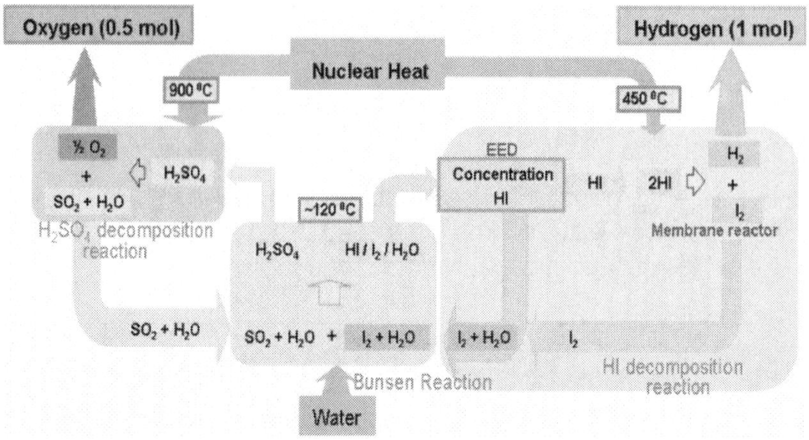

Figure 2. Schematic Diagram of the SI Cycle[1]

3. Cost Estimates

For the evaluation of the economic potential of the nuclear hydrogen production plant using VHTR coupled with SI cycle, preliminary total capital cost and the levelized unit hydrogen production cost(LUHC) were estimated.

The four prismatic modular reactors coupled with SI cycle (PMR, 4 x 200MWth) were chosen for the preliminary cost estimates. The specifications of the nuclear hydrogen production plant are described in Table 1.

Table 1. Specifications of Nuclear Hydrogen Production Plant

Design Parameters	Specification
Reactor Reference Plant	GT-MHR
SI Reference Plant	H2-MHR: SI-Based Plant
Thermal Output	4 x 200 MWth
Operational Period	60 years
Reactor Outlet Temperature	950 ℃
Plant Availability	90 %
Capacity Factor over Life	90 %
Fuel Cycle	Open

The hydrogen production rate for four, 200MWth prismatic modular reactors is calculated to be about 2.40 kg/s (68,000 tons/year), assuming a plant capacity factor of 0.9 and thermal efficiency of 48 %.

To calculate the capital cost for hydrogen production plant of VHTR with SI cycle, capital costs for the GT-MHR and for the H2-MHR based on SI plant were used as a reference, respectively. The cost/size scaling equations are applied to calculate the corresponding capital costs and some input data were modified to reflect country-specific factors for hydrogen production plant of VHTR with SI cycle.

Although the conventional scaling method for sizing and increase of design temperature was used in these calculations, it was no doubt that difference in design specifications between VHTR and GT-MHR gave rise to excessive assumptions in these calculations.

For the LUHC calculation, G4-Excell-based Calculation of Nuclear Systems (G4-ECONS) Version 2.0 was used. The general ground rules and assumptions follow those of G4-ECONS.

The capital costs for nuclear reactor system and SI plant are calculated and summarized in Table 2.

Table 2. Summary of Capital Costs for Nuclear Hydrogen Production Plant

Account Description		Costs ($M)
Reactor System	Buildings & Structures	68.3
	Reactor Plant Equipment	153.6
	Electrical Equipment	21.5
	Water Intake and Heat Rejection Plant	18.5
	Miscellaneous Plant Equipment	14.5
	Reactor Indirect Costs	71.7
SI Plant	Bunsen Reaction	43.9
	H_2SO_4 Decomposition	81.9
	HI Decomposition	168.6
	Vapor Recompression Equipment and Auxiliary Equipment	162.0
	Iodine Inventory	28.0
	SI Indirect Costs	70.5
Primary and Secondary HTS Capital costs		163.4
Capitalized Support, Operation, Supplementary Costs		389.4
Overall Contingency		87.1
Total Plant Capital Investment		1542.9

The cost of the turbo-generator was excluded while the costs of an intermediate heat exchanger, circulators and piping were included. Most of non-fuel operational annual recurring costs were considered and the capital costs for chemical inventory for SI plant were included. The costs include all direct and indirect costs, plus interest during construction. The reactor operating costs include all fuel cycles costs (fuel, conversion, enrichment, fabrication, waste disposal and decommissioning), normal operation costs, maintenance costs and high purity water cost.

The O&M cost for nuclear hydrogen production plant are presented in Table 3. The total annual O&M costs include a fixed and variable cost for maintenance materials, supplies and expenses, and offsite technical support plus overall administrative and general costs.

Table 3. Summary of Annualized O&M Costs for Nuclear Hydrogen Production Plant

Account Description	Costs($M/yr)
Nuclear Reactor O&M Costs	16.8
SI Plant O&M Costs	37.0
Nuclear Fuel Costs	40.2
Total Plant O&M Costs	94.0

With the exception of electricity costs, the O&M costs given in Table 3 are appropriately scaled up, based on the results of GA by Brown [1] in which the cost estimate was done for the four 600MWth prismatic modular reactors coupled with SI hydrogen production plant. The hydrogen production system consumes a lot of electricity mostly for shaft work associated with the SI process. The cost was about 188.0 $M/yr on a Korean specific basis for four, 200MWth prismatic modular reactor.

4. Results and Discussion

The total capital and operating costs, and O&M cost for the nuclear hydrogen production plant were estimated and used as an input to the G4-ECONS program to calculate the LUHC. The economics of nuclear hydrogen production plant was estimated based on LUHC. The costs for nuclear hydrogen production plant are summarized on an annual basis in Table 4. As shown in Table4, the LUHC was estimated to be 3.26 $/kg H_2.

Table 4. Summary of SI-based Hydrogen Production Costs ($M/yr)

Account Description	PMR ($M/yr)	Percentage of Total Cost
Reactor Capital Charges	50.3	
SI Plant Capital Charges	35.9	
Reactor O&M Costs	16.8	
SI Plant O&M Costs	37.0	
Nuclear Fuel Costs	40.2	
Electricity Costs	219.9	
Total Annual Costs	400.1	
Hydrogen Production Costs	3.26 $/kg	

Parametric studies were performed to determine the sensitivity on the LUHC of the major factors such as construction time, fixed charge rate and electricity cost.
The results of the study are summarized in Table 5.

Table5. Results of parametric studies

Parameters		Hydrogen Production Cost ($/kg)
Construction Time(Month)	36	3.27
	48	3.28
	60	3.30
Fixed Charge Rate(%)	5.3	3.26
	12.6	4.15
	15.9	4.60
Electricity Costs ($/kWhr)	0.035	2.58
	0.06	3.26
	0.09	4.08

As shown in Table 5, the fixed charge rate and electricity costs have a significant influence on hydrogen production cost. However, the effect of plant construction time is much less than those of fixed charge rate and electricity costs and results in only 0.03 $/kg ~ 0.06 $/kg increase in hydrogen production cost.

ACKNOWLEDGEMENTS

This study has been carried out under the Nuclear Research and Development Program financially supported by the Korean Ministry of Education, Science and Technology.

References

[1] L. C. Brown et al, et al, 2003 "High efficiency generation of hydrogen fuels using nuclear power," Final report to DOE, General Atomics report GA-A24285.
[2] J. H. Chang et al., A study of a Nuclear Hydrogen Production Demonstration Plant, Nuclear Engineering and Technology, Vol. 39, No. 2, pp. 111-122.
[3] M. B. Richards et al, 2006, "H2-MHR pre-conceptual design report: SI-based plant," Final report to DOE, General Atomics report GA-A25401.
[4] Gen-IV EMWG Members, 2007, "Cost estimating guidelines for Gen-IV nuclear energy systems," Gen-IV International Forum.
[5] K. A. Williams and K. Miller, 2006, "A user's manual for G4-ECONS Version 1.0: A generic EXCEL-based model for computation of the projected Levelized unit electricity cost (LUEC) from Generation IV reactor systems," Gen-IV International Forum.

Materials with Hierarchical Porosity for Energy Storage and Conversion

Andreas Stein, Jane Davidson,
Nicholas Petkovich, Yuqiang Qian, Stephen Rudisill, Luke Venstrom, Anh Vu
University of Minnesota, Minneapolis, MN 55455, U.S.A.

Introduction

Materials containing pores in size ranges from micropores to macropores offer interesting features for a number applications involving energy storage and conversion, such as batteries, fuel cells, hydrogen storage, and sunlight–to–fuel conversion. Depending on the specific application, they can provide large surface areas for reaction, interfacial transport, or dispersion of active sites; they can provide nanostructured features which enhance reactivity, alter materials properties, or shorten diffusion paths; they can act as host materials to stabilize other active components; or, in the case of porous carbons, they can provide electrically conductive phases as well as intercalation sites. However, the higher reactivity brought about by nanostructured features can also lead to decreased stability, particularly in applications where phase changes may alter the structure of the material or where high temperatures are employed. This talk will focus on two redox systems (lithium-ion batteries and sunlight-to-fuel conversion), in which templated porous materials provide a platform for either electrical energy storage or light-to-chemical energy conversion. In the first system the role of pore architecture in carbon-based electrodes will be discussed. In the second system, we investigate the role of porosity in ceria-based materials of interest for solar thermal splitting of water or carbon dioxide to produce hydrogen or carbon monoxide fuels, respectively.

Experimental Section

Three-dimensionally ordered macroporous (3DOM) carbon monoliths and 3DOM carbon with mesoporous walls (3DOM/m C) were synthesized by colloidal crystal templating and nonionic surfactant templating as decribed previously.[1] Tin, tin oxide, sulfur or $LiFePO_4$ were incorporated in the porous carbon materials by solution impregnation, vapor phase impregnation or direct triconstituent syntheses.[2,3] 3DOM CeO_2 and 3DOM $Ce_{0.8}Zr_{0.2}O_2$ powders were synthesized by colloidal crystal templating from $CeCl_3·7H_2O$ and $ZrOCl_2·8H_2O$ precursors. These samples were compared to a mesoporous CeO_2 sample prepared by thermal decomposition of $Ce(NO_3)_3·6H_2O$ in air at 723 K (DS CeO_2) and a commercial polycrystalline ceria sample (Alfa-Aesar, 99.9%). In-situ ceria reduction with hydrogen and oxidation with water or CO_2 were carried out in a fixed bed, vertical tubular flow reactor. Additional details about syntheses, characterization and test reactions are provided in references 1–4.

Results and Discussion

Improved rate capabilities for lithiation/delithiation are observed for hierarchically porous carbon electrodes. In composites with tin or tin oxide for anodes of lithium ion batteries, these structures maintain electrical contact between tin-based particles, even when those particles undergo significant volume changes during cycling, and hence the composite anode maintains good capacities over multiple cycles (Figure 1). Composites of hierarchically structured carbon with poorly conducting but otherwise desirable electrode materials (like sulfur or $LiFePO_4$) can be used to overcome limitations in electrical conductivity of those materials, increasing the choice of useful electrode materials.

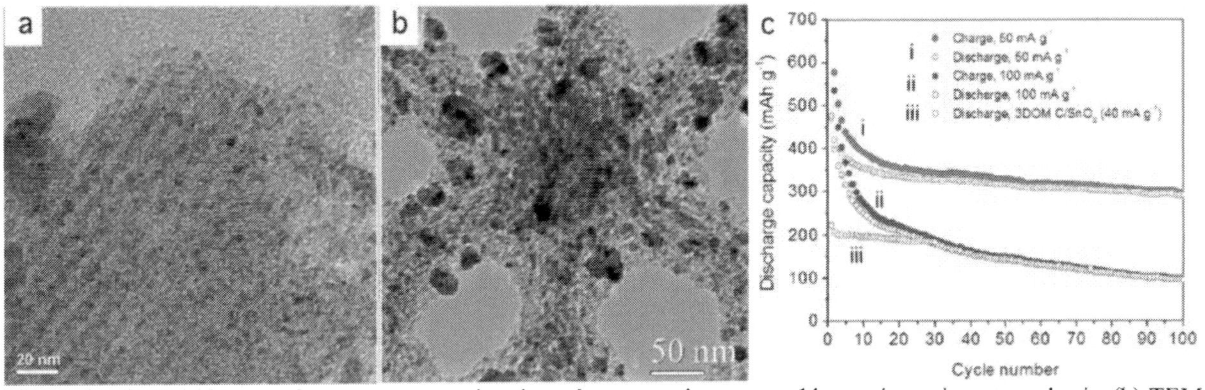

Figure 1. (a) TEM image of a mesostructured carbon-tin composite prepared by a triconstituent synthesis. (b) TEM image of a carbon-tin oxide composite structure with hierarchical porosity. (c) Lithiation charge and discharge curves of 3DOM/m C/SnO_2 (i), 3DOM C coated with SnO_2 (ii) and 3DOM carbon (iii) as function of cycle number.[2]

In reactions targeting the production of chemical fuels (H_2, CO) from water and carbon dioxide, the heterogeneous conversion processes benefit from added surface area in 3DOM CeO_2 samples that facilitate the oxidation cyle. While overall productivity does not depend strongly on sample morphology (given sufficient reaction time), the addition of porosity improves peak fuel production rates and reactivities (Figure 2). However, a mesoporous ceria sample displaying an even higher surface area (DS CeO_2) quickly lost activity as the material sintered during the reaction. The loss in surface area due to sintering of 3DOM CeO_2 was less and could be improved further by doping of ceria with zirconia, which stabilized the porous structures at the high reaction temperatures. The interplay between composition and morphology of these materials, thermal stability, and conversion efficiencies will be discussed.

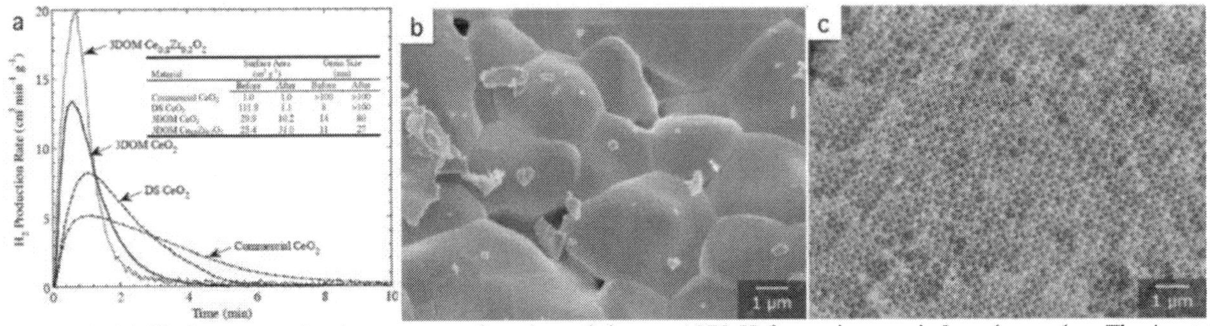

Figure 2. (a) Hydrogen production rate as a function of time at 1073 K for various ceria-based samples. The inset tabulates specific surface areas and grain sizes of the various ceria-based samples before redox cycling and after redox cycling at 1073 K. The SEM images show (b) a chemically cycled commercial CeO_2 sample and (c) a cycled 3DOM $Ce_{0.8}Zr0._{2}O_2$ sample.[4]

References

1. Wang, Z.; Stein, A. *Chem. Mater.* **2008**, *20*, 1029-1040
2. Wang, Z.; Fierke, M. A.; Stein, A. *J. Electrochem. Soc.*, **2008**, *155*, A658-A663.
3. Vu, A.; Stein, A. *Chem. Mater.* **2011**, *23*, 3237-3245.
4. Venstrom, L.; Petkovich, N.; Rudisill, S.; Stein, A.; Davidson, J. H. *J. Sol. Energ.-T. ASME* **2011**, submitted.

(106b) Investigation of Charge Storage Mechanism of Nanostructured Metal Nitrides and Carbides Electrodes for Electrochemical Capacitors

Priyanka Pande, Paul G Rasmussen and Levi T. Thompson, Department of Chemical Engineering, University of Michigan, Ann Arbor, MI

Electrochemical capacitors (ECs) offer aunique combination of high power and energy densities filling the gap betweenconventional capacitors and batteries. These devices store energy via doublelayer (i.e. electrostatic mechanism) and/or pseudo-capacitance mechanisms thatinvolve charge transfer processes [1]. Early transition metal nitrides andcarbides are promising candidates for use as electrode materials due to theirhigh electronic conductivities, surface areas (up to 200 m^2/g) andelectrochemical stabilities [2,3]. For example, vanadium nitride has beenreported to have the highest capacitance up to 1340 Fg^{-1} [4].However the charge storage mechanism for these materials is not wellunderstood. In this paper we will report results from ion isolation experiments,impedance spectroscopy, charge-discharge and chronotentiometry and suggest chargestorage mechanisms that reconcile the results.

Nanostructured V, Mo, W nitrides andcarbides were synthesized via temperature-programmed-reaction of their oxideprecursors with anhydrous NH_3 or 15% CH_4/H_2followed by passivation in 1% O_2/He at room temperature to form aoxygen-rich passivation layer preventing bulk oxidation on exposure to air[2,3]. Physical characterization was performed using BET surface area analysis,scanning electron microscopy and X-ray diffraction. The stability of these materials in aqueous KOH and $H_2SO_4$was determined by cyclic voltammetry. The species contributing to chargestorage were identified by isolation of the electrolyte ions and pairing themwith inactive counter ions. H^+ and SO_4^{2-}, wereisolated as $H^+BF_4^-$ (tetrafluoroboric acid) and$[(C_2H_5)_4N^+]_2\ SO_4^{2-}$ (tetraethylammoniumsulfate, $(TEA)_2SO_4$), while K^+ and OH^-ions were isolated as $K^+(CF_3SO_3)^-$(potassium triflate, K-triflate) and $(C_2H_5)_4N^+OH^-$(tetraethylammonium hydroxide, TEA-OH). Cyclic voltammetry experiments werecarried out in each of these electrolytes. In order to establish the chargetransfer reaction chronopotentiometry experiments were carried out in varyingelectrolyte concentrations.

The redox behaviorconfirmed that most of the charge storage was via the pseudocapacitancemechanism. The ion isolation experiments indicated that OH^- and H^+were the primary species contributing to charge storage in VN in KOH (figure 1)and Mo_2N in H_2SO_4 systems, respectively. Thechronopotentiometry results (figure 2) suggest that the charge storage reactionfor VN was:

$$VN_2O_3 - 2OH^- \leftrightarrow [VN_2O_3(OH)_2]^- - e^-$$

Results from impedancespectroscopy and three-electrode charge-discharge experiments will also bepresented

REFERENCES

[1] SimonP, Burke A, The ElectrochemicalSociety: Interface, Spring(2008) 38.

[2] Cladridge J B, York A P E,Brungs A J, Green Malcolm L H, Chem. Mater.12(2000) 132.

[3] WixomM R, Tarnowski D J, Parker J M, Lee J Q, Chen P -L, Song I, Thompson L T, Mat.Res. Soc. Symp. Proc. 496 (1998) 643.

[4] Choi D, Kumta P N, Electrochem. Solid-StateLett. 8 8 (2005) A418.

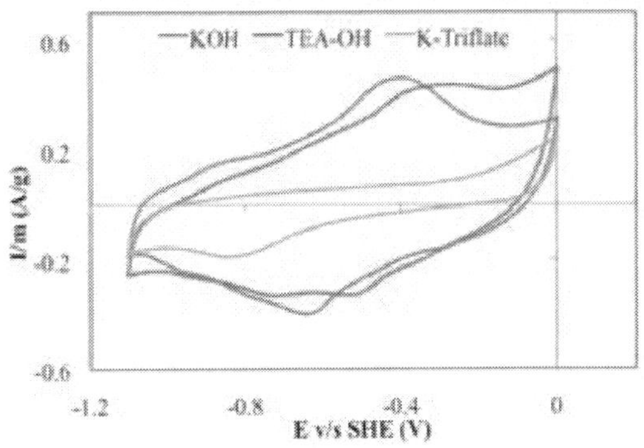

Figure 1: Cyclic voltammogram for VN in various electrolyte solutions at a scan rate of2mVs^{-1}.

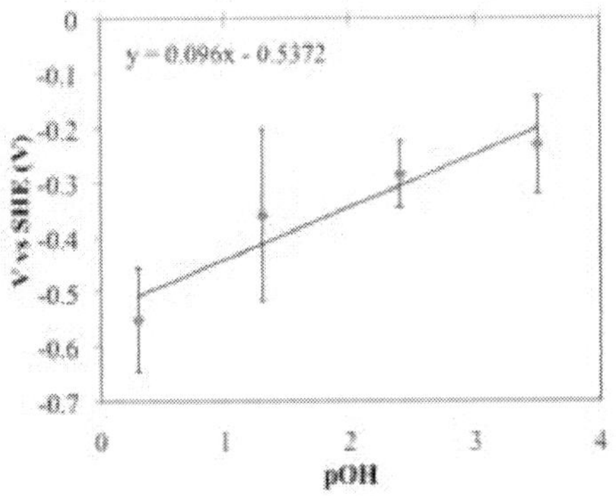

Figure 2: pOH *vs*Voltage for VN in KOH. Slope of 0.096 indicates that there are 2 OH$^-$ ions reacting per electron transferred.

See more of this Session: Nanomaterials for Energy Storage II
See more of this Group/Topical: Topical 5: Nanomaterials for Energy Applications

(106c) High-Performance Electrochemical Capacitors Based On Hybrid Graphene/Metal Oxide Nanostructured Textiles

Guihua Yu[1], Liangbing Hu[2], Yi Cui[2] and Zhenan Bao[1], (1)Chemical Engineering, Stanford University, Stanford, CA, (2)Materials Science and Engineering, Stanford University, Stanford, CA

Large scale energy storage system with low cost, high power and long cycle life is crucial for addressing the energy problem when connected with renewable energy production. To realize large-scale applications of the energy storage devices, there remain several key issues including the development of low-cost, high-performance materials that are environmentally friendly and compatible with low-temperature and large-scale processing. In this talk, we will present a novel structure based on graphene, transition metal oxide and textiles where solution-exfoliated graphene nanosheets conformably coated on highly porous textiles function as three-dimensional conductors and electrodeposited metal oxide nanostructures function as active electrode materials. Electrochemical capacitors based on hybrid graphene/metal oxide based textiles in aqueous electrolytes exhibit very promising characteristics with a power density of ~110 kW/kg, and excellent cycling performance of >95% capacitance retention over 5000 cycles, which are substantially better than those previously reported for this oxide-based system. Such low-cost, high-performance energy textiles based on all-solution-processed graphene/metal oxide hierarchical nanostructures offer great promise in grid-scale energy storage device applications.

See more of this Session: Nanomaterials for Energy Storage II
See more of this Group/Topical: Topical 5: Nanomaterials for Energy Applications

(106d) MnO2 Doped Graphene As Electrodes in Supercapacitor

Lixin Wang[1], Kapila Wadumesthrige[1], Steven O. Salley[2] and K. Y. Simon Ng[3], (1)Chemical Engineering, Wayne State University, Detroit, MI, (2)Chemical Engineering and Material Science, Wayne State University, Detroit, MI, (3)Wayne State University, Detroit, MI

Abstract:

Graphene doped carbon nanotubes for high-energy supercapacitors

A supercapacitor is an electrochemical capacitor with a relatively high energy density and is characterized by fast discharge rates and low maintenance. These characteristics have already been explored in various applications such as consumer electronics, wind energy system, smart grid applications and electric vehicles. The primary purpose behind the burgeoning effort in supercapacitor research is to achieve greater energy densities at a lower cost. Vertical carbon nanotube (CNT) is a promising electrode material for supercapacitor production with a superior conductivity, large voltage window of stability, high surface area, steady mechanical strength and low cost. Capacitance values up to 135 F/g were reported with multi-walled carbon nanotubes. However, the relatively small inherent capacitance of CNTs confines their use to supercapcitor applications. In this study, using different methods, graphite will be doped into CNT walls to improve its capacitance. The doped CNTs will then be used in supercapacitor electrode applications and their performance tested in order to find a suitable method of improving overall capacitance. 200 F/g capacitance is expected after the optimization of doping materials and doping methods.

See more of this Session: Nanomaterials for Energy Storage II
See more of this Group/Topical: Topical 5: Nanomaterials for Energy Applications

(106e) Thermopower Wave Dynamics: Nanoscale Alternating Current Sources and Geometry Effects

Joel T. Abrahamson[1], Wonjoon Choi[2], Sayalee G. Mahajan[1], Nicole S. Schonenbach[1], Jungsik Park[1], Michael P. Walsh[1], Jae-Hee Han[1], Kourosh Kalantar-Zadeh[3] and Michael S. Strano[1], (1)Department of Chemical Engineering, Massachusetts Institute of Technology, Cambridge, MA, (2)Mechanical Engineering, Massachusetts Institute of Technology, Cambridge, MA, (3)School of Electrical and Computer Engineering, RMIT University, Melbourne, Australia

The nonlinear coupling between exothermic chemical reactions and a nanostructure with large heat conduction results in a self-propagating thermal wave guided along the nano-conduit in the direction of its greatest thermal diffusivity [1,2]. The resulting reaction wave pushes charge carriers along, creating a concomitant thermopower wave and electrical current in the same direction. For carbon nanotubes, examples of one-dimensional conduits, we have measured high power density (> 7 kW/kg) from such waves. However, two-dimensional conduits such as graphene should also provide new avenues for control of wave velocity and temperature due to additional degrees of control from geometry and boundaries. We have developed the theory and mathematics of such waves for one- and two-dimensional conduits.

For one-dimensional conduits, our models predict that certain values of the chemical reaction kinetics and thermal parameters produce oscillating reaction wave velocities with well-defined frequency and amplitude [3]. We demonstrate such oscillations experimentally using a cyclotrimethylene-trinitramine/multiwalled carbon nanotube system, which produces frequencies in the range of 400 to 5000 Hz. The propagation velocity oscillations and the frequency dispersion are well described by our theoretical calculations and are linked to oscillations in the voltage generated by the reaction. These thermopower oscillations may enable new types of nanoscale power and signal processing sources.

[1] W. Choi, S. Hong, J. T. Abrahamson, *et al. Nature Materials*, 9, (2010), 423.

[2] W. Choi, J. T. Abrahamson, J. M. Strano, M. S. Strano. *Materials Today,* 13, (2010), 22.

[3] J. T. Abrahamson, *et al. ACS Nano* 5, (2011), 367.

See more of this Session: Nanomaterials for Energy Storage II
See more of this Group/Topical: Topical 5: Nanomaterials for Energy Applications

To be submitted to AIChE 2011 Annual Meeting, October 16-21, 2011, Minneapolis, MN

The Influence of Swirl and O_2 addition on Flame Stability, NO Formation, Carbon Burnout, and Flame Temperature in a High Oxygen Participation Pulverized Coal Burner

Darrel Zeltner, Teri S. Draper, and Dale R. Tree
Brigham Young University

Chendhil Periasamy, Taekyu Kang, Yuan Xue, and Remi Tsiava
Air Liquide

Three swirl vanes with different angles were used in a 150 $kW_{th,}$ oxy-fuel, pulverized-coal burner. Concentrations and total flow rates for O_2 and CO_2 were also changed and investigated with respect to swirl number. Gas measurements were recorded and ash was collected and analyzed for burnout at each condition. Data showed that there is a general trend of decreasing NO and increasing burnout with increasing swirl number. Addition of O_2 to the center of the primary fuel stream caused an increase in NO and a decrease in burnout. The addition of CO_2 to the secondary stream increased NO but generally improved burnout. Increased NO can be attributed to condition which increased the oxygen concentration of the near burner region where volatiles were oxidized while decreased burnout could be attributed to conditions where coal had a short residence time in the near burner reverse flow region.

Line of Sight Soot Measurements in Staged and Unstaged Air and Oxy-Coal Flames

Curtis Stimpson, Trevor Blanc, and Dale R. Tree
Brigham Young University

Will Morris and Yost Wendt
University of Utah

Andrew Fry
Reaction Engineering International

ABSTRACT

Soot concentration has been measured by an in-situ, line-of-sight extinction method at three axial distances from the burner in a 40 kW_{th}, down-fired, pulverized coal reactor. The oxy-fuel combustor (OFC) located at the University of Utah, can be fired with air, O_2/CO_2, and O_2 with flue gas recycle (FGR). Soot measurements were taken with two coals, a PRB sub-bituminous and Skyline bituminous coal. The burner was run under staged and unstaged conditions. For staged conditions the near burner S.R. was nominally 0.9 or 1.0 followed by burnout oxidizer at SR 1.17. Unstaged combustion was conducted at a nominal SR of 1.17. The flames for all conditions were lifted from the burner allowing oxidizer entrainment prior to ignition. Soot concentration was seen to decrease with increasing SR as would be expected with the additional availability of oxygen. Both coals produced relatively low soot concentrations with the highest soot volume fractions measured being 2.0 E-07 for the Skyline coal. The Skyline coal produced more soot at a given SR than the PRB coal. A general trend of decreased soot with increased axial distance from the burner was found. In the cases where O_2 in CO_2 or O_2 in FGR was used as oxidizer, the flame had a longer lift-off length and produced slightly less soot. Trends in soot concentration can be explained by changes in the SR of the fuel/oxidizer mixture at the point of ignition with increased mixing producing less soot. The ratio of oxidizer to fuel at the point of ignition appears to be more significant than the chemical kinetic effects produce by the different oxidizer compositions.

INTRODUCTION

Oxycombustion is a technology that can enable the capture of CO_2 produced by coal-fired boilers, but requires additional insights in order to optimize and integrate the various systems required. One area which requires more research is heat transfer in the burner region. Radiation is the principle mode of heat transfer in the near burner region which is typically dominated by soot. It is therefore of interest to understand differences in soot formation between air and oxy-combustion. Several investigators have reported an increase in soot formation when shifting from air to oxy-combustion. Shaddix and Molina[1] reported larger soot clouds when Pittsburgh coal was burned in O_2/CO_2 mixtures compared to O_2/N_2 mixtures. Andersson et al.[2,3] measured radiation intensity from gas flames with air and oxy-combustion.

[1] Shaddix C.R., and Molina A., "Particle Imaging of Ignition and Devolatilization of Pulverized Coal During Oxy-Fuel Combustion," Proceedings of the Combustion Institute, 32(2):2091-2098, 2009.
[2] Andersson, K., Johansson, R., Johnsson, F. and Leckner, B., "Radiation Intensity of Propane-Fired Oxy-Fuel Flames: Implications for Soot Formation," Energy and Fuels, 22: 1535-1541, 2008.

They found an increase in soot concentration for oxy-combustion, but in propane measurements with recycled flue gas, soot was almost eliminated at high flow rates of recycled flue gas (RFG) (21% O_2 in the RFG) and was higher at lower flow rates of RFG (27% O_2 in the RFG) suggesting that burner dynamics had a greater influence on soot formation than kinetics. Soot concentration measurements in coal flames are rarely performed, especially in oxy-combustion. Previous measurements have inferred soot concentrations qualitatively from luminosity measurements. This work utilized a direct line-of-sight extinction measurement to compare soot concentrations in air and oxy-fired coal flames.

EXPERIMENTAL METHOD

Background

One frequently used method for measuring soot in flames is line-of-sight laser light extinction. If the complex refractive index of soot is known and assumptions are made regarding the soot shape or morphology, the average soot volume fraction along a given path can be determined. Measuring soot in a coal flame is more challenging than pure hydrocarbon flames because of the presence of other solid particles such as coal, char, and ash. Transmittance measurements of two laser beams at two separate wavelengths can be used to overcome this difficulty.

Soot particles, coal, char and ash all attenuate light volumetrically, and as such the transmittance of light through the particles is subject to Beer's law as shown in Eq. 1; κ is the extinction coefficient and L is the path length. For spherical particles, the extinction coefficient is the product of: 1. Q_{ext}, the fraction of incident light on a particle that is absorbed by the particle 2. The projected surface area of a particle, and 3. The number density (N/V) of the particles as shown in Eq. 2. The extinction coefficient is therefore the key parameter that must be determined to convert extinction measurements to soot volume fractions.

$$\tau = \frac{I}{I_0} = \exp(-\kappa L) \qquad \text{Eq. 1}$$

$$\tau = \frac{I}{I_0} = \exp\left(-Q_{ext}\frac{\pi d^2}{4}\frac{N}{V}L\right) \qquad \text{Eq. 2}$$

When determining the extinction coefficient, the size parameter $\alpha = \pi d/\lambda$ can be helpful in identifying simplifications which can be made. Table 1 shows the relationship between the size parameter and the extinction coefficient based on information from Kerker[4]. The dependence of Q_{ext} on wavelength is strong for small particles but becomes independent of wavelength for large particles. This difference in the dependence of the extinction coefficient on wavelength is utilized to separate the extinction due to coal, char, and ash from extinction due to soot particles.

---

[3]Andersson, K., Johnsson, F., "Flame and Radiation Characteristics of Gas-Fired O_2/CO_2 Combustion," Fuel, 86, 656-668, 2007.
[4]Kerker, M., (1969), *The Scattering of Light*, Academic Press, New York.

Table 1. Dependence of the extinction coefficient on the size of particles.

Size Parameter	Qext	Comment
$\alpha < 0.1$	Rayleigh Limit $Q_{ext} = 4\frac{\pi d}{\lambda}\left\|\frac{m^2-1}{m^2+2}\right\|$	Unagglomerated soot particles are in this size range
$0.1 < \alpha < 50$	Full Mie Solution $Q_{ext} = f(\lambda, m, d)$	Soot can be in the lower end of this size range
$50 < \alpha$	$Q_{ext} = 2.0$	Ash and coal particles are in this size range

The total extinction by the two groups of particles is given by Eq. 3 where the total transmittance is the product of the extinction due to soot and the extinction due to coal, char, and ash (cca).

$$\tau = \frac{I}{I_0} = \exp(-\kappa_{\lambda,soot}L)\exp(-\kappa_{cca}L) \qquad \text{Eq. 3}$$

Using two measured wavelengths, an equation for each measured transmittance can be written. Taking the natural log of each equation and subtracting one transmittance equation from the others produces Eq. 4. In this equation, the two terms $\kappa_{cca}L$ can be cancelled because there is no wavelength dependence of the coal char and ash; this leaves only the extinction coefficient $\kappa_{\lambda,soot}$ to be determined from the ratio of the measurements of transmittance at two wavelengths as shown in Eq. 5.

$$\ln\frac{\tau_{\lambda 1}}{\tau_{\lambda 2}} = \left(-\kappa_{\lambda 1,soot}L - \kappa_{cca}L\right) - \left(-\kappa_{\lambda 2,soot}L - \kappa_{cca}L\right) \qquad \text{Eq. 4}$$

$$\ln\frac{\tau_{\lambda 1}}{\tau_{\lambda 2}} = \left(-\kappa_{\lambda 1,soot}L\right) - \left(-\kappa_{\lambda 2,soot}L\right) \qquad \text{Eq. 5}$$

Hottel and Broughton[5] introduced an empirical correlation for soot transmittance as a function of wavelength where $\kappa_{\lambda,soot} = K/\lambda^{1.39}$. This leads to Eq. 6 where KL is the optical thickness. Note that Hottel and Broughton's correlation is not homogeneous, therefore care should be taken to use units of micrometers to express the wavelengths and path lengths.

$$\ln\frac{\tau_{\lambda 1}}{\tau_{\lambda 2}} = \left(\kappa_{\lambda 2,soot} - \kappa_{\lambda 1,soot}\right)L = KL\left(\frac{1}{\lambda_2^{1.39} - \lambda_1^{1.39}}\right) \qquad \text{Eq. 6}$$

Once the optical thickness KL is determined, the soot volume fraction can be calculated. There are numerous models that can be used to relate the measured KL of the soot to a soot volume fraction. All of these models require information on the soot particle size, shape, and complex index of refraction. Again, an empirical approach was selected. Utilizing the correlation from Hottel and Broughton combined with Eq. 1 and Eq. 2, the relationship between KL and soot number density N/V can be

[5]Hottel, H. and Broughton, F. (1932) Determination of True Temperature and Total Radiation from Luminous gas Flames. Industrial and Engineering Chemistry 4(2):166-175.

determined as shown in Eq. 7. Using the definition of soot volume fraction $f_v = N\pi d^3/6V$, f_v can be solved as shown in Eq. 8

$$\frac{KL}{\lambda^{1.39}} = Q_{ext}\frac{\pi d^2}{4}\frac{N}{V}L \qquad \text{Eq. 7}$$

$$f_v = \frac{2K}{3dQ_{ext}\lambda^{1.39}} \qquad \text{Eq. 8}$$

Substituting the expression for Q_{ext} from the Rayleigh limit in Table 1 into Eq. 8 and selecting an appropriate value for the refractive index[6] leads to the final expression used to calculate soot volume fraction as shown in Eq. 9.

$$f_v = \frac{K}{4\lambda^{0.39}} \qquad \text{Eq. 9}$$

The value produced is admittedly an approximation. The error[7] is estimated to be on the order of ±50%. The assumptions made are that the soot particles are uniformly distributed, spherical, and obey Hottel and Broughton's emittance correlation. As long as the soot morphology is relatively constant between measurements of various coals or operating conditions, the relative changes in measured soot should be valid. Because soot is produced by the recombination of stable hydrocarbons that have decomposed from the parent fuel, soot properties tend to be relatively constant between fuels.

Experimental Set-up

Line-of-sight measurements were obtained in the Oxy-Fuel Combustor (OFC) at the University of Utah. The OFC, shown in Figure 1, is a down-fired, 40 kW_{th}, pulverized-coal reactor. The burner consists of a central primary tube and an annular secondary flow with 30 deg. swirl vanes. The burner zone is 1.2 m in length and 0.60 m in diameter, is electrically heated, and has vertically oriented optical access windows. A purge tube was installed that would slide vertically up and down the window on both sides to keep the window free of deposits. The same laser extinction set-up was used as is reported by Tree and Stimpson[8]. A multi-wavelength argon ion laser was used producing a 40 mW beam at 514.5 nm, and a 6 mW beam at 457.9 nm. Light was collected in an integrating sphere where it was filtered and measured by photodiodes. Over-fire or staged oxidizer was added in the transition zone of the reactor for cases that were overall fuel rich in the burner zone. The OFC is equipped for continuous monitoring of O_2, CO_2, CO, NO_x, and SO_2.

Two coals were measured with three different oxidizers, at three locations, and at three different stoichiometric ratios as listed in Table 1. The three SR's allowed investigation of overall fuel rich, fuel lean, and stoichiometric conditions. Coal was burned in air, a mixture of O_2 and CO_2, and a mixture of

[6]Ferguson, C.R., Tree, D.R., DeWitt, D.P. and Wahiduzzaman, S.A.H. (1987) "Design, Calibration and Error Analysis of Instruments for Heat Transfer Measurements in Internal Combustion Engines," ASME Reprint from *Developments in Experimental Techniques in Heat Transfer and Combustion,* HDT-Vol. 71.

[7]Tree, D.R. and Peart, J.A., "Two Color Transmittance Measurements in a Pulverized Coal Reactor," *Proceedings of the Combustion Institute*, Twenty-Eighth Symposium (International) on Combustion, August 2000

[8] Tree, Dale R., Stimpson, Curtis, Fry, Andrew, "Line of Sight Soot Measurements in a Pulverized Coal Oxy Flame," The 35th International Technical Conference on Clean Coal and Fuel Systems, June 6-10, Clearwater, Florida, 2010.

O_2 and flue gas recycle. For the oxy combustion conditions, the target O_2 concentration in the mixture of O_2 and diluent was 27% (vol, wet). However, in practice the concentration of O_2 ranged between 26% and 30%. Coal was conveyed to the burner with primary air for the air combustion cases and with a mixture of O_2 in CO_2 for the oxy- and FGR-combustion cases. The flue gas was recycled from the OFC for the FGR cases. All secondary oxidizers were heated to 260 °C (500 °F).

The flames in all experiments were lifted from the burner. Transmittance measurements were taken near the base of the flame, the center of the flame, and the tip of the flame as determined by visual inspection. Measurement locations are given in Table 2.

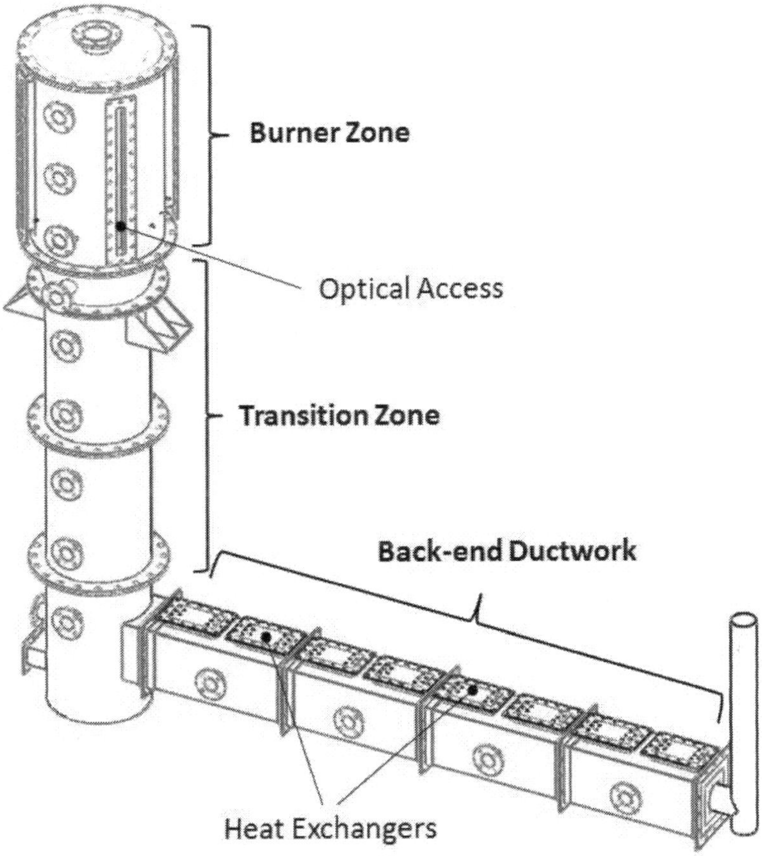

Figure 1. The University of Utah's Oxy-Fuel Combustor

Table 1. Test matrix showing coals and targeted operating conditions.

Coals	Stoichiometric Ratios	Oxidizers	Locations
Illinois #6	0.9	Air	Base
Skyline	1.0	Oxy (27% O_2 73% CO_2)	Center
PRB	1.15	FGR (27% O_2 in flue gas)	Tip

Table 2. Location (axial distance from burner) of soot measurements

Coal	Base (m)	Center (m)	Tip (m)
Illinois			
Air	0.21	0.53	0.82
Oxy	0.37	0.50	0.70
FGR	N/A	N/A	N/A
PRB			
Air	0.19	0.53	0.86
Oxy	0.36	0.51	0.67
FGR	0.36	0.51	0.67
Skyline			
Air	0.21	0.54	0.78
Oxy	0.33	0.51	0.70
FGR	0.34	0.51	0.70

RESULTS

The measured soot volume fractions for the Skyline coal are shown as a function of the overall burner stoichiometric ratio in Figure 2. The soot volume fraction is seen to decrease with increasing stoichiometric ratio (SR). Because soot is formed in fuel rich regions, it would normally be expected that a leaner mixture (higher SR) would produce less soot as is indicated by the data. A careful look at the data reveals that the green symbols (soot volume fraction of the flame tip) are generally below the blue symbols (soot volume fraction at the base of the flame) with the red symbols in-between. For the majority of the data, the soot concentration measured at the base of the flame is higher than the soot concentration at the tip and decreases with increasing distance from the burner. Differences due to the oxidizer (air vs oxy vs FGR) are not discernable in Figure 2 although it will be seen in a subsequent figure that there is a slight decrease in soot volume fraction for the oxy-fired and FGR-fired cases. The primary trend of decreasing soot with increasing SR therefore has two smaller trends superimposed on it: the first being a decrease in soot with increasing distance from the burner and the second being a decrease in soot when O_2/CO_2 or O_2/flue gas are the oxidizers.

The measured soot volume fractions for the PRB coal as a function of overall burner SR are shown in Figure 3. The same trend of decreasing soot with increasing SR can be seen in the PRB data although there are a few data points near SR 1.15 that are not in agreement with the remainder of the data. The trends relating to the measurement location or oxidizer composition are not discernable in the PRB data. The total amount of soot produced by PRB is about half that of the Skyline coal at a given SR. The lower value of soot for the PRB coal is near the measurement limit which is around 2.5 E-08 and making trends in the data more difficult to identify.

A plot of the average soot volume fraction at all positions and SRs for each oxidizer mixture is shown in Figure 4. The average soot volume fraction of the air combustion cases is seen to be the highest while the O_2/CO_2 mixture (oxy) and the O_2/flue gas mixture are smaller and similar in magnitude. The largest source of uncertainty was found to be the baseline laser intensity (I_0) which is the intensity of the beam without any particles in the path. The value of I_0 changed during the course of measurements due mostly to window fouling. The uncertainty was estimated based on how accurately this value could be determined and is displayed as error bars in Figure 4. The error bars suggest that the difference between soot volume fractions of the Skyline air and Skyline oxy are measurable, but differences between the PRB coals are too small to be conclusive.

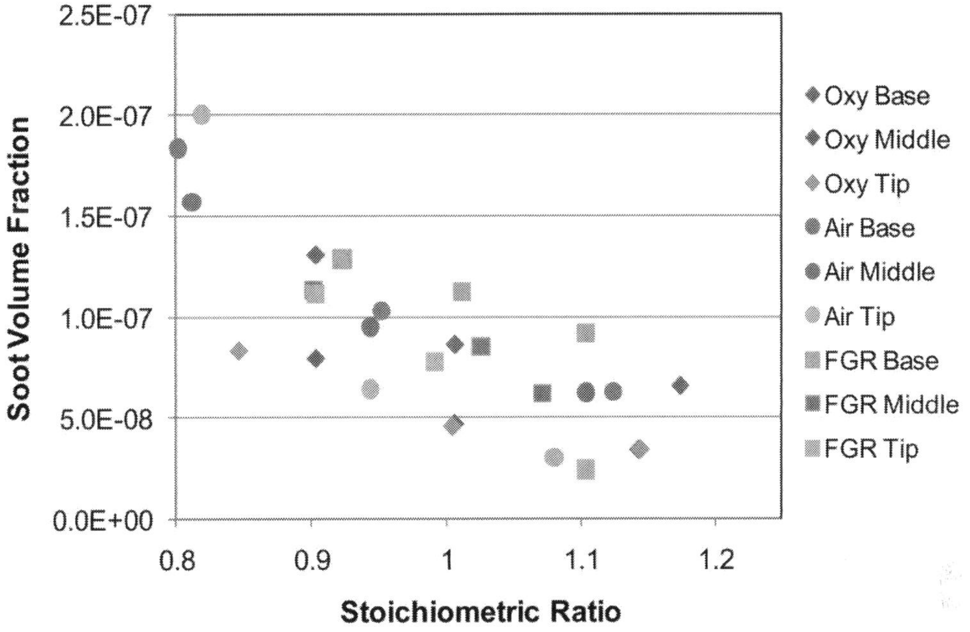

Figure 2. Soot volume fraction of Skyline coal as a function of overall stoichiometric ratio.

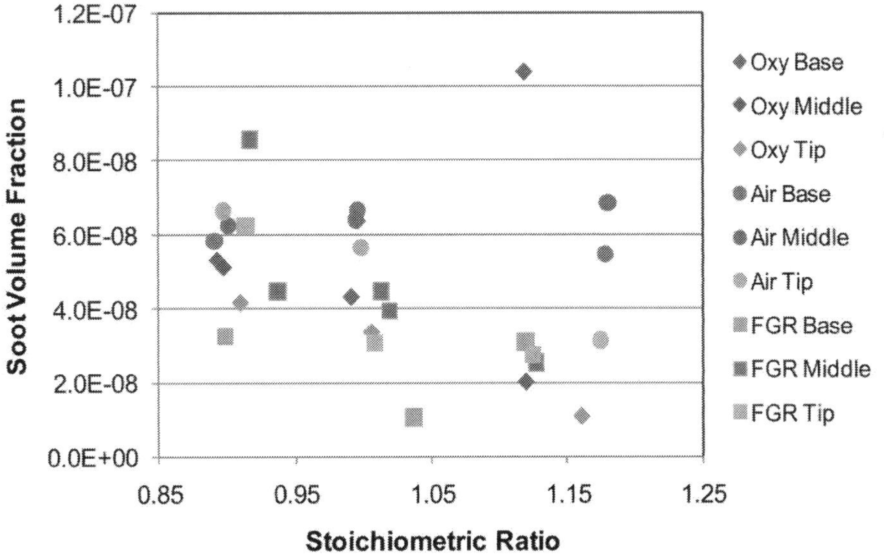

Figure 3. Soot volume fraction of PRB coal as a function of overall stoichiometric ratio.

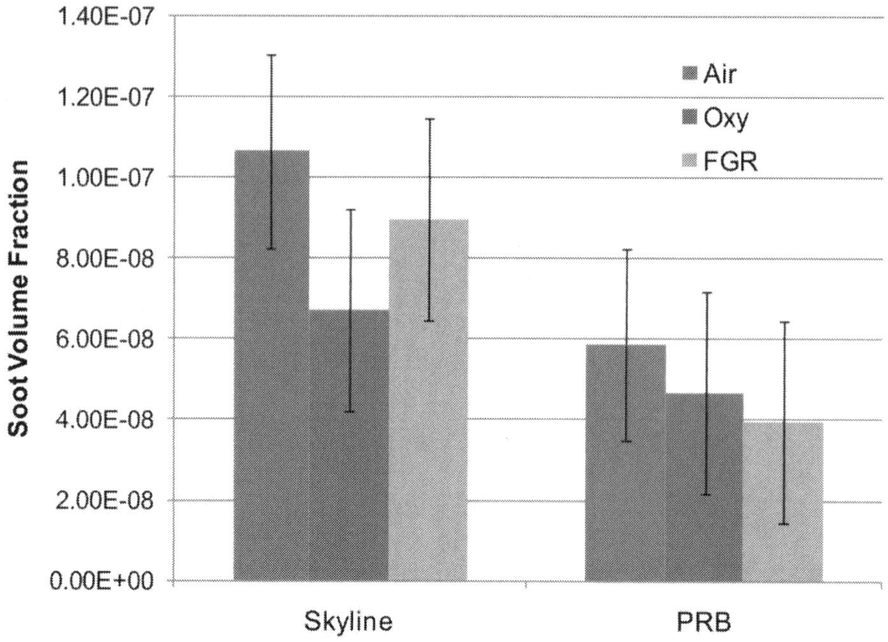

Figure 4. Average of all soot volume fractions for the Skyline and PRB coals for each oxidizer mixture.

Overall, the total soot volume fractions measured are relatively small when compared to other flames in the literature. Diesel flames can have a soot volume fraction of 1.0 E-05 while laboratory diffusion flames produce soot volume fractions of 1.0 E-06.[9] The low soot volume fraction may be attributed to the relatively high SR of the premixed mixture. Once there is enough oxygen to consume the volatiles, the formation of soot becomes less likely.

DISCUSSION OF RESULTS

A conceptual model is now introduced for soot formation based on soot formation in Diesel flames as described by Dec[10] and in pulverized coal flame structures as described by van der Lans et al.[11]. In order to understand soot formation in a coal flame, one must first consider the simplified schematic shown in Figure 5 of a type 1 coal flame[10]. In this flame the primary flow (coal and carrier gas) momentum is large enough to penetrate through the reverse flow recirculation zone and produces a lifted, long, relatively narrow, mixing limited flame. This flame type is representative of the observed flame behavior in the OFC.

Pulverized coal and primary oxidizer enter the combustion zone through a center primary tube. Because the coal is normally conveyed with air, a portion of the air required to oxidize the fuel is already premixed with the coal producing a fuel-rich, premixed mixture. The SR of this mixture in the OFC experiments was approximately 0.2 for all three oxidizer conditions. The primary fuel mixture must be heated before it will react; this occurs through both radiative heat transfer and mixing with hot combustion products. During heating, volatiles are released from the coal adding a gas phase fuel to the

[9] Tree, D.R., Svensson, K.I., "Soot Processes in Compression Ignition Engines," *Progress in Energy and Combustion Science Technology*, 33:3, pp 272-309, 2007.
[10] Dec, J. E. A Conceptual Model of DI Diesel Combustion Based on Laser-Sheet Imaging. SAE Paper 970873, 1997.
[11] van der Lans, R.P., Glarborg, P., Dam-Johansen, K., "Influence of Process Parameters on Nitrogen Oxide Formation in Pulverized Coal Burners," Progress in Energy and Combustion Science, 23, pp 349-377, 1997.

particles. When the temperature is high enough, the premixed volatile/oxidizer mixture will react causing an increase in temperature and a drop in oxygen concentration. If there is not enough oxygen in the mixture to burn all of the volatiles, the fuel-rich, hot products of combustion will remain and form soot. Soot is formed more readily from stable hydrocarbons such as the tars and aromatics but can also form from more reactive alkanes. The amount of soot formed will be a strong function of the stoichiometric ratio in this zone of premixed volatiles and oxidizer.

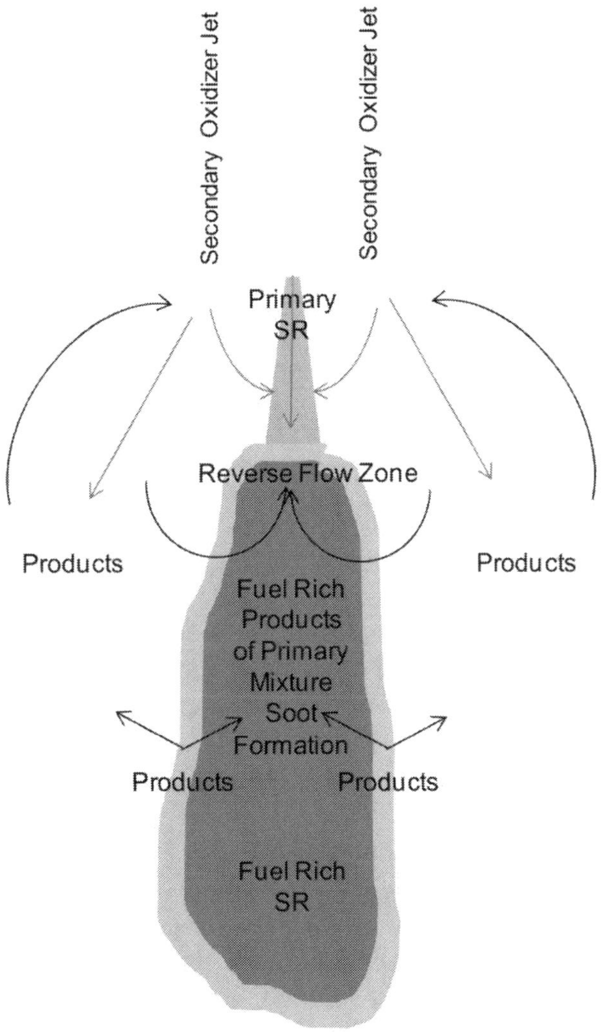

Figure 5. Schematic diagram of a lifted, pulverized-coal flame.

If a strong reverse flow recirculation zone is created, heating of the primary mixture will be rapid resulting in an attached flame near the burner. If slower heat transfer occurs, the flame may be lifted. A lifted flame allows more entrainment of the secondary oxidizer into the primary stream leaning out the fuel-rich mixture prior to ignition. The premixed fuel and oxidizer react rapidly to form rich premixed products of combustion. The rich products are surrounded by flamelets at the interface of the fuel and oxidizer regions. While only one large flame is shown in the schematic, an actual burner will produce many small flamelets of various lengths and shapes depending on the flow structures and turbulence produced by the burner flow dynamics. Each of these flamelets is supplied by the premixed fuel oxidizer mixture that reacts to form premixed products. A discussion of the existence of flamelet structures based on gas species measurements is found in Damstedt et al.[12]

Most of the soot will oxidize as it passes through these flamelets producing a luminous yellow color. Soot concentration can be increased by: 1. Reducing the SR of the mixture inside the fuel rich core of the flame 2. Burning a fuel (coal) that produces a greater amount of stable hydrocarbons that are more likely to produce soot or 3. Altering the kinetic mechanisms that produce and oxidize soot by changing the mixture composition. Soot is oxidized on the oxidizer side of the flame and is dependent on O_2 concentration and temperature.

The most prominent impact on soot formation in the OFC was an increase in overall SR causing a decrease in measured soot concentration. It is useful to define three different SRs in order to discuss soot formation. The first is the SR of the primary fuel/oxidizer. This value is easily calculated from the primary fuel and primary oxidizer flow rates and establishes a lower limit for the SR at the time of ignition as it can only increase with entrainment of the secondary oxidizer. The second SR of interest is the SR at the point of ignition or premixed SR. This SR represents the air fuel ratio when the volatiles are consumed and is the value most directly applicable to the amount of soot formed. Increasing the amount of oxygen entrainment prior to ignition will increase the premixed SR and decrease soot formation. The final SR is the overall SR of the burner zone and represents the total air and fuel delivered by the burner. When the overall SR is increased, the SR of gases entrained into the primary fuel and oxidizer becomes leaner causing the premixed SR to increase relative to a low overall SR. Thus, an increase overall SR should increase the premixed SR and decrease soot formation.

The second most prominent feature in the soot concentration data was a decrease in soot concentration with increasing distance from the burner. Since oxygen is being entrained or mixed into the flame with increasing distance from the burner, after reaching a peak where the soot is formed from the unburned volatiles, the soot is oxidized from that point downstream. The data suggest that the measurement point at the base of the flame is downstream of the initial soot formation zone and soot decreases in the downstream direction as it is oxidized.

The final less prominent observation is that soot decreased with clean (O_2 in CO_2) oxy-combustion and possibly with FGR oxy-combustion. When the O_2/CO_2 oxidizer was used, Table 2 shows that the base of the flame became even more detached from the burner. The base of the flame for air combustion was measured at 0.2 m but shifted to 0.36 m for oxy and FGR combustion. This increased lift-off length may cause increased oxidizer entrainment. In addition to the longer lift-off length, the oxy and FGR gas velocities are lower than air combustion which would increase the residence time prior to ignition for the oxy and FGR combustion flames. With additional oxygen entrainment prior to ignition, the oxy and FGR flames should produce less soot. It is not clear if this extra oxygen entrainment or chemical impacts produced by higher CO_2 concentrations are responsible for the observed decrease in soot.

The final observation is that PRB coal produced less soot in the OFC than the Skyline coal. There are two possible explanations for this result. The first is that PRB is a lower rank coal and will produce less tar and more light hydrocarbons than the Skyline coal. PRB should therefore produce less soot.

---

[12] Damstedt, B., Pederson, J.M, Hansen, D., Knighton, T., Jones, J., Christensen, C., Baxter, L., and Tree, D., "Biomass Cofiring Impacts on Flame Structure and Emissions," *Proceedings of the Combustion Institute*, V 31 II, Proceeding of the Combustion Institute – Thirty-First International Symposium on Combustion, 2007, p 2813-2820.

The second possibility involves mixing. Both PRB and Skyline coal have the same lift-off length suggesting that amount of oxidizer and product gas entrainment into the primary jet prior to ignition is similar. The PRB coal has a lower stoichiometric A/F ratio than Skyline; therefore, if they both start at the same SR in the primary tube and have the same amount of oxidizer entrainment, the PRB coal will have a higher SR at the point of ignition and produce less soot.

Interestingly, the amount of soot measured in a previous experiment using the same technique was higher for PRB coal than for Skyline[13]. In the previous experiments the flame was not lifted but attached. The entrained gas was therefore composed more of product gases and less of oxidizer gas. The product gases of PRB can be expected to contain less oxygen because of the rapid burnout and high reactivity of PRB coal compared to Skyline.

SUMMARY AND CONCLUSIONS

Soot concentration measurements have been obtained for two coals at three locations over a range of three stoichiometric ratios and three different oxidizers. The results show a decrease in soot concentration with increasing overall SR. The soot concentration was highest at the base and decreased slightly with increasing distance from the burner. Soot concentrations for Skyline coal were higher than for PRB coal with all soot concentrations being relatively low (below 2 E-07). The trends in soot concentration can be explained by a conceptual model of a coal flame with soot formation. The key parameter to soot formation in coal flames is the SR of the fuel oxidizer mixture at the time/point of ignition. The influence of changes in flow rate and burner flow dynamics created when changing from air-fired to oxy-fired would appear to have more impact on soot than any impacts caused by chemical effects of CO_2 on soot formation.

---

[13] Tree, Dale R., Stimpson, Curtis, Fry, Andrew, "Line of Sight Soot Measurements in a Pulverized Coal Oxy Flame," The 35th International Technical Conference on Clean Coal and Fuel Systems, June 6-10, Clearwater, Florida, 2010.

(113c) Effects of Biomass Particle Size During Cofiring Under Air-Fired and Oxyfuel Conditions

Melissa Holtmeyer, Benjamin Kumfer and Richard L. Axelbaum, Energy, Environmental & Chemical Engineering, Washington University in St. Louis, St. Louis, MO

Oxy-coal combustion enables carbon capture and storage (CCS) through the utilization of oxygen to produce a flue gas that is primarily CO_2. The flue gas can be recycled back to the combustion chamber to control temperatures and act as the primary fuel carrier. These changes to the combustion environment can impact flame structure and formation of NOx.

Although oxy-coal plants alone will not contribute to renewable energy targets, they may be able to if they are operated with biomass cofiring. When the flue gas is sequestered, CO_2 is effectively removed from the atmosphere because the CO_2 consumed by the biomass during growth is not released back into the atmosphere. In addition, the concentrations of SOx and mercury within the boiler can be reduced by cofiring.

However, biomass fuels are widely variable in composition, particle size, and nitrogen content, which can make utilization of these fuels challenging. In this work, a numerical study was conducted for cofiring of pulverized coal and sawdust under air-fired and oxyfuel conditions to investigate the effects of cofiring on flame length and nitric oxide (NO) formation. Previous experiments have shown an increase in nitrogen conversion to NO when cofiring under both air-fired and oxyfuel combustion, despite the fact that the sawdust cofired had less fuel-bound nitrogen. Computational Fluid Dynamics (CFD) is used to determine the cause of the increased NO conversion and to identify differences between air-fired and oxyfuel cofired flames. The simulations reveal that cofired flames have longer volatile-flame regions (the flame envelope), and this length is influenced by the increased volatile fraction and particle size associated with the biomass. Flame length theory for turbulent, non-premixed gaseous diffusion flames was found to be useful in interpreting the observed results in both air-fired and oxyfuel combustion. Large biomass particles that are not entrained in the near-burner region breakthrough the flame envelope, and this was shown to be detrimental to controlling NO formation. During oxy-cofiring combustion, particle breakthrough occurs at smaller diameter, leading to increased nitrogen conversion to NO when compared to air-fired conditions. This is a direct result of a decreased flame envelope length and elevated oxygen concentrations.

See more of this Session: Oxycombustion of Coal and Other Fuels II
See more of this Group/Topical: Energy and Transport Processes

(113d) The Fate of Sulfur In Oxy-Fired Fluidized Beds

Liyong Wang and Eric G. Eddings, Chemical Engineering, The University of Utah, Salt Lake City, UT

1. Introduction

SO_2emissions from coal combustion have several harmful effects on the environment,including their contribution to acid rain. A variety of approaches, includingwet and dry scrubbing and direct sorbent injection, have been developed tocontrol sulfur pollution. Direct dry sorbent injection is a relatively simpleand low-cost process. Until now, the fluidized bed has been one of the mostpopular furnaces for the application of direct sorbent injection. In addition,oxy-fuel combustion is a promising, practical method to reduce greenhouse gasemissions; thus, studies on the mechanisms related to the production of SO_2emissions under oxy-fuel conditions are important. The major limitation ofstudying SO_2 mechanisms in existing pilot-scale and industrial-scalefluidized beds is expense. Bench-scale fluidized beds, to some extent, canexamine fundamental mechanisms of sulfur under many experimental conditions dueto their low cost. However, very few previous studies on the mechanisms ofsulfur emissions in single coal particle fluidized beds have been conducted. Afew potential mechanisms of SO_2 capture by limestone are presentedin the literature[1-5].Little else can be confirmed except that the final product is $CaSO_4$[6].

Inthis work, the capture of fuel sulfur by sorbents in oxy-fired fluidized beds wasstudied by:

Performingbench-scale experiments to identify mechanisms and quantify rates;

Developinga detailed single-particle model to determine rate parameters and testmechanisms;

Performingpilot-scale experiments to verify bench-scale observations underindustry-relevant conditions.

2. Experimental Method

Abench-scale single-fuel-particle fluidized-bed reactor [7,8]is used to follow the transient evolution of sulfur in mixtures of $O_2$with N_2 or CO_2at the same conditions (mass flow rate, O_2concentration and fluidized bed temperature, etc.). In this way, we canexamine the effect of the presence of CO_2 on the formation andcapture of SO_2/SO_3 and other participating species.Illinois #6 coal was used as the fuel, due to its high sulfur content.

Thesingle-particle, coal-combustion experimental bench scale fluidized bed is shownin Figure1.It is a vertical, cylindrical, stainless-steel furnace with a combustionchamber length of 771 mm and an inner diameter of 44 mm. The column has a 2 mmthick perforated plate distributor with 60 holes. The temperatures of the bedmaterial, furnace wall and gas phase are measured by K-type thermocouples. Theeffluent is measured by a Magna-IR Spectrometer 550 FTIR.

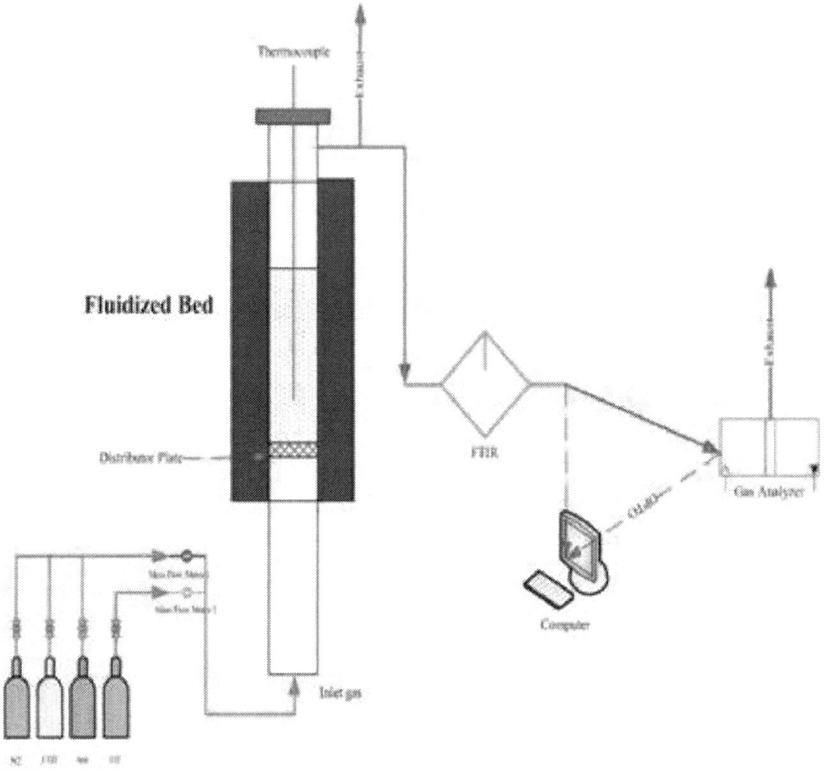

Figure1.Schematic setup of bench-scale fluidizedbed combustor.

3. Results

Thefate of sulfur in the bench-scale fluidized bed experiments were carried out inboth N_2 and CO_2 environments at three O_2concentrations (10%, 20% and 30%) and a range of bed temperatures (765 - 902C). The sulfur content in each single coal particle was not expected to beuniform; therefore, each experimental condition was repeated at least 5 timesand averages of the multiple runs and associated error bars are presented.

Figure2(left) shows the comparison of SO_2 emissions without limestone betweenoxy-fuel combustion and air combustion. It clearly indicates there was nodifference in the SO_2 emissions. As a conclusion, the effect of highconcentrations of CO_2 on SO_2 emissions appears to be the sameas N_2 in the bench-scale fluidized bed.

Figure 2(right) shows the comparison of SO_2 emissions with limestone betweenoxy-fuel combustion and air combustion. It is clear that the efficiency ofsulfur removal by limestone in oxy-fuel combustion is much lower than duringair combustion. Because the

CO_2 concentration is very high in thefluidized bed under oxy-fuel conditions, it is anticipated that there would bea great suppression of the decomposition of $CaCO_3$ (limestone) therebyreducing the efficiency of sulfur removal by limestone.

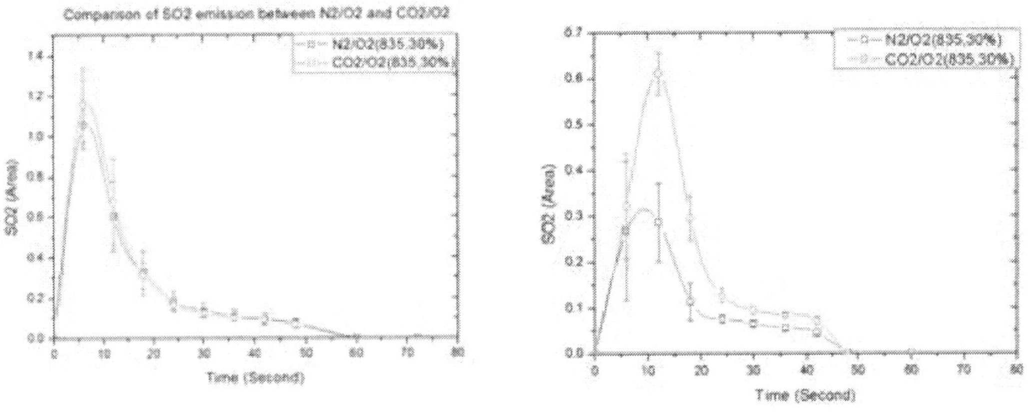

Figure 2.The fate of sulfur (T= 835 C) in either N_2or CO_2, without limestone (left) and with limestone (right) presentin a quartz fluidized bed.

In order to investigatelimestone sulfation behavior in detail, calibration gases of known SO_2 concentrationin either N_2 or CO_2were introduced as the sulfur source,instead of generating it from coal particle combustion.

Sulfation behavior oflimestone in O_2/N_2 or O_2/CO_2 isillustrated in Figure 3. As Figure3(left) shows, the reactor temperature had little effect on the degree of sulfationover the range considered in O_2/N_2. It is wellestablished that due to differences in molar volumes between $CaCO_3$and $CaSO_4$, significant pore plugging can occur that would limit thedegree of sulfation. If the overall rate is limited by particle diffusion, thenthe reaction would tend to take place near the particle surface. The poreswould become blocked after only a limited outer layer of the particle hadreacted to form $CaSO_4$. The low degree of sulfation is consistentwith this hypothesis. In addition, the very limited dependence on temperaturewould also indicate more of a diffusion-controlled regime.

The degree of sulfationfor the oxy-combustion condition is shown in Figure 3(right), and reveals a clear dependence on temperature. The sulfation mechanismof limestone in O_2/N_2 or O_2/CO_2 (T=874C)is shown in more detail in Figure 4.This figure provides a composite of the N_2 and CO_2atmospheres, and plots the degree of sulfation over time as determined from themeasured SO_2 concentration. As shown, the transient sulfationkinetic rate in O_2/N_2 is initially much higher comparedto O_2/CO_2; however, at longer times the degree of sulfationof limestone in O_2/CO_2 is higher. The crossover occurs attransition point A (t=2100s)for the temperature and conditions of thisexperiment, as noted in the figure.

It appears that the processof sulfation is controlled by a different mechanism in the presence of CO_2as compared with N_2, since very different sulfation behavior isobserved for the same operating conditions. It is likely that air-combustionand oxy-combustion conditions result in indirect and direct sulfationmechanisms. Also, it appears that the indirect sulfation mechanism is in adiffusion-controlled regime for the conditions evaluated, as there is verylittle temperature sensitivity and the conversion is low, consistent with theearlier discussion of pore plugging on the outside of the particle. The directsulfation mechanism may be in a kinetically controlled regime, since thetemperature dependence is much more pronounced. In O_2/N_2,if the kinetic rate for the SO_2-CaO (indirect) reaction isrelatively fast, then pore diffusion would limit the overall rate and poreswould become blocked very quickly, allowing only a limited outer layer of theparticle to react to form CaSO4 and limiting overall conversion. In O_2/CO_2,if the kinetic rate for the SO_2-$CaCO_3$ (direct) reactionis slow, then pores diffusion is not as limiting and the reaction to form $CaSO_4$could take place throughout more of the particle. This process would allow theformation of $CaSO_4$ over a larger surface area and allow greaterreaction time prior to pore plugging.

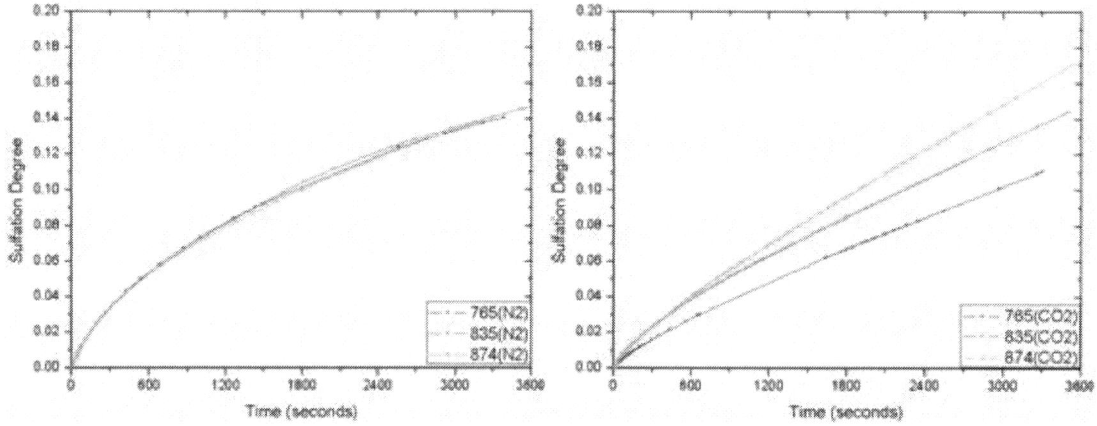

Figure 3. Limestone sulfation behavior at varioustemperatures in either O_2/N_2 (left) or O_2/CO_2(right).

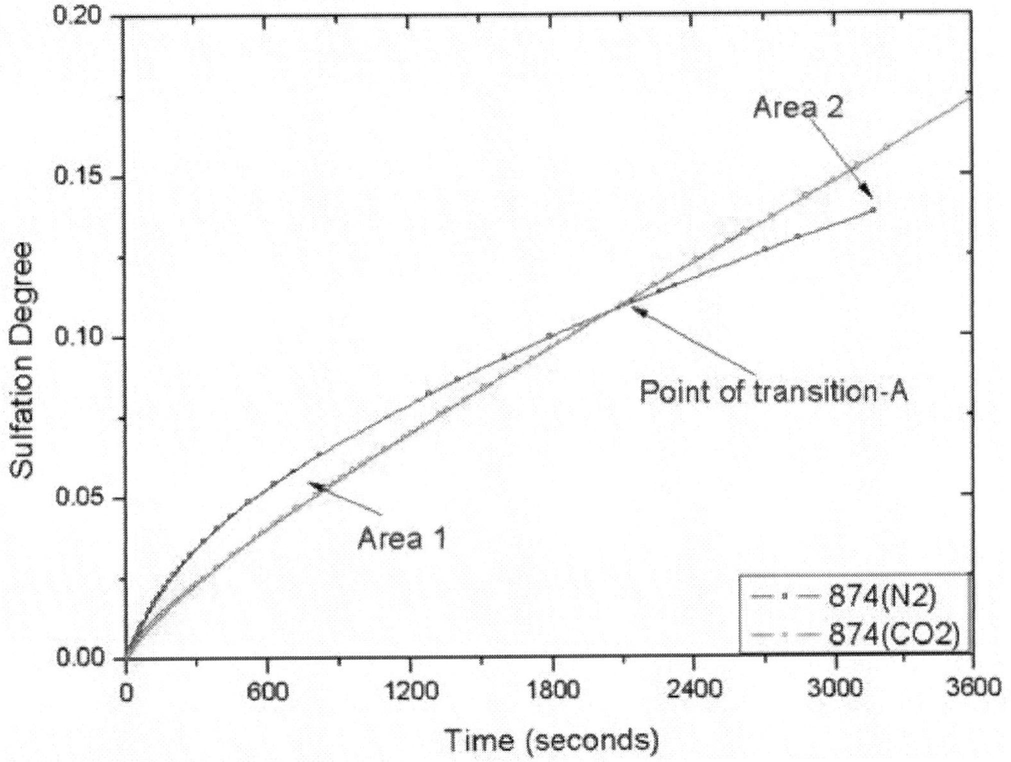

Figure 4. Sulfation behavior of limestone in either O_2/N_2 or O_2/CO_2at T=874C.

Reference:

1. Liu,H., et al., Sulfation behavior of limestone under high CO2 concentration in O_2/CO_2coal combustion. *Fuel*, 2000; 79(8): 945-953.

2. Silcox,G.D., J.C. Kramlich, and D.W. Pershing, A mathematical model for the flashcalcination of dispersed calcium carbonate and calcium hydroxide particles.*Industrial & Engineering Chemistry Research*, 1989; 28(2): 155-160.

3. Hajaligol,M.R., J.P. Longwell, and A.F. Sarofim, Analysis and modeling of the directsulfation of calcium carbonate. *Industrial & EngineeringChemistry Research*, 1988; 27(12): 2203-2210.

4. MichaelJ. H. Snow, J.P.L., Adel F. Sarofim, Direct Sulfation of Calcium Carbonate.*Ind. Eng. Chem. Res.*, 1988; 27: 268-213.

5. Tullin,C., G. Nyman, and S. Ghardashkhani, Direct sulfation of calcium carbonate: theinfluence of carbon dioxide partial pressure. *Energy & Fuels*,1993; 7(4): 512-519.

6. Hu,G., et al., Review of the direct sulfation reaction of limestone. *Progressin Energy and Combustion Science*, 2006; 32(4): 386-407.

7. Sanchez,A., E. Eric, and F. Mondragon, Fourier Transform Infrared (FTIR) OnlineMonitoring of NO, N2O, and CO2 during Oxygen-Enriched Combustion ofCarbonaceous Materials. *Energy & Fuels*, 2010; 24(9):4849-4853.

8. Can,G., P. Salatino, and F. Scala, A single particle model of the fluidized bedcombustion of a char particle with a coherent ash skeleton: Application togranulated sewage sludge. *Fuel Processing Technology*, 2007;88(6): 577-584.

See more of this Session: Oxycombustion of Coal and Other Fuels II
See more of this Group/Topical: Energy and Transport Processes

(113e) Application of a Multi-Grain Model: Single Large Coal Particle Oxy-Combustion

Liyong Wang and Eric G. Eddings, Chemical Engineering, The University of Utah, Salt Lake City, UT

In this paper, we present a mathematical framework for a multi-grain solid-gas reaction diffusion model. Typically, the application of single particle models is limited due to their assumptions. The shrinking-core model is the most frequently used model. It is a simplified model for gas-solid reactions, which assumes a sharp boundary between the un-reacted core and the formed product layer. It is often valid for non-porous particles but not for porous particles, unless the overall rate is kinetically controlled. However, very few general gas-solid reaction-diffusion models have been implemented for generalized gas-solid particle cases. Our mathematical framework is based on "multiple grains," instead of a "product layer." Our particle model is a general formulation that couples pore gas diffusion, external mass convection in the bulk gas, internal heat conduction, external heat convection, radiation with chemical reaction. It is based on the governing conservation equations including one energy equation and several mass equations whose numbers are determined by the number of species in the system (gas and solid). All of the mass conservation equations and the energy equation are coupled with reaction and heat generation. The solution of the governing equations is accomplished using Matlab. In addition, some numerical solution techniques to solve the stiff problem were used such as the implicit method and dynamic time stepping.

One application of our mathematical framework is single coal particle combustion under oxy-fired conditions in fluidized beds. Fuel particles in fluidized beds are significantly larger than in pulverized coal combustion, and thus the assumptions typically invoked for coal particle combustion models may not be applicable for all fuel particle sizes or over the entire lifetime of the particle. Initial model results illustrate the effect of particle size on combustion evolution: smaller particles favor kinetic control, while diffusion control occurs in the larger particles, as expected. For large particles, the O_2 penetration into the particle was limited to a limited region near the particle surface as would be expected for diffusion control. The evolution of CO/CO_2 was also examined and as the simulation results show, the process of CO converting to CO_2tendsto happen inside the particle for larger fluid-bed-sized particles (d=3000μm). However, the process of CO converting to CO_2 is more likely to occur outside the coal particle (gas phase) for pulverized coal-sized particles (d=50μm). The results of the simulation support the applicability of a shrinking core particle model for large particle coal combustion. The model framework presented here, however, provides a broader flexibility to address a range of particle sizes and temperature conditions.

See more of this Session: Oxycombustion of Coal and Other Fuels II
See more of this Group/Topical: Energy and Transport Processes

Rules of Thumb for Reactor Design and Process Conditions for Thermochemical Treatment of Chromated Copper Arsenate (CCA) Treated Wood Waste

Lieve Helsen, University of Leuven (K.U.Leuven), Heverlee, Belgium
Frederic Cuypers, University of Leuven (K.U.Leuven), Heverlee, Belgium

Extended abstract

Lab-scale pyrolysis experiments have been performed to gain insight in the mechanisms influencing the metal behavior during carbonization of chromated copper arsenate (CCA) treated wood chips. To this end a fixed bed low temperature pyrolysis reactor and a thermogravimetric (TG) analyser have been used. Besides CCA treated wood chips, mixtures of metal and organic model compounds (As_2O_5, As_2O_3, $CrAsO_4$, $CrAsO_4$ + $Ca(OH)_2$, As_2O_5 + glucose, As_2O_5 + lignin, As_2O_5 + activated carbon, As_2O_3 adsorbed on activated carbon, $CrAsO_4$ + glucose, $CrAsO_4$ + lignin, $CrAsO_4$ + $Ca(OH)_2$ + glucose, $CrAsO_4$ + $Ca(OH)_2$ + lignin) have been pyrolyzed in a controlled way. The aim of these model compound experiments is not to imitate the real CCA wood pyrolysis process, but to isolate some individual mechanisms (without being influenced by a variety of complex interactions) and to study the influence of different process parameters on these mechanisms. The parameters controlled during pyrolysis (and their corresponding range) are: temperature (330°C - 430°C), heating rate (5°C/min – 20°C/min), residence time (10 min – 40 min), reactor pressure (0 bar gage – 5 bar gage), heater gas flow rate (1000 Nl/h – 1800 Nl/h), oxygen concentration (1 vol% - 3 vol%) and steam concentration (5 wt% - 15 wt%).

Based on the experimental observations, reactor design guidelines are derived in order to identify an operating window that minimizes the volatilization of metals during thermochemical conversion of CCA treated wood, while ensuring a good quality of charcoal product.

The main observations related to metal and organic model compound experiments are:

1. The interactions between inorganic compounds (metals and minerals) and organic compounds play an important role during thermal decomposition. On the one hand the presence of organic compounds (wood, carbon, glucose, ...) promotes the reduction of As_2O_5 and subsequent volatilization of As_2O_3. On the other hand formation of complexes or agglomerates and adsorption effects may help to prevent arsenic loss below 400°C and may facilitate metal recuperation.
2. Arsenic adsorption on the solid residue may prevent the volatilization of As_2O_3, and may facilitate the formation of agglomerates of As_2O_3 with wood minerals, giving rise to thermally more stable micro-structures. Furthermore, in a fixed bed or moving bed reactor, small fractions of As_2O_3 that volatilize at the bottom of the reactor can be re-adsorbed by above lying wood or partially pyrolyzed wood layers, giving rise to thermally more stable compounds.
3. Mixtures of metal model compounds and respectively glucose and lignin behave in a different way. Glucose, which is a representative organic model compound for tar formation and secondary char formation, shows foam formation during pyrolysis which hampers the release of arsenic trioxide. Elevated pressure has a pronounced effect on the foam density and the generation of pyrolysis gas, resulting in almost 5 wt% higher arsenic retentions at 5 bar gage compared to 0 bar gage in glucose + As_2O_3 and glucose + As_2O_5 samples. Lignin, which is a representative organic model compound for

char production, shows no foam formation. The pyrolysis residue of lignin helps to stabilize arsenic trioxide.

4. Chromium arsenate forms thermally stable compounds with wood minerals consisting of Cr(III), As(V) and Ca(II), at temperatures below 310°C. Elevated pressure seems to have no significant influence on this process. Chromium arsenate, which had not formed stable compounds, decomposes to Cr_2O_3 (which may form volatile organo-chromium compounds) and As_2O_5, at temperatures below 310°C. Lignin helps to stabilize $CrAsO_4$, and thus counteracts this decomposition.
5. Once As_2O_5 is formed, it is readily reduced to As_2O_3 (following a radicallary mechanism which is favored by elevated pressure) in the presence of reductive pyrolysis vapors. Once As_2O_3 is formed, mass transfer limitations and adsorption on the pyrolysis residue (which are both enhanced by elevated pressure) play an important role in retaining arsenic in the pyrolysis residue.

The CCA wood experiments show that:

1. Percentages of metals volatilized depend strongly on the operating conditions of the process. Therefore, arsenic release can be limited by choosing the operating conditions very carefully.
2. The effect of residence time on arsenic and chromium retentions is limited for the ranges considered.
3. Heating rate mainly affects chromium retentions, primarily through a higher thermal lag of the inner particle temperature at higher heating rates.
4. Reactor temperatures lower than 400°C result in a carbon-like product, having the same matrix as the original wood, only porosity has increased. This allows the minerals and metals to migrate during the thermal process, and to be liberated afterwards since the solid structure does not collapse (in contrast to conventional carbonization processes).
5. Arsenic retentions decrease slightly with increasing temperature till 390°C, with a sharp decrease at higher temperatures, which may be attributed to accelerated desorption of arsenic trioxide at temperatures above 390°C. Chromium retentions show a less pronounced effect of temperature.
6. Arsenic and chromium retentions are affected in a different way by elevated pressure. Arsenic retentions increase about 5% at 5 bar gage compared to 0 bar gage, which is attributed to a higher mass transfer resistance. In the case of chromium, the effect is less clear; the increased mass transfer resistance might be counteracted by enhanced heat transfer at the early stages of pyrolysis.
7. Since pressure increase does not only influence the mode of occurrence of arsenic, but also the char structure and extent of tar reactions (elevated pressure enhances secondary reactions and heat transport), the metal behavior and ease of metal recuperation depends on the reactor pressure.
8. Oxygen concentration mainly affects the reactor temperature (a change of inlet oxygen concentration from 1 vol% to 3 vol% increases the peak temperature from 420°C to 550°C at the bottom of the reactor, and from 390°C to 420°C at the top of the reactor), which results in a decrease in arsenic retention with increasing oxygen concentration.
9. A higher gas flow rate and the addition of steam mainly result in a more homogeneous reactor temperature profile. The net impact on metal retention is limited.
10. Scanning electron microscopy coupled to energy dispersive X-ray analysis (SEM-EDXA) shows that during pyrolysis the metal and mineral compounds form agglomerates, which suggests a relatively easy (mechanical) separation between

metals/minerals and carbon. The mineral agglomerates (already existing in the initial wood before pyrolysis) may assist in the metal agglomeration process.

Although further research is needed to investigate the cross correlation between the various parameters, the experimental observations can be translated in following recommendations for reactor design and process conditions:

1. To ensure a good quality and brittle pyrolysis residue a residence time of at least 20 min and a process temperature of at least 370°C is necessary. However, the process temperature has to be limited to 390°C to limit metal release.
2. A more homogeneous wood input, with less fine particles, will result in a more homogeneous pyrolysis residue and reduced arsenic and chromium volatilization.
3. A longer residence time and lower temperature result in a similar wood conversion while reducing chromium and arsenic volatilization, however a lower plant capacity is obtained.
4. An operating pressure of 5 bar (or even higher) results in 5% higher arsenic retentions and mass retentions. Furthermore, elevated pressures allow to use lower heater gas temperatures to obtain the same wood conversion, resulting in even higher metal retentions.
5. Small gas flow rates are important with respect to:
    a. localizing the release of free and adsorbed water (drying) on the one hand and tar formation on the other hand in distinct phases
    b. minimizing the dilution of combustible vapors
6. Tar problems can be avoided by:
    a. preventing mixing of tar and water compounds (which would result in sticky tar compounds),
    b. facilitating secondary reactions (such as cracking of tar); repeated cracking of organic compounds (stimulated by stratification of pressure and temperature in the reactor column) eliminates the liquid phase.
7. The inlet oxygen concentration has to be strictly controlled below 1 vol% or even 0.5 vol% (at a heater gas flow rate of 1400 Nl/h and without steam addition), which increases both the arsenic retention and mass retention with 5% compared to an oxygen concentration in the range 1-3 vol%.
8. If oxygen control below 1 vol% is not feasible, the heater gas flow rate should be limited to 500 Nl/h to limit arsenic release. However, this could hamper the heating of the reactor bed. Nevertheless, heater gas flow rates should be lower than 1800 Nl/h to reach acceptable charcoal yields.
9. Steam addition results in better reactor temperature control, and as a consequence increased metal and mass retentions. Moreover, thermal lag is reduced which is beneficial with respect to process time. However, significant amounts of steam are needed (40 wt% steam at an initial oxygen concentration of 2 vol% and 70 wt% steam at an initial oxygen concentration of 3 vol% to limit the reactor bed temperature to 390°C).

These recommendations still need to be tested in an industrial scale reactor.

(118b) Biomass Pyrolysis and Gasification: Engineering Rules of Thumb

Michael J. Antal, Hawaii Natural Energy Institute, University of Hawaii, Honolulu, HI

In this presentation I review important findings recorded in the archival literature concerning the thermodynamics, kinetics, mechanisms, and catalysis of biomass pyrolysis and gasification that serve as rules of thumb for reactor design.

See more of this Session: Reactor Engineering for Biomass Feedstocks
See more of this Group/Topical: Sustainable Engineering Forum

(118c) Rule of Thumb for Design of Catalytic Pyrolysis of Biomass In Bubbling Fluid Bed Reactor

Foster Agblevor, Biological Engineering, Utah State University, Logan, UT and Ofei Mante, Biological Systsems Engineering, Virginia Polytechnic Institute and State University, Blacksburg, VA

The production of stable biomass pyrolysis depends strongly on both reactor design and the effective design of a condensation train to collect the pyrolysis products. There are several challenges in the design of the pyrolysis system including solids feeding, char separation from the pyrolysis vapors, condensation of the pyrolysis vapors and compression of the non-condensible gases. In the feeding of solid biomass into the reactor, the feedstock properties such is softening point are critical in the feeding. Bridging properties are also very important and that is dictated by the method of biomass preparation, either knife milling or hammer milling. Hammer milled and disc-milled feedstocks tend to bridge easily whereas knife milled material bridges less. The reactor operating temperature is strongly influenced by the composition of the feedstocks. Biomass feedstocks such as poultry litter which are high in ash tend to form clinker in the reactor and therefore reactor temperature control is very critical for successful operation. The condensation train; especially with recycling of the gases tend to create major problems especially if an electrostatic precipitator is used to condense the aerosols. For systems employing electrostatic precipitator, the cooling temperature in the condensers must be very low because any water vapor escaping the condensers tend to cause electrical short-circuiting in the unit and reduces efficiency of the aerosol capture. Catalytic pyrolysis tends to produce more water and therefore the cooling system must be very efficient. The compressor for the non-condensible gases must be able to handle dirty gases (large clearances) otherwise it is easily clogged with the pyrolysis oils that tend to degrade the performance of the compressor and requires frequent replacements.

See more of this Session: Reactor Engineering for Biomass Feedstocks
See more of this Group/Topical: Sustainable Engineering Forum

(118d) Rule of Thumb for Reactor Design and Process Conditions for Supercritical Water Gasification of Wet Biomass

Yukihiko Matsumura, Division of Energy and Environmental Engineering, Hiroshima University, Higashi-Hiroshima, Japan

The Rule of thumb for reactor design and process conditions for supercritical water gasification of wet biomass is introduced. The important aspects includes heat transfer and heating rate, reaction rate, plugging, gas composition, inorganic element behavior, and heat of reaction. The know-hows were developed by the laboratory scale, and pilot scale reactors.

See more of this Session: Reactor Engineering for Biomass Feedstocks
See more of this Group/Topical: Sustainable Engineering Forum

Rule of Thumb for Reactor Design and Process Condition for Biodiesel Production

Masato Kouzu, Doshisha University, Kizugawa, Kyoto, Japan

Introduction

For the last decade, the planet earth clearly tends toward warming due to the greenhouse gas emissions. One of solutions for the globally environmental problem is to utilize biomass resources, which are generated from CO_2 and H_2O with the aid of photosynthesis, as alternative fuel and chemicals. Plant oil is one of biomass resources, and is used as feedstock of alternative to petroleum diesel oil. The alternative fuel is extensively termed as "Biodiesel". Biodiesel has advantages of not only renewable nature mentioned above but also good fuel properties such as high flush point and excellent lubricity [1-2]. Moreover, it was reported that emissions of both carbon monoxide and particulate were reduced by fueling the engine with biodiesel [3-4]. In 2003, European Community decided to replace at least 5.75 % of the yearly consumed fossil fuels with biofuels, by the year 2010. This decision accelerated the use of biodiesel and its production has been constantly growth. The total of biodiesel yearly produced in the world was 7.75 milion metric tons in 2008 [5].

Biodiesel consists of fatty acids methyl ester (FAME), which was produced by transesterifying plant oil with methanol. For the industrial biodiesel production, plant oil is transesterified with the help of alkali catalyst dissolved in methanol. The alkali catalyst such as sodium hydroxide is so active that the yield of FAME is above 90 % for 0.1 h at 333 K under atmospheric pressure [6]. Glycerol that is by-produced from plant oil is separated from biodiesel by decantation. Then, purification to eliminate impurity such as soap from biodiesel is carried out.

In this presentation, the rule of thumb for reactor design and process condition for biodiesel production is introduced. The important aspect includes type of plant oil used as feedstock, an amount of methanol mixed with plant oil and quality of the produced biodiesel. Recently, much interest has been taken in transesterifying plant oil with the novel manners: enzymatic and heterogeneous catalytic reactions. Also, it is very important to consider the way for recycling glycerol of the by-product.

Type of plant oil

Plant oil consists of triacylglycerides having fatty acid moieties, and contains some minor components. Among the minor components, water and free fatty acids (FFA) inhibit the alkali-catalyzed transesterification that is applied to the industrial biodiesel production. Water brings about the undesirable hydrolysis of triglycerides. It is easy to limit the water content to the trace level by drying, because of the hydrophobic property of plant oil. FFA neutralizes the alkali catalyst. Moreover, soap produced by the neutralization causes formation of stable emulsion with the result that it is difficult to decant biodiesel from glycerol. An industrial way to eliminate FFA is the pre-neutralization with alkali hydroxide. For the pre-neutralization, aqueous solution of alkali hydroxide is stirred into the plant oil. Then, the soap sediment and the FFA-free oil separate out by decantation. Recently, much interest has been taken in pre-esterifying FFA into FAME with the help of acid catalyst such as sulfonated cation-exchange resin [7]. Waste cooking oil is the cost-advantageous feedstock for biodiesel production. In addition to FFA and water, polar compounds are impurities of the waste cooking oil. Heat for food preparation such as frying leads to thermolytic and oxidative reactions of triglycerides having unsaturated fatty acid chains, which produces polar compounds. Polar compounds seem to be of no negative effect on the base-catalyzed transesterification [8].

Furthermore, it should be noted that fatty acids composition of plant oil affects quality of biodiesel. As palm oil is abundant in triacylglycerides consisting of saturated fatty acid chains, the resultant biodiesel is inferior in the low-temperature properties such as cold filter plugging point (CFPP).

Methanol/oil molar ratio

An amount of methanol mixed with plant oil is the most important variable affecting the yield of FAME, because transesterification of plant oil with methanol is a reversible reaction. Due to the chemical equilibrium, excess of methanol is required for achieving the conversion of plant oil into biodiesel. The required methanol/oil molar ratio seems to be 6, while the stoichiometric ratio is 3 [6]. Although an increase in the methanol/oil molar ratio is effective in not only preventing the reverse reaction but also enhancing the conversion rate, a large amount of methanol mixed with the transesterified products makes it difficult to decant biodiesel from glycerol.

The efficient way to achieve the perfect conversion with the minimized amount of methanol is the two step transesterification. In this way, glycerol produced by the primary transesterification is removed in advance of the secondary transesterification. Due to the removal of glycerol, for the secondary transesterification, the chemical equilibrium shifts towards the forward reaction that is useful in maximizing the FAME yield.

Quality of biodiesel

As properties of biodiesel must meet the quality standard such as EN 14214 and ASTM D6751, purification is a very important operation for biodiesel production. Biodiesel collected after the decantation which is carried out after the transesterification contains soap deriving from alkali catalyst. Moreover, it is difficult to achieve the perfect removal of glycerol by only the decantation. Since these impurities are soluble in water, biodiesel is commonly purified by washing with water. As a large amount of soap leads to formation of stable emulsion, an amount of alkali catalyst employed for the transesterification should be optimized. Use of hot water is effective in preventing the emulsion formation. Water mixed with biodiesel for the purification can be eliminated perfectly by decantation followed by drying.

On the other hand, there are some water-free procedures to purify biodiesel. Cation-exchange resin serves adsorbent for eliminating soap and glycerol from biodiesel [9]. Also, it was reported that impurities of biodiesel was eliminated by utilizing amorphous magnesium silicate as the absorbent [10]. Some researchers studied vacuum distillation to collect highly-pure biodiesel [11].

Recycle of glycerol

It is essential for biodiesel production to recycle glycerol of the by-product. Glycerol collected after the transesterification followed by the decantation is contaminated with methanol, water and soap. Soap is decomposed with sulfuric acid, and the resultant alkali-sulfate is removed from stream of the glycerol purification. Fatty acids from soap are decanted from a mixture of glycerol, methanol and water. Finally, methanol and water are topped by heating in order to obtain the crude glycerol used as raw material of chemicals.

Also, it is important to study novel conversion routes of glycerol into chemicals, because intensive production of biodiesel has caused overproduction of glycerol that makes its market price plummeting. Corma et al. showed a variety of the novel catalytic reactions to convert glycerol into chemicals in their research paper

[12]. Hydrogenolysis into propanediol is one of the potential routes for promoting the recycle of glycerol as chemicals.

Catalyst selection

Sodium and potassium hydroxides are the alkali catalyst of commercial availability. Although sodium hydroxide is cheaper than potassium hydroxide, easy handling of glycerol that contains soap deriving from potassium hydroxide is advantageous to operation of biodiesel production process.

For the purpose of improving biodiesel production process, many researchers have focused their attentions on enzymatic and heterogeneous catalytic traansesterifications [13-14]. Both of them bring about not only the water-free purification of biodiesel but also easy recycle of glycerol. For the enzymatic transesterification, plant oil is converted into biodiesel with the help of lipase acting as bio-catalyst. Major characteristic of the enzymatic transesterification is mild reaction conditions: the transesterification plant oil is catalyzed at room temperature. Furthermore, the enzymatic transesterification is insensitive to FFA and water contained in the low-quality feedstock. For the practical use of the enzymatic process, the effect of immobilization on the catalytic activity of lipase has been investigated [15].

Zinc-aluminum mixed oxide was utilized as the heterogeneous catalyst for a biodiesel production process, which was termed as "Esterfif-H™" [16]. The heterogeneous catalyst seemed to be active at the temperature of 483-523 K under the pressure of 3-5 MPa. The catalytic activities of calcium oxide and sulfonated amorphous carbon are close to that of alkali catalyst [8, 17].

Reference

1. Dorado, M. P., E. Ballesteros, J. M. Arnal, J. Gomez and F. J. Lopez (2003), "Exhaust emissions from a diesel engine fueled with transesterified waste olive oil", *Fuel*, 82, pp. 1311-1315.

2. Lin, C. Y., H. A. Lin and L. B. Hung (2006), "Structure and properties of biodiesel produced by peroxidation process" *Fuel* ,86, pp. 1743-1749.

3. Karmee, S. K. and A. Chadha (2005), "Preparation of biodiesel from crude oil of *Pongamia pinaanta", Bioresour. Technol.*, 96, pp. 1425-1429.

4. Szybist, J. P., J. Song, M. Alam and A. L. Boehman (2007), "Biodiesel combustion, emissions and emission control", *Fuel Process. Technol.*, 88, pp. 679-691.

5. Knothe, G (2010), "Biodiesel: Current trends and properties", *Top. Catal.* 53, pp. 714-720.
6. Freedman, B. ,E. H. Pryde and T. L. Mounts (1984), "Variables affecting the yield of fatty esters from transesterified vegetable oils", *J. Am. Oil Chem. Soc.*, 61, 1638-1643.
7. Russbueldt, B. M. E. and W. F. Hoelderich (2009), "New sulfonic acid ion-exchange resins for the preesterification of different oils and fats with high content of free fatty acids", *Appl. Catal. A: Gen.*, 362, pp. 47-57
8. Kouzu, M., T. Kasuno, M. Tajika, Y. Sugimoto, S. Yamanaka and J. Hidaka (2008), "Calcium oxide as a solid base catalyst for transesterification of soybean oil and its application to biodiesel", *Fuel*, 87, pp. 2798-2806.
9. Berrios, M. and R. L. Skelton (2008), "Comparison of purification methods for biodiesel", *Chem. Eng. J.*, 144, pp. 459-465.
10. Kucek, K. T., M. A. F. Cesar-Oliveira, H. M. Wilhelm and L. P. Ramos (2007), "Ethanolysis of refined soybean oil assisted by sodium and potassium hydroxides", *J. Am. Oil Chem. Soc.*, 84, pp. 385-392.
11. Sakai, T., A. Kawashima and T. Koshikawa (2009), "Economic assessment of batch production processes using homogeneous and heterogeneous alkali catalysts", *Bioresour. Technol.*, 100, pp. 3268-3276
12. Corma, A., S. Iborra and A. Velty (2007), "Chemical routes for the transformation of biomass into chemicals", *Chem. Rev.*, 107, pp. 2411-2502.
13. Kulkarni, M. G. and Dalai A. K. (2006), "Waste cooking oil -An economical source for biodiesel: A review", *Ind. Eng. Chem.*, 45, pp. 2901-2913.
14. Yan, S., C. DiMaggio, S. Mohan, M. Kim, S. O. Salley and K. Y. Simon Ng (2010), "Advancements in heterogeneous catalysis for biodiesel synthesis", *Top. Catal.*, 53, pp. 721-736.
15. Oda, M., M. Kaieda, S. Hama, H. Yamaji, A. Kondo, E. Izumoto and H. Fukuda (2005), "Facilitatory effect of immobilized lipase-producing *Rhizopus oryzae* cells on acyl migration in biodiesel-fuel production", *Biochem. Eng. J.*, 23, pp. 45-51.
16. Bournay, L., D. Casanave, B. Delfort, G. Hillion and J. A. Chodorge (2005), "New heterogeneous process for biodiesel production: A way to improve the quality and the value of the crude glycerin produced by biodiesel plants", *Catal. Today*, 106, pp. 190-192.
17. Hara, M. (2010), "Biodiesel production by amorphous carbon bearing SO_3H, COOH and phenolic OH groups, a solid Bronsted acid catalyst", *Top. Catal.*, 53, pp. 805-810.

(118f) Rule of Thumb for Design and Analysis of Biomass Utilization System

Kazuhiro Mochidzuki, IIS, Univ. of Tokyo, Tokyo, Japan and Akiyoshi Sakoda, Institute of Industrial Science, University of Tokyo, Tokyo, Japan

This paper deals with the rule of thumb for design and analysis of overall biomass utilization system, including, the matching of available biomass and conversion technologies, the market and demand of biomass-based energy/products, the evaluation of material and energy balances throughout the system, the environmental impacts, and the sustainability. When laying a plot of biomass utilization, it is important to develop comprehensive strategy based on a reasonable scenario. Nevertheless every plot has different circumstances and viewpoints on it. There are endless varieties of factors regarding to design and analysis of the biomass systems. The problems spread across wide-ranged issues from agriculture/forestry to industry, from political/social sectors to scientific/engineering sectors, from local to worldwide, and so on. To establish a reasonable biomass system, it is necessary to make clear the whole picture of it, organizing the key factors and problems.

We have developed concepts and methods to design biomass utilization systems. Our approach includes the followings: 1) screening the key factors, 2) drawing up a scenario and its modeling, 3) matching the available biomass sources and utilization technologies, 4) analysis/optimization of material and energy balances throughout the whole system, and 5) evaluation of economical and environmental issues. In this paper our recent progress in this field is shown with some results of case studies.

See more of this Session: Reactor Engineering for Biomass Feedstocks
See more of this Group/Topical: Sustainable Engineering Forum

(173a) Silicon Nanostructures for Lithium Ion Battery Anodes

Justin Harris[1], Aaron Chockla[1], Matthew Panthani[1], Colin Hessel[1], Dariya Reid[1] and **Brian Korgel**[2], (1) Chemical Engineering, The University of Texas at Austin, Austin, TX, (2)Department of Chemical Engineering, The University of Texas at Austin, Austin, TX

Colloidal synthetic methods have been developed for a wide variety of nanocrystal and nanowire materials. Among these materials, silicon has been one of the most challenging to synthesize in in solution. The general approach is to feed molecular reactants into hot solvent environment to precipitate the nanomaterial of interest. In the case of silicon (Si), suitable reactants are needed that readily decompose in the solvent environment and the reaction must be controlled sufficiently to yield nanomaterials with the desired size and shape. Recently, we have been developing a variety of methods for the synthesis of Si nanomaterials, including amorphous and crystalline Si colloidal particles and Si nanorods and nanowires. These methods can yield significant quantities of Si nanomaterials and one potential application for them is as a replacement for carbon as the anode material in lithium ion batteries. Si has the ability to alloy with lithium ($Li_{4.4}Si$) to achieve very high charge storage capacity of as much as 4200 mAh/g, which is more than an order of magnitude higher than the storage capacity of carbon. Lithiation, however, leads to a massive volume expansion (~400%), which leads to pulverization. Si nanostructures have the ability to tolerate this volume expansion, while enabling significant lithiation. Recent data of the charge storage capacity of colloidal silicon nanoparticles and nanowires will be presented.

See more of this Session: Nanomaterials for Energy Storage III
See more of this Group/Topical: Topical 5: Nanomaterials for Energy Applications

(173b) Nano-Silicon/Graphene Composite Anode Modifications for Enhanced Performance Lithium Ion Batteries

Rhet Joseph De Guzman[1], Jinho Yang[2], Mark Cheng[2], Steven Salley[1] and K. Y. Simon Ng[1], (1) Department of Chemical Engineering and Materials Science, Wayne State University, Detroit, MI, (2) Department of Electrical and Computer Engineering, Wayne State University, Detroit, MI

Development of lithium ion batteries with improved capacity and cycle performance are in the forefront of efforts to increase the electrification of current-age vehicles. Lately, much attention has been focused in the use of silicon-based anode materials due to their low discharge potential and theoretical capacity (4,200 mAh/g) which is more than 10 fold of the traditional graphite anode. However, due to the increased accommodated lithium ions, silicon's volume varies by 400% causing pulverization and loss of electrical contact. Due to this, strategies have led to the use of nano-sized Si and incorporation of Graphene to be able to mitigate volume variation and therefore produce a stable anode material with improved capacity and cyclability. This project is aimed at the improvement of the dispersion of Si nanoparticles and the optimization of the graphene nanostructure producing a Si nanoparticle/graphene nanosheet composite anode demonstrating improved capacity retention and rate performance.

See more of this Session: Nanomaterials for Energy Storage III
See more of this Group/Topical: Topical 5: Nanomaterials for Energy Applications

(173c) Sn/SnOx /Graphene Nanocomposites: Synthesis and Anode Performance In Lithium-Ion Batteries

Mahbuba Ara[1], Dr. Steven Salley[2] and Dr.Simon Ng[2], (1)Chem. Eng., Wayne State University, Detroit, MI, (2)Chemical Engg., Wayne State University, Detroit, MI

The utilization of a Sn/SnO_x/graphene nanocomposite as anode material for lithium-ion batteries could potentially provide high reversible capacity as well as a reduction of capacity fade upon cycling, by increasing the Sn:Li_2O ratio in the anode matrix. A two step chemical synthesis route has been developed to produce a Sn/SnO_x/Graphene nanocomposite. In the first step, SnO_2 nanoparticles were prepared by a hydrolysis method and loaded on graphene nanosheets using an ultrasonic treatment to make SnO_2/graphene nanocomposites. In the second step, Sn/SnOx/graphene was prepared by the partial reduction of SnO_2/graphene in a 10% H_2 in Ar mixture at 700 °C. To get nanosized Sn/SnOx particles with size in the range of 5-10 nm, reaction time and temperature were varied to control the nucleation and to obtain a homogeneous distribution of Sn nanoparticles on the graphene sheets, while pH of the solution was optimized, as at a specific pH the surface charge of the particles affects particle organization on the surface. The small size and uniform distribution of Sn/SnOx nanoparticles are designed to reduce the volume expansion of Sn during cycling. Hence, the anode of lithium-ion battery endures lower strain and is less likely to be cracked upon cycling. Sn:C ratios of 1.5, 2.5 and 3.5 in the Sn/SnOx/graphene nanocomposites will be evaluated for a superior Li-storage performance with good cycle life and high capacity.

See more of this Session: Nanomaterials for Energy Storage III
See more of this Group/Topical: Topical 5: Nanomaterials for Energy Applications

(173d) A Facile Way to Fabricate MoO3 /Graphene Composite Anode for Lithium Ion Battery

Yun Xu, Qianglu Lin, Ling Fei, Joshua Hill, Shuguang Deng and Hongmei Luo, Chemical Engineering Department, New Mexico State University, Las cruces, NM

A Facile Way to Fabricate MoO_3 /Graphene Composite Anode for Lithium Ion Battery

Yun Xu, Qianglu Lin, Ling Fei, Joshua Hill, Shuguang Deng, Hongmei Luo

Department of Chemical Engineering, New Mexico State University, Las Cruces, New Mexico 88003

MoO_3 is capable of Li^{-+} insertion/extraction in excess of 6 Li^{-+} per formula unit, resulting in a significantly reversible capacity. Recently, chemically modified graphene materials have been used to form hybrid materials with SnO_2, Fe_2O_3, and Mn_3O_4 in attempts to improve electrode capacity and cycling stability. However, an anode combining MoO_3 with graphene has never been reported. MoO_3/ graphene composite was prepared through a facile two-step synthesis by homogeneous coprecipitation and subsequent reduction of the graphite oxide with hydrazine. The synergic effect of graphene and MoO_3 improves the electrochemical properties, like cycling performance and rate capability. SEM and TEM images show that MoO_3 uniformly dispersed on the surface of graphene. The synthesis approach presents a promising route to produce large amount of high performance electrode materials for Lithium ion batteries.

See more of this Session: Nanomaterials for Energy Storage III
See more of this Group/Topical: Topical 5: Nanomaterials for Energy Applications

(173e) Graphene-Encapsulated Hollow Fe3O4 Nanoparticle Aggregates As a High Performance Anode Material for Lithium Ion Batteries

Dongyun Chen[1], Ge Ji[1], Yue Ma[1], **Jim Yang Lee**[2] and Jianmei Lu[3], (1)Chemical & Biomolecular Engineering, National University of Singapore, Singapore, Singapore, (2)Department of Chemical and Biomolecular Engineering, National University of Singapore, Singapore, Singapore, (3)Chemical Engineering and Materials Science, Soochow University, Suzhou, China

Graphene-EncapsulatedHollow Fe_3O_4 NanoparticleAggregates as a High Performance Anode Material for Lithium Ion Batteries

Dongyun Chen[1,2], Ge Ji[1], Yue Ma[1], Jim Yang Lee[1] and Jianmei Lu[2]

-

[1]*Department of Chemical& Biomolecular Engineering, Faculty of Engineering, National University ofSingapore, 10 Kent Ridge Crescent, Singapore, 119260; Tel: (65) 65162186; E-mail:cheleejy@nus.edu.sg.*

[2]*Key Laboratory ofOrganic Synthesis of Jiangsu Province, Key Laboratory of Absorbent Materialsand Techniques for Environment, College of Chemical, Chemical Engineering andMaterials Science, Soochow University, Suzhou, China, 215123; E-mail: lujm@suda.edu.cn*

In this work we combined the unique propertiesof graphene sheets and a hollow assembly of nanoparticles to simultaneouslyprovide large reversible Li^+ storage capacity,good rate performance and long cycle life. Specifically graphene-encapsulatedordered aggregates of Fe_3O_4 nanoparticles withnearly-spherical geometry and hollow interior (G-HPM)were synthesized by a simple self-assembly process. TheHPM particles were first prepared by a one-pot hydrothermal synthesis. Theywere then modified by 3-aminopropyltriethoxysilane (KH550) to acquire a positivesurface charge, which enabled electrostatic assembly with negatively chargedgraphene oxide. Finally, a chemical reduction was used to reduce the grapheneoxide to graphene; finalizing the packaging of HPM particles in individualgraphene cages. This assembly could be carried out under relatively mildconditions; thereby minimizing the perturbations to the properties of HPM. Inour design, the open interior structure has adapted well to the volume changein repetitive Li^+ insertion and extraction reactions; and theencapsulating graphene connects the Fe_3O_4 nanoparticleselectronically. The composite particles showed a stable high specificreversible capacity of about 897 mA h g^{-1} which was nearly unvaryingover 50 cycles. These graphene-Fe_3O_4 composites are thereforea capable reversible Li^+ host with high storage capacity and goodcycle life which can be cycled at high rates.

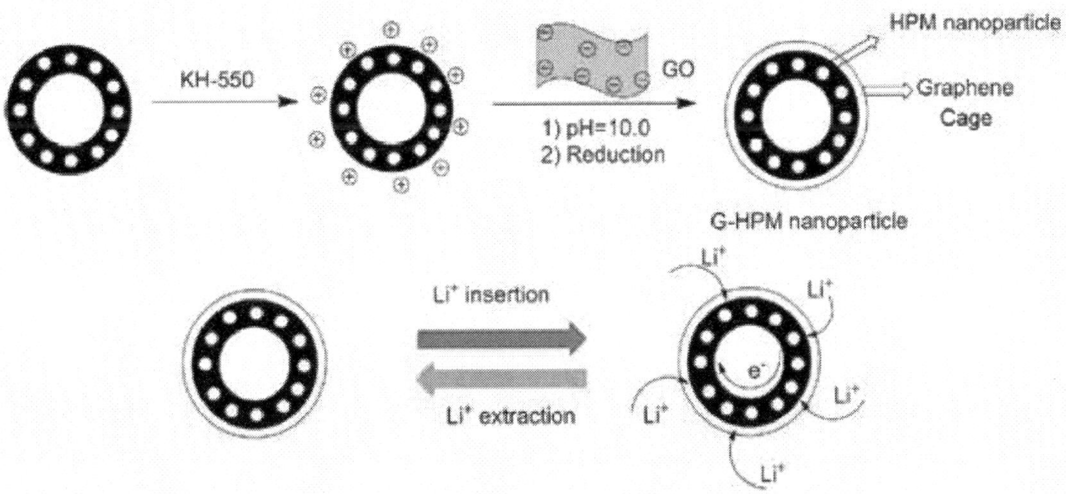

See more of this Session: Nanomaterials for Energy Storage III
See more of this Group/Topical: Topical 5: Nanomaterials for Energy Applications

(173f) An In Situ Method of Creating Metal Oxide-Carbon Nanocomposite As Anode Materials for Lithium-Ion Batteries

Zichao Yang, Jingguo Shen and Lynden A. Archer, Chemical and Biomolecular Engineering, Cornell University, Ithaca, NY

Transition metal oxides are attractive as anode materials for lithium-ion batteries (LIBs) due to their high lithium intercalation capacity and generally high natural abundances and their use on the nanoscale is being actively pursued. However, there are still challenges associated with the use of these metal oxides, including the pulverization problem and limited electrical conductivity of many metal oxides. It has been shown that these problems may be alleviated by creating metal oxide-carbon composites, which improve upon the mechanical flexibility and the electrical conductivity of the material.

There are different approaches for creating the carbon composites such as carbon nanopainting and attaching the metal oxide nanoparticles to carbon substrates. The current work describes an alternative method of creating metal oxide-carbon nanocomposites, in which the metal oxide nanoparticles are created *in situ* and embedded in a porous carbon matrix, which is made through calcination of a polyacrylonitrile-based polymer. Fe3O4 is used as a model compound.

Characterization techniques such as X-Ray diffraction, transmission electron microscopy and Raman spectroscopy were used to investigate the physical and chemical properties of the composite. Cyclic voltammetry of the composite shows lithium intercalation at 1.2V and 0.8V, which may be attributed to the reduction of Fe^{3+} to Fe^{2+} and Fe^{2+} to Fe^{0}, respectively. Very stable capacity is observed over >110 cycles at charging rate of 1C. Performance at other charging rates up to 5C is also shown and is seen to be stable. The contribution of factors including electrical conductivity, mechanical properties and porous nature of the carbon matrix in enhancing the electrochemical performance is investigated. The application of the method to other related materials such as MnO, Sn and Si is discussed.

See more of this Session: Nanomaterials for Energy Storage III
See more of this Group/Topical: Topical 5: Nanomaterials for Energy Applications

(173g) Binderless Cobalt Oxide Nanoparticle Films Through Electrophoretic Deposition, for Li-Ion Batteries

Don-Hyung Ha, Mohammad A. Islam and **Richard D. Robinson**, Materials Science and Engineering, Cornell University, Ithaca, NY

Nanoscalematerials for Li-ion battery electrodes have received great attention due totheir high surface-to-volume ratio and reduced Li-ion diffusion length whichresults in faster charge/discharge. Nanomaterialsalso show the ability to accommodate strain through volume expansion without pulverizationor capacity fading. In this work we have developed a hollow cobalt oxide NP filmfor Li-ion battery electrodes that require no additional support materials suchas carbon black or polymers to establish good electrical pathways. Thecolloidal cobalt NPs are prepared by decomposition of organometallicprecursors in non-aqueous conditions. The NP film is formed by using electrophoretic deposition (EPD). During EPD, a film isformed on both the cathode and anode but the film on the anode is slightlythicker than that of the cathode, which indicates that more particles arenegatively charged. The EPD deposited films are calcinedto form hollow cobalt oxide (Co_3O_4) NPs that act as the batteryanode.

Battery cycling of these anodes show lowdegradation and are near the theoretical capacity limits for this material. Theseresults are surprising since control experiments made from dropcastfilms form poor battery contacts under the same calcination and battery-test conditions.

See more of this Session: Nanomaterials for Energy Storage III
See more of this Group/Topical: Topical 5: Nanomaterials for Energy Applications

GASIFICATION OF ORANGE RESIDUE CHARS IN CO_2

Astrid Altamar[1], Suélen Amorim[1], Valquiria Balvedi[1], Regina F.P.M. Moreira[1] and Humberto J. José[1]
[1]Laboratory of Energy and the Environment. Federal University of Santa Catarina. Florianópolis, SC. Brazil.

Gasification of chars from orange residues and wood sawdust were studied in order to compare their reactivities using thermogravimetric analysis (TGA). Char samples were obtained by pyrolysis in a tubular reactor of quartz at a temperature of 900 °C and heating rate of 250 °C/min. The gasification of the chars using CO_2 was done on a thermogravimetric analyzer isothermally at temperatures up to 950 °C, and non-isothermally at temperatures up to 980 °C. Several morphological changes due to vaporization during pyrolysis were identified by comparing the SEM images of the chars and the corresponding biomass.

The biomass particle size was below 150 μm, and the sample size low enough to reduce heat and mass transfer side effects, thus ensuring that the reaction kinetics is the limiting step. The reactivity in CO_2 of both biomasses increased with conversion. However, differences in reactivity were observed between the two types of biomass. The char of orange residue exhibited a higher reactivity than wood sawdust. This could be due to the catalytic effect of the inorganic content, specifically potassium, which was found in the orange residue ash but not in the sawdust ash. The activation energies found for the wood sawdust char gasification were 197.4 and 187.7 kJ/mol for the isothermal and non-isothermal methods respectively, while for the orange residue char were 126.5 and 129.3 kJ/mol.

1. INTRODUCTION

Biomass waste from forest and industrial residues are an important source of energy, in particular orange residues can be considered in regions with high orange production like Brazil, where orange production was 2.02×10^7 Ton in 2010, which is 37.2% of the world production [1].

Biomass residues must be conditioned in order to be used in energy conversion devices, which require high energy content and homogeneity in their feed streams. Several techniques for biomass conditioning have been developed, being gasification one of the most promising because of its relative simplicity and high performance.

Gasification is a process where the carbon present in pyrolyzed samples reacts with a stream containing carbon dioxide, producing carbon monoxide and hydrogen. The kinetic study of the gasification is performed by thermogravimetric analysis, with the advantage of doing both isothermal and non-isothermal kinetic evaluation.

Isothermal kinetic analysis has been the traditional method used to find the reaction order and parameters such as activation energies and preexponential factor in chemical reactions. However, recently the benefits of doing this determination by non-isothermal experiments has been shown by measuring the weight change of solid samples while they react in an environment where the temperature is controlled with a dynamic gradient temperature called the heating rate (HR). [2-5].

Nomenclature			
c_{CO2}	concentration of carbon dioxide (mol/m^3)	R_i	reactivity of char (s^{-1})
Ea	activation energy (kJ/mol)	R_{50}	reactivity at 50% conversion (s^{-1})
HR	heating rate (K/s)	R	ideal gas law constant (0,008314) (kJ/(mol K))
km	reaction rate constant (m^3 kg^{-1} s^{-1})	t	time (s)
km,0	pre-exponential factor (m^3 kg^{-1} s^{-1})	T	temperature (K)
m_C	mass of carbon (kg)	T_0	initial temperature (K)
$m_{C,0}$	initial mass of carbon (kg)	TG	thermogravimetry
r	reaction rate (kg s^{-1})	X_c	Conversion of carbon
Mc	molecular weight of carbon (kg/mol)		

Results from Jess et al. [6] show that this method gives information not only about chemical reactions, but also about the mass transfer effect in a reaction by performing a single experiment.

2. THEORETICAL BACKGROUND

2.1. Isothermal Analysis

The gasification of lignocellulosic biomasses is considered according to the kinetic expression:

$$r = -\frac{dm_C}{dt} = k_m c_{CO_2} m_C Mc \tag{1}$$

Where r is the reaction rate, expressed as the weight loss in time, essentially from the carbon remaining in the pyrolyzed sample.

The reaction rate constant is expressed as:

$$\mathrm{k_m = k_{m,0} e^{(-E_A/RT)}} \tag{2}$$

The procedure used for the data analysis is the same as the one used by other researchers [7]. Based on the homogeneus reaction model the reactivity of the char is defined as:

$$\mathrm{R_i = -\frac{1}{m_C}\frac{dm_C}{dt}} \tag{3}$$

A representative reactivity value at 50% conversion R_{50} for each test was calculated. For isothermal analysis, R_{50} can be expressed as:

$$\ln(\mathrm{R_{50}}) = \ln\left(\mathrm{k_{m,0} c_{CO_2}}\right) - \frac{\mathrm{E_A}}{\mathrm{RT}} \tag{4}$$

Subsequently, values of preexponential factor and activation energy may be estimated from the intercept and slope, respectively, of a plot ln R_{50} versus 1/T.

2.2. Non-Isothermal Analysis

For the non-isothermal experiments the Eq. (1) can be re-arranged to a better form for the data provided by the thermogravimetric analyzer. Introducing the heating rate,

$$HR = \frac{dT}{dt} \tag{5}$$

The expression becomes:

$$-\frac{dm_C}{dT}\frac{HR}{m_C} = k_{m,0}c_{CO_2}M_C e^{(-E_A/RT)} \tag{6}$$

And in terms of the conversion X_C, calculated as

$$X_C = 1 - \frac{m_C}{m_{C,0}} \tag{7}$$

The following expression is obtained:

$$\ln\left(\frac{dX_C}{dT}\frac{HR}{(1-X_C)}\right) = \ln\left(k_{m,0}c_{CO_2}M_C\right) - \frac{E_A}{RT} \tag{8}$$

So, the plot of the term in the left side versus 1/T must be a straight line where the slope and intercept are used to calculate the activation energy and the preexponential factor respectively. The reason why it is possible to calculate these terms in a single experiment is that being the time dependent variable (HR) a constant, it is possible to directly tie the conversion to the temperature change, which leads to the direct kinetic parameters determination.

Eq. (6) can be integrated to obtain an expression for the weight of the sample as a function of temperature when a constant heating rate is applied,

$$-\int_{m_{C,0}}^{m_C}\frac{dm_C}{m_C} = ln\frac{m_{C,0}}{m_C} = \frac{k_{m,0}c_{CO_2}M_C}{HR}\int_{T_0}^{T} e^{(-E_A/RT)}dT \tag{9}$$

There is no analytical solution to the integral in the right side, but it can be solved through series expansion, according to:

$$\int_{T_0}^{T} e^{(-E_A/RT)}dT \approx \frac{RT^2}{E_A}e^{-\frac{E_A}{RT}}\left[1 + \frac{2!}{-E_A/RT} + \frac{3!}{(-E_A/RT)^2} + \cdots\right]\Bigg|_{T_0}^{T} \tag{10}$$

For activation energies greater than 100 kJ/mol, the terms corresponding to T_0 are negligible when the temperature difference $T-T_0$ is more than 200 K, and the series expansion can be reduced only to the first two terms as a good approximation, so the integral is calculated as:

$$\int_{T_0}^{T} e^{(-E_A/RT)}dT \approx \frac{RT^2}{E_A}e^{-\frac{E_A}{RT}}\left(1 - \frac{2RT}{E_A}\right) \tag{11}$$

So the expression for the sample weight leads to:

$$TG = \frac{m_C}{m_{C,0}} = \exp\left[\frac{k_{m,0}c_{CO_2}M_C}{HR}\frac{RT^2}{E_A}\left(1-\frac{2RT}{E_A}\right)e^{-\frac{E_A}{RT}}\right] \quad (12)$$

This expression is useful to determine the accuracy of the prediction of the model when compared to the experimental data obtained from the TGA.

3. EXPERIMENTAL

Wood sawdust and orange residues from the southern Brazilian region of Santa Catarina state were used in this study. The proximate and elemental analysis and the chemical composition of their ashes are shown in Table 1. The biomass samples were pyrolyzed in a tubular reactor of quartz. Approximately 1 g of biomass was loaded into the reactor which was then the reactor was inserted into the furnace. The biomass samples were placed at the center of the furnace and heated from ambient temperature to 900 °C using a heating rate of 250 °C/min under nitrogen flow (400 ml/min). The samples were kept at 900 °C for 15 min to complete the pyrolysis. An inert atmosphere (N_2) was maintained until the reactor had cooled down to the room temperature to prevent reaction with air. The char residue was stored under inert conditions in a desiccator for later use. The diameter of the char of biomass was smaller than 150 µm. Table 2 present its chemical composition.

Table 1
Analysis biomass samples (wt%, db)

	Wood sawdust	Orange residue		Wood sawdust	Orange residue
Proximate analyses			**Ash analyses**		
Fixed Carbon	20,13	29,26	Fe_2O_3	4,44	6,17
Volatile matter	79,44	66,38	CaO	31,27	16,11
Ash	0,43	4,36	MgO	11,64	4,18
Elemental analyses			Na_2O	1,67	0,16
C	55,06	46,40	K_2O	10,44	31,48
H	7,11	10,09	SiO_2	15,69	10,17
N	0,21	1,70	Al_2O_3	12,30	3,84
O (by difference)	37,14	37,39	TiO_2	3,94	0,25
			P_2O_5	2,74	8,46
			MnO	2,06	0,07
			SO_4	3,02	2,89
			Others	0,79	14,88

The kinetic study of the gasification was done by both, isothermal and non-isothermal experiments. The isothermal results were used as a reference to validate the non-isothermal ones. The isothermal and non-isothermal gasification experiments were carried out by means of a DTG60/60H Shimadzu thermogravimetric analyser (TGA) operating at atmospheric pressure. A constant amount (3-4 mg for wood sawdust and 5-6 mg for orange residue) was placed in the alumina crucible and the temperature was measured by the thermocouple located close to the sample. Carbon dioxide was used as the gasifying agent with a flow rate of 100 ml/min. Non isothermal runs proceeded at heating rates of 10, 20, 30°C/min from 35 to 980°C. Isothermal runs, were carried out at 800, 850, 900, 950°C, a char sample was kept at gasifying temperature to ensure complete carbon conversion. All weight measurement experiments were corrected with blank runs under the same heating conditions.

Table 2
Analyses of the char

	Char wood sawdust	Char orange residue
Elemental analysis (wt%, db)		
C	75,40	67,79
H	1,77	1,82
N	0,34	1,60
O (by difference)	22,22	28,55

Scanning electron microscopy (SEM) analysis of the raw material and char particles was employed to determine the effect of the pyrolysis on the microstructure. SEM analysis was performed using an JEOL JSM-6390LV.

4. RESULTS AND DISCUSSION

4.1. SEM ANALYSIS

Samples were analyzed by SEM, as may be seen in Figures 1 and 2, where the formation of cavities along the fibers is evident. This is the direct result of the materials volatilization.

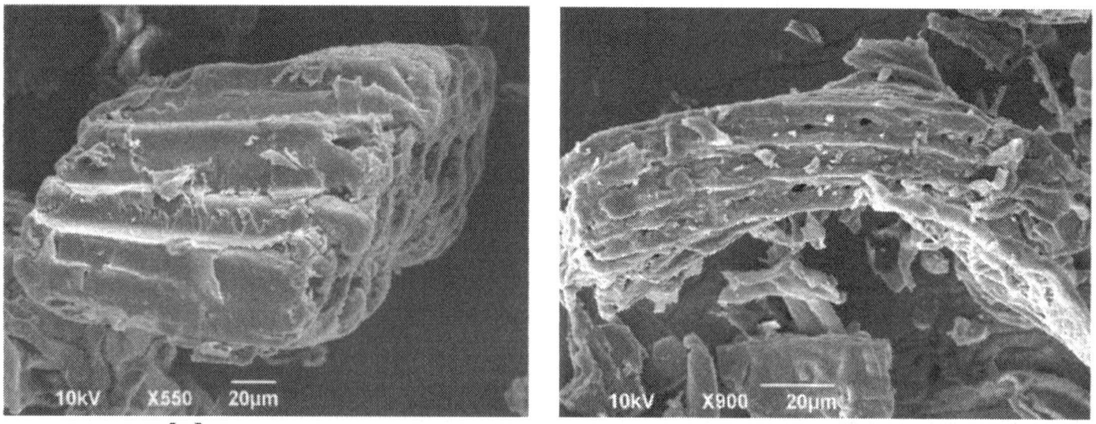

(a) **(b)**

Figure 1. SEM images of wood sawdust (a) raw material, (b) pyrolyzed at 900°C

(a) (b)

Figure 2. SEM images of orange residue (a) raw material, (b) pyrolyzed at 900°C.

4.2. KINETIC ANALYSIS

4.2.1. Isothermal Experiments

Behavior of the gasification reaction of wood sawdust and orange residue chars in CO_2 are shown in Figure 3. It can be seen that the orange residue exhibits more reactivity than the wood sawdust char.

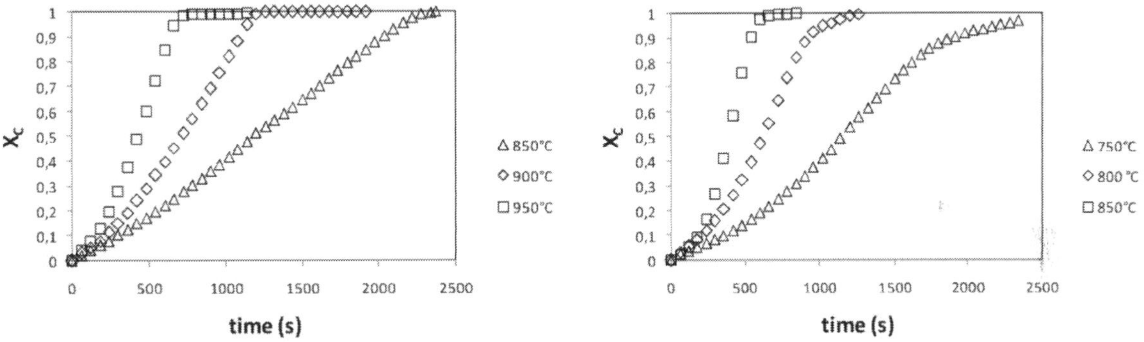

(a) (b)

Figure 3. Conversion of (a) wood sawdust and (b) orange residue chars in CO_2 at different temperatures.

To determine the activation energy, the methodology proposed by Ollero et al. [7], as described in section 2.1 was applied. Figure 4 shows the reactivity of wood sawdust in carbon dioxide as a function of conversion. The activation energy was determined by regression analysis and it was found to be 197.4 kJ/mol at 850-950 °C. This value is in agreement with literature. [3, 8-10].

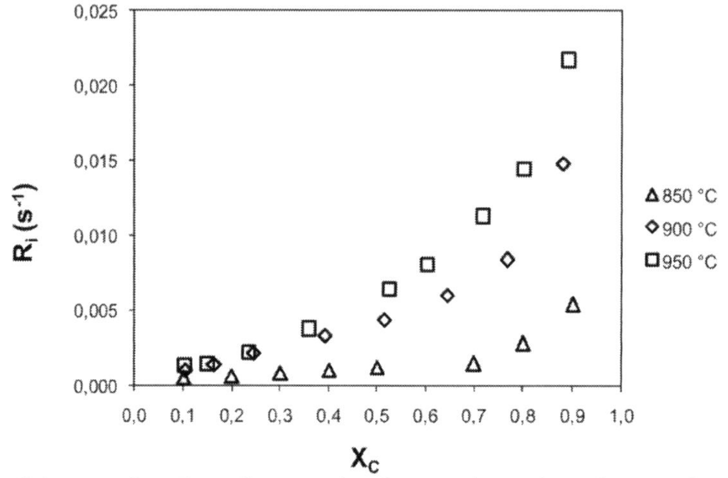

Figure 4. Reactivity as a function of conversion for wood sawdust char gasification in CO_2.

Figure 5 shows the reactivity as a function of conversion for the orange residue char. Using the same methodology as the one used with wood sawdust, the activation energy found to be 126.5 kJ/mol in the range 750-850 °C. [2, 7, 11, 12].

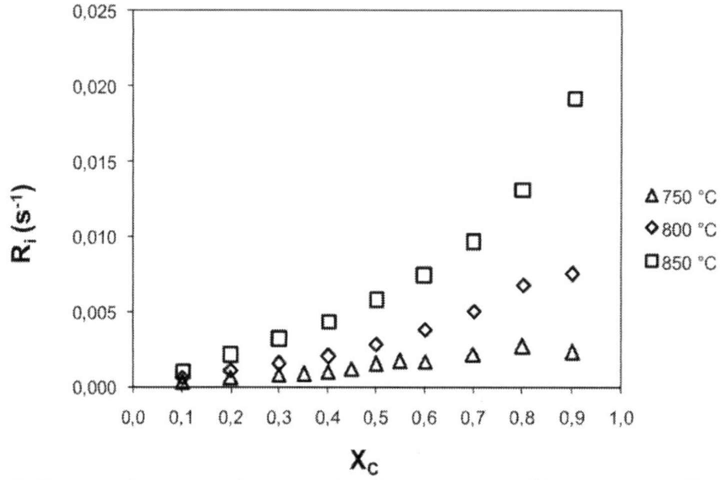

Figure 5. Reactivity as a function of conversion for orange residue char gasification in CO_2.

4.2.2. Non-Isothermal Experiments

Figure 6 shows the experimental results for the conversion of the chars in CO_2 at a heating rate of 20 K/min. It can be seen that the gasification for the orange residue char takes place at lower temperature than the one for the wood sawdust char.

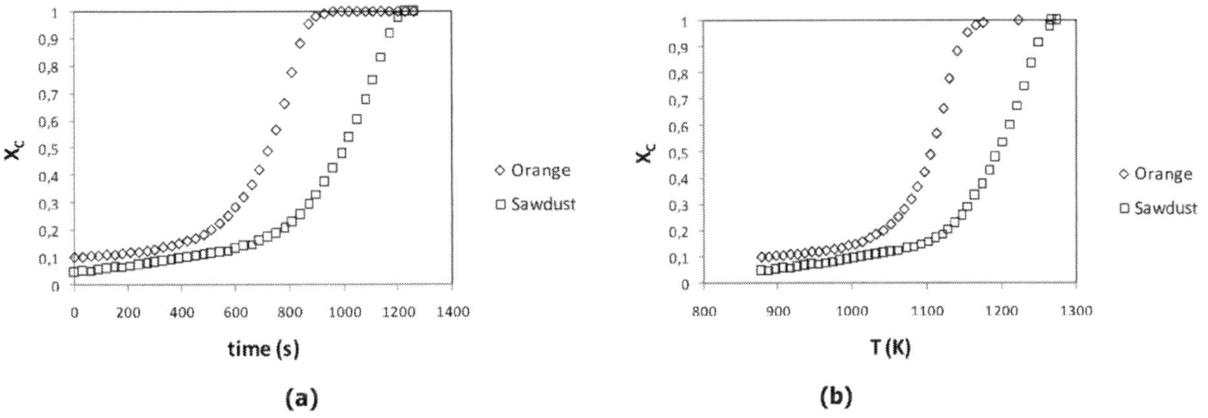

Figure 6. Conversion versus time (a) and temperature (b) of chars from orange residue and wood sawdust in CO_2 at a heating rate of 20 K/min.

As described in section 2.2, the plot of $\ln\left(\frac{dX_C}{dT}\frac{HR}{1-X_C}\right)$ vs 1/T (Figure 7), for the wood sawdust char, allows the calculation of the activation energy for the gasification. In this case the value obtained is 187.7 kJ/mol, which is close to the obtained for the isothermal method.

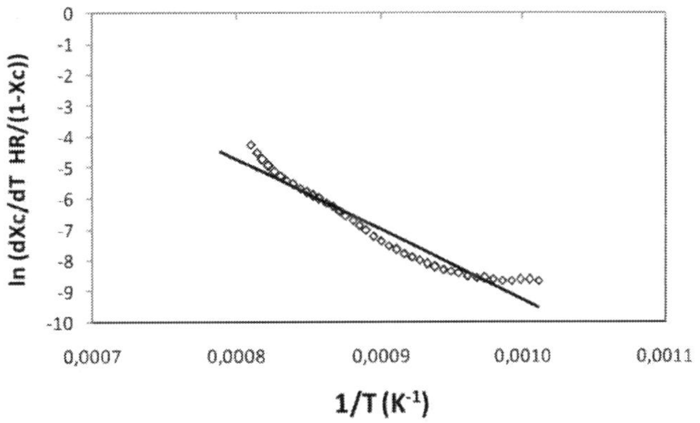

Figure 7. Plot to determine the activation energy of wood sawdust char gasification in CO_2 at heating rate of 20 K/min.

Figure 8 shows the values of the experimental and simulated thermogravimetric analysis as a function of temperature, that is, the fraction of the initial mass of the sample obtained from the TGA data, compared to the calculated values using Eq. (12). For conversions between 7.3% and 79.3%, the mean error associated in this range is 2.9%, calculated as the root mean square error (RMS).

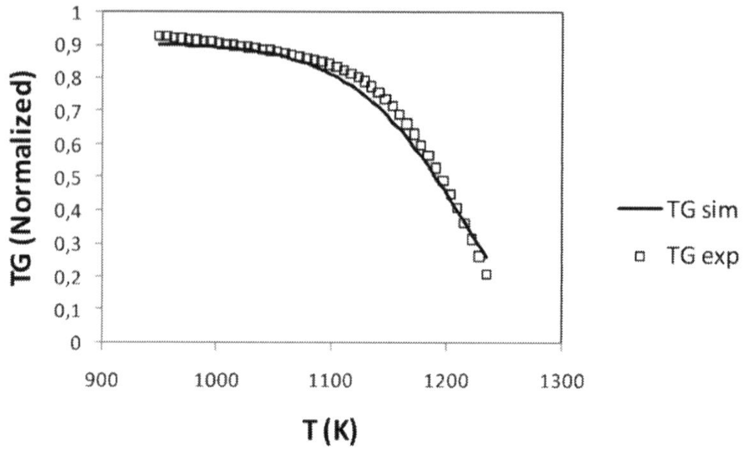

Figure 8. Comparison of experimental and simulated TG values for wood sawdust at heating rate of 20 K/min.

Figure 9 shows the plot used to estimate of the activation energy for the orange residue char, the obtained value was 129.3 kJ/mol. The tendency of the slope to the increase at higher temperatures indicates that there are several chemical reactions with different activation energies. The reactions with lower energies occur at lower temperatures, as a result, these low energy reactants disappear, leaving behind a sample with higher energy reactants, which becomes evident in the non-isothermal analysis with the increase of the slope. On the other hand, this tendency also confirms that there is no mass transfer limitation on the reactions.

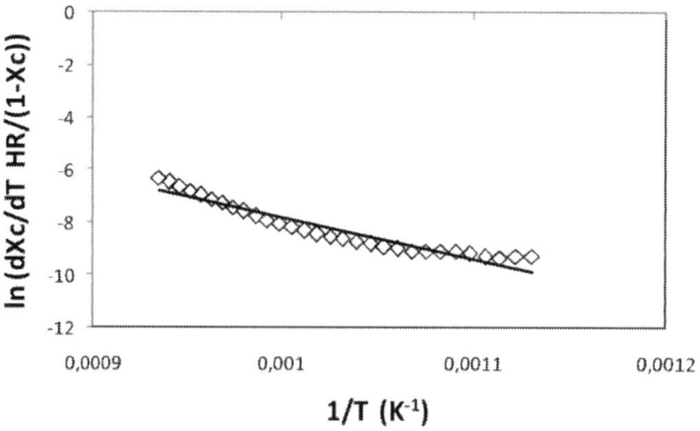

Figure 9. Plot to determine the activation energy of orange residue char gasification in CO_2 at heating rate of 20 K/min.

The comparison of the experimental and simulated values for the thermogravimetric analysis is shown in Figure 10. For the range of conversion between 10.0 and 54.0%, the associated error for the predicted values of the thermogravimetric analysis is 2.8%.

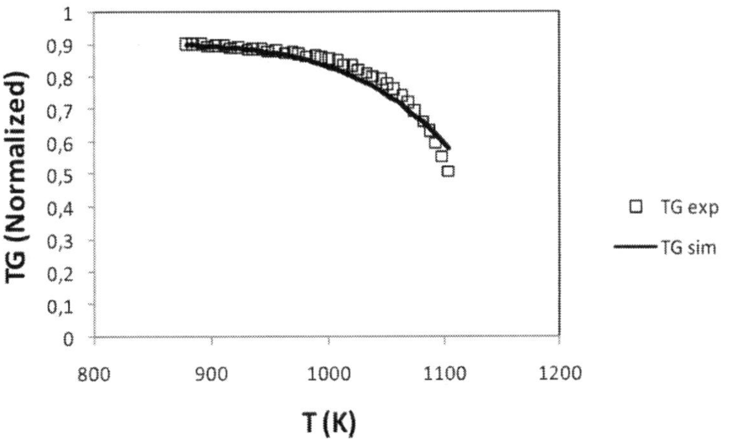

Figure 10. Comparison of experimental and simulated TG values for orange residue at heating rate of 20 K/min.

The activation energies calculated for the wood sawdust and orange residue chars are shown in Table 3. It can be seen that there is good agreement between the values obtained for the isothermal and non isothermal methods in each case, and there is consistency with literature.

Table 3
Activation energy calculated for wood sawdust char and orange residue char gasification

Wood Sawdust Char				
Method	**Ea, kJ/mol**	**T (K)**	**X_C**	**Error (%)**
Isothermal	197.4	1123-1173		
Non Isothermal	187.7	905-1235	0.07-0.79	2.9
Orange Residue Char				
Method	**Ea, kJ/mol**	**T (K)**	**X_C**	**Error (%)**
Isothermal	126.5	1023-1123		
Non Isothermal	129.3	880-1110	0.10-0.54	2.8

5. CONCLUSIONS

The orange residues char exhibit higher reactivity than wood sawdust due to its higher mineral content, which acts as catalyst for the gasification reactions. The use of these orange residues as biomass for gasification is an interesting alternative either directly or mixed with wood sawdust in order to improve the efficiency of the process.

The results obtained in this study show that the non-isothermal kinetic analysis is a powerful and reliable tool for the analysis of multiple chemical reactions, allowing the determination of the kinetic parameters, as well as the identification of possible interference from mass transfer diffusion.

6. ACKNOWLEDGEMENTS

The authors acknowledge CAPES and CNPq, Brazil, for financial support.

7. REFERENCES

(1) http://www.fas.usda.gov/psdonline/psdHome.aspx. Visited on 15 august 2011.

(2) Huang, Y., Yin, X., Wu C., Wang, C., Xie, J., Zhou, Z., Ma, L., Li, H. (2009), "Effects of metal catalysts on CO2 gasification reactivity of biomass char", *Biotechnology Advances*, Vol 27, pp. 568–572.

(3) Khalil, R., Várhegyi, G., Jäschke, S., Grønli, M. and Hustad, J (2009). "CO2 Gasification of Biomass Chars: A Kinetic Study", *Energy & Fuels*, Vol. 23, pp. 94-100.

(4) Sima-Ella, E., Yuan, G., Mays, T. (2005), "A simple kinetic analysis to determine the intrinsic reactivity of coal chars", *Fuel*, Vol 84, pp. 1920–1925.

(5) Vamvuka, D., Karouki, E. and Sfakiotakis, S. (2011), "Gasification of waste biomass chars by carbon dioxide via thermogravimetry. Part I: Effect of mineral matter", *Fuel*, Vol. 90. pp. 1120-1127.

(6) Jess, A., Andresen, A.-K., (2010), "Influence of mass transfer on thermogravimetric analysis of combustion and gasification reactivity of coke", *Fuel*, Vol 89, pp. 1541–1548.

(7) Ollero P, Serrera A, Arjona R, Alcantarilla S. (2003), "The CO2 gasification kinetics of olive residue", *Biomass Bioenergy*, Vol 24, pp. 151–61.

(8) Barrio, M., Hustand, J.E (2001), "CO2 gasification of birch char and the effect of CO inhibition on the calculation of chemical kinetics", in: Bridgwater AV, editor. *Progress in thermochemical biomass conversion*. Oxford: Blackwell Science Ltd., pp 47-60.

(9) Moilanen A, Saviharju K., (1997), "Gasification reactivities of biomass fuels in pressurized conditions and product gas mixtures", in: Bridgwater AV, Boocock DGB, editors. *Developments in thermochemical biomass conversion*. London: Blackie A & P, pp. 828–37.

(10) Risnes H, Sorensen LH, Hustad JE., (2001), "CO2 reactivity of chars from wheat, spruce and coal" in: Bridgwater AV, editor, *Progress in themochemical biomass conversion*. Oxford: Blackwell Science Ltd., pp. 61–72.

(11) Gómez-Barea, A., Ollero, P. and Fernández-Baco, C. (2006), "Diffusional Effects in CO_2 Gasification Experiments with Single Biomass Char Particles. 1. Experimental Investigation", *Energy & Fuels*, Vol. 20, pp. 2202-2210.

(12) Marquez-Montesinos F, Cordero T, Rodriguez-Mirasol J, Rodriguez JJ. (2002), "CO2 and steam gasification of a grapefruit skin char", *Fuel*, Vol 81, pp.423–429.

(199b) A Numerical Study On Bio-Oil Gasification Using A Multicomponent Approach

Lei Zhang and **Song-Charng Kong**, Mechanical Engineering, Iowa State University, Ames, IA

A Numerical Study on Bio-Oil Gasification Using A Multicomponent Approach

Bio-oil,also called pyrolysis oil, is a complex biorenewable fuel produced from fastpyrolysis of biomass. Since bio-oil is carbon dioxide neutral, it can be anideal replacement for petroleum fuel to be used in many kinds of combustion devices.It can also be gasified to produce synthesis gas, which in turn can be used forpower generation or synthesized to produce liquid fuels. The objective of thisstudy is to investigate the gasification/vaporization characteristics ofbio-oil drops under gasfier and combustor conditions. In this study, numerical methodsconsidering the actual composition of bio-oil are mainly used in the simulationof gasification/vaporization processes. Literature using experimentalapproaches such as Gas Chromatography/Mass Spectrometer has shown that bio-oil componentscan be categorized into several groups, including carbohydrates, furans,phenols, guaiacols, syringols, water, etc. By analyzing the composition ofbio-oil produced from different sources, a group of ten dominant components areidentified from the above groups. Using group contribution methods, thecritical and physical properties of these dominant components are estimated. Simplifyingbio-oil as a mixture of these dominant components, a multi-component dropvaporization model using a discrete component approach is developed in thesimulation of gasification/vaporization. Using the vaporization model, thevaporization histories of single bio-oil drops with different initialcompositions, which are the mass fractions of the components determined by thefeedstock and pyrolysis process, are calculated. In addition, the dynamics ofbio-oil spray is also studied numerically, and the distributions of the vaporsof various bio-oil components in the reactor are also predicted.

See more of this Session: Advances In Gasification Research
See more of this Group/Topical: Energy and Transport Processes

(199c) Realistic Impurity Generation Via Coal Gasification

Robert R. Mota, Chemical Engineering, University of North Dakota, Grand Forks, ND

Coal gasification is more efficient and environmentally friendly way to utilize coal than conventional coal combustion systems. However, there are some disadvantages to using coal as an energy source. Gasification comprises of many reactions, of which the two primary reactions; pyrolysis and steam gasification are endothermic. During the gasification process, impurities are transformed into gas, liquid and solid phases. The degrees to which these materials are transformed are dependent upon the impurity type and association in the fuel. During gas cooling the vapor phases condense and react with other impurities in the gas phase and with heat transfer surfaces such as gas cooling systems. Likewise, the liquid droplets will solidify into particles. These interactions are important in studying their effect on syngas cooling systems. Previous research[1] has shown that certain impurities have a high potential of condensing in the syn-gas cooler. The impurities found were lead, germanium, tin, antimony, bismuth, titanium, cadmium, zinc and arsenic. The focus of this research will be to find ways to mitigate the negative effects these impurities have on syn-gas coolers and hot gas cleanup by designing a small scale, fluidized bed, gasifier and characterizing the product gas through a syn-gas cooling system.

See more of this Session: Advances In Gasification Research
See more of this Group/Topical: Energy and Transport Processes

(199d) Comparison of 1D Modeling Methods for Entrained Coal Gasification

Moon-Kyeong Hwang, Mechanical Engineering, Pusan national university, Busan, South Korea

Comparison of 1D modeling methods for entrained coal gasification

Moonkyeong Hwang, Minsung Kim

School of Mechanical Engineering, Pusan National University, 30 Jangjeon-dong, Gumjeong-ku, Busan, 609-735, South Korea

Chunghwan Jeon, Youngjune Chang, Juhun Song*

Pusan Clean Center, Pusan National University, 30 Jangjeon-dong,

Gumjeong-ku, Busan, 609-735, South Korea

*Corresponding Author: jxs704@pusan.ac.kr

Telephone: +82-51-510-7330

FAX: +82-51-582-9818

Abstract

In the past, a number of one-dimensional modeling work have been made to predict the local gasification behavior of pulverized coals entrained at the normal operating conditions of tubular gasifier. In this work, three different modeling methods to compute the coal gasification at the gasifier exit in simpler manner are developed and compared. These include thermodynamic method, reaction kinetics method and well stirred reactor (WSR) method. With these models, important variables such as effect of oxygen addition and coal type on the gasification are evaluated in terms of coal conversion, the compositions and temperature of synthesis gas products at the exit.

Keywords

Coal gasification, 1D modeling, thermodynamic method, reaction kinetics method, WSR

See more of this Session: Advances In Gasification Research
See more of this Group/Topical: Energy and Transport Processes

(199e) Rigourous Modeling of An Entrained Flow Gasifier Considering a Heuristic Recirculation Model and a Detailed Energy Balance

Job Kasule[1], Richard Turton[1], Debangsu Bhattacharyya[1] and Stephen E. Zitney[2], (1)Department of Chemical Engg., West Virginia University, Morgantown, WV, (2)Collaboratory for Process and Dynamic Systems Research, National Energy Technology Laboratory, U.S. Department of Energy, Morgantown, WV

Coal-fired gasifiers are the centerpiece of integrated gasification combined cycle (IGCC) power plants. The gasifier produces synthesis gas that is cleaned and sent for combustion in a gas turbine. Several mathematical models have been developed to study the physical and chemical processes taking place inside the gasifier. Such models range from simple one-dimensional (1D) steady-state models to sophisticated dynamic 3D computational fluid dynamics (CFD) models. In this work, a 1D model of a single-stage, entrained-flow, General Electric Energy (GEE)-type gasifier has been developed.

In our model, mass, momentum and energy balance equations for solid and gas phases are considered. The model includes a number of heterogeneous and homogeneous chemical reactions along with devolatilization and drying of the slurry feed. The unreacted-core shrinking model is used for calculating the rates of the heterogeneous reactions. A detailed model of the radiative heat transfer has been developed considering interactions between all internal surfaces (side wall, top, and bottom surfaces) of the gasifier and the solids along with the interactions between the surfaces themselves. Unlike models developed in many previous studies, no arbitrary wall temperature profile is assumed but an energy balance across the wall including the heat loss from the gasifier wall to the environment is carried out. In the slurry-fed gasifiers, the recirculation near the inlet of the gasifier is promoted for rapid mixing of the slurry feed with a part of the hot reaction products. This vigorous mixing results in a significant rise in temperature that helps in evaporating the water and devolatilizing the coal. The recirculation is achieved by appropriately designing the feed burner, and feeding the oxygen through a swirling annular injector. In the current model, a heuristic recirculation model has been developed and the conservation equations have been appropriately modified.

The above formulation results in a highly nonlinear system of differential algebraic equations (DAE) that is solved using a Newton-type method in Aspen Custom Modeler® (ACM). The model is validated with the data from experimental studies and pilot plants. The paper includes species concentration profiles and temperature profiles of the gas and solid phases. A number of parametric studies are carried out to gain more insight into the gasifier performance as the inlet and operating conditions change. The results include the effects of the recirculation ratio on the initial heat up of the slurry feed as the type of coal and its quality are changed. For various inlet conditions, the model can be utilized to predict the profiles of a number of key variables which are very difficult, if not impossible, to measure in an actual gasifier.

See more of this Session: Advances In Gasification Research
See more of this Group/Topical: Energy and Transport Processes

Process Design and Validation of Operation in a Coal-to-Methanol Gasification Plant using Probabilistic Methodologies

Raul E. Ayala and Judeth B. Corry, GE Energy, 1333 W. Loop South, Houston, TX 77072

After a chemical process has been developed, designed, and launched at commercial scale, validation of the large scale results against design assumptions is essential. Typically, the engineer encounters factors that affect the quality of the data collected at a plant, both from a variability standpoint as well as from a data availability perspective, or lack thereof. This paper describes Six Sigma and probabilistic methodologies for uncertainty reduction used to ensure successful validation plans, data collection strategies, and methods for data analysis in order to validate a large-scale commercial system. The strategies will be illustrated using the GE designed and validated operation of a coal-to-methanol gasification facility in China using the aforementioned error-reducing strategies.

In a GE coal-to-methanol plant, coal is first ground as slurry and used as a feedstock to the entrained gasifier, where the slurry is converted to a gas mixture primarily of carbon monoxide, hydrogen, carbon dioxide and water to produce the raw syngas stream. After conditioning, the syngas is used as a reactant for the production of methanol in the downstream portion of the plant. A slag stream is the byproduct of the gasification process that contains the inert ash originally in the raw coal.

The GE validation process is used to ensure the test is executed in a manner to ensure high quality results. This process includes preparing for the test with reviews of the test plans, checklists, sample points, safety reviews, testing equipment required, and execution details as well as specifics required to define readiness such as "steady state" which can often be challenging in the field. During a field test, the test preparation and checklists provide the needed guidance at this time. After the test is completed, the team reviews the test execution to determine if it delivered the desired information along with the location of the data for future use. Lessons learned are also captured so that future tests benefit.

As it is challenging to simultaneously obtain discrete physical samples from all existing streams and vessels in a commercial plant, one requires a variety of statistical tools to make predictions about the non-measured point locations in various sub-systems or (black box) control volumes. This set of statistical tools makes up the overall probabilistic (or stochastic) approach to determine the overall plant performance analysis, which is the goal of the validation process.

Data from three major black box subsystems in the plant were used to carry out our probabilistic analysis: slurry preparation area, gasification island area, and slag handling area. Six Sigma techniques were used to achieve the data reconciliation across the subsystems by measuring time averages (means), local variability (standard deviations), and certainty in the instrumentation calibration and estimate of measurement drift (MSA), before doing a minimization of residuals across key parameters. The five-day validation test period was called the Baseline Test, and digital data from the distributed control system (DCS) was combined with daily chemical analysis of discrete physical samples (solid, liquid, and gases) to obtain snapshots of the overall plant operation. An in-house method of minimization of residuals in the mass and energy balances provided the means to data reconciliation across subsystem tie streams.

Once mass and heat balance reconciliation was achieved by minimizing residuals, three critical parameters were used to estimate the overall efficiency of the gasification and syngas production plant: total syngas production rate (SG flow rate), specific oxygen consumption (SOC, vol O_2/ vol ($CO+H_2$)), and specific coal consumption (SCC, mass dry coal/ vol ($CO+H_2$)). In addition, percent carbon conversion, another of the most difficult plant performance parameters to accurately quantify in a plant, was also estimated with improved precision using our probabilistic (stochastic) methodology than by standard deterministic methodologies.

Overall, results showed that our validation models enabled us to help existing customers improve the performance of their plants and enhance our designs for new projects. The expected error in the validated averages for carbon conversion was reduced over the test period because of the higher confidence in the measured and calculated values in each of the black-box subsystem control volumes. The results and recommendations provided to the customer defined a clear path to systematic improvements of plant operation in the future.

(219a) Theoretical Studies of Ion Transport In Solar Thermal Water Splitting by Mixed Metal Ferrites In Traditional and Hercynite Cycles

Christopher Lawrence Muhich[1], Charles Bruce Musgrave[2] and Alan W. Weimer[2], (1)Department of Chemical and Biological Engineering, University of Colorado, Boulder, Bouler, CO, (2)Chemical and Biological Engineering, University of Colorado, Boulder, Bouler, CO

We have used quantum simulations based on density functional theory within the periodic boundary condition approach to study ion transport occurring during solar thermal water splitting (STWS) by mixed metal ferrites. STWS occurs by cycling metal oxides through a high temperature reduction step and a lower temperature oxidation step where H_2O oxidizes the oxide and produces hydrogen gas. We investigate the transport of ions from the reduced/oxidized metal oxide interface to the surface in the traditional as well as hercynite STWS cycles and the effects of doping on the ion transport activation barrier. We have found that, in addition to lowering the reduction temperature, dopant ions has the added benefit of accelerating ion movement through the reduced phase. We compare the kinetics of ion transport to determine the differences in the traditional and hercynite cycles. The effects of magnetization were also studied to provide insight into similar processes taking place below the Currie temperature.

See more of this Session: Fundamental Research In Transport Processes I
See more of this Group/Topical: Engineering Sciences and Fundamentals

(219b) Oxidation Behavior of Copper Nanocrystals

Katherine P. Rice, Mark P. Stoykovich and Aaron E. Saunders, Chemical and Biological Engineering, University of Colorado, Boulder, CO

Nanoparticles of noble metals such as gold, silver, and copper interact with visible light in unique ways, giving rise to collective electron oscillations called the localized surface plasmon resonance (LSPR). The LSPR activity of copper has not been widely studied because of copper's propensity to oxidize. We have characterized the optical response of copper nanocrystals in oxygenated solvents to monitor in-situ and in real time the formation of a monolayer of copper oxide. Further, we obtained data on oxygen diffusivities throughout the nanoparticles based on thermogravimetric analysis. The electronic interactions between the solvent electrons and the copper nanoparticles can significantly quench the LSPR until a monolayer of copper oxide forms, and differing diffusion rates for copper and oxygen over time leads to hollow copper oxide nanoparticles.

See more of this Session: Fundamental Research In Transport Processes I
See more of this Group/Topical: Engineering Sciences and Fundamentals

(219c) Surface Morphological Stability of Crystalline Solids Under Simultaneous Electromigration, Thermomigration, and Stress-Driven Surface Diffusional Mass Fluxes

Dwaipayan Dasgupta, Georgios I. Sfyris, M. Rauf Gungor and Dimitrios Maroudas, Chemical Engineering, University of Massachusetts Amherst, Amhest, MA

Surface morphological instability underlies various reliability problems of technologically important materials, which have a broad range of applications in microelectronics and nanotechnology. For example, the competition between elastic strain energy and surface energy is responsible for surface morphological instabilities in stressed elastic solids, such as the well-known Asaro-Tiller or Grinfeld (ATG) instability under conditions that promote surface diffusion; under such conditions, it has been demostrated that surface cracking can occur through the rapid evolution of a planar surface of a stressed elastic solid to a cusped surface, with smooth tops and deep crack-like grooves. In recent studies, we have shown that applying on a stressed elastic conductor an external electric field of sufficient strength and proper direction can inhibit the ATG instability. Nevertheless, the surface morphological stability of crystalline solids under the simultaneous action of applied mechanical stresses, electric fields, as well as imposed temperature gradients that are commonly present in materials processing and in service has not been examined systematically.

In this presentation, we report a detailed analysis of the morphological stability of planar surfaces of electrically and thermally conducting stressed elastic crystalline solids under the simultaneous action of an electric field and an imposed temperature gradient using linear stability theory and self-consistent dynamical numerical simulation. The applied electric field and temperature gradient are directed parallel to the crystalline solid surface so that they drive surface mass fluxes due to surface electromigration and surface thermomigration. The solid is under uniaxial tension, also directed parallel to the surface, which is responsible for stress-driven surface diffusion. Our analysis is based on a fully nonlinear surface transport model that accounts for curvature-driven surface diffusion, surface electromigration, surface thermomigration, and stress-driven surface diffusion along with surface diffusional anisotropy. Our self-consistent dynamical simulations combine Galerkin boundary-integral computations of elastic displacement fields, temperature fields, and electrostatic potentials with a front tracking method for monitoring surface morphological evolution.

We report results on the effects of material properties (surface diffusional anisotropy parameters) on the morphological stability of <111>-, <100>-, and <110>-oriented planar surfaces of face-centered cubic (fcc) metals under the simultaneous action of the multiple external fields. We determine the surface morphological stability domain boundaries and the critical values of the applied electric field strength and temperature gradient required to stabilize the planar surface morphology against surface cracking through an ATG instability. Special emphasis is placed on exploring the synergistic or competing effects on the surface morphological response of the simultaneously applied thermal and electric fields, aiming at the optimization of the electric-field strength and temperature gradient requirements for planar surface stabilization. The results of our analysis are used to derive systematic surface engineering rules, expressed by the dependence of the critical electric-field strength and simultaneously imposed temperature gradient for stabilization of the surface morphology of the stressed solid as a function of the surface diffusional anisotropy parameters.

See more of this Session: Fundamental Research In Transport Processes I
See more of this Group/Topical: Engineering Sciences and Fundamentals

(219d) Dissolution In Fractured and Porous Rocks - the Cave Formation Paradox and Other Instabilities

Anthony J.C. Ladd, Department of Chemical Engineering, University of Florida, Gainesville, FL and Piotr Szymczak, Department of Physics, University of Warsaw, Warsaw, Poland

It has long been realized that limestone caves are solutional in origin; atmospheric CO_2 forms a weakly acidic solution which dissolves the surrounding limestone as it percolates through the fracture network. Under the simplest assumptions - uniform flow and linear kinetics - the concentration of reactant deacays exponentially with distance into the fracture, making it apparently impossible for long conduits to develop. Nevertheless, limestone caverns extend for kilometers; Mammoth cave in Kentucky has nearly 400 miles of passages. So how does the dissolution get so deep? The answer until recently has been described in terms of changes in chemical kinetics; in natural calcite the reaction rate decreases by orders of magnitude near saturation. Paradoxically this promotes dissolution since the undersaturated solution can penetrate deeper into the fractured rock.

Although this is an appealing and widely accepted resolution of the cave formation paradox, it turns out to be incomplete. About 12 years ago, Hanna and Rajaram[1] showed by computer simulation that a fracture does not necessarily open uniformly across its width, but can develop localized regions of dissolution. This insight was subsequently confirmed by laboratory experiments[2]. Recently, Piotr Szymczak and I realized that there is a universal instability in the equations for fracture dissolution, so that a dissolution front is *always* unstable[3], even under the most idealized circumstances: i.e. a fracture modeled as two parallel plates, laminar flow, and linear reaction kinetics at the fracture surfaces. This generic instability provides a more effective means to promote dissolution than changes in chemical kinetics and has a profound effect on how long it takes for breakthrough (when the fracture opens along its whole length) to occur. Dissolutional breakthrough is an important component in assessing the safety of CO_2 sequestration sites.

In this presentation I will outline the equations governing the dissolution of fractures and also of a uniform porous media. They are similar to the standard equations for heat and mass transfer but because material dissolves there is an additional coupling to changes in permeability. A planar dissolution front in an entirely uniform fracture or porous matrix is unstable to infinitesimal perturbations and inevitably breaks up into highly localized regions of dissolution. Our results suggest that there is an inherent wavelength to the erosion pattern in dissolving fractures, which depends on the reaction rate and flow rate, but is independent of the initial roughness. I will compare the instability in fractured rocks with the better-known instability in dissolving porous media.

1. R. B. Hanna and H. Rajaram, Water Resources Res., 34:2843, 1998.
2. R. L. Detwiler, R. J. Glass and W. L. Bourcier., Geophys. Res. Lett., 30:1648 (2003).
3. P. Szymczak and A. J. C. Ladd, EPSL, 201:424-432, 2011.

See more of this Session: Fundamental Research In Transport Processes I
See more of this Group/Topical: Engineering Sciences and Fundamentals

(219e) Self-Consistent Equations Governing the Dynamics of Nonequilibrium Colloidal Systems

Shuangliang Zhao, UC Riverside, Riverside, CA and **Jianzhong Wu**, Chemical and Environmental Engineering, UC Riverside, Riverside, CA

A self-consistent theoretical procedure is proposed to derive the governing equations for the dynamic properties of nonequilibrium colloidal systems within the framework of the probability theory. Unlike alternative methods in the literature, the self-consistent procedure completely decouples dynamic variables from thermodynamic functions introduced for equilibrium systems. The intrinsic characteristics of a nonequilibrium system is described by the one-body temporal- and spatial-dependent dynamic variables, including the particle density profile, the local momentum, the kinetic energy or dynamic temperature, and by various forms of the two-body position and momentum correlation functions. Within appropriate constraints related to the initial/boundary conditions of a nonequilibrium system, the governing equations for the time evolution of these dynamic functions are obtained by maximizing the information entropy, i.e., the time-evolution equations for the dynamic variables correspond to a probability distribution in the reduced phase space that best represents the known information. It is shown that the dynamic equations are in parallel to and fully consistent with the statistical description of equilibrium systems. With certain assumptions, the self-consistent procedure can be reduced to various conventional theories of nonequilibrium processes.

See more of this Session: Fundamental Research In Transport Processes I
See more of this Group/Topical: Engineering Sciences and Fundamentals

(219f) Heatline Analysis of Natural Convection In Rhombic Enclosures with Isothermally Hot Side or Bottom Wall

Tanmay Basak, Chemical Engineering, Indian Institute of Technology Madras, Chennai, India and R. Anandalakshmi, Chemical Engineering, IIT Madras, Chennai, India

Heatline method is used to analyze natural convection via differentially heated enclosure convection [Case 1: hot and cold side walls with horizontal adiabatic walls] and Rayleigh-Benard convection [Case 2: hot and cold horizontal walls with vertical adiabatic walls] for rhombic enclosures with various inclination angles, φ. An accurate prediction of the flow structure and heat distribution in such configurations are of great important due to its significant engineering applications such as ventilation of rooms, cooling of electronics devices or air flow in buildings. Simulations are performed for the range of Rayleigh number, $Ra = 10^{3}-10^{5}$ for various inclination angles ($\varphi=30^\circ$, 45°, 60°, 75° and 90°) using Galerkin finite element method. Interesting features of heat flow patterns are visualized by heatlines for various φs in both cases. At $Ra = 10^3$, heatlines and isotherms are less distorted and flow circulation is very weak at $\varphi=30^\circ$ in both cases. Increase in φ ($\varphi=90^\circ$) shows more distorted heatlines with closed loop heatline cells due to increase in flow strength compared to $\varphi=30^\circ$ at $Ra = 10^3$ in case 1 whereas heatlines and isotherms are found to be parallel and orthogonal to adiabatic walls, respectively indicating pure conduction dominant heat transfer with stagnant fluid condition for $\varphi=90^\circ$ in case 2 at $Ra = 10^3$. At $Ra = 10^5$, strength of fluid and heat flow increases for all φs due to enhanced convection effect and $\varphi=90^\circ$ shows maximum magnitude of streamfunction (ψ_{max}) and heatfunction (Π_{max}) values in both cases. It is found that, both ψ_{max} and Π_{max} values are comparatively higher in case 2. Both cases are compared based on local (Nu) and average Nusselt numbers ($\overline{Nu}$) and those are adequately explained based on heatlines. It is found that $\overline{Nu}$ is independent of φ at $Ra = 10^3$ in both cases. Also, $\overline{Nu}$ increases with Ra and shows its maximum at $Ra = 10^5$ for all φs in both cases. It is also shown that, $\varphi=30^\circ$ shows low heat transfer rate in case 1 compared to case 2 whereas $\varphi=90^\circ$ shows high heat transfer rate in case 1 compared to case 2. Heat transfer rate is almost similar for $\varphi=45^\circ$ in both cases. Overall, average heat transfer rate is maximum for case 1 at $\varphi \ge 45^\circ$, eventhough ψ_{max} and Π_{max} values are high in case 2.

See more of this Session: Fundamental Research In Transport Processes I
See more of this Group/Topical: Engineering Sciences and Fundamentals

(219g) Soret and Dufour Effects On Double Diffusive Convection In a Square Cavity Totally Filled with Porous Material Using the Multigrid Method

Maximilian S. Mesquita, Departamento de Engenharias e Computao, Universidade Federal do Esprito Santo UFES/CEUNES, So Mateus, Brazil, Marcelo J.S. De Lemos, Departamento de Energia - IEME, Instiruto Tecnolgico de Aeronutica - ITA, So Jos dso Campos, Brazil, Luzia Aparecida Tofaneli, Departamento Regional da Bahia, Centro Integrado de Manufatura e Tecnologia., SENAI -, Salvador, BA, Brazil and Fernando F. Fachini, Laboratrio de Combusto e Propulso - LCP, Instituto Nacional de Pesquisas Espaciais - INPE, Cachoeira Paulista, SP, 12630-000, Brazil

This work studies the Soret and Dufour effects on the flow due to natural double diffusion convection in a square cavity totally filled with a porous medium saturated with Newtonian fluids with constant wall temperature and concentration. The effects of the Dufour parameter, Soret parameter, Lewis number, and buoyancy ratio on the heat and mass transfer characteristics have been studied. The local Nusselt number tends to decrease as the Dufour parameter is increased. The effect of the Dufour parameter on the local Nusselt number becomes more significant as the Lewis number is increased. Moreover, an increase in the Soret number leads to a decrease in the local Sherwood number and an increase in the local Nusselt number.

See more of this Session: Fundamental Research In Transport Processes I
See more of this Group/Topical: Engineering Sciences and Fundamentals

(231a) A Novel SHF Process for Conversion of AFEX-CS to Ethanol Featuring High Ethanol Productivity, Enzyme Recycling, and Yeast Cells Reuse

Ming Jie Jin, Christa Gunawan, Nirmal Uppugundla, Shishir Chundawat, Bruce Dale and **Venkatesh Balan**, Chemical Engineering and Material Science, Michigan State University, East Lansing, MI

There are two major problems in current cellulosic ethanol production process: low ethanol productivity and high enzyme loading. Based on the current Separate Hydrolysis and Fermentation (SHF) process, it takes at least 96h to perform enzymatic hydrolysis to convert AFEX-CS to fermentable sugars at 6% glucan loading. It will then take another 96-168h to ferment those sugars to ethanol. By simply integrating enzymatic hydrolysis and fermentation, which is SSCF (simultaneously saccharification and co-fermentation), it still requires at least 168h to reach comparable results as SHF. Here we will present a novel SHF process. In this process, the easily hydrolyzed part of biomass is hydrolyzed first to fermentable sugars in 24h. The hydrolysate was then separated from residue solids and fermented to ethanol in 24h using high yeast cell inoculation. The residual solids (difficultly hydrolyzed part of biomass (DHPB)), which bound enzymes, are then transferred to the next cycle of hydrolysis. This way, enzymes are largely reused and enzyme loadings in the next cycles can thus be reduced. The yeast cells after 24h fermentation are also recovered and reused for the next cycle of hydrolysate fermentation. The time for each cycle hydrolysis/fermentation is 24h/24h. After five cycles, the difficult hydrolyzed part is collected and enzymatic hydrolysis is performed without adding additional enzymes. Since enzymes are accumulated to a relatively high concentration and part of the decomposition products are removed with liquid during the five cycles, the hydrolysis of DHPB would be easier to achieve high sugar conversion. Profiles of enzyme recycled in each step were characterized.

See more of this Session: Integrated Processes for Biochemical Conversion of Renewable Feedstocks to Fuels and Chemicals I
See more of this Group/Topical: Sustainable Engineering Forum

(231b) Concentration of a Solids-Free Lignocellulose Hydrolysate for Very High Gravity Fermentation: Inhibition Effects

Steven J. Schneiderman, Todd J. Menkhaus and Patrick C. Gilcrease, Chemical and Biological Engineering, South Dakota School of Mines and Technology, Rapid City, SD

A major problem in the efficient conversion of lignocellulose hydrolysates to ethanol is low ethanol concentrations in the fermentation broth due to low initial sugar concentrations. This is undesirable because low ethanol concentrations lead to increased separation costs downstream. Distillation is energy intensive, and constitutes more than 50% of the total energy consumption of a typical ethanol plant. An obvious solution is to start with a high initial sugars concentration, which is referred to as a very high gravity (VHG) fermentation. By increasing the glucose concentration in the hydrolysate from 40 to 100 g/l, the distillation energy consumption would be reduced by 60%. In the first step of our integrated process, polyelectrolyte flocculants facilitate the efficient removal of biomass solids from a pine wood hydrolysate. Once the hydrolysate is solids-free, an evaporation or reverse osmosis step can be readily used to remove water and concentrate the fermentable sugars. VHG fermentations present some challenges of their own, including higher energy costs associated with concentrating the sugars, simultaneous concentration of some or all of the fermentation inhibitors, and osmotic stresses on the fermentation organism at higher substrate concentrations. Because some of the inhibition effects are synergistic, this could become more problematic at VHG conditions.

The goal of this study was to investigate the potential problems associated with fermenting a concentrated lignocellulose hydrolysate, and to determine operating conditions that maximize productivity and minimize costs for the combined sugars concentration-fermentation-distillation process. A batch culture design of experiments was used to quantify the inhibition effects of five common inhibitors on the productivity of our *S. cerevisiae* D_5A strain. The inhibitors studied were glucose, ethanol, furfural, HMF, and acetic acid at concentration ranges of 25 – 300 g/l, 0 – 150 g/l, 0 – 5 g/l, 0 – 5 g/l and 0 – 10 g/l, respectively. Our experimental results were used to regress kinetic model parameters, and define an acceptable operating window for VHG fermentation. The hydrolysate concentration step (evaporation/reverse osmosis) was then tailored to minimize total separation costs (sugars concentration and ethanol distillation) while maintaining a high level of ethanol productivity. Finally, the fermentation of a concentrated pine hydrolysate at optimum conditions was demonstrated at the bench scale.

See more of this Session: Integrated Processes for Biochemical Conversion of Renewable Feedstocks to Fuels and Chemicals I
See more of this Group/Topical: Sustainable Engineering Forum

(231c) Separate Processing of Hemicellulose and Cellulose to Optimize Ethanol Production From Corn Stover

Alexandre Chapeaux[1], Nancy Dowe[1] and Daniel J. Schell[2], (1)National Bioenergy Center, National Renewable Energy Laboratory, Golden, CO, (2)National Renewable Energy Laboratory, Golden, CO

Separate processing ofhemicellulose and cellulose to optimize ethanol production from corn stover

Alexandre Chapeaux, NancyDowe, Daniel J. Schell

Using a series of thermochemical (pretreatment)and biochemical steps, lignocellulosic biomass is converted to sugars that subsequentlycan be made into fuels and chemicals. One such process uses heat and dilute sulfuric acid to hydrolyzes the hemicellulosic fractionof the biomass to monomeric and oligomeric sugars.The removal of hemicellulose and to a lesser extent lignin from the biomass significantlyimproves the performance of the next step in the process, enzymatic hydrolysisof cellulose to glucose. However, enzymatic cellulose hydrolysis can beperformed in the presence or absence of the hemicellulosic sugars producedduring pretreatment. It is well known that removing the hemicellulose sugarsand other compounds produced during pretreatment improves enzymatic celluloseconversion yields. In previous work, we assessed the potential of a processthat removes the hemicellulosic sugars prior to enzymatic cellulose hydrolysis byexploring the interaction between biomass solids loading, yields, and processcosts. However, this process is more expensive because of the additional costfor solid-liquid separation equipment required to remove th-richsugar liquor from the cellulosic solids.

In this study, the goal was to deepen ourunderstanding of the advantages and disadvantages of the separatehemicellulose/cellulose process. The feedstock was dilute-acid-pretreated cornstover produced in a 200 kg dry biomass/d continuous horizontal reactor. Thepretreated slurry was separated into a liquor stream (hemicellulosic sugars)and a washed cellulosic solid stream. Enzymatic cellulose hydrolysis was investigatedat four enzyme loadings (10, 20, 30 and 40 mg protein/g cellulose) and atinsoluble solids concentrations from 9% to 21% (w/w). The resulting glucose wasfermented to ethanol using several yeasts and a glucose-xylose co-fermentingbacterium, *Zymomonas mobilis* 8b. The sugars in the liquorstream were fermented to ethanol with three strains of co-fermenting *Z. mobilis* after dilution of the streamwith water to lower sugar concentrations. Finally, the overall performance ofthe process was assessed after partitioning water between the two phases in away that optimizes overall ethanol yields. We then compared these results to aprocess configuration in which the streams are not separated. In the latterprocess, enzymatic cellulose hydrolysis is performed in the presence of thebackground hemicellulosic sugars produced during pretreatment of the biomass.

See more of this Session: Integrated Processes for Biochemical Conversion of Renewable Feedstocks to

(231d) Ethanol and Furfural Production From Corn Stover Using Hybrid Fractionation Process with Zinc Chloride and Simultaneous Saccharification and Fermentation (SSF)

Chang Geun Yoo[1], **Tae Hyun Kim**[1] and Monlin Kuo[2], (1)Agricultural and Biosystems Engineering, Iowa State University, Ames, IA, (2)Department of Natural Resources Ecology and Management, Iowa State University, Ames, IA

Two-stage hybrid fractionation process was investigated to produce cellulosic ethanol and furfural from corn stover. Zinc chloride ($ZnCl_2$) was used to selectively solubilize hemicellulose in the first stage. During the second stage, the remaining treated solids were converted into ethanol using commercial cellulase and *Saccharomyces cerevisiae* or recombinant *Escherichia coli*, KO11.

Hybrid fractionation process recovered >90% of glucan, ~90% of xylan, and ~75% of lignin at optimal reaction conditions: 1st stage: 5% acidified $ZnCl_2$, 7.5 ml/min, 150 °C (10 min) and 170 °C (10 min); 2nd stage: Simultaneous saccharification and fermentation (SSF). Furfural yield from hemicellulose hydrolysates was 58 %. SSF of $ZnCl_2$ treated solid with *Saccharomyces cerevisiae* resulted in 69-98 % of theoretical maximum ethanol yields based on glucan present in the treated solids. After fermentation, solid residues contained mostly lignin. Based on total lignin in untreated corn stover, the lignin recovery yield was ~75%.

See more of this Session: Integrated Processes for Biochemical Conversion of Renewable Feedstocks to Fuels and Chemicals I
See more of this Group/Topical: Sustainable Engineering Forum

(231e) Fischer Tropsch Synthesis Via Biomass Derived Synthesis Gas

Syed Ali Gardezi, Chemical Engineering, University of South Florida, Tampa, FL, Babu Joseph, Chemical and Biomedical Engineering, Clean Energy Research Center, University of South Florida, Tampa, FL and John T. Wolan, University of South Florida, Tampa, FL

Due to the recent surge in energy costs and uncertainties in fuel prices, there is significant interest in seeking alternative sources of energy particularly from renewable such as biomass. Indirect liquefaction of biomass (via the thermo-chemical conversion to syngas followed by liquefaction via Fischer-Tropsch Synthesis or FTS) offers a commercially viable route for meeting the challenge of producing renewable fungible liquid fuels. Apart from biomass, FTS is also the critical step for converting natural gas and coal to liquid fuels.

A FTS technology has been developed for producing middle distillates (diesel and aviation fuel) by using pine chips as the main feedstock. In this "two step" process, pine chips were subjected to an entrained flow gasification process developed by Pearson et al. [1]. The resulting tar free gas is comprised mainly of carbon monoxide, hydrogen, carbon dioxide, methane and some organic contaminants such as benzene, toluene, and naphthalene along with substantial amount of water. For further liquefaction this gas was cleansed using multiple adsorbents. An inline "Shaw Moisture Meter" was used to continuously monitor the moisture content of the feed to the liquefaction process.

For the production of liquid fuels, eggshell cobalt catalysts were used. Cobalt catalysts have several advantages e.g. they produce linear paraffinic hydrocarbons and minimize undesirable water gas shift reaction. This eggshell design is a well known strategy for producing a narrow hydrocarbon distribution. In this regard previous research work conducted by Iglesia et al. was also consulted [2]. A bench scale fixed bed reactor was used for the conversion of syngas. Modeling of the reactor startup and of adjustment of the reactor packing ensured successful operation resulting in a high conversion and narrow distribution of hydrocarbons in the range of diesel and aviation fuel.

References :

[1] Pearson et al., U.S. Patent 4,857,076, issued Aug 15, 1989.

[2] E. Iglesia, S. L. Soled, J. E. Baumgartner, S. C. Reyes, *J. Catal.* **153**, 108 (1995).

See more of this Session: Integrated Processes for Biochemical Conversion of Renewable Feedstocks to Fuels and Chemicals I
See more of this Group/Topical: Sustainable Engineering Forum

(231f) Integrations of Biosurfactant Production Into Advanced Biorefineries

William Colonna[1], **Mustafa E. Marti**[1], Michelle Pynn[2], Gabriel Reznik[2], Kevin Jarrell[2], Buddhi Lamsal[3] and Charles E. Glatz[1], (1)Chemical and Biological Engineering, Iowa State University, Ames, IA, (2)Modular Genetics, Inc., Woburn, MA, (3)Food Science and Human Nutrition, Iowa State University, Ames, IA

Several classes of biosurfactants are produced by microorganisms. One such surfactant is FA-Glu, which is related to the natural biosurfactant, surfactin. FA-Glu consists of a long-chain β-hydroxy fatty acid coupled to glutamic acid. It is produced by a strain of *Bacillus subtilis* genetically-engineered by Modular, Inc. It has several potential applications including use as a dispersant for oil spills, where biodegradability is desirable.

Economic production and eventual application will be favored if substrate costs can be reduced and low-cost recovery methods developed. Several such opportunities arise in advanced biorefineries. Product value would be further enhanced if such production could achieve favored "green" rating by avoiding solvents other than water and simple alcohols and effect emulsification and low toxicity could be demonstrated. This report focus on production considerations.

Production of FA-Glu by fermentation on glucose is compared to results using lower cost substrates available in advanced biorefineries, including fiber components of soy and glycerol. FA-Glu yields from soybean hulls (SBH) ranged from 29-51% of that from glucose on a substrate weight basis.. When SBH polysaccharides were hydrolyzed (~65%) by pre-treatment with carbohydrases, FA-Glu levels were only ~21-37% of those from glucose. The cells consumed the enzymatically-released sugars; however, very little of this was converted into FA-Glu. Addition of glucose to the enzymatic SBH hydrolysate improved FA-Glu yields, but only slightly. The low FA-Glu yields from SBH could be due to inhibitors released from the hulls by the enzymes or inhibitory substances in the enzyme preparations.

Glycerol, a by-product of biodiesel production, proved to be a suitable carbon source for FA-Glu production. Cells grown on 2% w/v glycerol produced ~25% more FA-Glu than cells grown on 2% glucose. Glycerol is abundant and low in cost. Its use as a feedstock for production of microbial biosurfactants would provide a means for efficient utilization of soy by-products and improving the ag economy.

The FA-Glu was purified by collecting the foam from the aerated broth, clarification by centrifugation or filtration, and adsorption and elution from an anion exchange column to a purity of 97.9%. Final product color can be improved by liquid-liquid extraction, but this may compromixe "green" status. A precipitation based alternative process is being examined as a means of eliminating solvent use.

(258a) Mechanistic Modeling of Simultaneous Enzymatic Saccharification of Cellulose and Xylan Using Continuous Distribution Kinetics

Ambarish Nag[1], Andrew J. Griggs[2], Jonathan J. Stickel[2], Michael A. Sprague[1] and James J. Lischeske[2], (1)Computational Science Center, National Renewable Energy Laboratory, Golden, CO, (2) National Bioenergy Center, National Renewable Energy Laboratory, Golden, CO

Cost effective production of biofuels and biochemicals often involves enzymatic saccharification, which has been the subject of intense research and development. In this work, we develop a mechanistically based kinetic model for the enzymatic hydrolysis of biomass composed of cellulose, xylan, and lignin. The prototypical biomass particle in our model has a cellulosic core that is partially covered by xylan and lignin that impede the access of glucanase enzymes to the cellulosic core. Our model hence incorporates both glucanase and xylanase enzymatic saccharification of surface xylan and the cellulosic core. For degradation of insoluble cellulose chains we consider the actions of endoglucanase (endo-β-1,4-glucanase,) and exoglucanase (exo-β-1,4-cellobiohydrolase). Likewise for insoluble xylan we consider the actions of endoxylanase (endo-β-1,4-xylanase) and exoxylanase (xylan β-1,4-xylosidase), the latter enzyme hydrolyzing soluble xylan oligomers as well. The enzyme system also includes beta-glucosidase for the hydrolysis of soluble glucan oligosaccharides. The kinetics for the depolymerization of cellulose and xylan chains are described using population-balance equations, which are capable of describing the evolution of the distribution of chain sizes during hydrolysis, without solving equations for all chemical species present in the reacting mixture. This approach affords computationally efficient simulations, enabling the use of the model for process design and optimization. In combination with experiments, the kinetic model is used to study the transformations of the cellulose and xylan chain-size distributions during hydrolysis by an enzyme cocktail of all the enzyme species.

See more of this Session: Reaction Kinetics and Transport Fundamentals for Biomass Conversion
See more of this Group/Topical: Sustainable Engineering Forum

(258b) Kinetic Modeling of Cellulose Hydrolysis with First Order Inactivation of Adsorbed Cellulases

Zhuoliang Ye and R. Eric Berson, Chemical Engineering, University of Louisville, Louisville, KY

The rate of glucose release during enzymatic hydrolysis of cellulose slows down significantly as the reaction proceeds, and the mechanism is not yet clear. It is, therefore, desirable to develop a kinetic model that accurately describes the enzymatic hydrolysis reaction. In this work, a model based on Langmuir binding kinetics that also incorporates first order inactivation of adsorbed cellulase was developed that accurately predicts rates and yields during the hydrolysis reaction for two different cellulose substrates. Substrate concentrations of 20g/L and less were tested in order to remove any potential effects due to product inhibition or mass transfer limitations. A key result of the modeling suggests that Vmax decreases over time due to deactivation of the adsorbed enzyme. Rate constants associated with product formation and enzyme inactivation from our model yield are comparable to reported values elsewhere. The model predicted results within 10% of experimental results for both Solka Floc and Sigmacell substrates. Key model parameters were similar for the two substrates: the apparent hydrolysis rate (k_2) and inactivation rate (k_f) were 33 min^{-1} and 0.38 h^{-1}, respectively, comparable to values reported elsewhere. The rate constants were within 5% for the two different substrates, further strengthening the value of the model and implying that a common but yet undefined rate-limiting step associated with loss of enzyme activity likely exists in the pathway of cellulose hydrolysis. It is proposed that this rate-limiting step may be either dissociation or isomerization of the enzyme.

See more of this Session: Reaction Kinetics and Transport Fundamentals for Biomass Conversion
See more of this Group/Topical: Sustainable Engineering Forum

(258c) Observing and Modeling Cellulosic Substrate Depolymerization by Commercial Enzyme Cocktails Using Confocal Fluorescence Microscopy

Jeremy S. Luterbacher[1], Jose M. Moran-Mirabal[2] and Larry P. Walker[2], (1)Department of Chemical and Biomolecular Engineering, Cornell University, Ithaca, NY, (2)Department of Biological and Environmental Engineering, Cornell University, Ithaca, NY

Understanding and modeling the depolymerization mechanisms of cellulosic and lignocellulosic substrates by cellulase cocktails is a critical step towards further optimizing the production of monosaccharides from biomass. The Spezyme CP cellulase cocktail combined with the Novo 188 b-glucosidase blend was used to depolymerize bacterial microcrystalline cellulose (BMCC), filter paper cellulose and untreated and H_2O-CO_2 pretreated lignocellulosic substrates (hardwood and switchgrass), which were immobilized on a glass surface. The enzyme mixture was supplemented with a small fraction of fluorescently labeled *T. Reseii* CBH I, which served as a reporter to track cellulase diffusion into and binding onto the internal physical structure of the cellulosic substrate. All reported experiments were conducted at 50°C, the optimal temperature for maximum hydrolytic activity of the enzyme cocktail. BMCC was observed throughout degradation by labeling it with a fluorescent dye. Native autofluorescence was used in a similar way to track filter paper and lignocellulosic substrates. This method allowed us to measure the binding of cellulases *in situ* and follow the temporal morphological changes of cellulose and lignocellulosic substrates during their depolymerization by a commercial cellulase mixture.

In the case of BMCC, a kinetic model was developed and fitted to fluorescence intensity data obtained through image processing. This model was successfully used to predict the bulk sugar concentrations that were liberated over time from BMCC in a separate experiment and measured by liquid chromatography. A more complete mass transfer and kinetic model is currently under development for the more complex substrates (filter paper and lignocellulosic biomass).

See more of this Session: Reaction Kinetics and Transport Fundamentals for Biomass Conversion
See more of this Group/Topical: Sustainable Engineering Forum

(258d) Effects of Mixing Quality On the Kinetics of Biomass Liquefaction

David M. Lavenson[1], Emilio J. Tozzi[1], Tina Jeoh[2], Michael J. McCarthy[3] and Robert L. Powell[1], (1) Chemical Engineering and Materials Science, University of Calfiornia Davis, Davis, CA, (2)Biological and Agricultural Engineering, University of Califonrnia, Davis, Davis, CA, (3)Biological and Agricultural Engineering, University of Calfiornia Davis, Davis, CA

Abstract: Biochemical conversion is a promising route to produce fuels and chemicals from cellulosic biomass. A key step in the process is biomass liquefaction, where enzymes are used to break down pretreated cellulosic particles. During this process, drastic changes in the mechanical properties of the cellulosic slurry cause a transition from a solid-like material to a pumpable liquid. A fast liquefaction is desirable to achieve reductions in equipment size and cost. To assess the effect of mixing quality on the speed of liquefaction, cellulases were added to static fiber beds in vials. Different initial mixing conditions were used. One of the conditions was a well-mixed slurry where the enzyme is uniformly distributed in the slurry. The opposite extreme condition was addition of a thin layer of enzymes on one end of the vial, representing a completely unmixed enzyme. Samples were incubated for 96h at 50C. The state of liquefaction of the fiber beds was assessed using a mechanical penetrometer test. Large differences in the force-displacement curves were observed between untreated control slurries and enzymatically treated slurries. More importantly among the treated ones, the liquefaction times were much shorter for well-mixed slurries than for poorly mixed ones. Liquefaction data was also compared with reducing sugars measured throughout the incubation period to correlate with product yields.

See more of this Session: Reaction Kinetics and Transport Fundamentals for Biomass Conversion
See more of this Group/Topical: Sustainable Engineering Forum

(258e) Coupled Fluid-Dynamics and Population-Balance Kinetic Models for Enzymatic Hydrolysis of Biomass

Michael A. Sprague[1], Jonathan Stickel[2] and Andrew Griggs[2], (1)Computational Science Center, National Renewable Energy Laboratory, Golden, CO, (2)National Bioenergy Center, National Renewable Energy Laboratory, Golden, CO

A population-balance kinetic model developed previously for the enzymatic hydrolysis of lignocellulosic biomass is extended to include fluid mixing and transport. The kinetic model is focused on the depolymerization of insoluble cellulose chains by an endoglucanase and an exoglucanase. Population-balance models combined with a continuous-distribution representation provide computational efficiency, which is critical when considering the addition of fluid-mixing effects. The resulting coupled equations are solved with a high-order open-source computational fluid dynamics (CFD) package. The CFD model is validated against experimental and analytical data for a simple vane mixer. Preliminary results examining the effects of fluid mixing on dilute enzymatic hydrolysis are presented and compared to existing experimental data. This work provides a foundation for future efforts to model pilot- and commercial-scale bioconversion-reactors.

See more of this Session: Reaction Kinetics and Transport Fundamentals for Biomass Conversion
See more of this Group/Topical: Sustainable Engineering Forum

(258f) Flow and Transport Properties In High-Solids Biomass Slurries

James J. Lischeske and Jonathan J. Stickel, National Bioenergy Center, National Renewable Energy Laboratory, Golden, CO

Economic biochemical conversion of cellulose to sugars necessitates minimal addition of water to biomass slurries during conversion, resulting in high-solids systems. The resulting slurries are effectively wet-granular materials. These materials presents difficulties for reactor engineering, as their flow and transport properties, along with the mechanical properties that govern them, are poorly understood. This work examines the nature of high-solids flow, along with the heat and mass transfer properties of biomass and related model materials. Methods were developed to measure the volume fraction of solid, liquid and gaseous phases of high-solids slurries. The effect of phase composition on flow was systematically examined in a rotating vessel. Qualitative observations, as well as accessible quantitative information about size distribution of aggregates, were made. Further, a model for measuring bed diffusion in biomass was developed, and experiments were performed to measure the bed diffusion coefficient in model materials and biomass. Finally, heat transfer in a packed bed of biomass was examined, and conductivity coefficients were calculated and related to phase composition.

See more of this Session: Reaction Kinetics and Transport Fundamentals for Biomass Conversion
See more of this Group/Topical: Sustainable Engineering Forum

(280a) Investigation of CuO-Based Oxygen Carriers As CLOU Materials for Solid Fuels

Mehdi Arjmand[1], Henrik Leion[1], Abdul-Majeed Azad[2], Tobias Mattisson[2] and Anders Lyngfelt[2], (1)Department of Chemical & Biological Engineering, Chalmers University of Technology, Gothenburg, Sweden, (2)Department of Energy and Environment, Chalmers University of Technology, Gothenburg, Sweden

Chemical-loopingcombustion (CLC) is a promising combustion technique with efficient CO_2-captureat low cost due to inherent gas separation. In a CLC process, a metal oxide iscirculated between two reactors, wherein they undergo changes from an oxidizedform in the air reactor to a reduced one in the fuel reactor. In the case ofchemical-looping with oxygen uncoupling (CLOU), the oxygen carrier releases oxygenin the gas phase in the fuel reactor as shown in the figure. This enables a highrate of conversion of char from solid fuels, as it eliminates the need for thegasification step needed in normal CLC with solid fuels.

Only certain puremetal oxides of Co, Cu and Mn have an equilibriumpartial pressure of oxygen suitable for CLOU. Due to its favorablethermodynamics (facile redox properties) and agile kinetics (high reactivity),CuO is a promising CLOU material, in addition to being an excellent oxygencarrier in the combustion of gaseous fuels.

In this work, theperformance of CuO-based oxygen carriers supported on ZrO_2stabilized by calcium oxide, magnesia and ceria will be presented to assesstheir suitability as oxygen carriers for solid fuel combustion. The samples wereprepared by freeze-granulation and the reactivity was examined in a laboratoryfluidized bed reactor under cyclic oxidizing and reducing conditions,simulating the CLOU process. Owing to the fast release of oxygen by CLOUparticles, the reaction rates were investigated using coke as the fuel ofchoice. The influence of reaction temperature and oxygen concentration on therate of reaction was also studied.

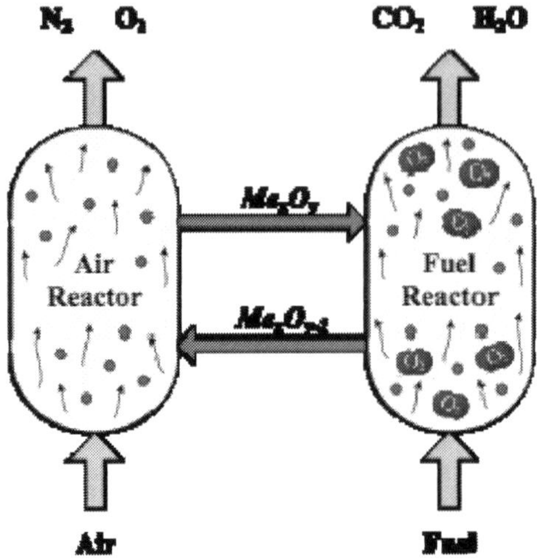

Schematicof the chemical-looping combustion via oxygen uncoupling (CLOU) process

See more of this Session: Chemical Looping Processes I
See more of this Group/Topical: Energy and Transport Processes

(280b) Oxygen Uptake and Release Kinetics of Cu/ Cu_2O/CuO On Al_2O_3 and ZrO_2-SiO_2 As Materials for Chemical Looping Combustion

Björn Brandt[1], Philipp Gruene[2], Frank Rosowski[2], Bob Farrauto[3] and Marco J. Castaldi[1], (1) Department of Earth and Environmental Engineering (HKSM), Columbia University in the City of New York, New York City, NY, (2)BASF SE, Ludwigshafen, Germany, (3)R & D, BASF Catalysts LLC, Iselin, NJ

One key parameter in ChemicalLooping Combustion (CLC) is the oxygen transport capacity of the CLC materialsbetween the two process reactors; in practice, it may be limited by acombination of many factors, including the metal content of the inert binder,the oxygen uptake and release kinetics and the thermodynamic equilibrium underthe process conditions in the two reactors, the respective residence times, therate of contaminant deposition and removal (such as coke, sulfur, or ashdeposits), and deterioration of the oxygen carrier. Such factors can have a significantimpact on the economics and viability of a CLC power plant, because it isexpected that many tons of CLC materials would be required in a practical CLCpower plant (the required inventory of CLC material can be estimated to be inthe range of 0.5 to 6 tons/MW power for typical CLC materials and processoperation conditions).

With our ongoing program, weaim for a systematic study of those interconnected aspects, and for obtaining amore quantitative description of CLC from the perspective of gas-surfacekinetics. As CLC materials, we have chosen copper oxide supported on two inertbinders: 1. ZrO_2-SiO_2(a high-performance material used in the field of catalysis), and 2. γ-$Al_2O_3$for comparison (a more economic, generic binder previously described inliterature); those materials are prepared by incipient wetness impregnationfollowed by a quantification of the prepared formulation by calibrated X-Ray fluorescence. The study is centered on TGA/DTA/DSC experiments, which allow for testing theoxygen uptake and release kinetics as a function of important processparameters; they will be amended with morphological characterizations (BET, Hg-porosimetry, electron microscopy, EDX elementaldispersion).

We will present experimentsobtained in the temperature range from 450 to 1000 C, using CLC materialswith Cu loadings ranging from ~ 5 to 50 wt.-%. Our first resultsindicate that the ZrO_2-SiO_2 material (17 wt.-% Cu)can be cycled many times (more than 30 reduction/oxidation cycles have been studied) with lessthan 2 % change of its oxygen uptake and release capacity. The experimentsshow that the material can be looped quantitatively between the fully oxidizedstate (CuO), and the fully reduced (metallic) stateusing methane as fuel at 750 C; at higher temperatures (starting around800 C), the material begins to spontaneously decompose to $Cu_2O$when the air is removed, likely due to thermodynamic driving forces; thus, thistemperature constitutes a significant upper boundary in order to maximize theoxygen transport capacity of the material in CLC. The observed oxygen uptakerates are $2.5x10^{-8}$ mole O_2 sec^{-1} (for 1mg of this CLC material), and varied by less than 4 % over atemperature rate from 550 to 750 C; this apparent lack of variationof the rate is probably due to extrinsic limitations of the oxygen uptake step– *e.g.* due to bulk or porediffusion processes. On the other hand, oxygen release rates under methaneexposure displays a strong temperature dependence; the reduction rate variesbetween $1.8x10^{-8}$ mole O_2 sec^{-1} at 750 C,and $3.1x10^{-9}$ mole O_2 sec^{-1} at 550 C (for1 mg of the material, respectively). Moreover, reduction rates are lower thanthe oxygen uptake rates at the same temperatures (26 % slower at750 C, 79 % slower at 550 C). These temperature findingsindicate that oxygen diffusion in the lattice is not the limiting factor forthe reduction rates. It will be discussed whether the observation of differentreduction rates is linked to the kinetics of the fuel activation on the CLCmaterials. Finally, to begin to understand the impact of a more realistic fuellooping cycle, tests done with simulated gasified coal including steam andppm-levels of H_2S will be shown.

See more of this Session: Chemical Looping Processes I
See more of this Group/Topical: Energy and Transport Processes

(280c) Investigation of Cobalt Ferrite and Nickel Ferrite for Chemical Looping Hydrogen Production

Victoria J. Aston[1], Anthony H. McDaniel[2], Brian W. Evanko[1], Jonathan R. Scheffe[3], Mark D. Allendorf[2] and Alan W. Weimer[4], (1)Chemical and Biological Engineering, University of Colorado at Boulder, Boulder, CO, (2)Sandia National Laboratories, Livermore, CA, (3)Department of Mechanical and Process Engineering, ETH Zurich, Zurich, CO, Switzerland, (4)Department of Chemical and Biological Engineering, University of Colorado at Boulder, Boulder, CO

The chemical looping hydrogen production process uses synthesis gas or methane to reduce a metal oxide material, forming a waste stream of CO_2 and H_2O from which CO_2 can be easily sequestered. The reduced metal oxide is then oxidized with steam to produce pure H_2 and regenerate the original metal oxide. This system requires a metal oxide material that achieves high H_2 and CO conversions to H_2O and CO_2 during the reduction step while maintaining fast oxidation and reduction kinetics through many cycles. In this study, the use of nano-scale cobalt- and nickel-ferrite ($CoFe_2O_4$ and $NiFe_2O_4$) based materials was investigated for chemical looping hydrogen production. Nano-thin films of ferrite were deposited on high surface area ZrO_2 support material using the atomic layer deposition (ALD) process to minimize sintering and diffusion limitations on reaction kinetics. The kinetics and cyclability performance of these thin film materials were analyzed and compared with the performance of conventionally made impregnation ferrite samples on the same high surface area substrates.

See more of this Session: Chemical Looping Processes I
See more of this Group/Topical: Energy and Transport Processes

(280d) Evaluation of Iron Based Oxygen Carriers for Chemical Looping Dry Reforming

Saurabh Bhavsar, Department of Chemical Engineering, University of Pittsburgh, US DOE-National Energy Technology Laboratory, Pittsburgh, Pittsburgh, PA, Michelle Najera, Department of Chemical Engineering, University of Pittsburgh, Mascaro Center for Sustainable Innovation, University of Pittsburgh, Pittsburgh, PA and Gtz Veser, Department of Chemical Engineering, University of Pittsburgh, US DOE-National Energy Technology Laboratory, Pittsburgh; Mascaro Center for Sustainable Innovation, University of Pittsburgh, Pittsburgh, PA

'Chemical Looping Combustion' (CLC) is an emerging clean combustion technology which offers an efficient and elegant route for fossil fuel combustion with inherent CO_2 capture. In CLC, an oxygen carrier (typically a metal) undergoes cyclic reduction and oxidation with fuel and air, respectively. By changing the oxidant, the process can be applied for purposes beyond combustion: Replacing air with CO_2 as oxidant results in dry reforming of the fuel, producing CO as result of the carrier oxidation. This process, Chemical Looping Dry Reforming (CLDR), hence results in utilization of CO_2, providing a potential alternate pathway to carbon mitigation via sequestration. However, since CO_2 is a weaker oxidant than air, significant challenges lie in the slow oxidation kinetics. While nano-sizing the oxygen carrier can be expected to increase carrier activity and result in higher carrier utilization, it also exacerbates the carrier stability issues at the demanding high-temperature cyclic operation conditions.

In the present study, iron was chosen as oxygen carrier for CLDR based on a prior thermodynamic evaluation of various metals. Different nanocomposite iron carriers, including Fe-BHA (barium hexaaluminate) and Fe@SiO_2 core-shell materials, were synthesized with the aim to combine high reactivity and high carrier utilization with good thermal stability. The carriers were characterized by TEM, SEM, XRD, and BET, and then evaluated in fixed bed reactor and TGA (thermogravimetric analysis) CLDR studies. Our results indicate that sufficiently small Fe nanoparticles show high CO_2 conversion activity down to temperatures well below 400°C, but that the oxide matrix required to render the carriers thermally stable can result in significant mass transfer limitations at high-temperature conditions (~800°C). Overall, our results indicate that CLDR could offer a novel way for efficient CO_2 utilization over a broad temperature range.

See more of this Session: Chemical Looping Processes I
See more of this Group/Topical: Energy and Transport Processes

The Effect of Fly Ash from Different Coal Sources on the Reactivity of an Iron-based Oxygen Carrier for Chemical Looping Combustion

Aurora Rubel*, Yi Zhang, James K. Neathery, and Kunlei Liu
University of Kentucky, Center for Applied Energy
2540 Research Park Drive
Lexington, KY 40511

Introduction

The University of Kentucky, Center for Applied Energy Research is investigating oxygen carriers (OC) for chemical looping combustion (CLC) with the goal of developing a pressurized chemical looping combustion/gasification process for generating electricity from coal. An important issue in the application of CLC to a solid fuel is the effect of ash on the performance of the OC. A previous study using a bituminous coal ash obtained from an electric utility suggested that the ash was not detrimental to the capacity and reactivity of two iron based OCs and iron oxide already present in the ash may have contributed additional OC activity[1]. With the wide variety of coals used to generate electricity world-wide, it is unlikely that all ashes will have the same effect on OCs. Recently four ashes obtained from bituminous, sub-bituminous, and lignite coals were used in a comparative study of their effect on the reactivity of an iron based OC to combust a high carbon beneficiated coal char under dry inert gas (argon) conditions in a TGA[2]. As expected the ash effect on the OC varied with the different ashes. The lignite ash did not affect the performance of the OC up to the highest concentration (75%) tested. The Bulgarian subbituminous which had a Fe_2O_3 concentration of nearly 14% caused a 20% reduction in the performance of the OC which was contrary to anticipated results. The Powder River Basin (PRB) ash produced a slight reduction in carbon removal ability as did the China bituminous ash. However, the results for the China ash which had 8.5% residual carbon indicated that the OC at the right carbon to OC ratio could also combust the residual carbon in the ash as well as the carbon in the fuel char. This study was undertaken to further investigate the impact on an iron based OC of these four different fly ashes in the presence of water vapor (WV) consistent with conditions which would exist during CLC in comparison to dry gas results.

Experimental

Four ashes were collected from power plants in Bulgaria (subbituminous), China (bituminous) and the US (Texas lignite and Powder River Basin subbituminous). The ashes were analyzed for CHNS and major and minor elements. The ashes were used as-received The four ashes were mixed with a 1:9 by weight mixture of char and OC so that the final mixture of char:OC:ash was 5:45:50. A 2 gram sample was heated in a TGA to 950°C in a 200 ml/min (STP) flow of Argon with and without 10% WV added. The sample was heated at 950°C for 5 hours. The OC was generated at the CAER by freeze granulation (FG) by procedures published elsewhere[3]. The FG OC used in this study was 50% Fe_2O_3 on an Al_2O_3 support. Control runs included the OC, char, and each ash run alone. A (1:9) char:OC mixture with no ash added was considered the base case. Aliquots of the before and after heating of the sample were submitted for carbon determinations to determine the percent carbon removed from the starting mixtures. Results were normalized to the base case of no ash added samples. Reaction rates were determined from the differential weight loss curves obtained by TGA.

Results

The ashes varied in chemical composition as determined from elemental and inorganic analyses (Table 1). Residual carbon and iron oxide contents varied from 0.07 to 8.52% and 4.22 to 13.71% respectively. These were considered the most important parameters but other oxides which might contribute oxygen (CaO, MgO, SO_3) were also present in high concentrations in some of the ashes. The residual carbon in all ashes except the China bituminous ash was less than 1%.

Table 1. Fly ash type and composition.

Ash	%C	%Fe_2O_3	CaO	MgO	SO_3
Bul Sub	0.78	13.71	7.48	2.52	2.78
Chi Bit	8.52	4.99	8.08	1.16	1.00
PRB	0.07	5.73	21.13	4.38	1.72
TX Lig	0.81	4.22	14.68	2.02	1.46

Heating the ashes under wet gases removed the 99, 97, 99, 58% of that carbon for the lignite, China , Bulgarian, and PRB ashes respectively (Figure 1). Under a flow of dry argon, comparable carbon removal from the ashes were 89, 15, 78, and 58% indicating the advantage of wet gases providing some gasification of the ash carbon. The effect of gasification on carbon removal was very evident by the increase in carbon removal from the char and char:OC controls from dry to wet gases of 1.2 to 62.3% and 56.7 to 86.8% respectively.

The benefits of the OC were improved carbon removal of 27.5% under wet gases and 55.5% under dry gases in comparison to the char alone control (Figure 2). Under wet gas conditions, the char:OC:ash mixtures all had carbon removals equal to or exceeding the zero ash base case indicating that none of the ashes affected the capacity of the OC to oxidize carbon. This was not true for the dry gas runs where the China, Bulgarian, and PRB ash all produced some decrease in the OC capacity.

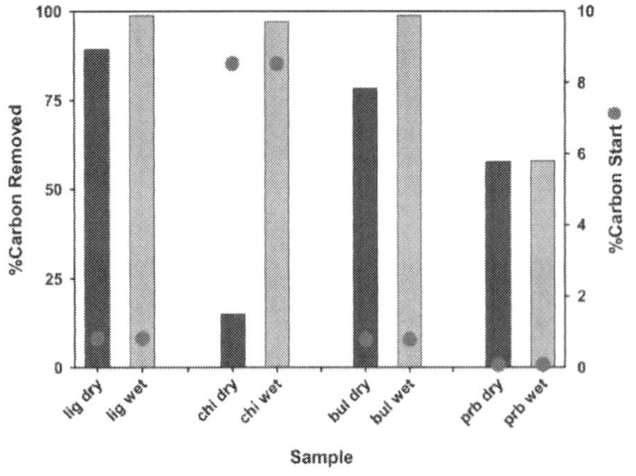

Figure 1. Carbon removed from ashes alone under experimental conditions.

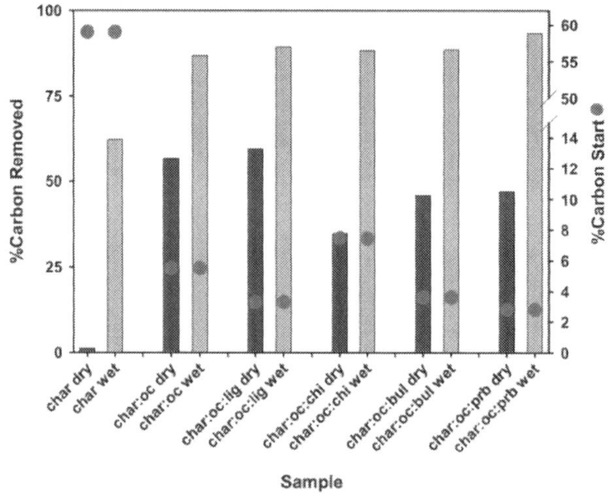

Figure 2. Carbon removed from char alone and char:FGOC:ash mixtures under experimental conditions.

The influence of the different ashes on the reaction rate of carbon removal from the char was determined from the TGA results. Rates under wet gas conditions were significantly slower than under dry gases. For instance, the maximum char gasification rate (no OC or ash) was -3.5 mg/g/min wet and -8.5 mg/g/min dry. The addition of OC improved the maximum char gasification rate for wet gases to -7.0 mg/g/min and the dry gas rate remained unchanged at -8.4 mg/g/min. Maximum carbon removal rates for the ash mixtures varied from a low of -6.5 mg/g/min for the lignite to a high of -9.9 mg/g/min for the PRB ash under dry condition and from a low of -3.9 mg/g/min for lignite to a high of -6.3 mg/g/min for the PRB ash mixtures. All ashes did reduce the maximum char removal rate under wet gases. Reductions were 45, 13, 16, 10% for lignite, China, Bulgarian, and PRB ashes respectively. Whereas under dry gases, all carbon removal rate except for the lignite were equal to or more than the base case of zero ash. Another outcome from this work was the fact that 50% or more of the carbon was removed during the first 10 minutes of dry gases and 20 minutes for wet gases. Since a CLC process will have short residence time in comparison to the 5 hours used in this study, faster initial reaction rates are important.

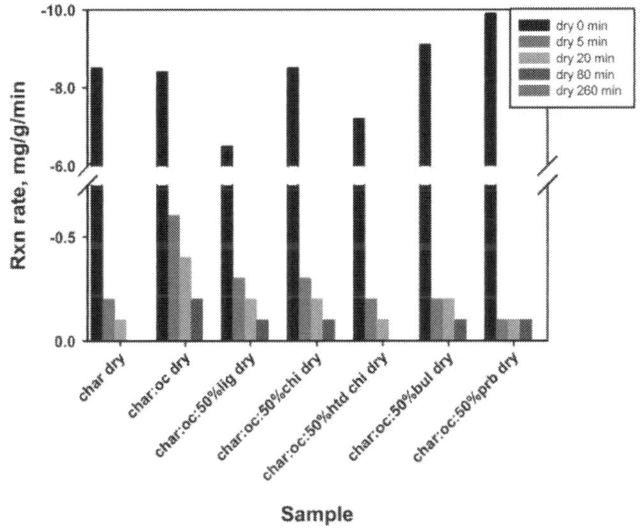

Figure 3. Carbon removal from char, char:FGOC, and char:FGOC: ash mixtures under dry gas experimental condition.

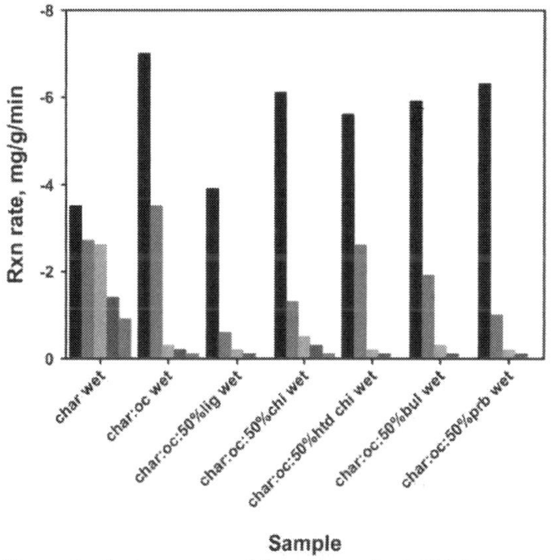

Figure 4. Carbon removal from char, char:FGOC, and char:FGOC: ash mixtures under wet gas experimental condition.

Conclusions

Both positive and negative results were obtained for the effect of ash with WV on the performance of the OC. The effect of each ash was dependent on the source coal. WV had a beneficial effect eliminating the detrimental effect of ash seen under dry conditions for the China, Bulgarian, and PRB ashes on the OC capacity to oxidize carbon. The OC had beneficial impact on carbon removal under both wet and dry gas. Carbon oxidation rates were significantly increased by the OC in wet gases over the rate of carbon gasification only. WV negatively impacted the OC's overall carbon removal rate over comparable dry runs with 50% removal of the carbon requiring 10 minutes under dry conditions and 20 under wet gas conditions. The conclusion was that in the presence of WV, ashes have no effect on the capacity of the OC to oxidize the carbon in the fuel but there was a reduction in the rate of oxidation.

References
1. Rubel, A.; Zhang, Y.; Liu, K.; Neathery, J. Effect of Ash on Oxygen Carriers for Chemical Looping Combustion of a High CarbonChar *Oil & Gas Science and Technology* [Online], 2011.
2. Rubel, A.; Zhang, Y.; Neathery, J.; Liu, K. In *Comparative Study of the Effect of Different Coal Fly Ashes on the Performance of Oxygen Carriers for CLC*, 7th International Symposium on Coal Combustion, Harbin, China, July 17-18, 2011; Harbin, China, 2011.
3. Zhang, Y.; Rubel, A.; Lui, K.; Neathery, J. In *Development of Freeze Granulated Oxygen Carriers for Chemical Looping Combustion of Solid Fuels*, 2010 Spring AIChE Conference, San Antonio, TX, March 21-25; San Antonio, TX, 2010.

(280f) Investigation of Oxygen Diffusivity within Oxygen Carrier In Chemical Looping Process

Siwei Luo[1], Zhenchao Sun[2], Fanxing Li[3], Deepak Sridhar[4], Liang Zeng[1] and Liang-Shih Fan[5], (1) William G. Lowrie Department of Chemical and Biomolecular Engineering, Ohio State University, Columbus, OH, (2)Chemical and Biomolecular Engineering, Ohio State University, Columbus, OH, (3) Chemical and Biomolecular Engineering, North Carolina State University, Raleigh, NC, (4)Chemical and Biomolecular Engineering, The Ohio State University, Columbus, OH, (5)William G. Lowrie Department of Chemical and Biomolecular Engineering, The Ohio State University, Columbus, OH

Chemical looping process is effective in converting carbonaceous fuels into hydrogen and electricity while generating a sequestration-ready CO_2 stream. The process scheme is enabled by the cyclic reduction and oxidation reactions of an oxygen carrier (OC). The properties of OCs are of critically importance for the performance of chemical looping process. Our previous experimental studies show that composite iron oxide synthesized by adding support materials to the pure iron oxide can serve as good OCs. Unlike the unsupported iron oxide which deactivates within the first few redox cycles, the composite iron oxide maintains good reactivity over a long time although pore volume decreases significantly in the first few cycles. This phenomenon indicates that traditional understanding on support effect, mainly based on morphological analysis, may not be accurate. In this study, we investigated the role of support through both marker experiments and periodic Density Functional Theory (DFT) calculations. Ilmenite ($FeTiO_3$) was used to model the simplified TiO_2 supported iron (II) oxide material. Both the oxygen diffusion barrier and vacancy formation energy in ilmenite ($FeTiO_3$) and wustite ($Fe_{1-x}O$) were calculated. The diffusivities of oxygen in both materials were estimated based on the calculation results. It was found that the energy barrier for oxygen anion diffusion in ilmenite was significantly lower than that in wustite. The calculated oxygen diffusivity in ilmenite was also higher. This finding was corroborated by the marker experiment which indicates notably enhanced oxygen anion conductivity in the supported iron oxide.

See more of this Session: Chemical Looping Processes I
See more of this Group/Topical: Energy and Transport Processes

THE STATUS OF CHEMICAL LOOPING PROCESSES DEVELOPED AT THE OHIO STATE UNIVERSITY

Presented by

Deepak Sridhar

William G. Lowrie Department of Chemical and Biomolecular Engineering

The Ohio State University

Columbus, OH, USA

CHEMICAL LOOPING AT OSU

OSU CCR Process Demonstration

OSU 25KW$_{th}$ Calcium Looping System for Hydrogen Production

OSU 25KW$_{th}$ Syngas Chemical Looping System for Hydrogen Production

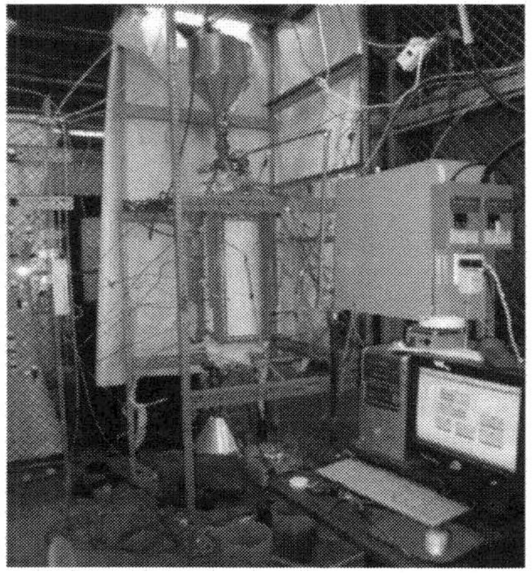

OSU 2.5KW$_{th}$ Coal Direct Chemical Looping System – Bench Unit

Syngas Chemical Looping (SCL) Process

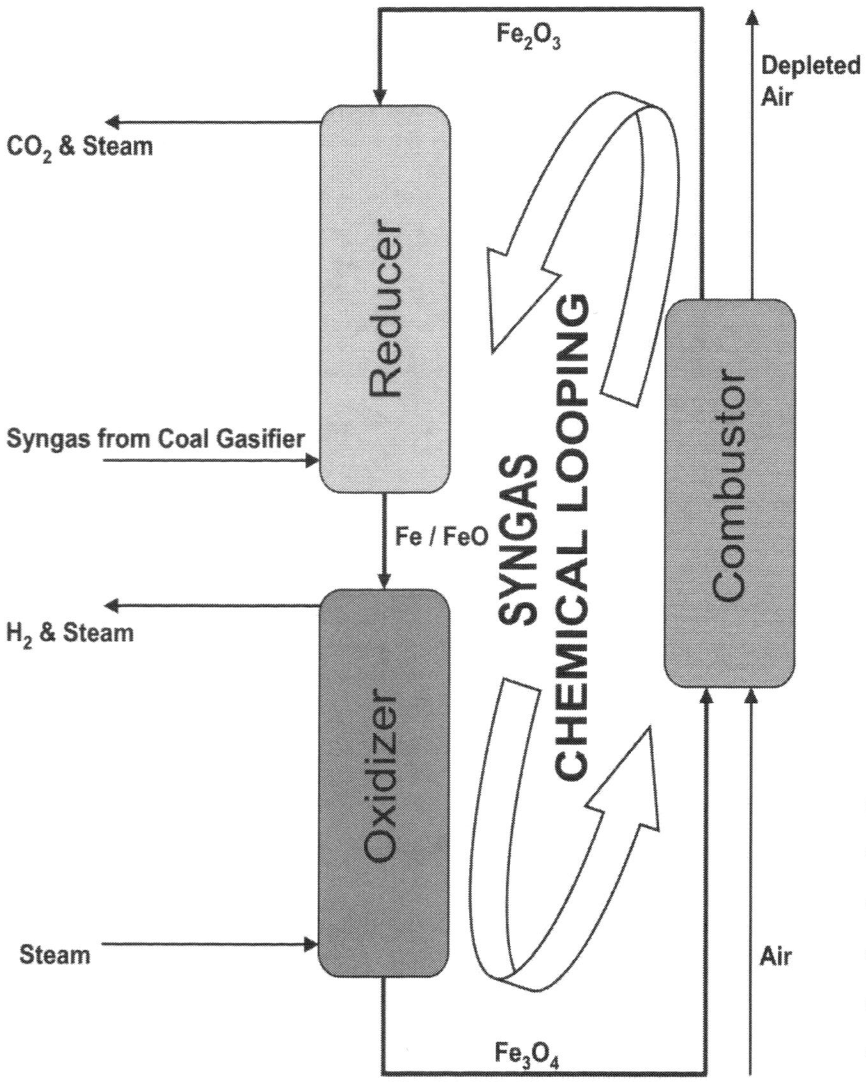

Main reactions:

Reducer	$C_xH_yO_z + Fe_2O_3 \rightarrow CO_2 + H_2O + Fe$
Oxidizer	$Fe + H_2O \rightarrow Fe_3O_4 + H_2$
Combustion Train	$Fe_3O_4 + O_2 \rightarrow Fe_2O_3 + Heat$

Overall reaction

$$C_xH_yO_z + H_2O/O_2 \rightarrow CO_2 + H_2/Heat$$

Possible Oxygen Carriers:

$Fe_2O_3 - Fe / FeO$

General Observations:

- 2 Moving Bed + 1 Entrained bed reactors
- Very High Fuel Conversion
- Near 100% in-situ CO_2 capture
- High Purity H_2 generation
- High Solid Conversion
- Low Solid Circulation Rate

Syngas Chemical Looping – Applications

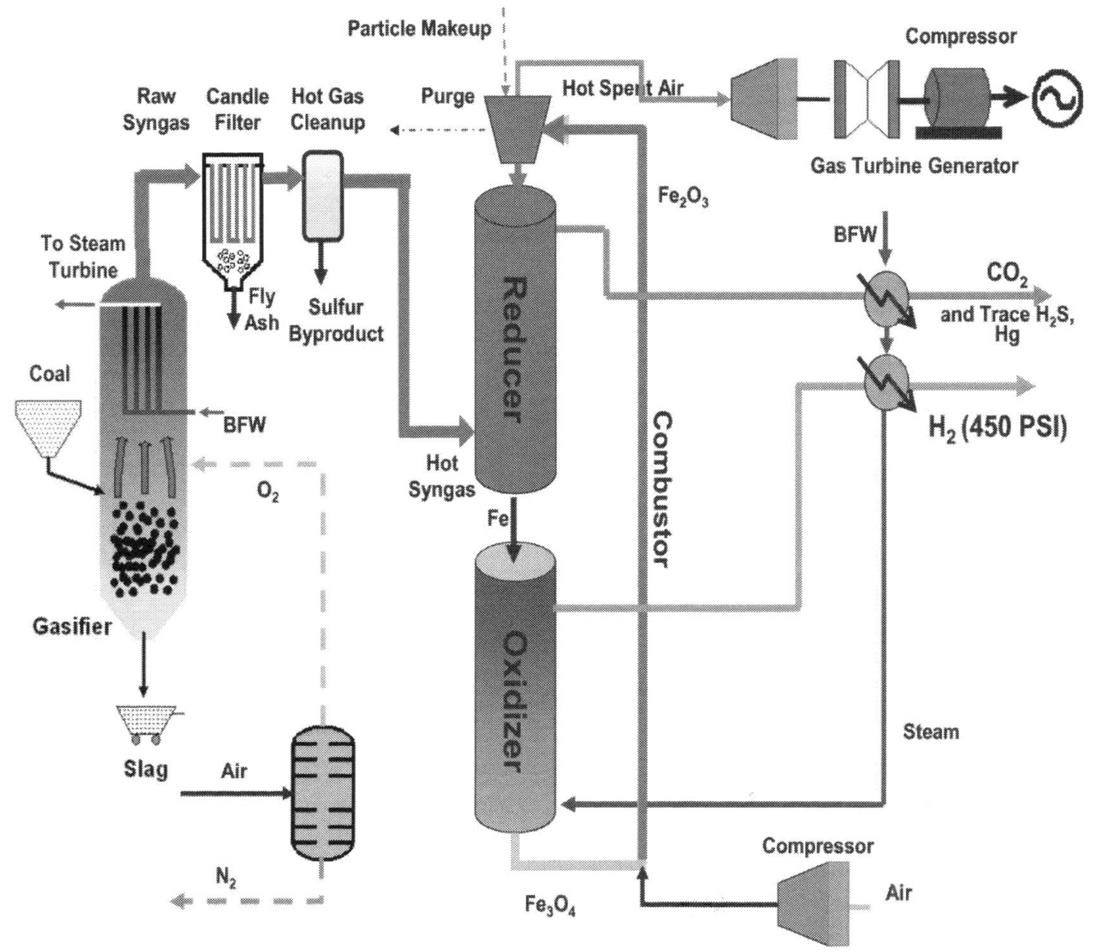

General Process Flow scheme

Product Flexibility

- Can co-generate any ratio of H_2 and Electricity from gaseous fuels
- Coupled with F-T process to generate liquid fuels efficiently

Chemical Looping Reactor Design

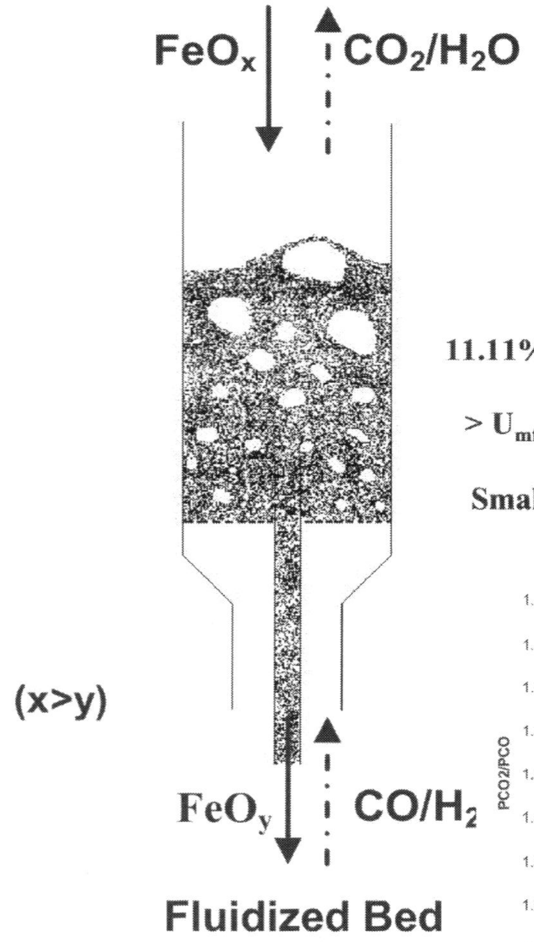

Fluidized Bed *v.s.* Moving Bed

11.11% ⟵ Maximum Solid Conversion ⟶ **50.00%**

> U_{mfv} ⟵ Gas Velocity ⟶ **< U_{mfv}**

Small ⟵ Particle Size ⟶ **Large**

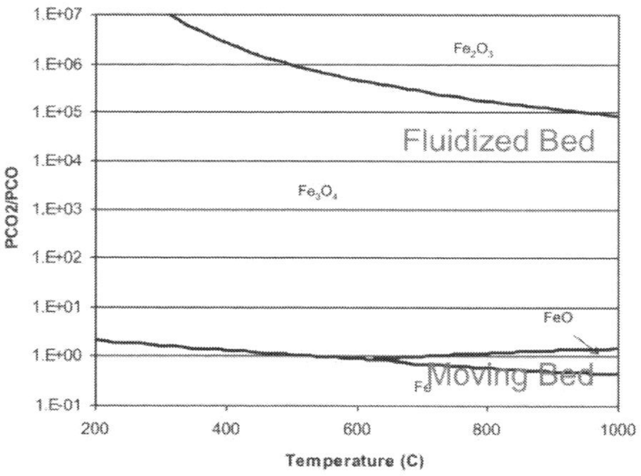

(x>y)

5

Syngas Chemical Looping Process Development

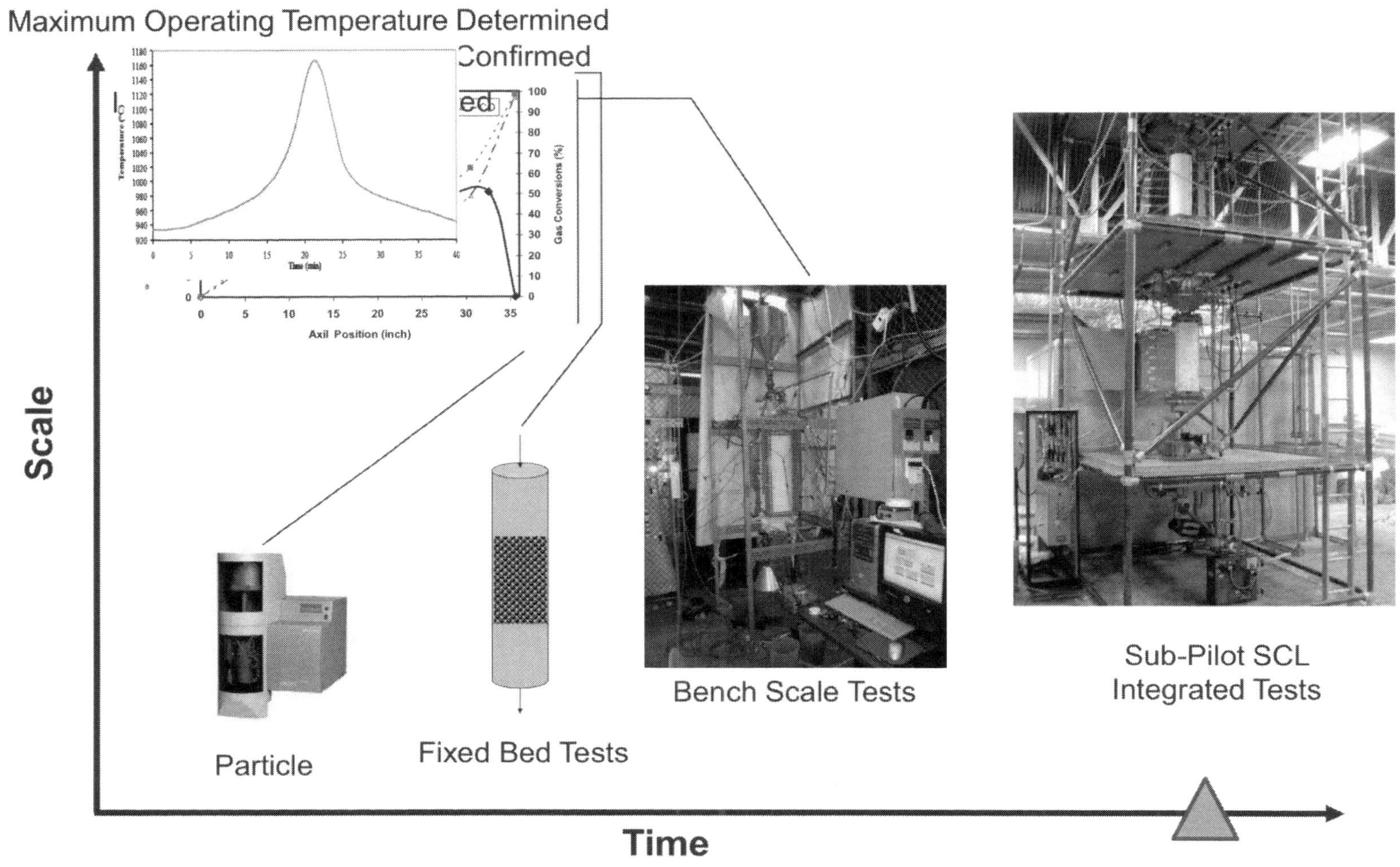

Techno-Economic Evaluation of the SCL process

	IGCC Process	SCL Process Electricity	CDCL Process Electricity	Conventional to Hydrogen Process	SCL Process Hydrogen	CDCL Process Hydrogen
Coal Feed (ton/hr)	132.9	132.9	132.9	132.9	132.9	132.9
Carbon Capture (%)	90	100	100	90	100	100
Hydrogen (ton/hr)	0	0	0	14.4	15.6	20.0
Net Power (MW)	321	365	435	2.1	26	3
Efficiency (%HHV)	32.1	36.5	43.5	57.8	64.1	79.0

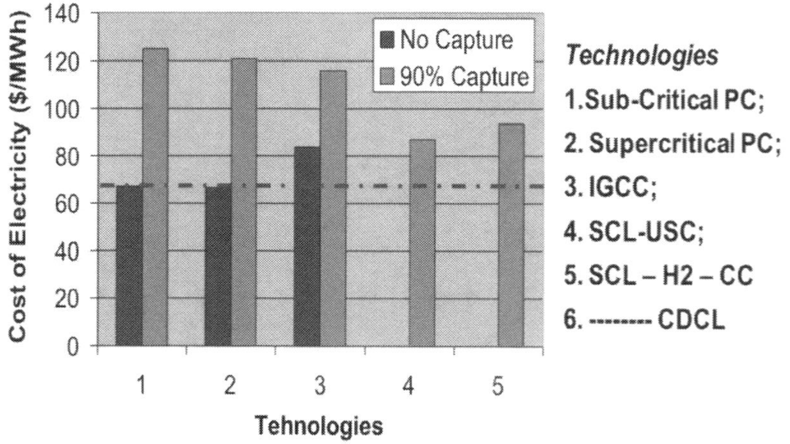

Process unit performance conditions and Assumptions used are similar to those adopted by the DOE baseline studies given in "Cost and Performance Baseline for Fossil Energy Plants", DOE/NETL-2007/1281.

- Improvement in Efficiency over conventional processes
- Lowest cost of electricity under a CO_2 regulated scenario
- SCL can pave the way to CDCL demonstrations in the future
- H_2 Production: SCL is 7 – 10% More Efficient Than Conventional Process (Analyzed by SAIC)
- Electricity Generation: SCL is 4.5 – 7% More Efficient Than IGCC with 100% CO_2 Capture (Analyzed by Noblis)
- When Integrated CTL, SCL can Increase the Liquid Fuel Yield by 10% with 19% Reduction in CO_2 Emission (Analyzed by Noblis)

SCL – Sub-Pilot Scale Unit Design

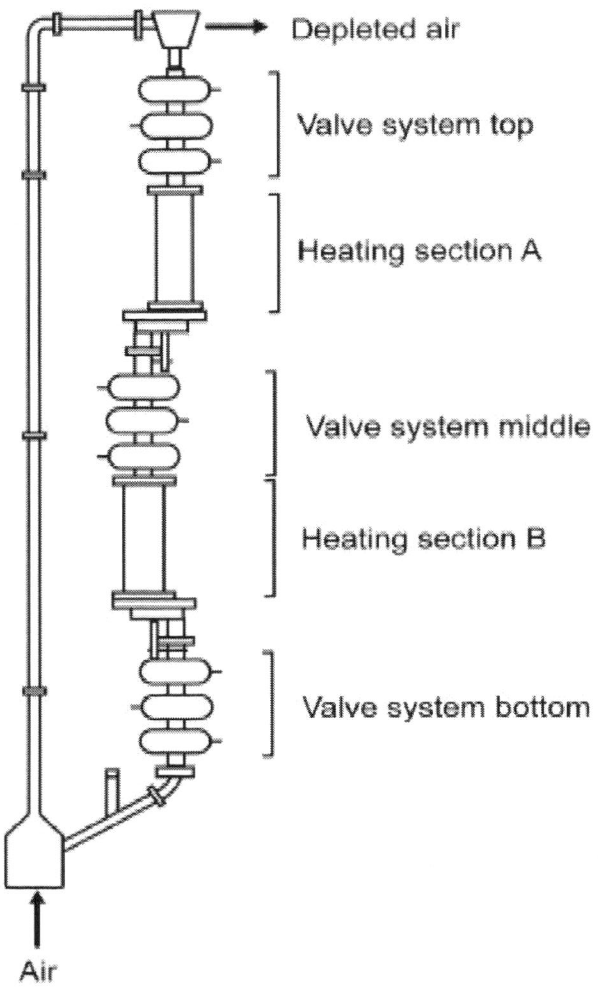

SCL – Sub-Pilot Scale Unit Features

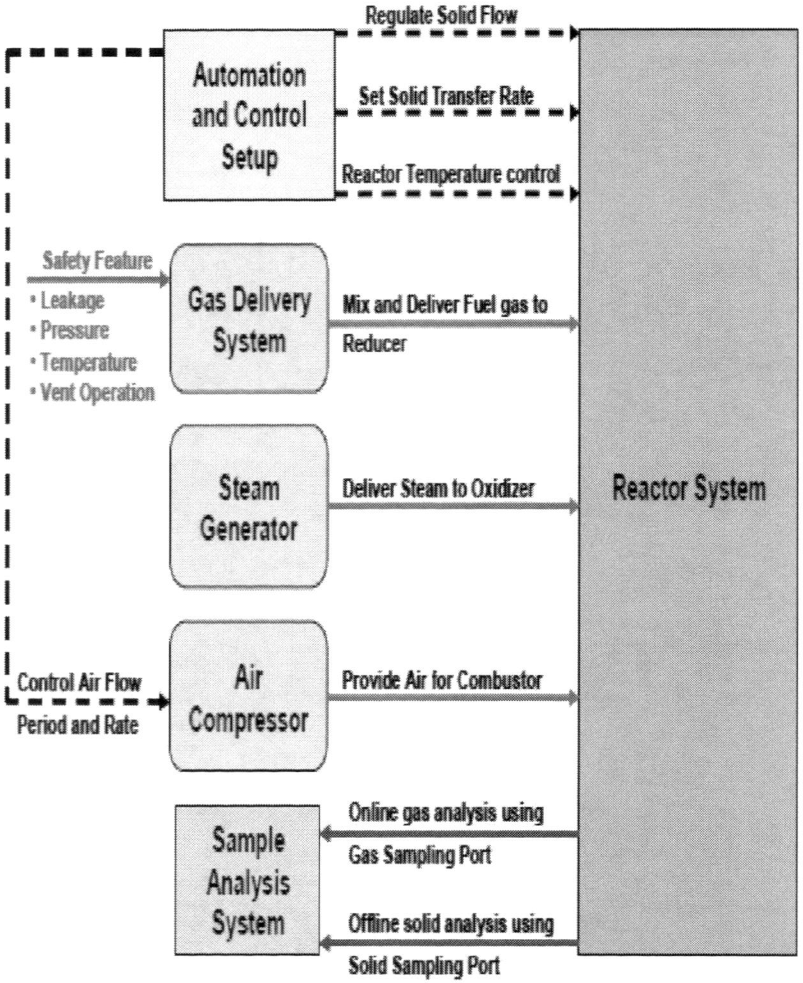

Salient Features:

- Completely automated process
  - Reactor temperature regulation
  - Valve operation
  - Air entrainment
- Regulated Gas flow
  - gas mixing panel could deliver any fuel combination
  - accurate steam and air flow
- Multiple Analyzers to obtain reliable data
  - Varian Micro GC
  - Sick Maihak Gas Analyser
- Safety features
  - Pressure Control
  - Temperature Control
  - Gas Leak Control
  - Vent Failure Control

Initial Sub-Pilot Scale Unit Operational Results

- 40+ hours of operational experience gained
  - 4 test runs
  - Final run was for 24 hours continuous
  - Average Syngas conversion – 98.9%
  - Average H_2 purity – 93.5%

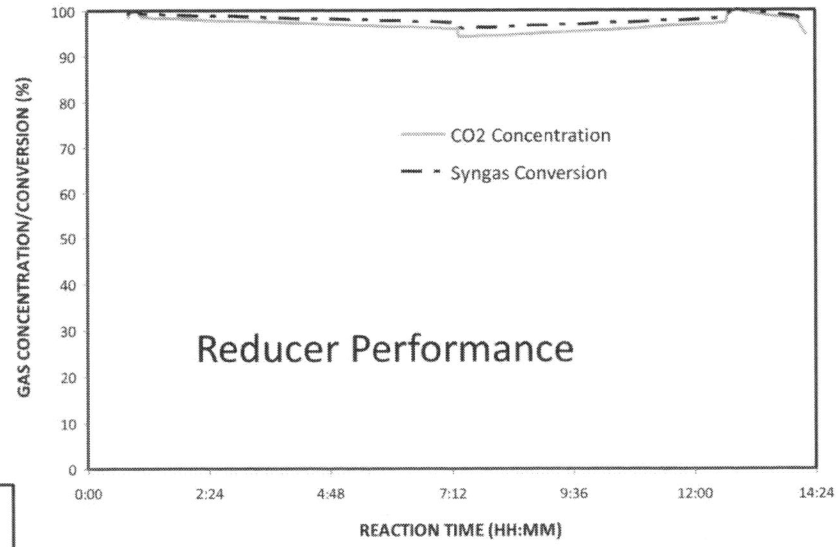

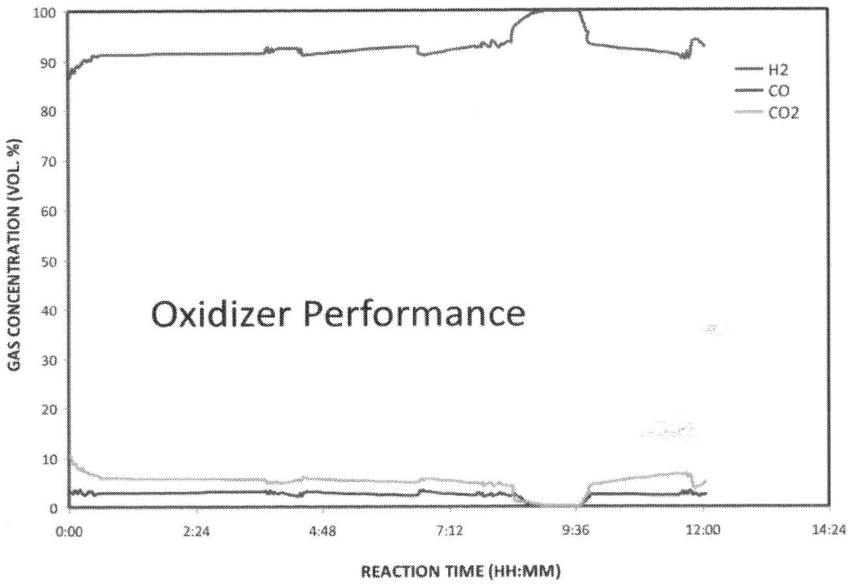

- Longer tests were not performed due to failure of the mechanical valves and solid feeder system
- Carbon deposition detected due to lack of gas pre-heating
- Attrition of 1% per cycle observed (more than double the inert condition attrition test results)

Sub-Pilot Scale Unit Performance Improvement

- **Longer Continuous Duration Testing required**
  - Changing from two point solid flow control to one point control
  - Reducing the number of mechanical valves used
- **Reduce Particle Attrition**
  - Change Combustor bend design
  - Use the improved oxygen carrier
- **Improved Reactor Performance**
  - Add pre-heating of inlet gases for all the three reactors

2008

2009

2010

Performance Results Modified Sub-pilot Unit

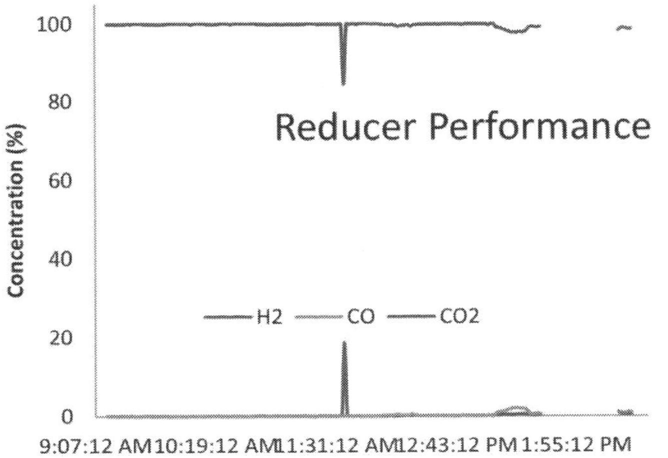

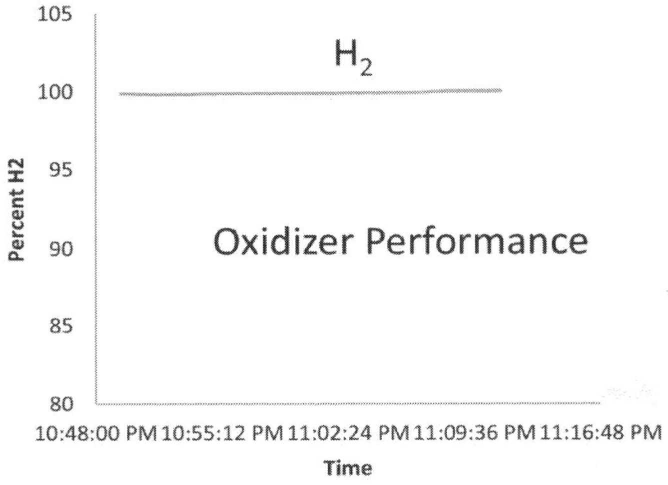

- Longer tests were not performed spanning more than 48 hours
- Carbon deposition was not observed
- Attrition of 0.53% per cycle observed (similar to the inert condition attrition test results)
- Syngas conversion of ~100% was observed.
- H2 purity of >99.9% was observed.
- Particle performance post the sub-pilot unit operation was similar to fresh particles

Accomplishments of the SCL Demonstrations

- 2.5 kW_{th} Demonstration Completed, > 99.5% Pure CO_2 and >99.95% Pure Hydrogen Obtained
- H_2 Production: SCL is 7 – 10% More Efficient Than Conventional Process (Analyzed by SAIC)
- Electricity Generation: SCL is 4.5 – 7% More Efficient Than IGCC with 100% CO_2 Capture (Analyzed by Noblis)
- When Integrated CTL, SCL can Increase the Liquid Fuel Yield by 10% with 19% Reduction in CO_2 Emission (Analyzed by Noblis)
- A 25 kW_{th} Sub-Pilot unit constructed and successfully operated, >99% in syngas conversion and >93.5% in H_2 purity obtained
- ***DOE/ARPA-E grant awarded, a 250 kW_{th} plant is being designed***

Successful Bench and Sub-pilot scale tests + Encouraging Techno-economic analysis

↓

Pilot Scale Design, Construction and Testing

Coal Direct Chemical Looping – Process Flow

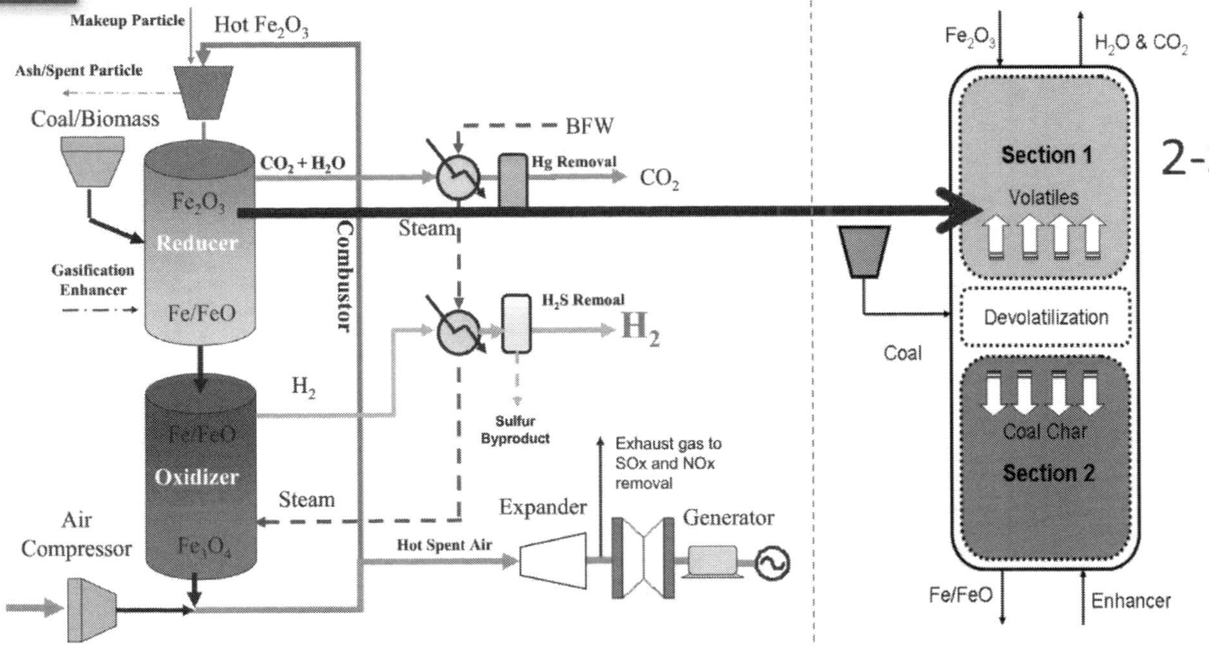

2-Stage Moving Bed

Counter-current flows

Gaseous volatiles

Coal char

More challenging reducer operation less developed compared to SCL due to handling of solids at different sizes : Pulverized coal (~μm) vs. OC (2-5mm)

- **No need for Capital/Energy Intensive Gasifier and Air Separation Unit**
- **Great Scheme for the Retrofit to PC**

Reaction Enhancer

– Initiates the solid-solid reaction

– Direct enhancer : CO_2, H_2O, O_2

- $CO_2 + C \rightarrow 2CO$
- $H_2O + C \rightarrow H_2 + CO$

– Indirect enhancer : H_2

- $3H_2 + Fe_2O_3 \rightarrow 2Fe + 3H_2O$

CDCL Current Status and Future Work

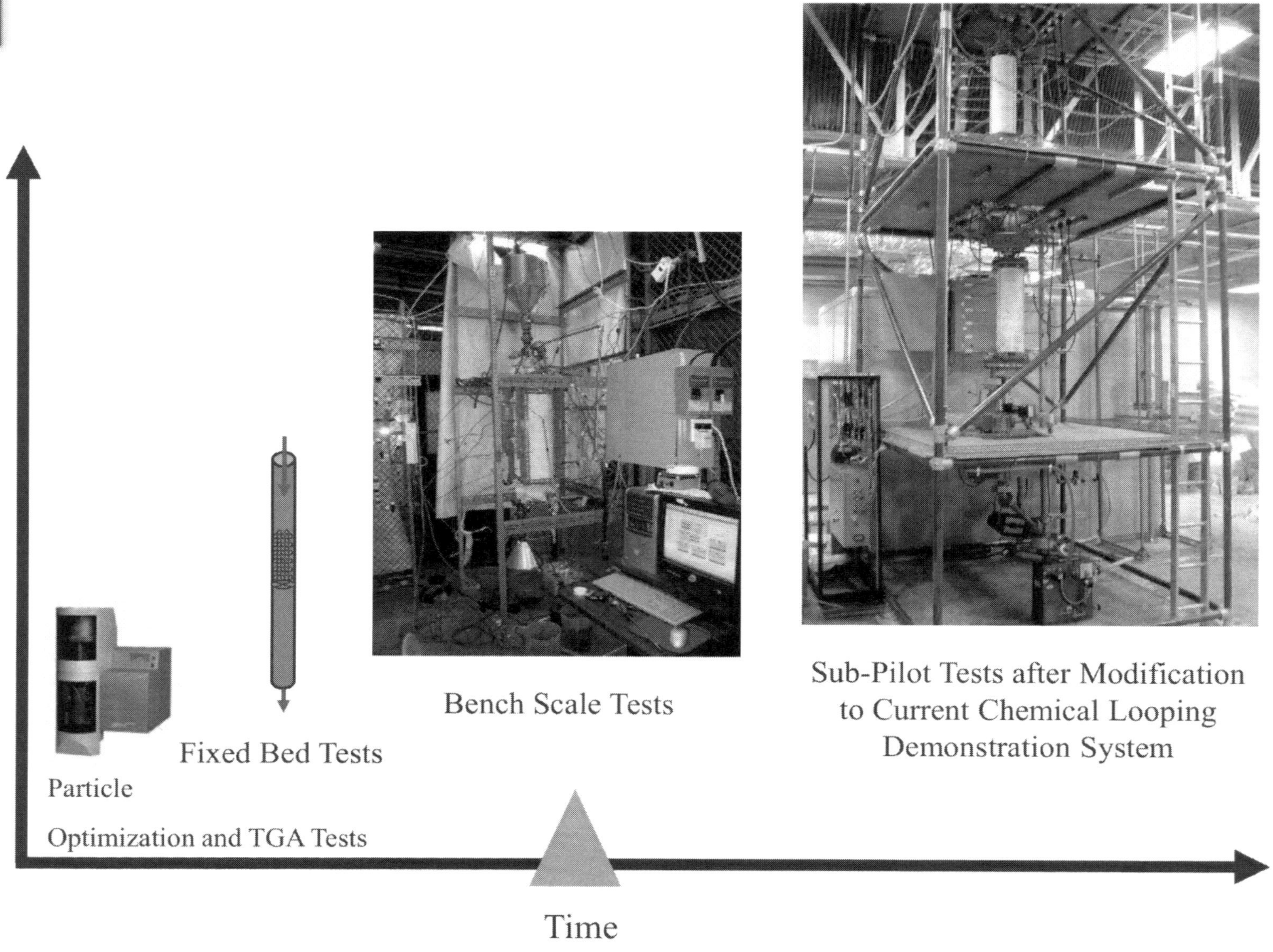

Coal Direct Chemical Looping – Key Results

Type of Fuel	Coal Volatile (CH_4)	Lignite Char	Ohio Coal Char	Anthracite Coal
Fuel Conversion (%)	99.8	94.9	90.5	95.5
CO_2 Concentration in Exhaust (% Dry Basis)	98.8	99.23	99.8	97.3

Particle Reactivity After Multiple Bench Tests

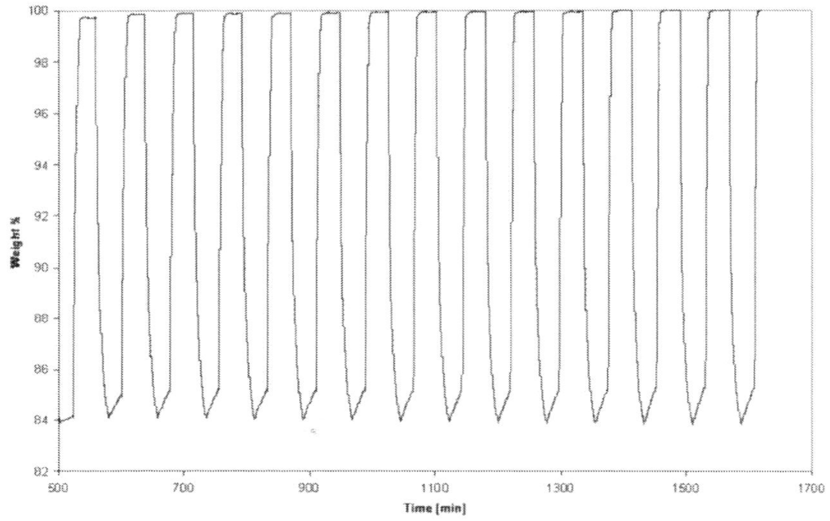

THE OHIO STATE UNIVERSITY

Comparison Among SCL, CDCL and Traditional Coal to Hydrogen/Electricity Processes

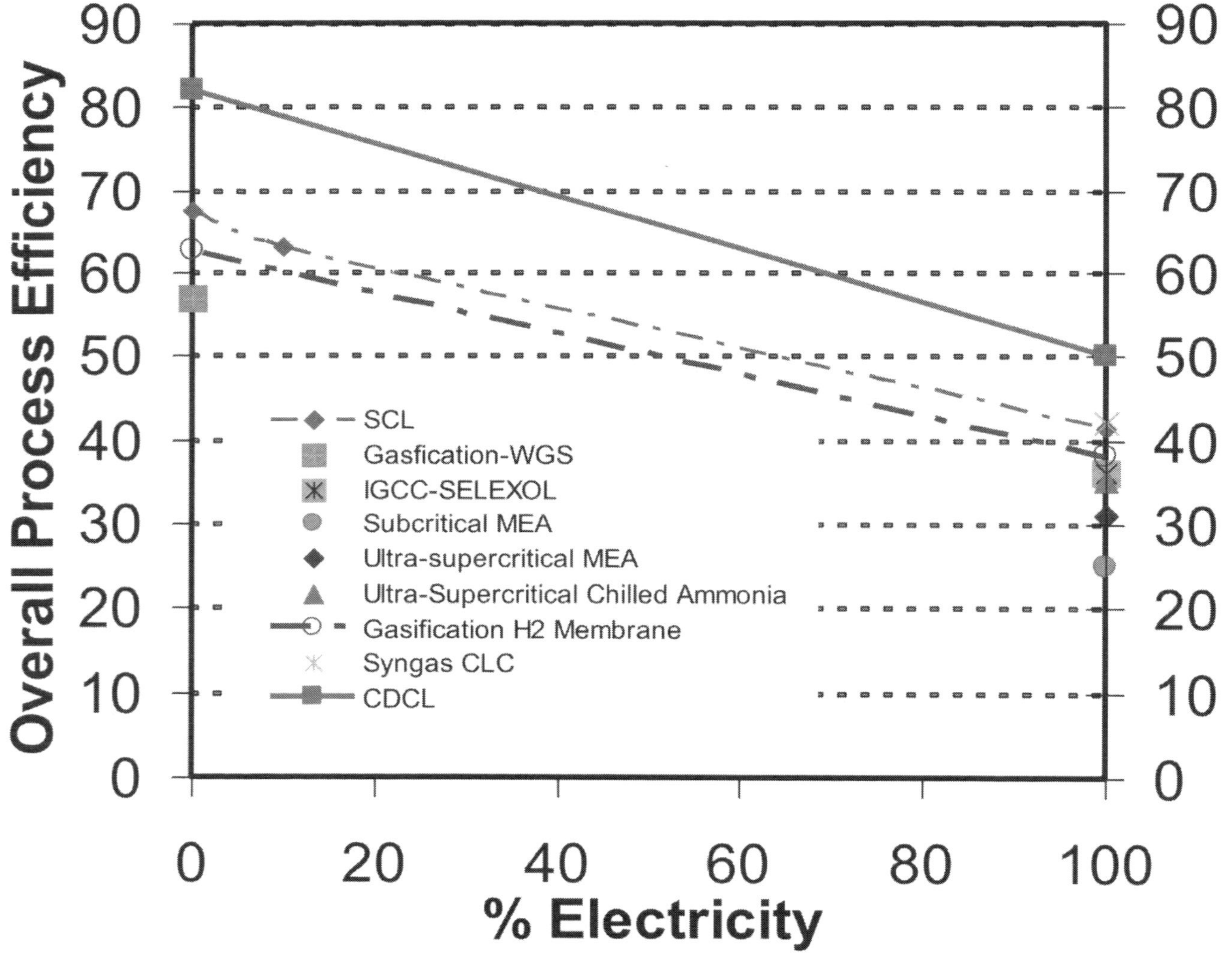

Team Work is Key

Dr. Liang-Shih Fan

Deepak Sridhar	Dr. Dawei Wang	Nihar Phalak
Ray Kim	Dr. Qiang Zhou	Niranjini Deshpande
Andrew Tong	Liang Zeng	William Wang
Elena Chung	Siwei Luo	Yao Wang
Mandar Kathe	Aining Wang	Zhenchao Sun
Samuel Bayham	Hui Yang	

Acknowledgements

Government Agencies and Industrial Corporations

- Department of Energy
- Ohio Coal Development Office
- Ohio Department of Development

Thank You

- Clear Skies
- Noblis/Metritek
- CONSOL Energy
- PSRI
- AEP
- Duke Energy
- Babcock & Wilcox
- Air Products
- Shell
- CRI/Criterion
- First Energy
- Carmeuse Lime and Stone
- LittleFord Day
- Specialty Minerals Inc.

The Fundamental Concepts of Transport and Reaction

Stuart W. Churchill
Department of Chemical Biomolecular Engineering
University of Pennsylvania

Introduction

A secure knowledge of the elementary but fundamental concepts of chemical engineering that evolved in the course of the past century, and that have been refined and re-refined, is essential practicing engineers. Most of these concepts were acquired while they were undergraduate and/or graduate students although they may have been reinforced in practice. Basic concepts are to be found in every course in but the word count imposed on the written version of this disquisition and the time-limit on its oral presentation allow only for a few illustrative examples in a representative set of three topics in a single context. Fluid mechanics, heat transfer, and chemical conversions in steady flow through a round tube were chosen because of my personal experiences in both industry and academia. Thermodynamics would instead have headed the list if relative importance were the criterion. I hope that the brief commentary that accompanies each of the chosen concepts will convince you that it is at least worthy of consideration for a comprehensive list.

Elementary concepts are here defined as those that are so concise that they can be memorized without drill, but whose importance belies their brevity. The identification and exposure of concepts that were once accepted as valid but that are now known to be false or inferior is as important as a list of the legitimate survivors, so examples of the false ones are also examined herein.

Two characteristics of chemical engineering that contribute to its vitality are its breadth in the sense of fluids, and its reliance on universalities. However, these very characteristics may encourage the application of concepts beyond their limits of validity. It is more difficult to persuade students and practicing engineers to discard false concepts than to accept new ones. Therefore teachers should be especially careful to avoid the introduction of false concepts, to identify those that have found their way into our textbooks, and to point out the limitations and errors introduced by idealizations and approximations. The designation of concepts that originated with one our idols as false or out-dated is painful and perhaps even blasphemous to some of you, but science and technology can progress only by doing so. It is is an aspect of the self-renewal that is essential to all professionals. I am incapable of identifying all false or over-extended concepts, and if I were, such a listing and discussion would further exceed the aforementioned limitations of the current format. Therefore the false concepts listed herein are also only illustrative. I hope my list will inspire others, and particularly the authors of textbooks, to search out, identify, and excise or exorcize those in their domain of expertise. A single listing is utilized for both the true and false concepts because, as is illustrated repeatedly, it is often difficult to draw a firm line between them.

We share an interest in fluid mechanics with mechanical, civil, environmental, petroleum, and nuclear engineers, and possibly with physicists; in heat transfer with mechanical engineers; and in chemical conversions with chemists. However, in almost all instances the primary interests of chemical engineering differ from the others in terms of geometry and/or by virtue of a broader range of fluids. Hence the ordering of concepts herein in terms of their importance in chemical engineering does not necessarily apply to these other fields.

Chemists and physicists by and large choose to live in a Lagrangian world, whereas chemical engineers necessarily live in an Eulerian world because of the economic advantages of continuous processing. As an example, our models and solutions for continuous tubular and stirred reactors are almost beyond the ken of chemists.

One common denominator of most of the concepts listed here is their simplicity in a mathematical sense. The first line of the well-known Shaker hymn of the 19th century, "Tis the gift to be simple," and subsequently added line, "Tis the gift to be taught and a richer gift to learn", rightfully praise simplicity. Simplicity implies and possibly evokes generality, but its primary merit in the context of technical concepts is the reduction of complex ideas and processes to ones that can readily be understood and possibly committed to memory. Many of these simple concepts were first conceived by the giants of science and engineering of the past, which gives them a historical context and luster and thereby makes them easy to remember.

Because the great majority of listeners and readers of this disquisition are familiar with these concepts, and if not, can find them in standard textbooks, the details of their derivation or conception are not presented here. Many of these same topics were described in somewhat more length in my plenary lecture "A Century of Transport -A Personal View", which may be found on the web or obtained on request from the author.

As a very important disclaimer before examining specific examples, the elemental concepts listed here provide at best a skeletal structure for fluid mechanics, heat transfer, and chemical conversions. They are not a substitute for the full accounts to be found in textbooks and discussed in the classroom.

What use is a mental list of concepts? When, as a student, I expressed my distaste for surprise (unscheduled), closed-book tests my professor said it was part of the preparation of an engineer who would be expected, if employed in industry, to respond to unexpected events, to unexpected questions in formal and informal conferences with collaborators, and, most importantly, to questions from one's superior, without access to a book and perhaps while standing. Likewise a teacher is expected to respond without reference to a textbook if their lecture is interrupted by a student. Today, that admonition would presumably include being prepared to respond without reference to an electronic device. That proved to be good advice to me. My off-the-cuff response in industrial settings had a decisive impact during my years in industry and is an every-day occurrence in the classroom.

One cannot commit to memory all of the important passages in textbooks and handbooks and in particular the information listed in tables, displayed in graphs, and summarized by equations. (As an aside, I had one teacher who had committed to memory the table of logarithms and was able to used those values for mental multiplications, but that is a freak skill and not a reasonable expenditure of effort for normal folks.)

All of us acquire a mental library of sorts from our studies and experiences without conscious effort. We rely on that resource in formulating a response in every-day situations as well as crises. I am merely suggesting a few formal additions and corrections. Most of the items in our mental libraries once had validity but many have become out-dated and should be deleted or replaced. I will try to identify widely cited but invalid concepts as well as those that are valid in some contexts but not in others.

Overall Concepts

The Eulerian frame of reference. The use of spatial coordinates rather than a temporal one could be said to be the defining concept of chemical engineering. Its adoption is a is natural consequence of the discovery that heat transfer and chemical conversions are carried out most effectively and efficiently in steady flow through a round tube. Its ubiquity overshadows its importance. Physicists and chemists, who live in a Lagrangian universe are helpless when faced with taking diffusion into account in a flowing system. **This is a keeper**

The concept of dimensional analysis by Rayleigh. All prior and subsequent methodologies and interpretations are false or redundant. Students struggle to master this concept but there efforts are repaid by the understanding it confers. **A keeper**

All other concepts of dimensional analysis. At best they are inferior to that of Rayleigh. **Discards**

The concept of correlation in terms of products of powers of dimensionless groups. This concept is legacy of a faulty analysis by Nusselt himself, in the very investigation that defined the dimensionless group named in his honor. Power dependences only occur in asymptotic relationships. It is my candidate for the most pernicious concept in the history of heat transfer. Rayleigh knew better at the time. **A discard**

Correlating equations in the form of power-means of asymptotes. **An under-utilized keeper**

The method of Hellums and Churchill for the simplification of mathematical models and the identification of similarity transforms. **A keeper**

Fluid Mechanics

Fully developed flow. This concept, which is a good approximation for a long tube and thereby very useful in a practical sense, implies invariant physical properties. It was a necessary precursor for the derivation of Poiseuille's law for the pressure drop in a round tube or parallel plate channel as well as for the various empirical expressions for turbulent flow in a round tube. Its introduction obligates the teacher to point out all real tubular flows encompass a regime of developing flow. **This a keeper but with the noted reservations.**

The mixed-mean velocity. **A keeper**

The parallel-plate channel. This is a hypothetical geometry and thereby an example of a "false" concept in a realistic sense. It can, however, be interpreted as an asymptote for a rectangular channel. Also, it shares some characteristics, such as a parabolic velocity distribution, with a round tube. This concept is over-used because of its simplicity, namely one-dimensionality and the absence of curvature. **This a mixed-bag**

The friction factor. This quantity is not a "fudge factor" but simply a convenient, defined combination of real (measurable) quantities. For example, an exact theoretical expression exists for the friction factor in fully developed laminar flow. **A keeper**

The 1/5-power dependence of the friction factor on the Reynolds number in turbulent flow in a tube as determined graphically by Blasius. The unfortunate consequences of this innocent error have not yet all been excised. **A long-overdue discard**

The power-law velocity distribution. One of the consequences of the graphical correlation of Blasius. **A discard**

Prandtl's law of the wall, namely $u^+ = \varphi\{y^+\}$. Although this concept has a speculative origin, it appears to be exact in an asymptotic sense. **A keeper**

Prandt's law of the center, namely $u_c^+ - u^+ = \varphi\{y/a\}$. Just as for its counterpart for the wall, this speculative concept appears to be exact in an asymptotic sense. **A keeper**

The law of the region of overlap, namely $u^+ = \text{A} + \text{B} \ln\{y^+\}$. This concept, just as its counterparts (and components) appears to have validity in an asymptote sense. **A keeper but with limitations**

Millikan's derivation of the law of the region of overlap. Perhaps the of the most ingenious speculations in all of fluid mechanics. ***A keeper***

Prandtl's derivation of the law of the region of overlap. ***A discard***

The concept of space-averaging by Reynolds. This is my choice as the greatest contribution of all time to the modeling of turbulent flow, although its validity is questioned by some. The concept itself meets the standard of brevity although its evolution and implementation do not. **A keeper**

Direct numerical simulation. When this concept was first proposed and implemented in the late 1980's it was expected to make all other modeling of turbulent flows obsolete but its promise remains unfulfilled because of computational limitations. **A disappointment**

Large eddy simulation. This attempt to extend *DNS* to higher Reynolds numbers is a failure because it incorporates a wall function. **A discard**

The eddy viscosity. This concept of Boussinesq has recently been shown to be a valid only in round tubes. **A keeper with reservations**

The mixing length. Totally without merit despite its origin with Prandtl and its application by von Karman. **A discard**

The local fraction of the shear stress due to turbulence. I immodestly cite this model as decisively superior to all other algebraic ones for turbulent transport in fully developed flow. **A keeper**

The k-ε model. This model never fulfilled its early promise but is nevertheless our only resource for developing flow. **A necessary keeper but with severe reservations**

The Navier-Stokes equations. When I asked the renowned George Uhlenbeck, my teacher in a graduate course in physics, why he never referred to the Navier-Stokes equations, he replied that he was unsure of their validity and preferred statistical mechanics as the starting point in fluid mechanics.

He subsequently wrote in the introduction of an invited article in *The Reviews of Applied Mathematics* that my question provoked him to "waste" two doctoral students in a vain attempt to derive the former from the latter. That article contains the following quotation that expresses his frustration in this regard. "Quantitatively, some of the predictions from these equations surely deviate from experiment, but the *very* remarkable fact remains *that qualitatively* the Navier-Stokes equations *always* describe physical phenomena sensibly.....The mathematical reason for this virtue of the Navier-Stokes equations is completely mysterious to me." **The N-S equations are keepers**

Heat Transfer

The heat transfer coefficient. The recognition by Newton that the rate of heat transfer is proportional to the area and the difference in temperature led to the replacement of four variables with one and is thereby the cornerstone of all predictive and correlative equations for convection. **A keeper**

Fully developed convection. This concept was elevated to a viable one by Seban and Shimazaki, who revealed its dependence on the thermal boundary condition. **A keeper**

The mixed-mean temperature. **A keeper**

The turbulent Prandtl number. The representation of the turbulent shear stress and the turbulent heat flux density as fractions of the total shear stress and heat flux density, respectively, revealed somewhat surprisingly to the speaker that this was a valid concept. **A keeper**

The eddy thermal conductivity. **A keeper only as a component of the turbulent Prandtl number**

The classical analogies between heat and momentum transfer. Although the similarities of the partial differential equations of conservation are the source of this concept, the classical implementations of Reynolds and Prandtl, and in particular that of Colburn, are all invalid functionally and highly inaccurate numerically. **Discards**

The analogies between heat and momentum transfer of Churchill and Zajic. These adaptations of the analogy of Reichardt are immodestly cited as superior to all others both functionally and and numerically. **Keepers**

The analogy between heat and mass transfer. This concept has been proven to be false by Hanratty and his collaborators but it persists in many textbooks. **A discard**

The small Prandtl number of liquid metals. The false concept that the Prandtl number of liquid metals is asymptotically small resulted in attempts to correlate data for convection in terms of $Pe = RePr$, which of course failed. **A long-overdue discard**

The effective thickness. The concept of representation of the rate of convection by an effective thickness for conduction is an example of a little known and often scorned but very useful concept devised by Langmuir in 1912 in the context of the estimating the heat loss from the filament of an electric light bulb. **A keeper**

The Lévêque solution. Although this solution is of limited utility, its derivation is a classic in itself . **A keeper**

The classical solution for transient temperature produced by thermal conduction in semi- infinite region following the imposition of a step in temperature at the surface, namely, $(T_w - T)/(T_w - T_\infty) = \mathrm{erfc}\{x/2(\alpha t)^{1/2}$. This solution has long been recognized by mathematicians, if not by engineers, to be erroneous because it predicts an instantaneous rise in temperature at an infinite distance. The explanation (see, Churchill and Brown) is the false postulate of incompressibility in the differential model, namely
$\rho c(\partial T/\partial t) = k\,(\partial^2 T/\partial x^2\,)$. **Discard**

The hyperbolic equation of conduction. Although this model, which was formulated as a correction for the failure of solutions for transient thermal conduction, has been shown by Churchill and Brown to be based on a totally false premise and to lead to false predictions, it has been utilized in hundreds of publications. Although the false solutions in print cannot readily be eradicated, the editors of technical journals should exclude any future ones. **Long over-due discard**

Chemical Conversions

Plug flow. This concept is the most pernicious one in all of chemical engineering. It occurs physically only when a semi-solid is pushed through a tube with a plunger. Plug flow is not approached as the Reynolds number increases; the velocity goes to zero at the wall, and the profile is more nearly parabolic than flat. The length of the reactor required for a specified conversion may be up to twice that so-predicted. The error in the prediction of the concentration of residual secondary products such as pollutants may be unbounded. Perhaps most serious of all, it precludes the consideration of external heat exchange. These shortcomings far over-balance the claimed gain in simplicity. **An absolute discard**

Perfect radial mixing. This concept produces the same solutions as plug flow but is physically conceivable as an asymptote for $Pr_T \to 0$ and $Sc_T \to 0$ and thus an upper bound for the conversion. The other shortcomings remain. **Mixed-bag. A keeper with severe reservations**

Solutions for fully developed flow with no radial mixing. The solutions produced by this concept are in serious error but constitute lower bounds for the conversion, and are thereby complementary to those for perfect radial mixing. **Mixed-bag. A keeper with severe reservations**

Rate mechanisms in terms of concentrations. This concept, which all undergraduates in chemical engineering must recognize as incompatible with their background in thermodynamics, particularly for gases, is deeply imbedded in all the textbooks of reaction engineering, but that does not relieve the faculty of their responsibility to say so or the journals to refuse to propagate this usage. **A discard**

Global models for combustion. This important class of reactions has long been known to proceed only by means of free-radical mechanisms. Furthermore, most of the critical free-radical mechanisms have been identified and quantified. This concept should be abandoned. **A discard**

A pseudo stationary state for the concentrations of free radicals in combustion. The concept of Christiansen of a pseudo stationary state for the concentrations of free radicals on a solid-state catalyst is one of the greatest in the history of chemistry in that it explains the previously puzzling observations of fractional and bilinear mechanisms. However, it is inapplicable for combustion because of the infrequency of the collisions between the molecules and free radicals. Fortunately, with the present state of computational software and hardware renders this approximation is no longer needed. **A discard**

A Simple Listing

The following concepts and recommended actions would have been discussed were there no constraints in time and space. You can carry out your own analysis and reach your own conclusion as to my categorization if you consider it worth while.

The boundary layer. **A mixed-bag**

The thin boundary layer. **A keeper**

Free streamlines. **A mixed-bag**

The orifice coefficient based on free streamlines. **A discard**

Creeping (Stokes) flow. **A mixed-bag**

Potential (inviscid) flow. **A mixed-bag**

Inertial flow. **A keeper**

The shear stress on the wall. **A keeper**

The equivalent length of straight pipe for pipe fittings. **A mixed-bag**

The equivalent diameter of non-circular channels. **A mixed-bag**

Integral boundary layer theory. **A discard**

Homogeneous turbulence. **A border-line keeper at best**

The criterion of Reynolds for the onset of turbulent flow in pipes. **A keeper**

The asymptotic expression derived by Murphree and others for the turbulent shear stress near the wall. **A keeper**

The asymptotic expression of Hinze and others for the turbulent shear stress near the center. **A keeper**

The friction factor law of Nikuradse for artificial (uniform) roughness. **A keeper**

The friction factor law of Colebrook for natural roughness. **A keeper**

The analogy of MacLeod for the velocity distribution in laminar and turbulent fully developed flow in round tubes and parallel-plate channels. **A keeper**

The adaptation of Ohm's law for thermal conduction and convection. **A keeper**

The heat transfer coefficient for radiation. **A keeper**

Black- and gray-body radiation. **Keepers**

Newton's law laminar film condensation. **Keeper as an asymptote**

The solution of Wilhelm and Kwauk for fluidized beds. **A keeper**

The marker-and-cell model for unconfined flow. **A keeper**

Specialized Concepts

The following are examples of important concepts for a special field (natural convection) that may not be important to every one

The "Boussinesq" approximation for natural convection. **A keeper with reservations**

Emmon's analogy for gravitational convection. **A keeper in a qualitative sense**

The false transient of Samuels for modeling natural convection. **A keeper**

The streakline of Ozoe for display of natural convection. **A keeper**

The integral transformation of Saville for free convection. **A keeper**

The use of the vector potential to model three-dimensional natural convection. by Aziz and Hellums. **A keeper**

The recognition that the universally utilized boundary condition for free convection from a vertical plate is physically unattainable. **This knowledge is a keeper**

The recognition by Ozoe that the streaklines in laminar natural convection are all double helices. **A keeper**

Summary

You may be thinking "Does he really expect us to memorize all this stuff?

The short answer is "Yes".

All of you listeners or readers have been exposed to these concepts and most of you already have them in your mental library. What I am proposing is that you review these concepts in your mind and books, and mull over my interpretations. That minimal but organized effort should be sufficient to rejuvenate your mental library in these particular topics. I hope that you will do the same in the aspects of chemical engineering that are of most interest to you.

(295b) Scaleout of Turbulent Drag Reduction From a Pipe to a Flat Plate

Preetinder S. Virk, Department of Chemical Engineering, Massachusetts Institute of Technology, Cambridge, MA

Drag reduction by wall-injected Polyox solutions was studied by axial pressure profile and flow visualization measurements at Reynolds numbers from 40000 to 140000 in an acrylic test pipe of ID 15.7 mm and L/D 430. Friction factors for the deionized water solvent followed the Prandtl-Karman law to within 0.2 $1/\sqrt{f}$ units. The P309 Polyox, of MW ~ 8×10^6, was injected in concentrated solutions, of C_{inj} 500, 1000 or 5000 wppm, through 6 equi-spaced circumferential ports to provided downstream polymer concentrations C_{line} from 0.1 to 100 wppm that exhibited eventual fractional drag reductions DR_{line} from 0.15 to the asymptotic maximum drag reduction MDR ~ 0.80 possible at the present Re. At low $C_{line} < 1$ wppm, flow enhancements $S'_{line} = (1/\sqrt{f_p} - 1/\sqrt{f_n})_{Re\sqrt{f}}$ increased almost linearly with increasing C_{line}, with the highest intrinsic slip $[S'] = \lim_{C \to 0} (S'/C_{line})$ ~ 8.0 $wppm^{-1}$, while at high $C_{line} > 10$ wppm the asymptotic maximum S'_{MDR} ~ 18 was approached. Polymer dissolution into the flow was assessed both from the initial increase in fractional drag reduction towards the eventual downstream DR_{line}, which provided a characteristic development distance $(L/D)_{dr}$, and from the final disappearance of the red-dyed strands of injected solution, which provided a characteristic dissolution distance $(L/D)_{pv}$. At all C_{line} and C_{inj}, the $(L/D)_{dr}$ ~ 100 were generally less than $(L/D)_{pv}$ ~ 100 to > 400. Finally, pipe flow drag reduction observations were "scaled out" to flat plate boundary layers by assuming that Type B drag reduction prevails identically in both internal and external flows, with $U_{CL} = U_o$, the free-stream velocity, and $R^+ = du_\tau/\nu$, with d thickness and u_τ friction velocity at the boundary layer trailing edge L. Scaleout of the present data to the classical HMS Highburton Polyox W301 ejection trials of Canham et. al (1971) suggests that the present Polyox P309 might have provided their reported drag reductions at roughly 1/20 th their polymer expenditure rate per unit of hull wetted area.

See more of this Session: Fundamental Research In Transport Processes II
See more of this Group/Topical: Engineering Sciences and Fundamentals

A FUNDAMENTAL CONVECTIVE-DIFFUSION BASED MODEL TO EXPLAIN INTERFACIAL MIXING IN MULTIPRODUCT PIPELINES

Anirudh R. Patrachari, Oklahoma State University, Stillwater, Oklahoma
Arland H. Johannes, Oklahoma State University, Stillwater, Oklahoma

Biography

Anirudh R. Patrachari is a PhD Student in Chemical Engineering at Oklahoma State University in Stillwater, Oklahoma. He holds a M.S. degree in Chemical Engineering from Oklahoma State University and a Bachelor's degree in Chemical Engineering from University of Mumbai in Mumbai, India. He is a member of various honor societies and professional organization, including The International Society of Automation and the honor society, Phi Kappa Phi. His current research interests include investigation of axial dispersion in petroleum pipelines, process modeling and simulation, and turbulent flow.

Arland "AJ" Johannes is Professor of Chemical Engineering at Oklahoma State University in Stillwater, Oklahoma. He holds a PhD in Chemical Engineering from the University of Kentucky, an M.S.E. in Civil Engineering from West Virginia University and a B.S. in Physics and Chemistry from Illinois State University. He has won numerous awards for teaching and research. His current research includes the study of transmix in pipelines and viscous heating in non-Newtonian fluids.

Abstract

Interfacial contamination of batches of refined petroleum products transported through multiproduct pipelines is a long standing problem in the petroleum industry. Determination of the extent of mixing depends on various factors including the fluid properties, operating conditions and the flow regimes. The convective-diffusion model along with a fictional parameter introduced by Sir G. I. Taylor, known as the axial dispersion coefficient is the commonly employed methodology to evaluate longitudinal mixing. However, the model is formulated with a uniform turbulent or laminar flow across the entire flow field. This study argues that the assumption of a uniform flow regime leads to inaccuracies in predicting the extent of contamination at the interface. This study incorporates both the turbulent and the viscous effects in predicting the interfacial contamination volume. Taylor's analysis of dispersion coefficients has been reviewed, and an improved model combining the weighted effects of the viscous boundary layer contributions and the turbulent core contributions to axial dispersion is proposed. The proposed model provides a possible explanation for the underlying physics of flow through straight pipes without pipe fittings and other devices. A comparison of the model predictions with literature data demonstrates the importance of considering both turbulent and viscous effects to achieve better overall prediction accuracy.

Nomenclature

Letters

%AAD	Percent absolute average deviation
%D	Percent deviation

a, b	Arbitrary constants
a_δ	Contribution coefficient
d	Pipe diameter
D	Molecular diffusion coefficient
f	Fanning friction factor
k	Interface cut-off multiplier
K_E	Effective axial dispersion coefficient
K_T	Turbulent axial dispersion coefficient
K_V	Viscous axial dispersion coefficient
L	Pipe length
n	Exponent in Prandtl's power law velocity profile
R	Pipe Radius
Re_d	Reynolds number, based on pipe diameter
Re_δ	Boundary layer Reynolds number
S	Contaminated length
$\bar{u}$	Average flow velocity across the pipe cross-section
u_0	Pipe centerline velocity
u_δ	Velocity at the boundary of the viscous region
$u_{\delta, corr}$	Corrected velocity at boundary of the viscous region
u_τ	Friction velocity
ΔP	Pressure drop

Symbols

δ	Viscous region thickness
δ_v	Viscous sub-layer thickness
ε	Eddy diffusivity
ν	Kinematic viscosity
ρ	Density

Introduction

Batching refers to the process of transporting batches or slugs of petroleum products successively through a pipeline. Large quantities of different types and grades of refined petroleum products are transported from the refinery to the distributers by this operation. Such pipelines commonly referred to as the "products pipelines" or "multiproduct pipelines". Swiftness of transport and operating economics make batching and multiproduct pipelines extremely popular in the petroleum transportation industry and almost all transportation pipelines today operate in this manner.[1]

The effectiveness of batching is limited by the operating conditions in the pipeline. An interface slug forms between the batches of products transported one behind the other without any separating medium. This mixed or contaminated slug has to be isolated into a separate storage tank at the distributor end as any mixing on either side might contaminate the pure products. The extent of contamination depends upon the operating conditions of the pipeline and the materials being transported; and it is of great economic significance as any error in estimation might lead to an entire batch/tank of product being off the desired specification. To provide an idea of the extent of mixing, Fowler and Brown[2] reported an estimate of a few hundred barrels would be contaminated sue to mixing within a 700 mile long 8" line, based on an analysis by Smiley on the Phillips production pipeline constructed in 1931.

Contrary to perception, contamination volume (or length) does not increase linearly with pipe length. The presence of a mixed interface between the two pure product slugs acts as a buffer and reduces the rate of dispersion at longer pipe lengths.[2] Consider an example of two products A and B being pumped sequentially through a pipe. An interface forms between the two products as soon as the second product (B) enters the line. Initially, the interface grows at a steady rate as the two pure products are in direct contact with each other. However, once the interface grows to a considerable extent, the pure products are no longer in contact with each other but are in contact with an interface slug consisting of a mixture of A and B. This interface can be considered to be a less contaminated version of B coming in contact with pure B (and vice versa), resulting in a reduction in the contamination growth rate.

A number of models for predicting contamination (transmix) growth rate have been developed.[1-20] Most of these studies build on Taylor's concept of axial dispersion due to convective-diffusion.[17,18] Taylor[17,18] suggested that velocity profile plays a significant role in the dispersion of matter along the axial direction and proposed a Fickian type diffusion model to determine the extent of axial dispersion.

Taylor[17] visualized axial dispersion to be a two-step process. The first step consists of the contaminant concentration profile getting skewed into a parabolic shape (in laminar flow) due to the velocity profile, with diffusion across the cross section smoothing out the radial concentration variations in the second step (Figure 1). Taylor[17,18] solved the convective diffusion equation using the velocity profile across the cross section of the flow system and obtained an estimate for the longitudinal dispersion coefficient to be used in the Fickian model. A parabolic velocity profile was employed for estimations in laminar flow[17] and the universal velocity profile was employed for turbulent dispersion calculations.[18]

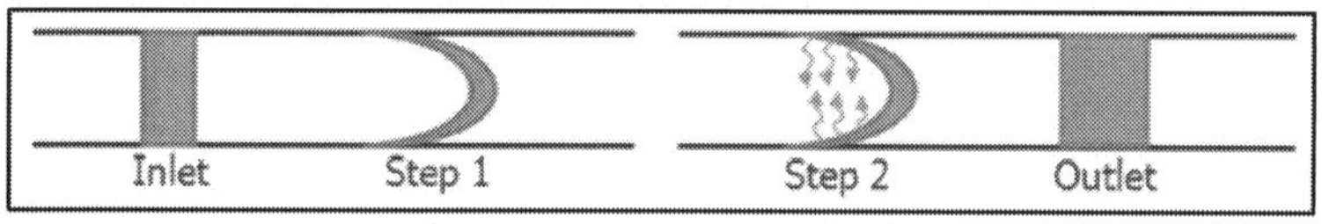

Figure 1. Taylor's proposed diffusion mechanism

Taylor's results for axial dispersion estimations[17,18] are in many cases inconsistent with the experimental observations[1,19-21], with the predictions being consistently lower than the experimental results. Taylor's analysis[17,18], though, is excellent in predicting the general trend in the process. Austin and Palfrey[1] showed through quantitative analysis that Taylor's expression[18] under-predicts axial dispersion coefficient by a huge extent at low Reynolds numbers (≤ 5000). Similar comparisons at higher Reynolds number (≥ 100000) showed better agreement with experimental data, though Taylor's analysis[18] still under-predicts the extent of contamination. Austin and Palfrey[1] suggested the inadequacies of Taylor's analysis[18] in incorporating "tailing" effects as the reason for the slight deviation at high Reynolds numbers.

Aris[3] pointed out that Taylor's expressions[17,18] were obtained assuming Reynolds analogy for mass and momentum transport, and hence, are only valid for systems with Schmidt number of 1. Aris[3] then corrected Taylor's expression[17,18] by adding molecular diffusivity to the axial dispersion coefficient and improved the predictions for low Schmidt number systems (gases). For Schmidt numbers higher than 1, which is the case with liquids, the effect of molecular diffusivity is negligible and hence can be ignored. This was shown to be true at higher Reynolds numbers using the data assembled by Levenspiel[11], and Flint and Eisenklam.[9,22] Many researchers[4,12,13,19,21-23] suggest Taylor's[17,18] use of the universal velocity profile and concentration independent diffusion coefficients as reasons for this eccentricity. Tichacek[19] evaluated the convective diffusion equation with different velocity profiles and showed that even a 3% variation in the velocity profile resulted in a deviation of around 50% in the dispersion coefficient. Flint and Eisenklam[22] reworked Taylor's analysis[17,18] for turbulent diffusion for axial dispersion in gases and pointed out similar observations.

The improved theoretical models however still exhibit similar inconsistencies in accurately predicting the experimental results, as observed in the original. Many different investigators have proposed empirical[1,5,6,10,15,16] and semi-empirical models[8,20,21] for estimating the axial dispersion coefficient. However, like their theoretical counterparts they are not consistent in replicating the experimental observations. Most of the empirical equations (except Austin and Palfrey[1]) are found to be accurate only under certain operating conditions, specifically for systems similar to the experimental setup used in the respective investigations and hence are not reliable. The empirical equations proposed by Austin and Palfrey[1] were based on their analysis of a large database consisting of laboratory scale, pilot scale and field data that were available in open literature and hence can be considered to be comprehensive and trustworthy. Empirical models perform better than the theoretical equations and are therefore the preferred analysis routine in most cases. They however, fail to explain the actual phenomenon governing axial dispersion. This investigation proposes a mathematical relation based on theory that explains the dynamics of axial dispersion systems. The proposed model is expected to improve on, or, in the least preserve the prediction accuracy of empirical models.

Theoretical Analysis

A theory that has been suggested to explain the deviations in Taylor's analysis[18] for low Reynolds number turbulent flow is the inattention to viscous effects in the near wall region. Austin and Palfrey[1] were one of the first researchers to suggest this hypothesis. They point to the similarity between the viscous boundary layer thickness equation (Equation 1) and their proposed empirical relation for axial dispersion in turbulent flow (Equation 2) to bolster their claim. They also argued the significance of viscous effects by pointing out to experimental results showing an increase in contamination length with increasing pipe diameter at the same Reynolds number.

$$\delta_V \propto d\,\mathrm{Re}_d^{-0.875} \tag{1}$$

$$S \propto \left(d \cdot L\right)^{0.5} \mathrm{Re}_d^{-0.9} \tag{2}$$

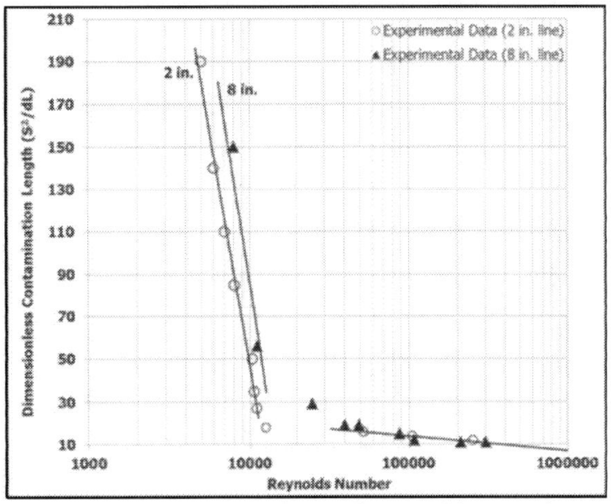

Figure 2. Dimensionless contamination length vs. Reynolds number (Adapted from Austin and Palfrey[1])

Austin and Palfrey[1] identified two distinct dispersion curves for Reynolds numbers in the turbulent regime (Figure 2). In the low Reynolds number region, the rate of axial dispersion decreases rapidly with increasing Reynolds number and beyond a certain critical value it flattens out and appears to be only slightly dependent on the changes in the Reynolds number. Based on this observation, they proposed the occurrence of two distinct mechanisms in the turbulent regime and suggested viscous dispersion to be one of the mechanisms.

The above observation can be explained by considering the effects of the viscous boundary layer. For fully developed laminar flow, the viscous boundary layer covers the entire cross section of the pipe. This changes as the flow enters into transition and then turbulence. The viscous layer thickness decreases as a hyperbolic function of the Reynolds number, with the pipe diameter as an asymptote for Reynolds number at the onset of turbulence and zero as the other asymptote at infinite Reynolds number. This phenomenon is very similar to the observations provided by Austin and Palfrey[1] and hence can be taken as a possible governing parameter in axial diffusion estimations.

Chatwin[7,21], and Chikwendu and Ojiakor[8] proposed multilayer models for flow in open channels, with each layer having different diffusion characteristics and velocity profiles. Chatwin[21] solved the convective diffusion equation using a linear velocity profile for the viscous sub-layer region (non-dimensional wall units, $y^+ \leq 5$) and the log law profile in the turbulent core region to obtain a 20% increase in Taylor's turbulent dispersion equation. However, the equations still under-predict the overall dispersion coefficient and Chatwin[21] attributed this to the incompetency in modeling the viscous effects.

Taylor[17,18] proposed independent theoretical equations to calculate axial dispersion coefficient in the laminar and turbulent regimes. As stated earlier, these expressions perform very well for fully laminar flow and very high Reynolds number turbulent flows; however their accuracy drops in the transitional low Reynolds numbers and moderate Reynolds number flow regimes, indicating a need for some sort of corrections in these regions. This investigation introduces a combined axial dispersion model, which unifies the two independent equations proposed by Taylor into a single equation so that both viscous and turbulent effects are collectively considered for axial dispersion estimations.

The model proposed in this paper derives its inspiration from the shear stress distribution across the cross section in turbulent flow and the pressure drop analysis proposed by Ergun[24] for turbulent flow through packed beds. The shear stress for turbulent flow can be segregated into two components a) shear due to laminar or viscous effects and b) shear due to turbulent Reynolds stresses (Figure 3). The viscous effects entirely constitute for the total shear at the wall, and its contribution falls rapidly as one proceeds away from the wall towards to turbulent core. One the other hand, the contribution of turbulent effects to the total shear stress is zero at the wall as all fluctuations go to zero at the wall, and they increase with increasing distance away from the boundary, reaching a maximum value at the pipe center. Thus, the total shear stress experienced by the flow depends on both the viscous and turbulent contributions. This has been widely accepted and confirmed through many experimental investigations.[25]

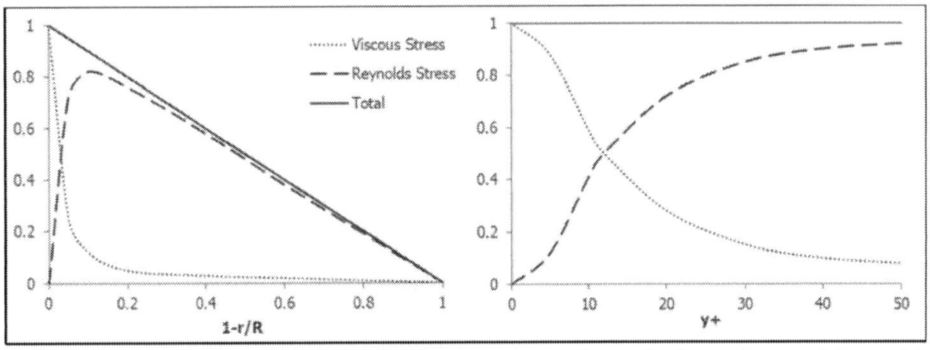

Figure 3. Profiles for viscous and Reynolds shear stresses in turbulent channel flow (Adapted from Pope[25])

Taylor's theories[17,18] indicate a clear dependence of axial dispersion on the shear stress distribution across the flow cross section. Hence, it would not be farfetched to suggest that viscous contributions will have a major effect on axial dispersion coefficient. The effects of viscous stresses would be pronounced in the lower Reynolds number region as the viscous boundary layer would be thicker in this region. Therefore, neglecting the viscous contributions would have a proportionally higher impact on calculations in this region as compared to higher Reynolds numbers. This behavior fits very well with the observation of higher deviation of Taylor's analysis[18] from experimental data in the low Reynolds number region.[1] Based on this analysis, Taylor's expression[18] can also be expected to under-predict axial dispersion coefficient for higher Reynolds number, though to a lesser extent. This also agrees with observations presented by other investigators.[1,7,9,21,22]

This shortcoming in the analysis can be resolved by selectively adding the contributions of the viscous and the turbulent stresses. Similar concepts have also been employed for pressure drop calculations. Pressure drop, like axial dispersion is also an outcome of shear due to a solid wall; and therefore, the distribution of shear stress across the flow cross-section would have a similar effect on both these quantities. Ergun[24] proposed a relation for pressure drop taking both the viscous and the turbulent effects into account (Equation 3).

$$\frac{\Delta P}{L} = a\nu\bar{u} + b\rho\bar{u}^2 \tag{3}$$

The first term in Equation 3 represents the contribution of viscosity to pressure drop, while the second term indicates the pressure drop resulting due to the momentum effects. On similar lines, the overall axial dispersion coefficient across a cross section of the flow can be determined as:

$$K_E = a_\delta K_V + (1 - a_\delta) K_T \tag{4}$$

Where, a_δ is a constant representing the contributions of the viscous dispersion coefficient (K_V) to the overall axial dispersion coefficient (K_E). The value for a_δ being determined based on the thickness of the region where viscous effects contribute significantly to K_E.

Model Equations

The expressions proposed by Taylor[17,18] for axial dispersion coefficient calculations in laminar and turbulent flows are used in the current model. The laminar dispersion coefficient is used to represent the contributions of the viscous stresses to the effective dispersion coefficient (Equation 5a) and the turbulent axial dispersion equation is employed to model turbulent dispersion (Equation 5b).

$$K_V = \frac{R^2 u_0^2}{192D} \quad (5a)$$

$$K_T = 10.1 R u_\tau = 3.57 d\bar{u}\sqrt{f} \quad (5b)$$

The contribution coefficient (a_δ) in Equation 4 is the fraction of the flow area dominated by viscous characteristics and is expressed as a ratio of the area occupied by the viscous region to the total cross sectional area. The viscous region occupies the annular space extending from the pipe walls to a certain thickness into the flow and the viscous layer thickness (δ) expressed in terms of the non-dimensional wall units (y^+) can be written as:

$$\delta = \frac{y^+ \nu}{u_\tau} \quad (6)$$

A value of $y^+ = 5$ is commonly accepted for the viscous sub-layer thickness.[25] However, the region denoted by this value does not necessarily cover the entire region of appreciable viscous effects. It is a commonly accepted fact that the turbulent and viscous contributions to the total shear stress are nearly equal at y^+ of about 12 and that viscous contributions are known to be appreciable even at higher values of y^+ (about 30~50).[25] Therefore, a value of $y^+ = 30$, which is generally accepted as the thickness of the buffer layer, is selected for the current model.[25]

Taylor's expression[17] for K_V is based on the entire flow cross-section and the maximum or centerline flow velocity. This expression therefore needs to be modified before it can be used for calculating the dispersion in the viscous region of the cross section. Equation 5b can be rewritten using the maximum velocity in the viscous region of the flow as:

$$K_V = \frac{R^2 u_\delta^2}{192D} \quad (7)$$

Where, u_δ refers to the velocity at the boundary of the viscous region ($y^+ = 30$). This velocity can be calculated from Prandtl's power law velocity profile (Equation 8).

$$u_\delta = u_0 \left(\frac{\delta}{R}\right)^{1/n} \quad (8)$$

Equation 5a with the above modification would then represent the dispersion coefficient for fluid flowing with a velocity (u_δ) through a pipe of diameter (2R). This equation would therefore overestimate the value of K_V and requires modification to appropriately represent the dispersion in the annular region. This can be done by correcting the velocity from Equation 8 using a ratio of the boundary layer Reynolds number (Re_δ) to the overall flow Reynolds number (Re_d). Equation 8 can be therefore be revised to give:

$$u_{\delta,corr} = u_0 \left[\frac{(2n+1)(n+1)}{4n^2}\right]\left(\frac{\delta}{R}\right)^{\frac{2}{n}+1} = \frac{\bar{u}}{2}\left[\frac{(2n+1)(n+1)}{2n^2}\right]^2 \left(\frac{\delta}{R}\right)^{\frac{2}{n}+1} \quad (9)$$

The u_δ in Equation 7 can then be replaced with $u_{\delta,corr}$ (Equation 9) to represent the final form of the dispersion coefficient due to the viscous effects. The derivation for Equation 9 is provided as an Appendix to this paper.

$$K_V = \frac{R^2 u_{\delta,corr}^2}{192D} \quad (10)$$

Taylor's[18] expression for axial diffusion in turbulent flow can be used in its original form in Equation 4, with the friction factor (f) being calculated using the Swamee-Jain[26] or any other commonly used equations. It can be argued that using the entire pipe diameter in the equation for turbulent dispersion would result in over predicting the turbulent contributions to the overall dispersion. However, this increase would be negligible considering that the viscous region thickness (δ) would be much smaller compared to the diameter and hence would have no appreciable effect on the calculated effective dispersion coefficient.

Statistical Analysis and Experimental Data

Lab scale, pilot scale, and field experiment data from published literature[1,2,6,10,14-16] were assembled (Table 1) for validating the performance of the proposed dispersion model. The data consisted of axial dispersion measurement for pipes 0.124" to 40" in diameter and 5 ft. to 2270000 ft. in length using water-water (with solute) and kerosene-gasoline interfaces. The information on physical properties such as density and viscosity for the fluids employed were in most cases specified by the investigators; and for the rest, information obtained from Perry's Chemical Engineers' Handbook[27] were used.

Table 1. Experimental data matrix

Source	Diameter (in.)	Length (ft.)	Reynolds Number	# of Data Points
Austin and Palfrey[1]	20 – 40	355 – 1230000	128000 - 970000	12
[1]Fowler and Brown[2]	0.124 – 0.313	5 – 105.6	2220 - 19800	27
[1]Birge[6]	4 – 12	140000 – 2270000	67000 - 538000	8
[1]Hull and Kent[10]	8 – 10	72000 – 970000	24000 - 300000	6
[1]Smith and Schulze[16]	2 – 12	2500 – 2033803	14160 - 591000	62
[1]Sjenitzer[14]	8 – 10	3000 – 246000	8000 - 52000	7

[1]Part of the data taken from Austin and Palfrey[1]

Most sources express the axial dispersion data in terms of measured contamination lengths at the pipe exit. Therefore, the axial dispersion coefficient obtained from the proposed model is converted into its contamination length form for comparison. This can be done by using the effective axial dispersion coefficient (Equation 4) to solve the transient Fick's diffusion equation to give:

$$S = k\sqrt{\frac{K_E L}{\bar{u}}} \tag{11}$$

Where, "S" is the contaminated length, "L" is the pipeline length, and "k" is a constant multiplier dependent on the interface concentration cut-off points ($k \approx 6.57$ for the case of 1-99% interface cut-off). The contamination lengths reported in literature are based on different interface concentration cut-off basis in each investigation. They therefore need to be converted to a common basis for impartial comparisons. Austin and Palfrey [1] performed such corrections using a ratio of inverse error function of the differences in purities concerned. Similar calculations were performed in the current investigation using an interface cut-off concentration of 1-99%.

Another item of note concerns data for the 2" pipeline.[1,15] Smith and Schulze[15] employed a loop-like facility for their experiments, which consisted of a complicated network with multiple bends and pumps. The effect of the bends and pumps in particular cannot be quantified and therefore the data obtained from this source cannot be directly used for comparison with the model calculations. However, this issue can be resolved by correcting the reported lengths to be applicable to straight pipes. Smith and Schulze[16] provide a set of Reynolds number dependent contamination length curves standardized in terms of the pipe diameter. These contamination curves can be used to approximate the contamination lengths for shortened pipe lengths, which would be more indicative of mixing in straight pipes. In spite of this, some of the experimental results from the loop experiments were included in the analysis in this work to demonstrate the effects of pumps and fittings on contamination.

The values for K_E (and contaminated length) were evaluated using the model equations (5 – 10) for each of the 122 data points listed above. These values were then converted to percent deviation (%D) form using Equation 12a. Owing to the large quantity of experimental data, it is impossible to list the comparison for each case individually. Therefore, all comparisons have been expressed in terms of percent absolute average deviation (%AAD), segregated based on pipe diameters.

$$\%D = \frac{Estimated - Experimental}{Experimental} \times 100 \quad (12a)$$

$$\%AAD = \frac{1}{n}\sum |\%Deviation| \quad (12b)$$

Results and Discussion

In accordance with the procedures outlined by previous researchers, the physical properties of equi-volume mixtures of the leading and trailing fluids were employed for determining the Reynolds number and for subsequent calculations. Table 2 lists the %AAD values for the calculations using the proposed model and the Austin and Palfrey[1] model.

Table 2. Comparison of model performance for pipes of varying diameters

Diameter (in.)	0.124	0.313	2	4	6	8	10	12	20	[1]40
%AAD	64.2	62.5	12.4	8.9	4.1	12.2	7.7	6.3	22.2	0.6
%AAD[1]	8.3	8.2	28.8	7.2	1.8	15.0	7.9	4.1	17.3	1.7

[1]Only 1 data point

The proposed model provides reasonable agreement with experimental results for all diameters 2" and above except for the 20" pipe case. However, the proposed model overestimates the amount of contamination in the 0.124" and 0.313" pipelines. One possible explanation for this discrepancy might be the pipe lengths used for the experiments. Taylor[17,18] conducted the axial dispersion analysis assuming that an equilibrium exists between the axial convective and radial diffusive transport. Such a condition requires that the interface reside in the pipe for a certain amount of time as specified by the equation (Equation 13) given below for laminar flow systems.[17] Taylor's equations[17,18] are only valid for interface residence times greater than Equation 13 and the equation would over predict axial dispersion coefficient otherwise.

$$T = \frac{R^2}{3.8^2 D} \quad (13)$$

Levenspiel[11] by means of statistical analysis showed that Taylor's laminar dispersion equation over-predicts the results for Fowler and Brown's laminar flow data[2] because of the short pipe lengths employed in the experiments. A similar argument can be proposed for axial dispersion in turbulent flow. It is assumed that Equation 13 also holds for turbulent flow, albeit with the diffusion coefficient being replaced with eddy diffusivity (ε). It is common knowledge that eddy diffusivity increases with increasing Reynolds number and therefore the value for T would decrease with increasing Reynolds number. The proposed model being a variance of Taylor's analysis, the above proposed theory should probably also apply to this model. Then, for a pipe with constant dimensions, the model prediction must improve considerably with increasing Reynolds numbers. Such a trend is observed in the model predictions for the turbulent flow data provided by Fowler and Brown.[2] Figure 4 shows the percent average deviation for the experimental data obtained in a 20.6 ft. long 0.313" diameter glass tubes.

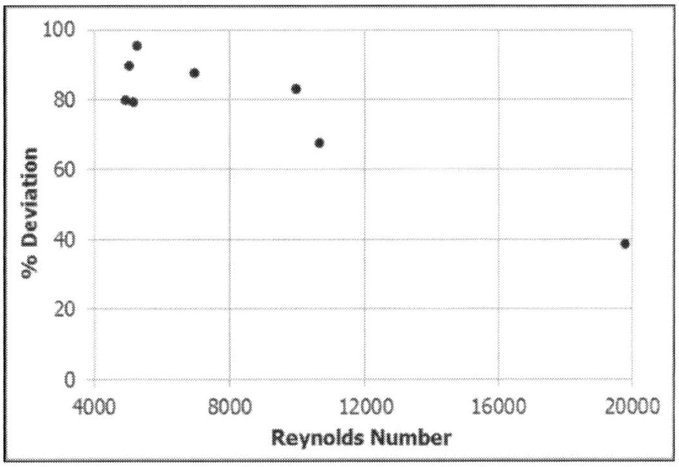

Figure 4. % Deviation vs. Reynolds number for 20.6 ft. long, 0.313" diameter pipe

Based on the above analysis, it can be concluded that the 0.313" experimental setups are possibly shorter than that required for the proposed model to be applicable. Similar explanation could be provided for the 0.124" lines. However this could not be quantitatively verified due to lack of reported measurements from experiments with the same pipe length.[2] Many investigators have proposed transient dispersion models for such scenarios.[28-30] In these models, the axial dispersion coefficient is determined as a fraction (time dependent) of Taylor's axial dispersion coefficient calculated using Equation 5. Such an analysis was out of the scope of this investigation; though, trial and error estimations were carried out and dispersion fractions of 0.35 and 0.31 were determined for the 0124" and 0.313" cases respectively. Table 3 lists the %AAD with these modifications and also compares the performance of the proposed model against Taylor's turbulent dispersion model.[18]

Table 3. Comparison of proposed model performance against Taylor's model[18]

Diameter (in.)	0.124	0.313	2	4	6	8	10	12	20	[1]40
%AAD	5.7	8.2	12.4	8.9	4.1	5.4	7.7	6.3	22.2	0.6
%AAD [18]	71.3	49.0	58.5	11.6	6.0	20.6	25.9	6.3	23.6	0.7

[1]Only 1 data point

The improvements are easily discernable for all the diameters under investigations. It should be noted that Taylor's model[18] under-predicted the axial dispersion coefficient for all the data points, with higher deviations observed at lower Reynolds number. The proposed model on the other hand is fairly unbiased over a large Reynolds number range (Figure 5).

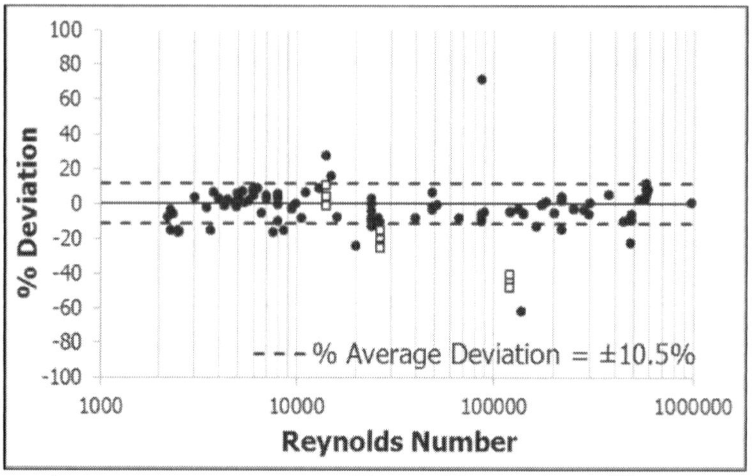

Figure 5. Model performance over a range of Reynolds numbers □ Smith and Schulze [15,16] 2" diameter loop experiment data, ● Experimental data from all other sources

The total %AAD for all the 122 data points including the data from the loop experiments of Smith and Schulze[15,16] falls around 10.5%. While, the %AAD excluding the loop experiment data is 9.3%. Apart from the loop experiment, some data points in the higher Reynolds number range (around 150,000) also lie outside the 10.5 %AD area. These belong to the 20" pipe experiments, indicating the poor performance of the model for this system. The proposed model performs fairly evenly through the entire range, with around 70 % of the data points falling within the %AD of 9.5% (greater than 80% excluding loop experiment results and 20" pipe results). The model predictions for the loop experiments are represented using transparent squares in Figure 5. An obvious trend can be observed in the loop experiment predictions, with the model performance declining progressively as the Reynolds number deviates from about 13,000. The presence of pipe fittings in the experimental setup would result in enhanced mixing at higher Reynolds numbers.

It is noted that the model consistently underperforms for the 20" experiments. An analysis of the calculations indicates that the contribution of the viscous dispersion coefficient for this system is an order of magnitude less than the turbulent coefficient, which is reasonable considering the high Reynolds numbers employed. This suggests that Taylor's[18] predictions for turbulent dispersion coefficient themselves are lower than those indicative of the experimental results. The Austin and Palfrey[1] empirical model does not improve the predictions by a huge margin and they attribute the disparity to the presence of complicated pipe networks and valve switching at the inlet and exit. This seems a reasonable suggestion, as complicated pipe networks, fittings, pipeline debris, and valve switching tend to increase the flow friction factor, which will result in increased turbulent dispersion coefficient (Equation 5b). This hypothesis was tested for the 20" pipe diameter case by increasing the wall roughness, which results in higher friction factors. It was found that increasing the wall roughness did result in higher turbulent dispersion coefficient and therefore higher contamination lengths, thereby reducing the percent deviation from the experimentally measured results. An equivalent roughness value of 0.03 ft. corresponding to riveted steel reduced the %AAD to around 6.3%, with reduced bias.

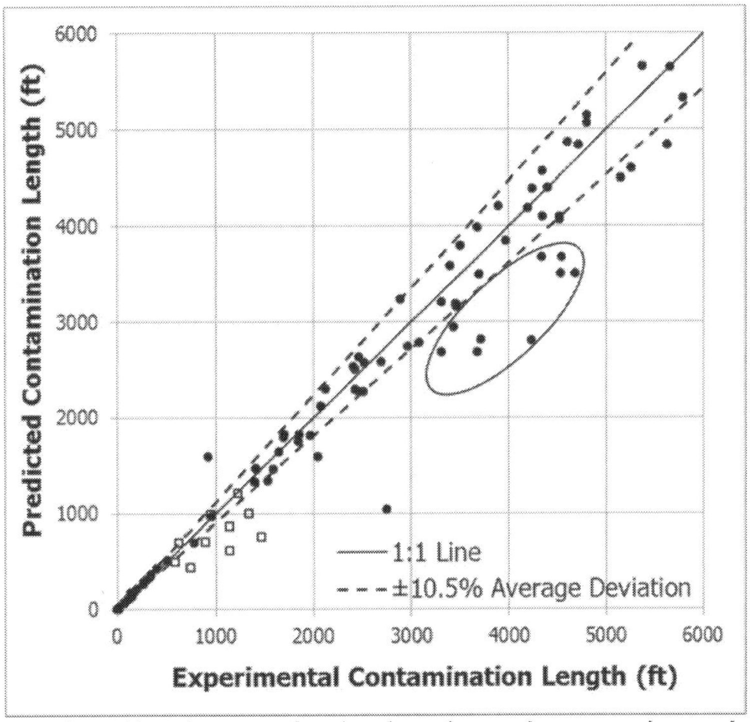

Figure 6. Comparison of predicted contamination lengths against experimental results □ Smith and Schulze [15,16] loop experiment data, ● Experimental data from all other sources

Figure 6 shows a comparison of the predicted contamination lengths and the experimentally measured lengths. As discussed in the previous sections, contamination lengths increase with increasing pipe lengths. Therefore, Figure 6 as an indicator of the model's performance for increasing pipe lengths. The data is fairly evenly distributed across the 45° line (Slope ~ 0.928 for all data points) indicating that the model is reasonably impartial for increasing pipe lengths. The experimental results from the loop experiments are clearly higher than the predicted values and therefore lie outside the 10.5% deviation lines. Expect for the data for the 20" pipeline (circled out in Figure 6) and a couple of other points (possibly outliers), most lie close to the area covered by the 10.5% deviation lines indicating the robustness of the model. A slope greater than 1 in Figure 6 suggests that the model under-predicts the extent of contamination for higher pipe lengths. However, this is because of the discrepancies with the 20" pipeline data, excluding which results in a more favorable slope of 0.983.

Conclusions

This investigation was aimed at providing an insight into effects of the viscous forces on axial dispersion, while suggesting possible methodologies for incorporating the same in a mathematical form. Experimental results and comparisons with the theory of wall bounded shear flows reveals that viscous effects play a significant role in axial mixing during batched fluid transportations. A straightforward model has been proposed for predicting the axial dispersion coefficient (extent of contamination) for flow through straight pipes of circular cross-section without considering the effects of pipe fittings and transport devices. The model builds on the basic theory of dispersion (Taylor[17,18]) by combining the convective-diffusion transport equation and boundary layer theory.

One possible technique for integrating the viscous and the turbulent effects into a single equation is to selectively add the contributions from the two effects. The resultant increase in the axial dispersion coefficient was more evident in the transitional, low Reynolds number turbulent flow and it decreased rapidly at higher Reynolds numbers. This is very much in concurrence with the experimental observations. A value of 30 non-dimensional wall units has been proposed for approximating the thickness of the viscous region. It should be noted that the proposed model is a variation of Taylor's equations[17,18] (Equation 5) and the assumptions in Taylor's analysis[17,18] also apply to the current model. Therefore, the model would not be applicable for short pipe lengths where the convective and diffusive transports have not reached equilibrium.

As expected, the model performs much better than Taylor's turbulent dispersion analysis for low Reynolds number turbulent flows and its performance is also comparable with that of the Austin and Palfrey[1] empirical model. The predictions exhibit little or no bias in the contamination predictions over the range of Reynolds number and pipe dimensions investigated, signifying the robustness of the model.

References

1. Austin, J. E. and J. R. Palfrey (1964), "Mixing of miscible but dissimilar liquids in serial flow in pipelines," *Proceeding of Institution of Mechanical Engineers,* 17 (15), pp. 377-395.
2. Fowler, F. C. and G. G. Brown (1943), "Contamination by successive flow in pipe lines," *Transactions of the American Institute of Chemical Engineers,* 39, pp. 0491-0516.
3. Aris, R. (1956), "On the dispersion of a solute in a fluid flowing through a tube," *Proceedings of the Royal Society of London, Series A, Mathematical and Physical Sciences,* 235 (1200), pp. 67-77.
4. Atesman, K. M., L. V. Baldwin and R. D. Haberstroh (1971), "The dispersion of matter in turbulent pipe flows," *Journal of Basic Engineering,* 93 (4), pp. 461-474.
5. Aunicky, Z. (1970), "The Longitudinal Mixing of Liquids Flowing Successively in Pipelines," *The Canadian Journal of Chemical Engineering,* 48, pp. 12-16.
6. Birge, E. A. (1947), "Contamination control in products pipelines," *The Oil and Gas Journal,* 46, pp. 176.
7. Chatwin, P. C. (1973), "Calculation illustrating effects of the viscous sub-layer on longitudinal dispersion," *Quarterly Journal of Mechanics and Applied Mathematics,* 26 (4), pp. 427-439.
8. Chikwendu, S. C. and G. U. Ojiakor (1985), "Slow-zone model for longitudinal dispersion in two-dimensional shear flows," *Journal of Fluid Mechanics,* 152, pp. 15-38.
9. Flint, L. F. and P. Eisenklam (1970), "Dispersion of matter in transitional flow through straight tubes," *Proceedings of the Royal Society of London, Series A, Mathematical and Physical Sciences,* 315, pp. 519-533.
10. Hull, D. E. and J. W. Kent (1952), "Radioactive tracers to mark interfaces and measure intermixing in pipelines," *Industrial and Engineering Chemistry,* 44 (11), pp. 2745-2750.
11. Levenspiel, O. (1958), "Longitudinal mixing of fluids flowing in circular pipes," *Industrial and Engineering Chemistry,* 50 (3), pp. 343-346.
12. Maron, V. I. (1978), "Longitudinal diffusion in a flow through a tube," *International Journal of Multiphase Flow,* 4, pp. 339-355.
13. Rachid, F. B. F., J. H. C. d. Araujo and R. M. Baptista (2002), "Predicting mixing volumes in serial transport in pipelines," *Journal of Fluids Engineering,* 124, pp. 528-534.
14. Sjenitzer, F. (1958), "How much do products mix in a pipeline?," *Pipeline Engineer,* 30, pp. 31-34.
15. Smith, S. S. and R. K. Sulze (1948), "Interfacial mixing characteristics of products in products pipeline - Part 1," *The Petroleum Engineer,* 19 (13), pp. 94-104.

16. Smith, S. S. and R. K. Sulze (1948), "Interfacial mixing characteristics of products in products pipeline - Part 2," *The Petroleum Engineer,* 20, pp. 330-337.
17. Taylor, G. I. (1953), "Dispersion of soluble matter in solvent flowing slowly through a tube," *Proceedings of the Royal Society of London, Series A, Mathematical and Physical Sciences,* 219 (1137), pp. 186-203.
18. Taylor, G. I. (1954), "The dispersion of matter in turbulent flow through a pipe," *Proceedings of the Royal Society of London, Series A, Mathematical and Physical Sciences,* 223 (1155), pp. 446-468.
19. Tichacek, L. J., C. H. Barkelew and T. Baron (1957), "Axial Mixing in Pipes," *AIChE Journal,* 3 (4), pp. 439-442.
20. Udoetok, E. S. and A. N. Nguyen (2009), "A disc pig model for estimating the mixing volumes between product batches in multi-product pipelines," *Journal of Pipeline Engineering,* 8, pp. 195-201.
21. Chatwin, P. C. (1971), "On the interpretation of some longitudinal dispersion experiments," *Journal of Fluid Mechanics,* 48, pp. 689-702.
22. Flint, L. F. and P. Eisenklam (1969), "Longitudinal gas dispersion in transitional and turbulent flow through a straight tube," *Canadian Journal of Chemical Engineering,* 47 (2), pp. 101-106.
23. Krantz, W. B. and D. T. Wasan (1974), "Axial dispersion in the turbulent flow of power-law fluids in straight tubes," *Industrial and Engineering Chemistry Fundamentals,* 13 (1), pp. 56-62.
24. Ergun, S. (1952), "Fluid flow through packed columns," *Chemical Engineering Progress,* 48 (2), pp. 89-94.
25. Pope, S. B. *Turbulent lows*: University Press, Cambridge; 2000.
26. Swamee, P. K. and A. K. Jain (1976), "Explicit equations for pipe-flow problems," *Journal of Hydraulics Division,* 102 (5), pp. 657-664.
27. Perry, R. H. and D. W. Green. *Perry's Chemical Engineers' Handbook.* 7 ed: McGraw-Hill; 1997.
28. Gill, W. N. (1966), "Analysis of axial disperison with time variable flow," *Chemical Engineering Science,* 22, pp. 1013-1017.
29. Gill, W. N. and R. Sankarasubramanian (1970), "Exact Analysis of Unsteady Convective Diffusion," *Proceedings of the Royal Society of London, Series A, Mathematical and Physical Sciences,* 316 (1526), pp. 341-350.
30. Soltanieh, M. and S. Sadraei (1991), "A simplified model for prediction of Time-dependent axial disperison coefficient," *Chemical Engineering Science,* 46 (1), pp. 301-305.

Appendix

Viscous Region Velocity Correction

The flow velocity at the edge of the viscous region can be determined using Prandtl's power law velocity profile as:

$$\frac{u_\delta}{u_0} = \left[1 - \frac{r_\delta}{R}\right]^{1/n} = \left(\frac{\delta}{R}\right)^{1/n} \tag{14}$$

Applying the Reynolds number correction as discussed in the **Model Equations** section yields the viscous layer velocity to be used in the viscous dispersion equation (Equation 10).

$$\frac{u_{\delta,corr}}{u_0} = \frac{u_\delta}{u_0}\frac{\mathrm{Re}_\delta}{\mathrm{Re}_d}$$

Where Re_δ refers to the Reynolds number based on the boundary layer thickness and Re_d is the standard flow Reynolds number based on the pipe diameter.

$$\therefore \frac{u_{\delta,corr}}{u_0} = \frac{u_\delta}{u_0}\frac{(\delta u_\delta / v)}{(d\bar{u}/v)}$$

$$\therefore \frac{u_{\delta,corr}}{u_0} = \frac{u_\delta}{u_0} \cdot \frac{\delta u_\delta}{d\bar{u}}$$

Substituting Equation 14 into the above expression then yields:

$$\frac{u_{\delta,corr}}{u_0} = \frac{1}{2}\left(\frac{\delta}{R}\right)^{1+\frac{1}{n}}\left(\frac{u_\delta}{\bar{u}}\right) \tag{15}$$

We know that the average velocity can be related to the centerline velocity using Equation 16.

$$\frac{\bar{u}}{u_0} = \frac{2n^2}{(2n+1)(n+1)} \tag{16}$$

Using this relation in Equation 15 results in:

$$\frac{u_{\delta,corr}}{u_0} = \frac{1}{2}\left(\frac{\delta}{R}\right)^{1+\frac{1}{n}}\left(\frac{u_\delta}{u_0}\right)\left(\frac{(2n+1)(n+1)}{2n^2}\right)$$

Re-substituting Equation 14 in the above equation gives:

$$\frac{u_{\delta,corr}}{u_0} = \frac{1}{2}\left(\frac{(2n+1)(n+1)}{2n^2}\right)\left(\frac{\delta}{R}\right)^{1+\frac{2}{n}}$$

$$\therefore \frac{u_{\delta,corr}}{\bar{u}} = \frac{1}{2}\left(\frac{(2n+1)(n+1)}{2n^2}\right)^2\left(\frac{\delta}{R}\right)^{1+\frac{2}{n}} \tag{17}$$

Viscous Region Area Fraction

The three variables involved in calculating the effective axial dispersion coefficient are the viscous region area fraction (a_δ), Taylor's viscous (K_V) and turbulent (K_T) axial dispersion coefficients. The turbulent dispersion coefficient is used without any changes. However, the viscous dispersion coefficient is corrected to apply just to the viscous region instead of the entire pipe cross-section. The relationship required to calculate the area fraction to be used in Equation 4 is listed below.

$$a_\delta = \frac{A_\delta}{A_d} = \left[1 - \frac{\delta}{R}\right]^2 \tag{18}$$

(295d) Flow Characterization In a Taylor-Couette Bioreactor In the Presence of Mobile Scaffolds

Jian Qiao and Chi-Hwa Wang, Department of Chemical and Biomolecular Engineering, National University of Singapore, Singapore, Singapore

In recent years, Taylor vortex flow is commonly encountered in many practical applications such as reaction, filtration and extraction. Taylor vortices can be formed by subjecting a viscous liquid to a shear stress within an annular space between two rotating cylinders and a stationary bottom surface. Taylor-vortex flow can be applied in a bioreactor to culture cells that are seeded in degradable porous scaffolds. In this study, the effect of floating scaffolds on the vortices inside a Taylor-Couette system is investigated using a Couette apparatus with an aspect ratio (Γ) of 6.06 and a radius ratio (η) of 0.613. A laser-based non-invasive technique known as Particle Image Velocimetry (PIV) is applied to characterize the fluid flow field in the system. The results are substantiated by FLUENT simulations for better understanding on the flow pattern of the fluid. We investigate on the specific problems such as how the vortex structure can be affected and how the flow field is developed with the presence of floating scaffolds inside a wide gap and a low aspect ratio of the Taylor-Couette bioreactor. The interactions between scaffolds of various shapes and the vortex structure are also studied in order to understand the influence of each scaffold towards the overall flow field inside the Taylor-Couette bioreactor system.

See more of this Session: Fundamental Research In Transport Processes II
See more of this Group/Topical: Engineering Sciences and Fundamentals

(295e) Effect of Side and Bottom Wall Temperature Non-Uniformities On Multiplicity of Steady Solutions for Mixed Convection In A Square Cavity

Tanmay Basak, Chemical Engineering, Indian Institute of Technology Madras, Chennai, India and **Madhuchhanda Bhattacharya**, Independent Researcher, Chennai, India

Galerkin penalty finite element method with 28x28 bi-quadratic elements has been used to investigate the role of bottom and side wall temperature non-uniformities on the existence of multiple steady solutions for mixed convective flows within a square enclosure with thermally insulated moving top wall. Here, non-uniformities of bottom and side wall temperatures are varied such that side walls always remain colder than bottom wall with continuity of temperature maintained at the bottom corners. Variations of bottom and side wall temperature non-uniformities are parametrically represented by thermal aspect ratio (0 \leqslant \Theta \leqslant 1), which is the ratio of temperature gradient of side walls to the total temperature gradient of bottom and side walls. Side walls remain isothermally cold with entire temperature gradient confined at the bottom wall at \Theta = 0. On the other hand, side walls drive entire temperature gradient at \Theta = 1, where bottom wall remains uniformly hot. Any intermediate non-uniformity of bottom and side walls in proportion to \Theta and 1 - \Theta, respectively can be simulated by varying \Theta from 0 to 1. Present work traces possible steady solution branches in the parameter space of \Theta at constant Grashof and Prandtl numbers (Gr =105, Pr = 10) and for three different Reynolds numbers as Re = 1, 10 and 100 representing almost natural to truly mixed convective flows. Bifurcation diagrams of various steady branches at Re = 1, 10 and 100 are presented in terms of trajectories of average bottom wall Nusselt number along \Theta. It has been shown that there exists a critical thermal aspect ratio, which differentiates the region of unique and multiple solutions. Steady multiple solutions appear for \Theta > \Theta_{cr}, where \Theta_{cr} = 0.626, 0.6446 and 0.218 for Re = 1, 10 and 100, respectively. This shows that a certain non-uniformity of side wall temperature is essential for existence of multiple solutions. We have obtained six steady branches for Re = 1 resulting in eleven steady solutions at \Theta = 1. Number of steady branches decreases to three for Re = 10 and 100. Correspondingly, number of steady solutions at \Theta = 1 also decreases to five at Re = 10 and 100. Transition of flow and temperature profiles along each branches are also shown, which illustrate a sequence of formation, merging, splitting, disappearance an reappearance of secondary cells near bottom wall of the enclosure along each trajectory.

See more of this Session: Fundamental Research In Transport Processes II
See more of this Group/Topical: Engineering Sciences and Fundamentals

(295f) Laminar MIXED Convection IN A DOUBLE LID-Driven Square Cavity Totally FILLED with POROUS Material

Maximilian S. Mesquita, Departamento de Engenharias e Cincias Exatas, Universidade Federal do Esprito Santo UFES/CEUNES, So Mateus, ES, Brazil, Marcelo J.S. De Lemos, Departamento de Energia - IEME, Instiruto Tecnolgico de Aeronutica - ITA, So Jos dso Campos, Brazil, Fernando F. Fachini, Laboratrio de Combusto e Propulso - LCP, Instituto Nacional de Pesquisas Espaciais - INPE, Cachoeira Paulista, SP, 12630-000, Brazil and Luzia Aparecida Tofaneli, Departamento Regional da Bahia, Centro Integrado de Manufatura e Tecnologia., SENAI -, Salvador, BA, Brazil

A numerical investigation is conducted to study the effect of porous material characteristic on heat and mass simultaneous transport in a lid-driven cavity filled with a fluid saturated porous medium. The left and right vertical walls of the cavity are insulated while both the top and bottom horizontal walls are kept at constant but different temperatures. The top horizontal wall is moving on its own plane at a constant speed while all other walls are fixed. The governing equations are solved using finite volume approach along with the SIMPLE algorithm. Numerical solutions are obtained for a wide range of parameters. The effects of the controlling parameters involved in the heat transfer and mass characteristics are studied in detail.

See more of this Session: Fundamental Research In Transport Processes II
See more of this Group/Topical: Engineering Sciences and Fundamentals

(295g) A Simple Approach of Mass Transport On Gas-Liquid Interface by Statistical Rate Theory

Xiaohua Lu, Rong An, Hanbing He, Yuanhui Ji, Yudan Zhu, Suoying Zhang and Gulou Shen, State Key Laboratory of Materials-oriented Chemical Engineering, Nanjing University of Technology, Nanjing, China

In many multi-phase chemical processes, mass transport at the interface is the rate-determining step, and the corresponding transport behavior can affect the system performance directly. Thus the effects of the interface on the material performance or the process have to be taken into account. The description for the transport behavior at solid-liquid interface has successfully provided theoretical guidance for chemical process. However, for the description of the mass transport at the gas(vapor)-Liquid interface, it is difficult to calculate the interfacial area accurately, which is a key factor for building accurate transport rate model. The interfacial area at gas(vapor)-liquid interface can be generally assumed as liquid cross-sectional area, which is different from the actual interfacial area. Therefore, it is necessary to introduce a nanobubble theory to predict the interfacial area at vapour-liquid interface.

With Statistical Rate Theory (SRT), the mass transport at gas(vapor)-liquid interface can be simplified as that of the two processes linking in gas(vapor)-phase, which is, mass transport from liquid into gas (vapor) phase and then that from gas(vapor)-liquid interface into bulk phase. It is necessary to utilize the nanobubble theory for predicting the interfacial transport area, as a key factor for mass transport at gas (vapor)-liquid interface. Afterwards, with the nanobubble theory and mass conservation, the transport area at the gas(vapor)-liquid interface can be determined by considering nanobubble diameter related to interfacial tension which is one of the main factors to control the transport rate. The nanobubble size distribution function can be also determined with the theory to further predict the transport rate with experimental kinetic data. Using the chemical potential gradient $\Delta\mu$ as driving force and then with parametric regression, the transport behavior at interface could be analyzed to determine the rate-controlling factor.

See more of this Session: Fundamental Research In Transport Processes II
See more of this Group/Topical: Engineering Sciences and Fundamentals

(358a) Chemical Looping Combustion Development and Perspectives

Florent Guillou[1], Thierry Gauthier[1], Hlne Stainton[2] and Sbastien Rifflart[2], (1)IFP Energies Nouvelles, Solaize, France, (2)Total Gaz Energies Nouvelles, Paris, France

Chemical Looping Combustion is presently considered as one of the most promising CO_2 capture technology in terms of efficiency penalty and cost reduction. *TOTAL* and *IFP Energies nouvelles* are collaborating for the development of chemical looping combustion applications. In order to develop appropriate process solutions several batch and continuous experiments were conducted on a wide range of fossil feedstocks (liquid, gaseous, solid) over the past years in order to serve as the basis of design for a life size pilot on coal. Several tests have been conducted at different temperatures ranging between 800°C and 1000°C with several oxygen carriers.

Further to these encouraging results the technology is being taken to the next development stage via a several MW pilot. This larger pilot is at engineering stage, and it is planned to start construction on a selected industrial site by early 2012.

This paper presents the most recent results of this joint development program and also some projections for CLC process efficiencies as well as technico-economical evaluation for a full scale industrial plant in comparison with alternative routes including post combustion and oxy-combustion. Finally the perspective of the full technology development plan is introduced.

See more of this Session: Chemical Looping Processes II
See more of this Group/Topical: Energy and Transport Processes

(358b) Design of a Chemical-Looping Combustion System Using Solids and Gaseous Fuels

Suman K. Hajra, TBA, Pittsburgh, PA and Thomas J. O'Brien, Computational Science Division, National Energy technology Laboratory, Morgantown, WV

Climate changing is a critical concern through out the world and it is due mostly to the burning of fossil fuels in the power plants. Nearly 80% of energy production is from the combustion of fossil fuels and green house gas such as carbon dioxide emits into the atmosphere on the combustion of all fossil fuels. Chemical looping combustion (CLC) is a unique technology for carbon capture and is for the separation of carbon dioxide from nitrogen and excess air. In this process fuel is burnt using oxygen rich metal oxide. The oxygen carrier particles circulates in a reactor system composed of a fuel reactor, where oxygen carrier particles are reduced by the fuel, and an air reactor, where the particles are regenerated to the oxidized state. Method of determination of coal molecular formula will be presented. Coal combustion and heat of formation will be discussed. Mass and energy balance of the CLC reactors will be presented for both solid and gaseous fuels. A simplified mathematical relation is developed for solids circulation in the CLC reactors. Parameters such as nature of oxygen carrier, carrier oxidation state, carrier conversion, and fuel rate will be discussed.

See more of this Session: Chemical Looping Processes II
See more of this Group/Topical: Energy and Transport Processes

(358c) Gasification and Combustion of Biomass In Chemical Looping Combustion Over A Ni Based Oxygen Carrier

Mohammad R. Quddus, Chemical and Biochemical Engineering, The University of Western Ontario, London, ON, Canada, Mohammad M. Hossain, Chemical Engineering, King Fahd University of Petroleum & Minerals, Dhahran, Saudi Arabia and Hugo I. de Lasa, Chemical & Biochemical Engineering, University of Western Ontario, London, ON, Canada

This study reports the development of a Ni based oxygen carrier and the investigation of an Integrated Gasification (IG) and Combustion of solid biomass as a fuel for Chemical-Looping Combustion (CLC). The aim is to develop a highly reactive, stable and fluidizable oxygen carrier (OC): a cobalt (Co) promoted Ni/γ-Al_2O_3. Every OC impregnation step was accomplished using incipient wetness impregnation. The OC was modified first by stabilizing the γ-Al_2O_3 with a 5% La[1] addition. The resulting La-γAl_2O_3 was then impregnated with 1wt% Co. Finally, a total of 20% Ni was loaded to the OC precursor via successive impregnations of 2.5wt% Ni. The synthesized OC displays excellent reducibility (94%) and stability, as demonstrated in successive TPR/TPOs and H_2 chemisorptions. TPR profiles for this OC consistently show essentially complete reduction at 700 °C. These results are encouraging given that they demonstrate that the oxygen in the OC is very close to the expected amount. These results also show that the OC reducible phase is primarily NiO with small $NiAl_2O_4$ fractions. The addition of Co helps the formation of easily reducible NiO species minimizing Ni-support interactions and formation of $NiAl_2O_4$[2,3,4]. It was also found that the small TPR peak assigned to $NiAl_2O_4$ remained unchanged over repeated TPR/TPO cycles, with this suggesting that $NiAl_2O_4$ always remains as a small fraction in the OC. Pulse chemisorption results were valuable to further confirm the stable behavior of the sample in consecutive redox cycles, showing a well dispersed state of Ni crystals as well as the absence of Ni crystal agglomeration.

The prepared OC was used for integrated gasification and combustion of biomass in the CREC Fluidized Riser Simulator[5]under the expected conditions (turbulent fluidized bed and high temperature) of an industrial scale CLC unit. Glucose was employed as the model compound for biomass. The product analysis of IG-CLC (Integrated Gasification-Chemical looping combustion) experiments shows that H_2O, CO_2, CO and H_2 are the main products under the studied reaction conditions. A small amount of CH_4 was also detected with no glucose found in the formed products. The combustion of the gasified biomass was further confirmed by comparing products composition with the ones of thermal gasification. Results showed 7 % CH_4, 35 % CO and 44 % H_2 and 15 % CO_2 for thermal gasification and 7 % CH_4, 8 % CO and 0.5 % H_2 and 86 % CO_2 for the IG-CLC with OC/glucose close to the stoichiometric amounts for complete combustion. It can be concluded from these findings that during the IG-CLC most of the chemical species produced during gasification are burnt in the subsequent combustion step involving both product gases and OC. As a result, the CO_2 fraction in the IG-CLC product significantly increases. The only exception is CH_4, whose composition remains at almost the same level, both in thermal and IG-CLC runs. Thus, it appears that the CH_4 formed during the gasification step neither reacts with the oxygen carrier to give CO_2 and H_2O nor reforms with H_2O to produce H_2 and CO. In order to further investigate this matter, subsequent IG-CLC experiments were carried out using both an excess and a deficient stochiometric amount of OC under the same reaction conditions. It was shown, that the product analysis of these experiments displays significant variations of H_2, CO and CO_2 with the varying amounts of the OC., and with the CH_4 remaining close to a constant value. Thus, product selectivity using a supported metal oxide is closely associated with the degree of the reduction of the oxygen carrier. At the beginning of the reaction (or short contact times), the fully oxidized oxygen carrier favors the total oxidation of the gasification products forming CO_2 and H_2O. As the reaction proceeds, the partially reduced oxygen carrier starts catalyzing the reaction products ($CO+H_2$). Consequently, the methane in the oxygen deficient runs decreased due to the reforming reaction while CO and H_2 remained unreacted. Regarding the stability of the oxygen carrier, repeated combustion-regeneration experiments show that NiO/Al_2O_3 particles display excellent reactivity and stability during the cyclic process.

References: (1) Hossein M., Quddus M., de Lasa H., "Reduction Kinetics of Lanthanum Modified Ni/γ-Al2O3 Oxygen Carrier for CLC" Ind.Eng.Chem.Res. Vol.49, N 21, 11009-11017 (2010); (2) Hossein M., de Lasa H., "Reduction Kinetics of Co-Ni/γ-Al_2O_3 involved in a chemical-looping combustion cycles" Chem.Eng.Sci.,65 98-106 (2010); (3) Hossein M., de Lasa H., "Reduction and oxidation kinetics of Co-Ni/Al_2O_3 oxygen carrier involved in a chemical looping combustion cycles", Chem.Eng.Sci, Vol.65 98-106 (2009); (4) Hossein M., de Lasa H., "Chemical Looping Combustion (CLC) for Inherent CO2 Separations- A Review", Chem.Eng.Sci., Vol.63, 4433-4451 (2008); (5) de Lasa H., "Riser Simulator for Catalytic Cracking Studies", US Patent 5,102,628,(1992).

See more of this Session: Chemical Looping Processes II
See more of this Group/Topical: Energy and Transport Processes

(358d) Application of the Law of Additive Reaction Times to the Analysis of Oxidation Kinetics In the Air Reactor of a Chemical Looping Combustion System

Asad H. Sahir[1], JoAnn S. Lighty[1] and Hong Yong Sohn[2], (1)Chemical Engineering, University of Utah, Salt Lake City, UT, (2)Department of Metallurgical Engineering, University of Utah, Salt Lake City, UT

Chemical Looping Combustion (CLC) is one of the promising technologies for CO_2 capture which has the potential to reduce the energy penalty. A CLC system consists of two interconnected fluidized beds which act as a fuel and an air reactor, with a re-circulating metal/metal oxide system acting as the carrier of oxygen between the two beds.

Insights to the design and optimization of an air reactor in a CLC process could be obtained by analyzing the oxidation kinetics by the Law of Additive Reaction Times. The law offers an approximate closed form solution for a first order system in which structural changes on reaction can be neglected. It takes into account the pore diffusion of gaseous species in the interior of the porous pellet. The scenario is in contrast to the usage of shrinking-core models for porous solids, which assumes the occurrence of the reaction along a well-defined sharp reaction interface that progresses into the solid as the reaction occurs. This law provides a convenient and realistic means for analyzing air reactors in CLC systems for the oxidation process.

In this presentation, the results of applying The Law of Additive Reaction Times to the reported experimental data on the oxidation kinetics of copper in an air reactor of a CLC system will be discussed. It will also highlight the utility of a criterion provided by the law which helps in understanding the relative importance of chemical reaction, pore diffusion and external mass transfer that make up the overall reaction process.

See more of this Session: Chemical Looping Processes II
See more of this Group/Topical: Energy and Transport Processes

(358e) Comparison of Bench-Scale Fixed-Bed Units for Chemical-Looping Combustion

Zhiquan Zhou, Monica Navarro and George M. Bollas, Department of Chemical, Materials & Biomolecular Engineering, University of Connecticut, Storrs, CT

Abstract:

Theobjective of this work is to develop a detailed dynamic model for chemical-loopingcombustion in a fixed-bed reactor. The key properties desired of this model arethat it can be practically implemented in different fixed-bed reactors usingthe same reaction kinetics network. Experimental data from published fixed-bed chemical-loopingreactors are used for model development and validation.

Fixed-bedreactors are typically used to evaluate the properties of oxygen carriers inrepeated oxidation/reduction cycles [1-4]. Fuel and air are sequentially fed tothe reactor to simulate chemical-looping combustion oxidation and reductioncycles. Inert gas is fed between cycles to isolate reduction and oxidationsteps.

Comparisonof the results from various laboratories reveals inconsistencies regarding theapparent kinetics, and process time scales. Hence, it is useful to use a dynamicPFR model to perform a model-based comparison, including the most importantreactions, and time-space effects on volumetric flow, pressure drop, reactionextent and selectivity.

Reactionsinvolved in the chemical-looping reduction step are listed below:

$$2NiO + CH_4 \xrightarrow{Ni} CO_2 + 2H_2 + 2Ni \qquad CH_4 + NiO \xrightarrow{Ni} Ni + 2H_2 + CO$$

$$NiO + H_2 \rightarrow Ni + H_2O \qquad CO + NiO \rightarrow Ni + CO_2$$

$$CH_4 + H_2O \overset{Ni}{\leftrightarrow} 3H_2 + CO \qquad C + CO_2 \overset{Ni}{\leftrightarrow} 2CO \qquad CO + H_2O \overset{Ni}{\leftrightarrow} H_2 + CO_2$$

$$CH_4 + CO_2 \overset{Ni}{\leftrightarrow} 2H_2 + 2CO \qquad C + H_2O \overset{Ni}{\leftrightarrow} H_2 + CO \qquad CH_4 \overset{Ni}{\leftrightarrow} C + 2H_2$$

$$CH_4 + 2H_2O \overset{Ni}{\leftrightarrow} 4H_2 + CO_2$$

Kineticsfor the first four reduction reactions are taken from Iliuta et al. [5].Kinetics for steam reforming [6,7], dry reforming [8,9], CH_4pyrolysis [10,11], water gas shift [6,7], reverse methanation [6,7], reverseBoudounard reaction [12], and carbon gasification by steam [12] are taken from theliterature.

Figure 1presents model predictions of the profiles of H_2 and CO selectivity,oxygen carrier conversion and CH_4 to carbon conversion for the fixed-bedreactor of Ref [5]. An interesting observation is that there is an optimalpoint at relatively low reaction times

(~10 s) and in the mid-zone (~3.5 mm) ofthe reactor, where the process reaches to a H_2 and CO selectivityextremum, at high OC conversions (~65%) and negligible carbon formation (~0).

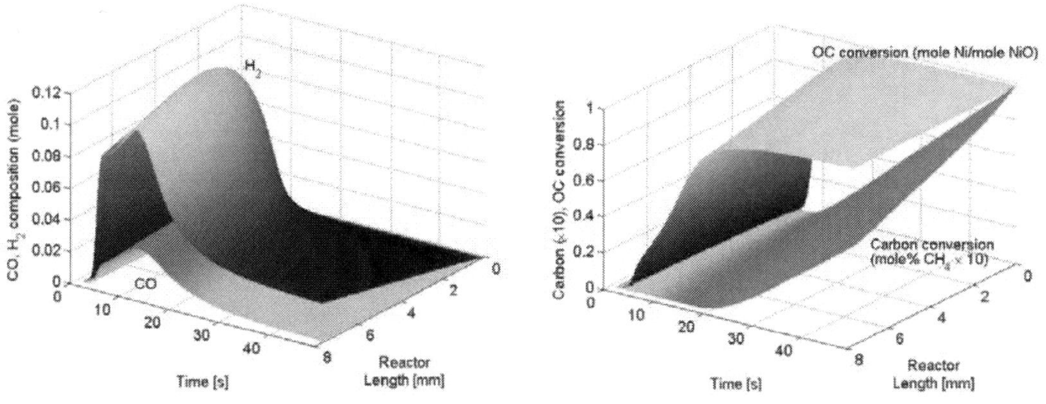

(a) (b)

Figure1:*Prediction of spatial and time profiles in chemical-looping reduction usingNi/Al_2O_3 as oxygen carrier at 800 C in a fixed-bedreactor; (a) H_2 and CO selectivity and (b) oxygen carrier and $CH_4$to C conversion.*

A thoroughreview of the kinetics of chemical-looping reactions is carried out, focusingmainly on Ni- and Fe-based oxygen carriers. Comparison of experimental data withthe results from the dynamic model allowed to identify the most importantchemical-looping reaction, to select a finite number of kinetic networks, andto obtain kinetic parameters capable of describing yields and selectivities ofdifferent experimental setups.

In thispresentation, the optimized kinetic network along with the fixed-bed model willbe presented. After implementing this model into different fixed-bed designs, acomparison of the predicted results and the experimental results for thesedifferent reactors will be illustrated and discussed.

Acknowledgement:NSF Career AwardNo. 1054718

References:

[1]Ryu H-J, Bae D-H, Jin G-T. *Korean Journal of Chemical Engineering.*2003;20(5):960-966.

[2]Adnez J, Garca-Labiano F, Diego LF de, et al. *Industrial & EngineeringChemistry Research.* 2006;45(8): 2617-2625.

[3]Corbella BM, Diego LF de, Garca-Labiano F, Adnez J, Palaciost JM. *Environmentalscience & technology.* 2005;39(15):5796-803.

[4]Ryden M, Lyngfelt a, Mattisson T, et al. *International Journal of GreenhouseGas Control.* 2008;2(1):21-36.

[5]Iliuta I, Tahoces R, Patience GS, Rifflart S, Luck F. *AIChE Journal.*2010;56(4):1063-1079.

[6]Xu J, Froment GF. *AIChE Journal.* 1989;35(1):88-96

[7]Hou K, Hughes R. *Chemical Engineering Journal.* 2001;82(1-3):311-328.

[8]Ginsburg JM, Pia J, El Solh T, et al. *Industrial & EngineeringChemistry Research.* 2005;44(14):4846-4854.

[9]Becerra AM, Iriarte ME, Castro-luna AE. *Reaction Kinetics and CatalysisLetters.* 2003;79(1):119-125.

[10]Borghei M, Karimzadeh R, Rashidi A, Izadi N. *International Journal ofHydrogen Energy.* 2010;35(17): 9479-9488.

[11]Snoeck J-W, Froment GF, Fowles M. *Journal of Catalysis.*1997;169(1):250-262.

[12]Snoeck J-W, Froment GF, Fowles M. *Industrial & Engineering ChemistryResearch.* 2002;41(17):4252-4265.

See more of this Session: Chemical Looping Processes II
See more of this Group/Topical: Energy and Transport Processes

(358f) Calcium Looping Process (CLP) for High Purity Hydrogen Production From Coal: Results From Sub-Pilot Scale Carbonator Testing

Nihar Phalak, Shwetha Ramkumar, Niranjani Deshpande, Yao Wang, William Wang and L.-S. Fan, The Ohio State University, Columbus, OH

CO_2 capture using calcium sorbents has wide ranging applications in post-combustion and pre-combustion systems, and has the potential of developing into a viable technology. In a typical calcium-based process, the carbonation of calcium oxide (CaO) by reaction with CO_2 from flue or fuel gas results in the formation of calcium carbonate ($CaCO_3$) and the reverse reaction (calcination) regenerates the CaO releasing a sequestration-ready pure stream of CO_2. Low cost of the naturally occurring CaO precursors and high temperature CO_2 capture capability make this process economically attractive for application in coal-based energy systems.

The Calcium Looping Process (CLP) is one of the clean coal technologies being developed for the production of hydrogen (H_2) and electricity from coal-derived syngas. It integrates the water-gas shift (WGS) reaction with in-situ carbon dioxide (CO_2), sulfur, and halide removal in a single-stage reactor. In the CLP, a regenerable calcium-based sorbent is used to chemically absorb CO_2, sulfur, and halide impurities from the synthesis gas during the production of hydrogen. The removal of CO_2 drives the water-gas shift reaction forward via Le Chatelier's principle, obviating the need for WGS catalyst and enabling the production of high-purity hydrogen. The spent sorbent is heated in a calciner to regenerate CaO for reuse in the process and to release a concentrated CO_2 stream that can be dried and sequestered. Overall CO_2 emissions from the process are essentially zero. The regenerated sorbent is reactivated in a hydrator, to reverse the effect of sintering and improve the recyclability of the sorbent, before being reintroduced into the hydrogen production reactor – carbonator. The high temperature operation of the three CLP reactors allows the utilization of high quality heat for electricity production. This significantly reduces the energy penalty associated with CO_2 capture.

Experiments conducted in a laboratory-scale fixed bed reactor have shown that production of high purity H_2 with in-situ CO_2 and contaminant removal is possible at low steam to carbon ratios (~1:1) and moderate pressures (3-5 atm). Based on these results, a sub-pilot scale fluidized bed carbonator has been designed and constructed at The Ohio State University (OSU). Enhanced hydrogen production has been investigated using synthetic syngas mixtures in this reactor and these results will be presented. A simplified model has also been developed to predict the performance of the combined reaction in the fluidized bed carbonator. The predictions of this model are compared to the experimental results.

See more of this Session: Chemical Looping Processes II
See more of this Group/Topical: Energy and Transport Processes

(358g) Thermodynamic Analyses of Solar Thermochemical CO2 Capture Via Calcination/ Carbonation Cycle

Leanne Matthews and Wojciech Lipinski, Department of Mechanical Engineering, University of Minnesota, Minneapolis, MN

The effect of solid and gas phase heat recovery on total energy required for an ideal solar-driven carbon dioxide capture cycle based on calcium carbonate/calcium oxide calcination/carbonation is studied. The effect of varying the carbonation and calcination reaction temperatures on total energy requirements is also investigated. The results are compared to values for minimum thermodynamic work of separation and work available from several common hydrocarbon fuels. For a calcination temperature of 1000°C and a carbonation temperature of 400°C, the addition of solid heat recovery decreased the cycle energy requirement by less than 10% for all considered carbon dioxide concentrations in the input gas stream. The addition of gas phase heat recovery reduced the energy requirements by up to 50%, with the largest decrease occurring at low carbon dioxide concentrations.

See more of this Session: Chemical Looping Processes II
See more of this Group/Topical: Energy and Transport Processes

(364a) Impact of Lignin Extraction and Cellulose III Conversion On the Enzymatic Digestibility of Corn Stover

Leonardo da costa Sousa, Shishir Chundawat, Nirmal Uppugundla, Bruce Dale and **Venkatesh Balan**, Chemical Engineering and Material Science, Great Lakes Bioenergy Center, Michigan State University, Lansing, MI

The benefits of delignification on plant cell wall digestibility by hydrolytic enzymes are well described in the literature. The presence of lignin increases non-productive binding of cellulases, resulting in enzyme inhibition and decrease in the sugar yields. Moreover, lignin acts as an anti-microbial agent, reducing cell viability and ethanol yields.

Recently, additional benefits by modifying the crystal structure of native cellulose (cellulose I) to cellulose III were also reported. After this modification, it was possible to double the rate of enzymatic hydrolysis of cellulose. However, this increase was observed for purified cellulose, native from algae (Cellulose Iα), which has a different crystal configuration than the cellulose native from plants (Cellulose Iβ). Therefore, it is not evident that this modification is possible to be obtained in plant cell walls and that it can represent the same kind of benefits observed for purified cellulose. From the literature, cellulose III is obtained by contacting cellulose I to supercritical ammonia or different types of amines in anhydrous conditions. Additionally, water is known to prevent the conversion of cellulose I to cellulose III, however the water concentration limit that allows cellulose III to be formed is not yet clearly defined. For biorefinery applications, plant biomass will contain a certain degree of moisture that could prevent stable cellulose III formation and therefore, it is important to verify in what extent is possible to convert native cellulose to cellulose III in plant biomass.

In this perspective, our goal was to test the benefits of cellulose III conversion under non-anhydrous conditions, using corn stover, and couple this modification with lignin extraction in a single stage process – Extractive AFEX (E-AFEX). These benefits were compared with the traditional non-extractive ammonia based pretreatment method - AFEX.

See more of this Session: Developments In the Pretreatment of Lignocellulosics for Bioconversion I
See more of this Group/Topical: Sustainable Engineering Forum

(364b) The Effects of Cationic Polyelectrolyte On the Pretreatment and Enzymatic Hydrolysis of Corn Stover

Shaowen Ji and Ilsoon Lee, Chemical Engineering and Materials Science, Michigan State University, East Lansing, MI

Lignocellulose is the most abundant renewable biomass and the potential substrate for ethanol production. The costly pretreatments and large consumption of enzymes are needed to reach high cellulose conversion. Lignin, as one of major cell wall components, is considered as the barrier to enzymatic hydrolysis by blocking cellulase from productively being adsorbed. The removal or effective modification of lignin is required for efficient pretreatment process. In this study, the effects of cationic polyelectrolyte, poly-(dimethyldiallylammonium chloride) (PDAC), on alkaline pretreated corn stover were investigated. Structural study by Scanning Electron Microscopy (SEM) imaging revealed particles morphology on cell walls of pretreated corn stover. These particles were confirmed to contain lignin by Confocal Laser Scanning Microscopy (CLSM) and Transmission Electron Microscopy (TEM) imaging. Chemical compositions of pretreated corn stover at different alkaline and polyelectrolyte concentrations were analyzed. Higher polyelectrolyte concentration in the pretreatment resulted in higher lignin retention. Evaluation of enzymatic hydrolysis showed the enhancement of cellulose and hemicellulose conversion after polyelectrolyte pretreatment. PDAC was proposed to facilitate cellulose disruption and lignin redistribution via its interaction with lignin, which significantly modified biomass morphological properties, such as increased porosity and improved accessibility of cellulose surface.

See more of this Session: Developments In the Pretreatment of Lignocellulosics for Bioconversion I
See more of this Group/Topical: Sustainable Engineering Forum

(364c) Novel Lignocellulosic Biomass Pretreatment Using a Modified Taylor-Couette Mixer for Biofuels and Biomaterials

Wei Wang and Ilsoon Lee, Department of Chemical Engineering and Materials Science, Michigan State University, East Lansing, MI

Bioethanol derived from lignocellulosic biomass can be a good candidate for energy alternative, not only because lignocellulosic biomass is abundant and sustainable, but also bioethanol production forms carbon cycle in a more timely fashion compared to fossil fuel, which leads to reduction of green gas emission. Lignocellulosic biomass pretreatment, which is one of the essential steps in bioethanol production, accounts for the majority portion of the overall bioethanol production cost. Hence an effective and efficient pretreatment method is highly demanded. Although this research area has been deeply investigated several decades ago, it has not attracted tremendous attention until recently. In our research, cost-effective and breakthrough biomass pretreatment was conducted with a novel modified Taylor-Couette mixing system, aiming to overcome the natural recalcitrance of lignocellulosic biomass in a short processing time and facilitate a better accessibility of enzymes to polysaccharides. Compositional analysis proved significant removals in both lignin and hemicellulose and rich glucan content left in the residue only after 2 minutes or less. Parameters like mixing speed, retention time, biomass loading and corresponding enzyme digestibility will be presented.

See more of this Session: Developments In the Pretreatment of Lignocellulosics for Bioconversion I
See more of this Group/Topical: Sustainable Engineering Forum

(364d) Modeling and Scale up of CO2-Water Pretreatment of Guayule Biomass

Ehsan Moharreri and J Richard Elliott, Chemical and Biomolecular Engineering, The University of Akron, Akron, OH

A carbon dioxide – water pretreatment process was developed for guayule biomass. The guayule biomass was mixed with water and then loaded to a reactor where it was exposed to a biphasic environment consisting of CO2 rich vapor phase and water rich liquid phase. Then, the biomass was enzymatically hydrolyzed and the sugar content was measured at the end. The effects of temperature, pressure, moisture content, accessible volume and holding time in the pretreatment step were investigated. One hundred and twenty (120) experimental runs were carried out in the temperature range of 145 C to 210 C, pressure was between 3.4 MPa to 34 MPa, the moisture content ratio was between 0.6 to 6, the accessible volume was between 0.2 to 0.1 L and the holding time was between 20 to 50 minutes. A model was fit to the experimental data based on dissolved CO2 in the liquid phase using CO2-H2O phase envelop and a temperature dependant constant. The model fits to one hundred and twenty (120) data points with an average absolute percent deviation(AAD%) of 13. Then, the same process was implemented to a pilot scale vessel with a 1 liter volume. The process was carried out in batch and semi-continuous methods and twenty (20) data points were produced to confirm the model. The model predicted the pilot plant runs with a 15% average absolute percent deviation (AAD%). In the lab scale of 190 C, 26 MPa, moisture ratio of 1.5, accessible volume of 0.22L, biomass load of 15 grams and 35 minutes in holding time resulted in 78% glucose yield and 66% hemicelluloses yield. In the pilot scale of 200 C, 10MPa, moisture ratio of 2, accessible volume of 0.88, biomass load of 60 grams and 35 minutes in holding time resulted in 78% total sugar yield.

See more of this Session: Developments In the Pretreatment of Lignocellulosics for Bioconversion I
See more of this Group/Topical: Sustainable Engineering Forum

(499g) Effect of Transition Metal Catalyst On Lignin Oxidation During Alkaline Hydrogen Peroxide Pretreatment

Zhenglun Li, Department of Chemical Engineering and Material Science, Michigan State University, East Lansing, MI and David Hodge, Chemical Engineering and Materials Science, Michigan State University, East Lansing, MI

Biomass lignocellulose can be used as feedstock for bioethanol production, and a pretreatment is necessary for achieving high ethanol yield and economical feasibility. As a method to increase enzyme digestibility of biomass by selective delignification, alkaline hydrogen peroxide (AHP) pretreatment has been proposed and studied. The lignin recalcitrance in lignocellulose biomass can be removed by H_2O_2 oxidation and the enzymatic hydrolysis efficiency of the polysaccharides can be greatly improved. However, a big issue with AHP pretreatment is the high cost of hydrogen peroxide during the pretreatment, and the selectivity of the oxidant need to be improved, i.e. to achieve highest pretreatment efficiency with low level of H_2O_2 loading.

In 1890s, Fenton discovered that iron ion can catalyze the oxidation of organic substrate by hydrogen peroxide. Since then, research has been done on the application of transition metal catalysts in catalyzing the phenol contaminant in wastewater and pulp bleaching. It is desired that catalyst can enhance selective electron transfer from lignin substrate to the oxidant and improve the selectivity of delignification reactions to unproductive side reactions.

See more of this Session: Developments In the Pretreatment of Lignocellulosics for Bioconversion I
See more of this Group/Topical: Sustainable Engineering Forum

(364f) Progress Toward Biological Pretreatment - Understanding Plant Cell Wall Degradation by Termite and White-Rot Fungi

Shulin Chen[1], Jing Ke[2], Dhrubojyoti D. Laskar[2] and Jijiao Zeng[2], (1)Department of Biological Systems Engineering, Washington State University, Pullman, WA, (2)Washington State University, Pullman, WA

In an effort developing more efficient and environmentally benign lignocellulosic biomass processing technologies, the authors' laboratory has been studying plant cell wall degradation process in termite and white-rot fungi. This presentation summarizes the results so far obtained from various experimental investigations of the lab. First, termite guts provide both aerobic and anaerobic environments, the former is required for lignin degradation. The termite guts also have close to neutral pH range, which suggest existence of lignin degrading enzymes that are different from that of fungi. The organization of the gut system allows the conversion of biomass in termite to occur as a sequential process in which the pretreatment of the biomass occurs in the foregut and midgut, whereas the hydrolysis of the cellulose occurs mainly in the hindgut. There are synergies between termite and its symbiont in deconstructing plant cell walls. The chewing action of the termite increases the surface area for enzymatic actions, the enzymes produced by the termite tissues modifies the bonds that forms the barriers for cellulase access and symbionts produce the enzymes that are mainly responsible for cutting the cellulose polymer to sugar units. The results shows the actual changes of lignin structure as the biomass material pass through the termite guts. The lignin structure changes are more modification than degradation in nature. Equally import is the deconstruction of the association of lignin with other components of plant cell walls such as hemicelluloses that occurs in the foregut and midgut. It becomes apparent by comparing with other biological systems that the enzyme systems and their actions in biological plant cell wall degradation systems are more complex than what was previously known. A generality has also been observed in that most of the biological systems studied tended to reduce G unit from the original lignin structure. Additionally, termite is a more effective plant cell wall degradation system than white rot-fungus because of the chewing action of the termite and the better-controlled environment in the termite guts. Termite is also more effective than the cow stomach as the latter lacks capability of lignin modification. Developing an artificial system that can surpass the natural biological system in major performance criteria may offer the ultimate solutions to lignocellulosic biomass processing and utilizations.

See more of this Session: Developments In the Pretreatment of Lignocellulosics for Bioconversion I
See more of this Group/Topical: Sustainable Engineering Forum

DESIGN CHALLENGES AND MODELING FOR AN EFFICIENT LIQUID HYDROGEN STORAGE TANK FOR AUTONOMOUS SYSTEMS

*Richard O. Stroman and Michael W. Schuette**
Naval Research Laboratory, Washington DC
**Global Strategies Group (North America) Inc., Washington DC*

Introduction

Small autonomous systems are rapidly adding capability and taking on new roles due to advances in navigation, controls, communications and sensing. Despite these improvements, however, the mission duration remains limited by the amount of useful energy available. There are two ways to provide more energy: raise the conversion efficiency or store more energy at the outset.

In recent years some vehicles have used fuel cells to convert fuel to electrical power for the propulsion plant and payload, because they tend to be more efficient than combustion engines in the 0.5 – 5 kW power range. One such vehicle is the Naval Research Laboratory's Ion Tiger, a 15.9 kg unmanned air vehicle (UAV) with a proton exchange membrane fuel cell (PEMFC) power system. PEMFCs convert H_2 to electrical power with about 50% efficiency, which in combination with the high specific energy of H_2 (33,400 Wh/kg) makes them ideal for long endurance systems.

The Ion Tiger demonstrated a 26 hr flight using 0.5 kg of gaseous H_2 (GH_2) stored at 5000 psi in a 22 L, 3.8 kg pressure vessel. [1] The pressure vessel and regulator together had to weigh less than 4 kg to satisfy the aircraft weight budget, which dictated a minimum storage efficiency of 12 wt% H_2 when combined with the goal of 24+ hr flight. Higher storage efficiency is necessary to increase the Ion Tiger flight endurance beyond 1 day, and our experience suggests that GH_2 is not a solution. Increasing the H_2 pressure would provide more H_2 storage in the same volume, but would require a disproportionately heavier pressure vessel and decrease storage efficiency. Furthermore, pressures above 5000 psi can require heavier materials not susceptible to H_2 embrittlement. Therefore, we opted to explore liquid H_2 (LH_2) as an alternative to GH_2.

The mass density of LH_2 is 62.6 g/L at 80 psi and 25 K, which is nearly three times larger than that of GH_2 at 5000 psi and 298 K (23.0 g/L). [2] The greater mass density of LH_2 could in principle provide nearly three times the storage efficiency of the present GH_2 system. LH_2 has been largely abandoned for most fuel cell applications because boil off cannot be eliminated entirely, so long term storage is problematic. For UAVs, however, LH_2 presents an interesting opportunity because they can be launched immediately after fueling and then continuously burn fuel in flight. The high mass efficiency of an LH_2 tank may enable unique small, long endurance electric UAVs. The major challenges when designing an LH_2 tank for an aircraft are to build a lightweight container with very low heat leak from ambient, and to accommodate transients in the fuel withdrawal rate.

Three aspects of the LH_2 storage tank design problem are described here: design of the insulation, design of the vent tube assembly (VTA), and the selection of an operating scheme. Validation of the insulation and VTA design models is also addressed.

Design Considerations

The tank design process was dominated by two (non-rigorous) optimization goals: maximize fuel storage and minimize venting. Fuel storage is directly related to the system weight – if the system weighs less, more of the weight budget can be allocated to fuel. Venting is related to the rate of heat leak from ambient – a lower rate of heat leak is less likely to raise tank pressure to the point at which it must be vented. Therefore the practical design goals were to minimize weight and minimize the heat leak rate.

There are additional reasons to minimize weight and heat leak rate. Lighter weight aircraft tend to require less propulsion power, be more maneuverable, and carry heavier payloads. With the rate of heat leak from ambient minimized, that the rate of boil off can be controlled over the widest possible range to match the withdrawal rate. Reducing the heat leak rate provides more control flexibility, "turn down" capability and endurance.

A dewar consisting of two nested vessels separated by a vacuum space is used in most high performance applications where exceptionally low heat leak is required. The vacuum eliminates conduction and convection heat transfer between the inner and outer vessels, and multi-layer insulation (MLI) is often added to reduce the rate of radiation heat transfer. Two heat conduction paths between the inner and outer vessels are largely unavoidable: the inner vessel supports and the structure(s) for filling and venting the inner vessel. The first design task is to determine the amount of MLI and the vacuum quality required by estimating the rate of radiation heat leak as a function of those parameters. The second is to design the vent tube by estimating the rates of conduction and radiation heat leak it introduces.

The fuel consumption rates for a typical Ion Tiger flight are shown in Table 1. Ideally the rate of LH_2 boil off would vary to match the rate of withdrawal, but because it is determined by the temperature difference between the LH_2 and ambient, it is nearly constant. To accommodate transients in the withdrawal rate one can allow the tank pressure to fluctuate, use a heater to "trim" the rate of boil off, or both. The third design task is to determine how large a heater is required, how often it should turn on, and how large an open space (filled with GH_2) above the LH_2 is required to damp transients in the withdrawal rate by allowing the tank pressure to vary.

Table 1. Fuel flow and heating rates required in a typical Ion Tiger flight.

Phase	H_2 Flow Rate [g/s]	Heat Rate, Vaporize LH_2 [W]	Heat Rate, Warm GH_2 [W]
Warm Up	8.33×10^{-4}	0.34	3.0
Climb Out	1.17×10^{-2}	4.78	41.4
Cruise	5.67×10^{-3}	2.32	20.1
Landing	8.33×10^{-4}	0.34	3.0

The total heat leak design goal was 90% of the heat rate required to boil off LH_2 at cruise power, shown as 2.32 W in Table 1. This limit kept the design goal feasible while providing some turn down capability. Early estimates of the relative rates of heat leak through each path led to the detailed heat leak budget in Table 2.

Table 2. Heat leak budget.

Path	Transfer Mode	Rate [W]
Supports	conduction	0.025
Insulation	radiation	1.889
VTA	conduction and radiation	0.175
Total	all	2.089

Design of the Insulation

A 1-D steady state heat transfer model was developed to determine how many radiation shields would be necessary to satisfy the heat leak budget. The model accounted for radiation heat transfer among layers of double aluminized Mylar insulation, using the Stefan-Boltzmann law [3-5] and absorbtivity data provided by the manufacturer. Conduction through the rarefied gas (assuming He at a pressure of 1×10^{-5} torr) [6] and through the Dacron net spacer were also included. Boundary conditions were the temperatures of the inner and outer vessels, assumed to be 25 K and 298 K, respectively. A nonlinear equation solver (the *fsolve* function in MATLAB) was used to determine which temperature profile through the stack of radiation shields would conserve energy, and the radiation flux through the stack. The number of MLI layers was increased until the desired radiation flux was reached.

Comparison to experimental data showed that the model consistently underpredicted the rate of heat transfer, and hence the number of MLI layers required. The discrepancy was attributed to insufficient fidelity in the spacer heat conduction model – the spacer reduces the insulation effectiveness by providing a conduction path, and the rate of conduction was underpredicted. The rate of heat conduction through the spacer layers depends on the contact area between the woven spacer and the radiation shields, and on the flux area, neither of which were well characterized. Attempts to fit these parameters were unsuccessful, so a published empirical model [7] was used instead. The empirical model predictions compared well with a test rig built for model validation (see Validation section) and confirmed that considerable effort is necessary to produce an accurate physics based MLI model. The empirical model should be sufficient for most designs.

Design of the Vent Tube Assembly

A 1-D steady state center differencing model was developed to estimate the rate of heat transfer through the VTA as a function of tube material, tube length, wall thickness and gas flow rate. The VTA heat leak model accounted for conduction through the tube wall, convection to the H_2 gas and radiation exchange between the tube and outer vessel (see Figure 1).

The tube and gas channel were broken into discrete volumes with the rates of heat transfer among them determined by Fourier's Law for conduction, Newton's Law of Cooling for convection and the Stefan-Boltzmann law for radiation. [3-5] Constant temperature boundary conditions were imposed at each end of the tube and at the gas entrance to the tube. A nonlinear equation solver (the *fsolve* function in MATLAB) was used to determine which tube and gas temperature profiles $T(x)$ conserved energy in each discretization. The resulting tube temperature profile was used to determine the heat flow rate into the inner vessel.

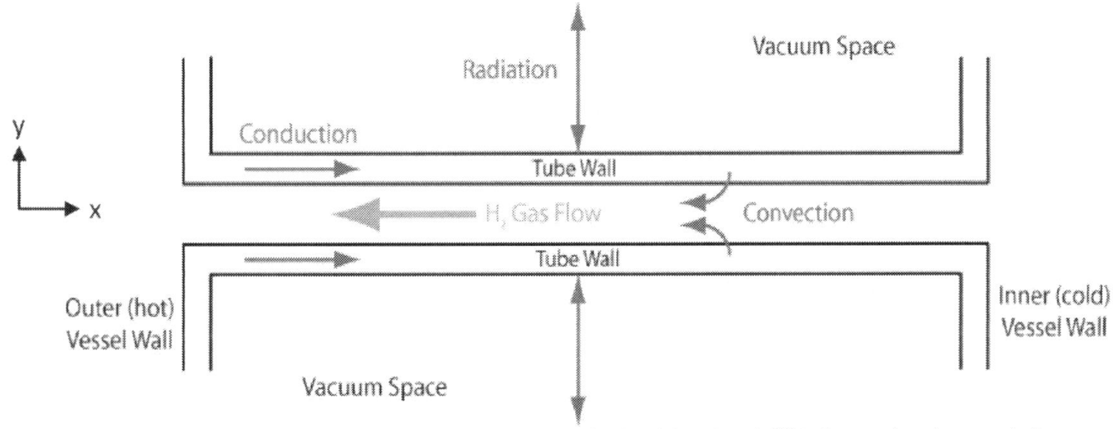

Figure 1. Heat transfer modes included in the VTA heat leak model.

The gas and tube thermal conductivities k varied considerably through the 25 – 298 K temperature range experienced by the tube, so the model interpolated among data from an online NIST database [8] to determine $k_{tube}(T)$ and $k_{H2}(T)$. The gas velocity (and hence the rate of convective heat transfer) was calculated in each discretization by accounting for the variation in gas density. The gas density was determined as a function of temperature and pressure by interpolating among data which functioned as an empirical equation of state. [2] It was assumed that the VTA was positioned in the vacuum space between the MLI and outer vessel, so it absorbed radiation from the outer vessel directly and via reflection from the MLI. This geometry was approximated by treating the VTA as if it were enclosed by a black body, i.e. assuming that the MLI was a perfect reflector. A typical radiation absorbtivity for each tube material was chosen from the literature, but the calculated temperature profiles showed only a small dependence on the value selected because the net rate of radiation heat transfer was small. The numbers of tube and gas discretizations were increased until the result ceased to vary with the number of discretizations.

An example result from the VTA heat leak model is shown in Figure 2, where the heat leak rate due to various tube lengths and gas flow rates have been plotted. Each curve in Figure 2 assumes the same tube material, inside diameter and outside diameter.

The importance of heat leak being intercepted by the gas through convection is clear – for each gas flow rate, there is a length beyond which the entire conduction and radiation heat leaks are intercepted. Given a maximum allowable rate of vent tube heat leak, expected gas flow rate and tube geometry, results such as in Figure 2 can be used to determine the minimum tube length necessary. It can also be used to explore different tube materials and combinations of inside diameter and gas flow rate.

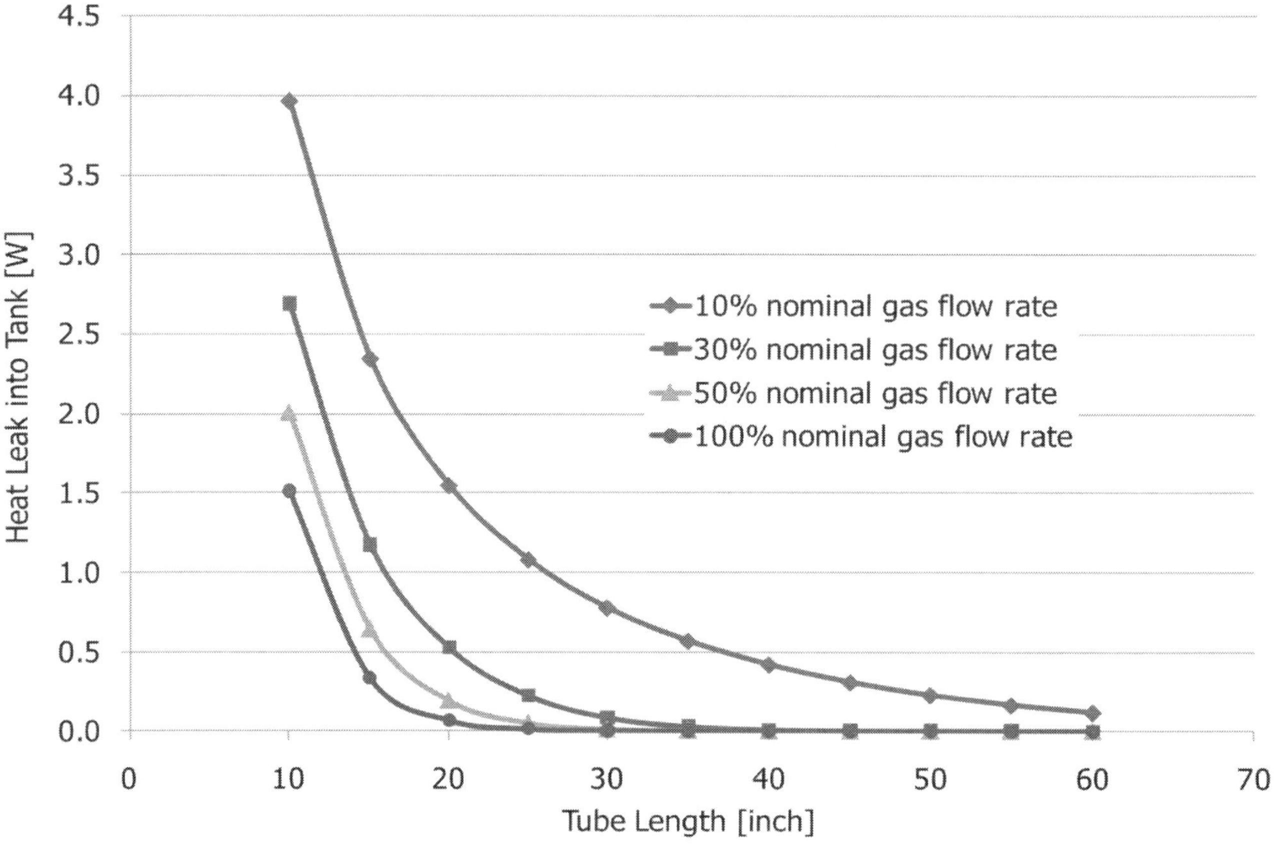

Figure 2. Predicted VTA net heat leak rate as a function of tube length and GH_2 flow rate.

A safety concern related to the VTA is the possibility of a catastrophic loss of vacuum which would cause rapid LH_2 vaporization, and if the gas cannot escape, rising tank pressure. In order to avoid a boiling liquid expanding vapor explosion (BLEVE) the VTA must vent GH_2 at the failure boil-off rate without exceeding the burst pressure of the inner vessel. The failure heat flux and boil off rate were estimated by assuming that gaseous N_2 from the air would condense on the outside of the inner vessel. This established temperature boundary conditions at the inside (25 K) and outside (77 K) surfaces of the inner vessel wall, which allowed an estimate of the heat flux to be calculated. Note that neglecting the thermal resistance associated with the LH_2 boiling made the heat flux estimate conservative (high).

The method outlined by [9] was used to estimate the maximum flow rate through tubes of various diameter and length. In each case the gas pressure in the inner vessel was assumed to be 75% of the inner vessel burst pressure. For the chosen tube length and inside diameter, the predicted gas flow rate was sufficient to avoid a BLEVE when compared to the predicted failure boil off rate.

Validation of Heat Transfer Models

A test rig was built to validate the model; the rig consisted of two concentric aluminum cylinders separated by a vacuum space. The inner cylinder was wrapped with MLI and separated from the outer cylinder by low thermal conductivity supports. The inner cylinder was filled with liquid N_2 (LN_2) and the total rate of heat transfer to the inner cylinder was estimated from the rate of LN_2 boil off.

The total rate of heat leak was measured for three different boundary temperatures: 207±1 K, 298±4 K and 387±3 K. The relative amounts of conduction and radiation heat transfer were estimated by assuming a dependence on $(T_h - T_c)$ for conduction and $(T_h^4 - T_c^4)$ for radiation. The rate of radiation heat transfer agreed with the empirical model to within 10% using a degradation factor of 7.4, which describes the deviation from ideal behavior introduced by seams and penetrations. In all of the measurements, the flow rate of LN_2 was sufficient to ensure that the heat leak rate was dominated by radiation heat transfer, confirming the expectation that MLI plays a major role in the design.

Load Based Transient Analysis

Fuel consumption rate data from GH_2 powered Ion Tiger flights were analyzed to estimate the transient performance required of the LH_2 tank. All aircraft have varying propulsion power requirements which depend on vehicle mass, drag and the way in which it maneuvers, but because the Ion Tiger is relatively small, the propulsion power also varies substantially with weather and thermal phenomena. Due to these variations, it is important to accurately estimate the range and typical rate of change in the fuel withdrawal rate.

To accommodate periods of low fuel consumption rate, a pressure relief valve (PRV) must be included in the tank design. Tank pressure increases when the boil off rate exceeds the consumption rate; when the pressure reaches the upper limit for the inner vessel, the PRV opens to release GH_2. Having a larger open space above the liquid allows more LH_2 to boil off before the PRV must open, thereby decreasing the amount of lost fuel, but this comes at the cost of storing less LH_2 at the outset. Having a lower overall heat leak rate and using a heater in periods of high fuel consumption is preferred, because then a smaller gas volume can be used to damp transients. The smaller gas volume leaves more space for LH_2 storage.

The design goal was to maintain tank pressure between 40 and 80 psig; to do this a transient performance model was developed. The model used propulsion load data and heat leak rate as inputs and the tank fill fraction and pressure as initial conditions. It integrated the difference between the LH_2 boil-off rate and the rate of fuel consumption, and then used an empirical equation of state [2] to estimate the tank pressure at each time step. When the pressure reached the upper limit, the tank vented, and when the pressure reached the lower limit, the heater turned on. Deadband of ½ the pressure range was assumed for the heater controller – the heater turned off at the nominal tank pressure. Model outputs were the predicted tank pressure, GH_2 blow-off rate and heater duty cycle as functions of time.

Figure 3 shows the predicted tank pressure as a function of time given an Ion Tiger fuel consumption curve, total heat leak rates of 2.0, 2.1 and 2.2 W, no PRV or heater, and a gas volume of 4 L. Note that changes in the heat leak rate as small as ±100 mW integrated over 26 hr meant the difference between exceeding the pressure limits of the tank and starving the fuel cell. The same is true to a greater extent when integrated over a 72 hr flight. Figure 3 shows the clear need for a low heat leak rate from ambient, a PRV, and a heater for adequate control of the LH_2 storage system. The high heat leak case exceeded the 80 psia inner vessel pressure limit and the low heat leak case could not support the fuel consumption rate over the entire flight.

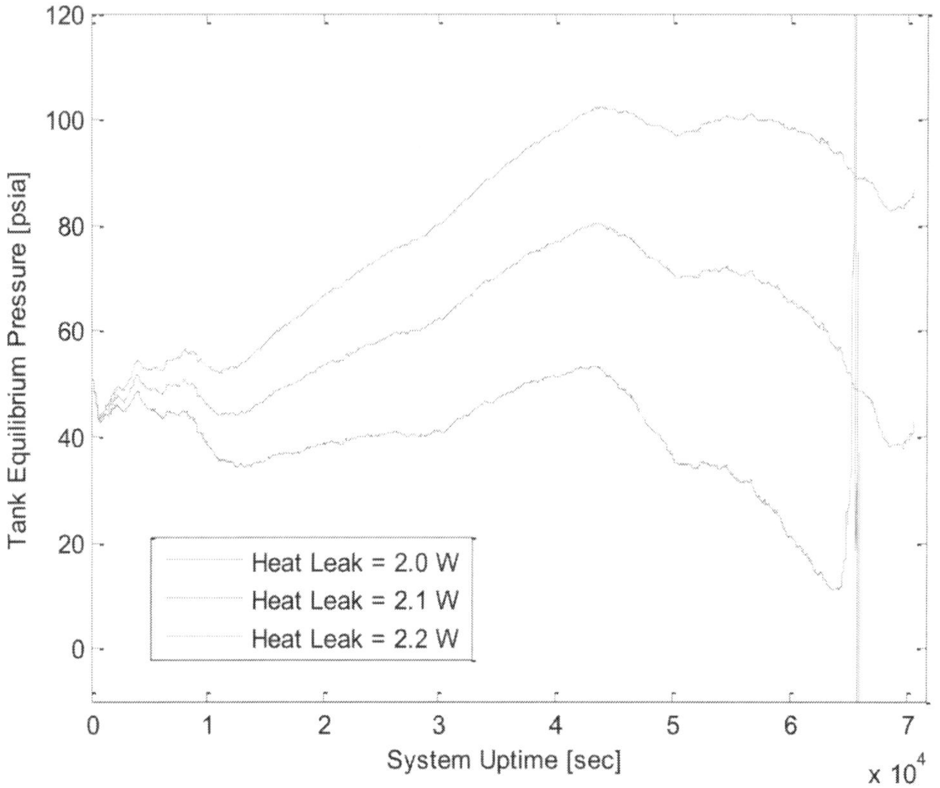

Figure 3. Predicted tank pressure assuming heat leak rates of 2.0, 2.1 or 2.2 W and a gas volume of 4 L. The large transient in the 2.0 W curve is an integration error due to the tank pressure falling below ambient pressure, which cause the equation of state to fail.

Assuming a total heat leak rate of 2.09 W, the transient performance model was used to compare two design scenarios: a large gas space and no heater vs. a small gas space with a heater. In the first case, the gas volume and PRV damp are the only means to damp transients, so the gas volume must be larger. In the second case, the heater supports the pressure during periods of high withdrawal rate, so the gas space can be smaller. The model predicted that adding a heater would improve tank pressure control and ensure it remained within the desired 60±20 psi range, as shown in Figure 4. The smaller gas space led to an increase in venting, but it was more than offset by larger volume of fuel stored at the outset. The tank with the heater was predicted to hold 18% more LH_2 at the end of the 26 hr simulated flight, as shown in Table 3.

Table 3. Tank performance in a simulated 26 hr flight, with and without a heater. * indicates predicted values.

Case	Initial fuel volume [L]	Total Heat Leak [W]	Fuel vented* [g]	Total Capacity* [g]
No heater, 4 L gas volume	18.0	2.09	0.0	1161
Heater, 0.5 L gas volume	21.5	2.09	19.4	1367

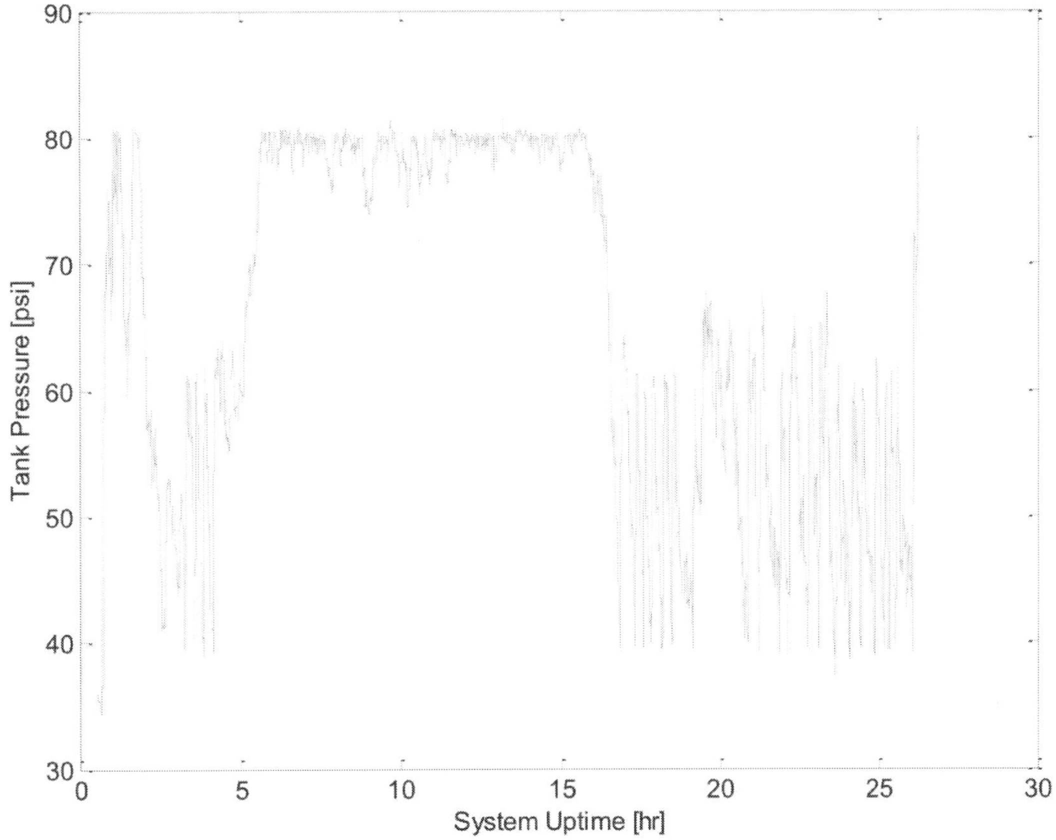

Figure 4. Predicted tank pressure with a heat leak rate of 2.09 W, 0.5 L gas volume, PRV set to 80 psia and heater activated at 40 psia.

Conclusions

The design of a LH_2 storage tank for small aircraft applications is challenging in large part because of the restrictions on weight and heat leak rate. For the Ion Tiger LH_2 storage tank design, these two restrictions drove most of the material selection, mechanical design and thermal design decisions. Steady state heat transfer and transient flow models can provide valuable insights into the performance of proposed designs, and guide the design process, once the basic design concepts have been established. As shown by the insulation modeling, however, validation of the models is essential and empirical models are sometimes preferred to physics based approaches.

Acknowledgements

The authors gratefully thank the Office of Naval Research for funding this work.

References

1. K. Swider-Lyons, R.O.Stroman, G. S. Page, J. F. Mackrell, J. A. Rodgers, M. W. Schuette. *The Ion Tiger Fuel Cell Unmanned Air Vehicle*. in *Proceedings of the 44th Power Sources Conference*. 2010.

2. *NIST Thermophysical Properties Database*. [cited 2011; Available from: http://webbook.nist.gov/chemistry/fluid/.
3. Incropera, F.P. and D.P. DeWitt, *Fundamentals of heat and mass transfer*. 5th ed. 2002, New York: J. Wiley. xix, 981 p.
4. Kays, W.M., M.E. Crawford, and B. Weigand, *Convective heat and mass transfer*. 4th ed. McGraw-Hill series in mechanical engineering. 2005, Boston: McGraw-Hill Higher Education. xxx, 546 p.
5. Brewster, M.Q., *Thermal radiative transfer and properties*. 1992, New York: Wiley. xxii, 543 p.
6. Barron, R.F., *Cryogenic systems.* 2nd ed. Monographs on cryogenics. 1985, New York; Clarendon Press. xvii, 507 p.
7. Gilmore, D.G., *Satellite thermal control handbook*. 1994, El Segundo, Calif.: Aerospace Corporation Press.
8. Marquardt, E.D., J.P. Le, and R. Radebaugh, *Cryogenic material properties database.* Cryocoolers 11, 2001: p. 681-687.
9. C. K. Liu, L.G.N., A. F. Manikowski, Jr. *Sizing an emergency venting system for a cryogenic dewar*. in *Advances in Cryogenic Engineering*. 1986. Cambridge, MA: Plenum Press.

(378b) Out-Gassing Study and Species Measurement for Cryogenic Carbon Fiber Hydrogen Storage Vessels

Joseph W. Reiter, Rajeshuni Ramesham and Alexander W. Raymond, Jet Propulsion Laboratory, Pasadena, CA

Out-gassingStudy and Species Measurement for Cryogenic Carbon Fiber Hydrogen Storage Vessels[1]

Joseph Reiter, Rajeshuni Ramesham, and AlexanderRaymond

JetPropulsion Laboratory, California Institute of Technology, Pasadena, CA 91109,USA

Cryo-adsorption andcryo-compressed hydrogen storage are two promising technologies for vehicularhydrogen storage. These low-temperature storagemethods have the benefits of increasing hydrogen density and materialadsorption capacity, which enhance both the gravimetric and volumetricperformance of these systems. In bothcases, carbon-fiber wrapped tanks appear to be preferable to metal pressurevessels for vehicular applications because of their higher gravimetriccapacity; however, maintaining adequate thermal isolation for sustaineddormancy at cryogenic temperatures is challenging with carbon compositevessels. Cryogenic pressure vessels aretypically surrounded by a vacuum jacket and multi-layer insulation (MLI.) The effectiveness of MLI depends on theextent to which high-vacuum is maintained in the vacuum jacket. Even moderate vacuum degradation can resultin significant gas conduction across the separated layers and will render theMLI ineffective. Carbon compositeout-gassing is studied in this preliminary investigation with relation to itseffects on vacuum insulation degradation.Species and total out-gassing rates will be presented on both agravimetric and on an areal basis using the sampleporous surface area provided by the BET (**Brunauer-Emmett-Teller**) method. Temperature dependence of the out-gassingrate will be deduced by conducting experiments at cryogenic operatingtemperature (77 K), room temperature (298 K), and elevated temperature (358 K). A discussion of the allowable out-gassingrates will be presented along with potential mitigation strategies.

Acknowledgments:The researchwork described in this paper has been carried out at the Jet Propulsion Laboratory,California Institute of Technology, Pasadena, CA. We acknowledge theDepartment of Energy (DOE) for supporting this effort.

See more of this Session: Hydrogen Storage Systems: Designs, Analysis, and Safety Aspects

(378d) Removal of Heat From a Hydrogen Storage Media by a Novel Microscale-Based Heat Exchanger

Christopher Loeb[1], **Agnieszka Truszkowska**[1], Bruce Hardy[2], Richard Chahine[3] and Goran N. Jovanovic[1], (1)Chemical, Biological and Environmental Engineering, Oregon State University, Corvallis, OR, (2)Computational Sciences, Savannah River National Laboratory, Aiken, SC, (3) Hydrogen Research Institute, Trois-Rivieres, QC, Canada

High volumetric storage density of hydrogen gas is a major challenge in the progress towards hydrogen-fueled automobiles. Of the various storage methods, pressurized gas tanks, liquid hydrogen storage, liquid chemical systems, and solid-state storage, activated carbon adsorbents show great potential. A reliable thermal management system within the storage tank is vital to provide quick, safe, and efficient charging and a controlled release of hydrogen gas on demand.

The current work investigates room temperature adsorption of nitrogen gas (model gas for hydrogen) in activated carbon bed and compares the temperature distribution within the storage media without thermal management, and with the addition of a microscale-based heat exchanger. Experimentally obtained temperature distribution data is used to assess the effect of charging time, and removal of generated heat within the system. A mathematical model is used to numerically simulate the experimental process.

The maximum height of activated carbon bed positioned between two heat exchange devices, while providing adequate thermal management, is investigated by varying the height of the adsorption bed and monitoring nitrogen-charging time and temperature distribution.

A novel microscale-based heat exchanger design takes advantage of 250 μm high features, which facilitate the flow of cooling fluid. The heat exchanger is fabricated by diffusion bonding of patterned stainless steel shims, resulting in an overall 250 μm thick resistance to thermal transport. The heat exchanger design also provides uniform gas distribution into the bed of storage media. The device's ability to provide a uniform temperature at the contacting surface is verified using infrared thermal imaging.

Experimental results indicate a dramatic improvement in charging time due to the fast and efficient removal of generated heat.

See more of this Session: Hydrogen Storage Systems: Designs, Analysis, and Safety Aspects
See more of this Group/Topical: International Congress on Energy 2011

(378e) Mitigating Strategies for Reactive Solids-Based Hydrogen Storage Systems

Joseph W. Pratt[1], Joseph G. Cordaro[2], Daniel E. Dedrick[3], Michael P. Kanouff[1] and Yehia (John) F. Khalil[4], (1)Thermal/Fluid Science and Engineering, Sandia National Laboratories, Livermore, CA, (2) Materials Chemistry, Sandia National Laboratories, Livermore, CA, (3)Hydrogen and Combustion Technology, Sandia National Laboratories, Livermore, CA, (4)Chemical Engineering Group, Physical Sciences Department, United Technologies Research Center (UTRC), East Hartford, CT

Solid-state materials such as metal hydrides and complex hydrides have been shown to be viable candidates for on-board hydrogen storage systems. However, some complex metal hydrides such as sodium alanate ($NaAlH_4$) in the powder form, exhibit unfavorable exothermic reactions when exposed to air or water. Mitigating this reactivity and its associated risks in the event of inadvertent exposure (e. g., tank rupture during a vehicular collision) is an important step for achieving widespread use and public acceptance of these technologies.

Several types of accident scenarios, exposures, and mitigating methods have been investigated in the past. This paper focuses on a specific scenario and mitigating strategy: exposure of the reactive solid to oxygen, and the effect of an active material-polymer composite mixture on its exothermic behavior.

Sodium alanate is selected as the representative of reactive solid hydrogen storage materials that exhibit unfavorable exothermicity, and is the "active material," although the mitigation methods developed are applicable to other similar materials. Previously, we reported on the use and effectiveness of polystyrene polymers synthesized (polymerized) along with the active material to create a composite. This paper presents our work on other polymers using the same methodology, comparing their effectiveness to the active material alone and to the previous work with polystyrene. Additive effectiveness is measured by the amount of heat released during controlled exposure to oxygen. Measurements are done both before and after subjecting the composite material to representative operating conditions of adsorption/desorption cycling, to understand the impacts (if any) of operating conditions on the stability and effectiveness of the composite material. SEM imagery is also used as a tool to better understand the method of mitigation and reasons for degradation (if any) of the composite. Additional tests were performed to investigate the impact of exposing the hydride (both in powder and powder compact forms) to vehicle fluids and road sprays and also to assess the sensitivity of the mitigated material to mechanical impact.

Daniel E. Dedrick, Joseph G. Cordaro, Michael P. Kanouff, Craig L. Reeder, Joseph, W. Pratt, and Y. F. Khalil, "Mitigation Technologies for Hydrogen Storage Systems based on Reactive Solids," presented at AIChE Annual Meeting, Salt Lake City, UT, Nov. 8-12, 2010.

(378f) Can a Liquid Hydrogen Storage Tank BLEVE?

Subramanya Nayak, Sergio Garcia, Victor Carreto and Sam Mannan, Chemical Engineering, Texas A&M University, College Station, TX

Abstract

A BLEVE (boiling-liquid expanding-vapor explosion) occurs when a tank containing liquefied gas ruptures, resulting in the instantaneous release of the vapor, boiling liquid, missiles (tank fragments), and blast wave[1]. The most common type of the BLEVE is caused when the external wall of the tank is exposed to a fire[1]. The continuous heat transfer from the fire to the tank wall increases the temperature and pressure inside the tank, until the tank fails. This sudden failure of the tank results in an explosive vaporization of its contents.

In passenger cars the LH_2 is stored as a cryogenic liquid at -240° C and at moderate pressures (20 to 30 bars)[2]. It is believed that the LH_2 will not BLEVE, since it is stored at such a low temperature[3]. However, LNG, which is also stored as a cryogenic liquid presents a potential for BLEVE[4]. Consequently, in this study we have theoretically investigated if the LH_2 storage tanks will BLEVE.

A theoretical model considering transient mass and energy transfer is developed and solved numerically[5]. It is assumed that the LH_2 tank is engulfed in a fire and the fire is caused due to spilled gasoline or hydrogen. Different tank materials and insulations are tested. The transient model has a capability to predict the time required for the LH_2 tank to BLEVE. This prediction will be useful for the fire fighters and emergency responders when dealing with LH_2 tank engulfed in a fire. Finally the consequences due to the BLEVE of LH_2 tank are also studied.

References

1. Abbasi, T. and S. A. Abbasi (2007). "The boiling liquid expanding vapour explosion (BLEVE): Mechanism, consequence assessment, management." Journal of Hazardous Materials 141(3): 489-519.
2. Rigas, F. and S. Sklavounos (2005). "Evaluation of hazards associated with hydrogen storage facilities." International Journal of Hydrogen Energy **30**(13-14): 1501-1510.
3. C. J. H. van den Bosch, R. A. P. M. W. (2005). Methods for the calculation of physical effects - due to the releases of hazardous materials (liquids and gases) - 'Yellow Book'. The Hague,

Committee for the Prevention of Disasters.
4. Pitblado, R. (2007). "Potential for BLEVE associated with marine LNG vessel fires." Journal of Hazardous Materials 140(3): 527-534.
5. Runnels, D. S. R. (2008). AFFTAC: Analysis of Fire Effects on Tank Cars.

See more of this Session: Hydrogen Storage Systems: Designs, Analysis, and Safety Aspects
See more of this Group/Topical: International Congress on Energy 2011

(388a) Metabolic Flux Analysis of *Escherichia Coli* MG1655 Under Octanoic Acid Stress

Yanfen Fu[1], Jong M. Yoon[1], Liam A. Royce[1], Maria Rodriguez-Moya[2], Ramon Gonzalez[2], Laura Jarboe[1] and Jacqueline V. Shanks[1], (1)Department of Chemical and Biological Engineering, Iowa State University, Ames, IA, (2)Departments of Chemical & Biomolecular Engineering and Bioengineering, Rice University, Houston, TX

Metabolic engineering has evolved to the point of fulfilling the dream of having our industrial chemicals produced renewably. Carboxylic acids (e.g., short chain fatty acids (SCFAs) such as octanoic acid (C8)) are one such chemical intermediate that can be produced with *Escherichia coli* engineered with a short chain thioesterase. However, C8 is toxic to *E. coli* growth at concentrations greater than 10 mM. To design a better host strain for the production of large amount of those SCFAs, metabolic flux analysis of *E. coli* grown under C8 stress was performed. When *E. coli* MG1655 was grown in M9 minimal media supplemented with 35 mM C8 under aerobic conditions, cell growth was inhibited by 55.1% ± 10.0% (mean ± standard deviation, n=4) compared to control (without C8). To study fluxes in central carbon metabolism, 10 g/L of a mixture of labeled and unlabeled glucose (25% uniform label glucose, 25% 1-label glucose and 50% natural glucose) was used as the sole carbon source for bacterial growth. Proteinogenic amino acid isotopomers were measured by 2D (^{13}C, ^{1}H) NMR. Notable differences of several amino acid isotopomer enrichments were observed between control condition and C8 stress condition, suggesting the precursor nodes of these amino acids in metabolic pathways are responding to the stress. More specifically, the difference suggested that the flux distribution among tricarboxylic acid (TCA) cycle, pyruvate node and α-ketoglutarate node and oxaloacetate node was changed. Under C8 stress, acetate production flux increased by ~50% and the flux from phosphoenolpyruvate to pyruvate increased by ~50%, while the TCA cycle pathway, the malic enzyme pathway, and the pentose phosphate pathway decreased by 25%, 90%, and 20% respectively. Downregulation of the TCA cycle can occur if an excess of NADH occurs, due to feedback inhibition of isocitrate synthase. Since our flux data shows that less NADH is produced via central carbon metabolism under C8 stress, an excess of NADH could occur if C8 stress inhibits regeneration of NAD^+ from NADH, via membrane disruption. Our transcriptome and proteomic data suggest that carboxylic acids may negatively impact the function and/or integrity of the cell membrane. Flux data combined with transcriptomic and proteomic analysis allow the development of hypotheses in context with the mechanism of C8 toxicity, which in turn helps to engineer strains with increased tolerance of C8.

See more of this Session: Microbial Engineering for the Synthesis of Bulk and High-Value Products
See more of this Group/Topical: Sustainable Engineering Forum

(388b) Metabolic Engineering of Rhizopus Oryzae for Enhanced Production of Fumaric Acid From Sugars

Baohua Zhang, WILLIAM G. LOWRIE Department of Chemical and Biomolecular Engineering, The Ohio State University, Columbus, OH and Shang-Tian Yang, The Ohio State University, Columbus, OH

Fumaric acid is a dicarboxylic acid used extensively in resins, food acidulants, and other applications including oil field fluids, esters etc. Some strains of *Rhizopus oryzae*, which is a filamentous fungus, can produce fumaric acid from sugars under aerobic conditions but the fumarate yield is relatively low due to the co-production of ethanol and other fermentation byproducts. In this work, metabolic engineering technique is applied to optimize the fermentation for enhanced fumaric acid production. Overexpression of native enzymes fumarase and pyruvate carboxylase showed no significant effect on the fumaric acid production, suggesting that the biosynthesis of fumaric acid is not limited by these two enzymes. On the other hand, overexpression of heterologous enzyme phosphoenolpyruvate carboxylase, which converts pyruvate with CO_2 fixation to oxaloacetate, led to increased fumaric acid production compared to the wild type control. The fermentation kinetics and activities of enzymes in the fermentation pathway of these *Rhizopus* mutants were studied and will be reported in this paper.

See more of this Session: Microbial Engineering for the Synthesis of Bulk and High-Value Products
See more of this Group/Topical: Sustainable Engineering Forum

ARTICLE IN PRESS

Metabolic Engineering ▪ (▪▪▪▪) ▪▪▪–▪▪▪

Contents lists available at ScienceDirect

Metabolic Engineering

journal homepage: www.elsevier.com/locate/ymben

Styrene biosynthesis from glucose by engineered *E. coli*

Rebekah McKenna, David R. Nielsen *

Chemical Engineering, Arizona State University, 501 E. Tyler Mall, ECG 301, Tempe, AZ 85287-6106, USA

ARTICLE INFO

Article history:
Received 18 March 2011
Received in revised form
2 June 2011
Accepted 14 June 2011

Keywords:
Styrene
Phenylalanine ammonia lyase
E. coli
Cinnamic acid
L-Phenylalanine
Aromatic

ABSTRACT

Styrene is a large volume, commodity petrochemical with diverse commercial applications, including as a monomer building-block for the synthesis of many useful polymers. Here we demonstrate how, through the *de novo* design and development of a novel metabolic pathway, styrene can alternatively be synthesized from renewable substrates such as glucose. The conversion of endogenously synthesized L-phenylalanine to styrene was achieved by the co-expression of phenylalanine ammonia lyase and *trans*-cinnamate decarboxylase. Candidate isoenzymes for each step were screened from bacterial, yeast, and plant genetic sources. Finally, over-expression of *PAL2* from *Arabidopsis thaliana* and *FDC1* from *Saccharomyces cerevisiae* (originally classified as ferulate decarboxylase) in an L-phenylalanine over-producing *Escherichia coli* host led to the accumulation of up to 260 mg/L in shake flask cultures. Achievable titers already approach the styrene toxicity threshold (determined as ~300 mg/L). To the best of our knowledge, this is the first report of microbial styrene production from sustainable feedstocks.

1. Introduction

Styrene is a versatile, large commodity chemical for which 60% of its global annual consumption supports the production of numerous, industrially important polymers, and co-polymers (SRI, 2010). In 2006, over 6 million metric tons of styrene were produced by U.S. manufacturers, a market that was valued at nearly $28 billion and projected to grow by 4.3% per year through 2010 (SRI, 2010). Today, all commercially available styrene is derived from the world's dwindling petroleum resources. Conventional styrene synthesis is achieved through the chemocatalytic dehydrogenation of petroleum-derived ethylbenzene (Wu et al., 1981) which requires over 3 metric tons of steam per metric ton of styrene produced. This exorbitant requirement renders styrene production as the most energy-intensive among commodity chemical production routes, consuming nearly 200 trillion BTU of steam for its domestic annual production alone (DoE, 2002). With that being said, the goal of this study was to engineer a biocatalyst capable of synthesizing styrene from renewable resources as a more sustainable and greener source of styrene and styrene-derived polymers.

Beyond its petrochemical origins, styrene has been observed as a trace metabolite in foods, in particular cheeses, where it acts as an aroma defect. For instance, the yeast *Penicillium camemberti* has been reported to be capable of synthesizing low levels of styrene from excess L-phenylalanine, but neither a defined pathway nor the requisite genes have thus far been elucidated (Pagot et al., 2007). Styrene is also known to be naturally synthesized by select plant species, including several trees in the *Styracaceae* family (including several *Styrax* sp.). Here, styrene is also synthesized from excess L-phenylalanine where it then subsequently accumulates as a minor constituent (< 0.55% of total dry weight) within benzoin resins (which are predominantly composed of benzoic acid) (Fernandez et al., 2005). Again, however, neither the enzymes nor genes associated with said pathway have been identified to date. Although it is possible to purify styrene from plant resins via distillation or liquid–liquid extraction (Clark, 1990), considering the extremely low productivity, poor net yields, and low inherent value of styrene, its potential, large-scale biological production by such a mechanism is rendered as completely uneconomical and unsustainable. A more sustainable and inexpensive approach, however, would involve the engineering of microorganisms that possess the unique ability to synthesize styrene directly from renewable resources.

In recent years, a variety of additional, novel synthetic routes have been proposed and engineered in microorganisms for the production (from renewable substrates such as glucose) of a number of other useful, functionalized monoaromatic compounds with structural similarity to styrene. For example, a biosynthetic pathway for the production of *p*-hydroxystyrene (a monomer used in synthesis of photo-resist polymers) from renewable sugars has been constructed using both *Escherichia coli* (Qi et al., 2007) and *Pseudomonas putida* (Verhoef et al., 2009) as host platforms. Meanwhile, both phenol (a precursor and monomer

* Corresponding author. Fax: +1 480 727 9321.
E-mail address: David.R.Nielsen@asu.edu (D.R. Nielsen).

1096-7176/$ - see front matter © 2011 Elsevier Inc. All rights reserved.
doi:10.1016/j.ymben.2011.06.005

Please cite this article as: McKenna, R., Nielsen, D.R., Styrene biosynthesis from glucose by engineered *E. coli*. Metab. Eng. (2011), doi:10.1016/j.ymben.2011.06.005

for phenolic resins) (Wierckx et al., 2005) and p-hydroxybenzoate (a precursor to parabens, which are used as preservatives) (Verhoef et al., 2007) have also been synthesized as individual products from glucose by engineered strains of *P. putida*. Interestingly, each of the above non-natural metabolites were derived using L-tyrosine (or its immediate precursor, 4-hydroxyphenylpyruvate) as a pathway precursor, thereby making them each phenolics (Boudet, 2007). To date, there remain few examples of engineered biosynthetic pathways for the production of non-phenolic, monoaromatic compounds using microbial biocatalysts. Moreover, there exist no previous reports regarding the development of a styrene biosynthetic pathway or the engineering of microbes capable of synthesizing styrene from renewable resources. With this in mind, the present study describes the *de novo* design and development of a functional styrene biosynthetic pathway and the engineering of *E. coli* strains capable of styrene biosynthesis from glucose.

The proposed styrene biosynthesis pathway utilizes endogenously synthesized (from glucose) L-phenylalanine as an intermediate precursor which is converted to styrene by a series of two enzymatic steps, as in Fig. 1. First, endogenously occurring L-phenylalanine is converted to *trans*-cinnamic acid (tCA) through its deamination, as catalyzed by phenylalanine ammonia lyase (PAL). Said activity and substrate specificity has been previously reported for a number of PAL isoenzymes that have been identified and characterized in yeast (Gilbert and Tully, 1982; Qi et al., 2007; Vannelli et al., 2007b), plants (Cochrane et al., 2004; Young et al., 1966), and (although less prevalently) bacteria (Moffitt et al., 2007a; Xiang and Moore, 2005; Xiang and Moore, 2006; Young et al., 1966). Among previously characterized PAL isoenzymes, considerable variability with respect to both substrate specificity and activity has been reported. The most commonly studied PAL isoenzyme in recombinant systems, including *E. coli*, remains the bifunctional PAL/TAL (TAL: tyrosine ammonia lyase, which also catalyzes the deamination of L-tyrosine to p-coumaric acid) of the yeast *Rhodotorula* sp. (Cui et al., 2008; Gilbert et al., 1985, 1983; Gilbert and Tully, 1982; Santos et al., 2011; Vannelli et al., 2007a). Meanwhile, a number of PAL isoenzymes have been also been studied from plant sources (Jones, 1984) where, for example, *Arabidopsis thaliana* has been characterized as possessing four distinct PALs (encoded by *PAL1*, *PAL2*, *PAL3*, and *PAL4*) (Cochrane et al., 2004). More recently, a number of prokaryotic PALs have been isolated and characterized, beginning with that which is encoded by *encP* from *Streptomyces maritimus* (Xiang and Moore, 2002, 2005). Soon after, two additional PAL isoenzymes were discovered in the cyanobacteria *Nostoc puntiformes* and *Anabaena variabilis* (Moffitt et al., 2007a; Xiang and Moore, 2005, 2006). Interestingly, each of these prokaryotic PALs was also found to be highly specific for L-phenylalanine, and thus do not also display TAL activity that is so common among yeast PALs.

The second step in the proposed styrene biosynthesis pathway involves the subsequent decarboxylation of tCA by a phenylacrylate decarboxylase (PADC) displaying *trans*-cinnamate decarboxylase activity to yield styrene as the final product. Several genes encoding PADC activity have been characterized and reported in the literature, including *pdc* from *Lactobacillus plantarum* and *padC* from *Bacillus subtilis* (Tran et al., 2008). It currently remains unknown, however, if their enzyme products could specifically decarboxylate tCA to produce styrene. Meanwhile, the yeast *Saccharomyces cerevisiae* has demonstrated the ability to synthesize styrene when supplied with exogenous tCA (Clausen et al., 1994); however, the underlying mechanism and genetic basis for said activity presently remains unclear. For instance, previous characterizations of *PAD1* have shown that its over-expression in *S. cerevisiae* results in increased resistance to tCA (as achieved by its conversion to styrene) (Larsson et al., 2001). However, it has also been demonstrated that the expression of *PAD1* alone is insufficient for achieving PADC activity (Clausen et al., 1994), and it was later speculated that the expression of a second enzyme was necessary to convert tCA to styrene (Jiang et al., 2005). Most recently, it was reported that tCA decarboxylase activity in *S. cerevisiae* maintains an essential dependence on the co-expression of both *PAD1* and *FDC1*, the latter a gene previously characterized as encoding ferulic acid decarboxylase (note that ferulic acid is also a phenylacrylic acid) (Mukai et al., 2010).

The present study describes the *de novo* design of a styrene biosynthetic pathway, as supported through the comprehensive screening of composite pathways enzymes from various genetic sources. The synthesis of styrene from glucose was ultimately achieved through the co-expression of PAL and tCA decarboxylase enzymes in an L-phenylalanine over-producing *E. coli* host platform.

2. Materials and methods

2.1. Chemicals

Chemicals used in this study include L-phenylalanine (98.5%, VWR, Westchester, PA), *trans*-cinnamic acid (99%, MP Biomedicals, Solon, OH), styrene (99%, Alfa Aesar, Ward Hill, MA), methanol (99.8%, VWR, Westchester, PA), and trifluoroacetic acid (99.5%, EMD, Darmstadt, Germany). All other chemicals used in this study are from Sigma-Aldrich (St. Louis, MO).

2.2. Strains and media

All strains, plasmids, and oligonucleotide primers used in this study are listed in Table 1. Custom oligonucleotide primers were synthesized by Integrated DNA Technologies (Coralville, IA). cDNA of *N. puntiformes* and *A. variabilis* were gifts from Prof. Bradley Moore (UCSD). Strains of *B. subtilis*, *L. plantarum*, and *S. cerevisiae*, as well as the plasmids pSTV28 and pTrc99A were all gifts from Prof. Kristala Prather (MIT). Genomic DNA was prepared from whole cells using the ZR Fungal/Bacterial DNA MiniPrep (Zymo Research, Irvine, CA) according to vendor protocols. Strains were

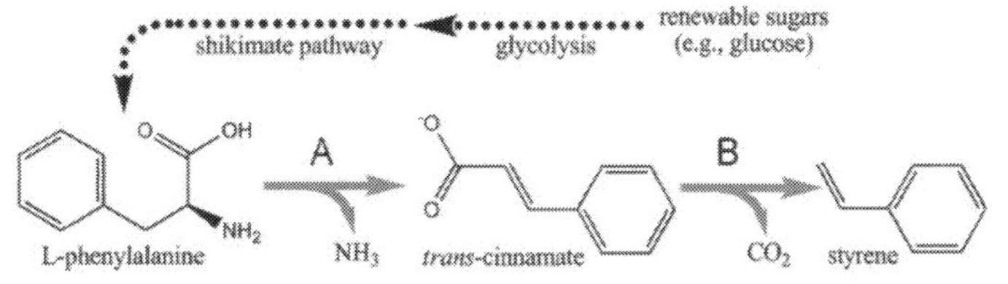

Fig. 1. Enzymatic pathway to convert the precursor L-phenylalanine to the product styrene via the intermediate *trans*-cinnamate. The two-step pathway from L-phenylalanine is achieved by the co-expression of one or more genes which encoded phenylalanine ammonia lyase (PAL) activity (A), and one or more genes which encoded *trans*-cinnamic acid decarboxylase (CADC) activity (B).

Please cite this article as: McKenna, R., Nielsen, D.R., Styrene biosynthesis from glucose by engineered *E. coli*. Metab. Eng. (2011), doi:10.1016/j.ymben.2011.06.005

Table 1
Strains, plasmids, and oligonucleotide primers used in the present study.

	Description	Source
Strains		
E. coli NST74	*aroH367, tyrR366, tna-2, lacY5, aroF394(fbr), malT384, pheA101(fbr), pheO352, aroG397(fbr)*	ATCC 31884
E. coli BL21Star(DE3)	*F-ompT hsdS*$_B$ *(r*$_B$*-m*$_B$*-) gal dcm rne131 (DE3)*	Invitrogen
E. coli NEB-10 beta	*araD139 Δ(ara,leu)7697 fhuA lacX74 galK16 galE15 mcrA f80d(lacZΔM15)recA1 relA1 endA1 nupG rpsL rph spoT1Δ(mrr-hsdRMS-mcrBC)*	New England Biolabs
S. cerevisiae W303	Source of *pad1* and *fdc1*	Prather Lab, MIT
L. plantarum	Source of *pdc*	Prather Lab, MIT
B. subtilis 3610	Source of *padC*	Prather Lab, MIT
Plasmids		
pTrc99A	P_{trc}, pBR322 ori, $lacI^q$, Amp^r	Prather Lab, MIT
pSTV28	P_{lac}, pACYC184 ori, Cm^r	Prather Lab, MIT
pSpal1At	*PAL1* of *A. thaliana* inserted into the *EcoRI* and *SphI* sites of pSTV28	This study
pSpal2At	*PAL2* of *A. thaliana* inserted into the *EcoRI* and *SphI* sites of pSTV28	This study
pSpalAv	*pal* of *A. variabilis* inserted into the *EcoRI* and *BamHI* sites of pSTV28	This study
pSpalNp	*pal* of *N. punctiforme* inserted into the *EcoRI* and *BamHI* sites of pSTV28	This study
pSencPSm	*encP* of *S. maritimus* inserted into the *EcoRI* and *BamHI* sites of pSTV28	This study
pTpad1Sc	*PAD1* of *S. cerevisiae* inserted into the *NcoI* and *XbaI* sites of pTrc99A	This study
pTpad1Sc-fdc1Sc	*FDC1* of *S. cerevisiae* inserted into pTpad1Sc	This study
pTfdc1Sc	*FDC1* of *S. cerevisiae* inserted into the *SalI* and *HindIII* sites of pTrc99A	This study
pTpadCBs	*padC* of *B. subtilis* inserted into the *BamHI* and *SbfI* sites of pTrc99A	This study
pTpdcLp	*pdc* of *L. plantarum* inserted into the *BamHI* and *SbfI* sites of pTrc99A	This study
pUN15-pal1	Clone U10120 containing AT2G37040 (*PAL1*) from *A. thaliana*	ABRC
pUN15-pal2	Clone U12256 containing AT3G53260 (*PAL2*) from *A. thaliana*	ABRC
Primers (5′→3′)		
F_pal_AV_EcoRI	TAAGAATTCAAGGGGATAAATAATGAAGACACTATCTCAAGC	
R_pal_AV_BamHI	ATTGGATCCTTAATGCAAGCAGGGT	
F_pal_NP_EcoRI	TAAGAATTCAAGGGGATAAATAATGAATATAACATCTCTACAAC	
R_pal_NP_BamHI	ATTGGATCCTTACGTTGACTTTAAGCT	
F_pal1_AT_EcoRI	TAAGAATTCAAGGGGATAAATAATGGAGATTAACGGGGCAC	
R_pal1_AT_SphI	ATTGCATGCTTAACATATTGGAATGGGAGCTC	
F_pal2_AT_EcoRI	TAAGAATTCAAGGGGATAAATAATGGATCAAATCGAAGCAATG	
R_pal2_AT_SphI	ATTGCATGCTTAGCAAATCGGAATCGGAG	
F_encP_SM_EcoRI	TAAGAATTCAAGGGGATAAATAATGACCTTCGTCATAGAGCT	
R_encP_SM_BamHI	ATTGGATCCTTAGTGCGCCGCCACG	
F_padC_BS_BamHI	AAAGGATCCCGACTAAGGGAGGATAAGATGGAAAACTTTATCGGAAG	
R_padC_BS_SbfI	ATACCTGCAGGATGTTTATTATAATCTTCCCGCG	
F_pdc_LP_BamHI	ATAGGATCCCTCTGGAGGCAGTTCTAATGACAAAAACTTTTAAAACACT	
R_pdc_LP_SbfI	ATACCTGCAGGCCAGAATGTTTCACGTGAA	
F_pad1_SC_NcoI	TTACCATGGAGGAACCTAGGCACACAATGGTCCTATTTCCAAGAAGAA	
R_pad1_SC_XbaI	ATTTCTAGATTACTTGCTTTTTATTCCTTCCC	
F_fdc1_SC_SalI	ATAGTCGACAGACATCAAAGGACGGTTCATGAGGAAGCTAAATCCAGCT	
R_fdc1_SC_HindIII	ATTAAGCTTTTATTTATATCCGTACCTTTTCCAAT	

routinely cultured in Luria-Bertani (LB) broth (supplemented with antibiotics as necessary). Cultures were assayed for their ability to synthesize L-phenylalanine, *trans*-cinnamate, and styrene by cultivation in phosphate-limited minimal media (herein referred to as "MM1") with glucose. MM1 was adapted from Qi et al. (2007), and is composed of glucose (nominally 15 g/L), $MgSO_4 \cdot 7H_2O$ (0.5 g/L), $(NH_4)_2SO_4$ (4.0 g/L), MOPS (24.7 g/L), KH_2PO_4 (0.3 g/L), K_2HPO_4 (0.7 g/L), and 5 mL/L ATCC Trace Mineral Supplement (Catalog No. MD-TMS) (EDTA (0.5 g/L), $MgSO_4 \cdot 7H_2O$ (3 g/L), $MnSO_4 \cdot 7H_2O$ (0.5 g/L), NaCl (1 g/L), $FeSO_4 \cdot 7H_2O$ (0.1 g/L), $Co(NO_3)_2 \cdot 6H_2O$ (0.1 g/L), $CaCl_2$ (0.1 g/L), $ZnSO4 \cdot 7H_2O$ (0.1 g/L), $CuSO_4 \cdot 5H_2O$ (0.01 g/L), $AlK(SO_4)_2$ (0.01 g/L), H_3BO_3 (0.01 g/L), $Na_2MoO_4 \cdot 2H_2O$ (0.01 g/L), Na_2SeO_3 (0.001 g/L), $Na_2WO_4 \cdot 2\,H_2O$ (0.10 g/L), and $NiCl_2 \cdot 6H_2O$ (0.02 g/L)).

2.3. Toxicity assays

To determine extent and effects of metabolite toxicity on *E. coli*, the impacts of the exogenous addition of tCA and styrene (at increasing final concentrations) to growing cultures was investigated. Seed cultures of *E. coli* NST74 were prepared in 5 mL of LB broth and grown at 32 °C overnight while shaking at 250 rpm. The seed culture (1 ml) was then used to inoculate 50 ml of LB broth in a 250 mL shake flask. Cultures were grown to an optical density (OD_{600}) of either ~0.6 or ~2.0, at which time either tCA or styrene was added to the flasks at an array of final concentrations between 0 to 1 g/L. Two different initial OD_{600} levels were used to investigate the potential effects of product toxicity as a function of cell density and/or growth stage. In all cases, culturing then resumed at 32 °C for another 6–8 h while cell growth, as determined by OD_{600} measurements, was periodically monitored using a UV/vis spectrophotometer (Beckman Colter DU800, Brea, CA).

2.4. Cloning of candidate genes encoding PAL activity from *A. variabilis*, *N. punctiforme*, and *A. thaliana*

All genes used in this study were PCR amplified using a BioRad iCycler system with Phusion DNA Polymerase (Finnzymes, Espoo, Finland) using custom oligonucleotide primers. PCR cycling and reaction conditions were standardized according to manufacturer instructions. Candidate PAL encoding genes were amplified from the cDNA of *A. variabilis* and *N. punctiforme*. Candidate PAL encoding genes from *A. thaliana* were derived from cDNA library plasmids containing the specific loci of interest (Table 1) obtained from the ABRC (Ohio State University, Columbus, OH). Whereas *A. thaliana* posses four distinct PALs, we focused on those encoded by *PAL1* and *PAL2* as these have displayed the greatest activity when expressed in recombinant *E. coli* (noting also that PAL3 was

Please cite this article as: McKenna, R., Nielsen, D.R., Styrene biosynthesis from glucose by engineered *E. coli*. Metab. Eng. (2011), doi:10.1016/j.ymben.2011.06.005

found to be of 'very low activity') (Cochrane et al., 2004). Amplified linear DNA fragments were purified using the Zyppy Clean and Concentrator kit (Zymo Research, Orange, CA). Purified fragments were treated by endonuclease digestion using appropriate restriction enzymes (all from New England Biolabs, Ipswich, MA). Amplified fragments containing the *pal* from both *N. punctiforme* and *A. variabilis* were digested with *BamHI* and *EcoRI* whereas fragments containing *PAL1* and *PAL2* from *A. thaliana* were digested with *EcoRI* and *SphI*. All digestions were performed at 37 °C according to manufacturer's protocols. The expression vector pSTV28 was similarly digested with either *BamHI* and *EcoRI* or *EcoRI* and *SphI*. Digested fragments were gel purified using the Zyppy Gel DNA recovery kit (Zymo Research, Orange, CA). Linearized fragments were ligated using T4 DNA ligase (New England Biolabs, Ipswich, MA) at 4 °C overnight. Chemically competent *E. coli* NEB10-Beta (New England Biolabs, Ipswich, MA) cells were used for all transformations. Transformants were selected by plating on LB solid agar containing 34 mg/L chloramphenicol and culturing at 37 °C overnight. The transformant pool was subsequently screened according to both colony PCR (employing the same primers as used in the initial amplification) and restriction digest mapping of the resultant plasmids to identify those clones harboring the successful construct. This approach resulted in construction of plasmids pSpalAv, pSpalNp, pSpal1At, and pSpal2At, as listed in Table 1.

2.5. Cloning of candidate genes encoding PADC activity from L. plantarum, B. subtilis, and S. cerevisiae

Candidate PADC encoding genes, including *pdc*, *padC*, *PAD1*, and *FDC1*, were amplified via PCR using genomic DNA templates derived from *L. plantarum*, *B. subtilis*, and *S. cerevisiae*, respectively. PCR amplified DNA fragments were purified before treatment by endonuclease digestion. Fragments containing *padC* and *pdc* were each digested with *BamHI* and *SbfI*. The *E. coli* expression vector pTrc99A was similarly digested with *BamHI* and *SbfI*. Meanwhile, the amplified fragment containing *PAD1* was digested with *NcoI* and *XbaI* whereas the *FDC1* containing fragment was digested with *SalI* and *HindIII*. The *E. coli* expression vector pTrc99A was similarly digested with either *NcoI* and *XbaI* (for the insertion of *PAD1*) or *SalI* and *HindIII* (for the insertion of *FDC1*). All digested fragments were gel purified then ligated at 4 °C overnight before their transformation into chemically competent *E. coli* NEB10-Beta. Selection was then achieved by plating transformed cells on LB solid agar containing 100 mg/L ampicillin and culturing at 37 °C overnight. After confirmation of the correct transformant, these works resulted in the generation of plasmids pTpadCBs, pTpdcLp, pTpad1Sc, and pTfdc1Sc, as listed in Table 1. The *SalI–HindIII* digested *FDC1* fragment was then also cloned into the same sites in the newly generated plasmid pTpad1Sc by an analogous protocol, resulting in the plasmid pTpad1Sc-fdc1Sc.

2.6. Assaying PAL activity in crude lysates of recombinant E. coli

Each of the newly created PAL harboring plasmids (namely pSpalAv, pSpalNp, pSpal1At, and pSpal2At) were individually transformed into *E. coli* BL21(DE3) (Invitrogen, Carlsbad, CA). Seed cultures of each of the resultant strains were prepared (in triplicate) in 5 mL LB broth supplemented with 34 mg/L chloramphenicol and cultured at 32 °C while shaking at 250 rpm overnight. 50 μl of each seed was then used to again inoculate 5 ml LB broth. These cultures were grown until reaching an OD_{600} of ~0.6, at which point each was induced by the addition of isopropyl β-D-1-thiogalactopyranoside (IPTG) at a final concentration of 0.2 mM. Induced cultures were incubated for an additional 6 h after which an equal number of cells (determined by OD_{600} measurement) were collected by centrifugation at 1400 × *g* for 4 min. The cell pellet was re-suspended in 900 μL distilled water. Cell lysis was achieved using the FastBreak Cell Lysis Reagent kit (Promega, Madison, WI) and the supernatant collected after centrifugation at 11,000 × *g* for 2 min. PAL activity was analyzed at room temperature in pH 7.5 50 mM Tris-HCl buffer containing 250 mM L-phenylalanine. Activity assays were initiated by the addition of 5 μL of crude cell lysate. The production of tCA was followed at 290 nm on a Beckman Colter UV/vis spectrophotometer for a total of 5 min at 20 s intervals. A molar extinction coefficient of 9000 M^{-1} cm^{-1} and a 1 cm path length were used to establish enzyme activity in terms of U mg^{-1} protein. The PAL protein content in each crude lysate was determined via first separation by SDS-PAGE using Mini-PROTEAN TGX 4–20% precast gels (Bio-Rad, Hercules, CA) and standard protocols. Concentration was then analyzed using the ImageJ software package (NIH, Bethesda, MD) and calibrated vs. Precision Plus Unstained Protein Standards (Bio-Rad, Hercules, CA).

2.7. Assaying PAL activity in recombinant E. coli whole cells

Seed cultures of *E. coli* BL21(DE3) harboring one of pSpalAv, pSpalNp, pSpal1At, or pSpal2At, were prepared in 5 mL LB broth and grown overnight. Shake flasks (250 mL) containing 50 mL of LB were inoculated with 1 mL of each seed culture. Cultures were grown until an OD_{600} of ~0.6 was reached, at which point the cultures were induced by IPTG addition at a final concentration of 0.2 mM. Cultures were then incubated for an additional 6 h (resulting in an OD_{600} of ~2) before an equal number of cells were collected and centrifuged at 1400 × *g* for 5 min. The pellet was washed once with PBS (phosphate buffered saline) at pH 7 before being re-suspended in 12 ml PBS buffer. Finally, the appropriate substrate, L-phenylalanine or L-tyrosine at a final concentration of 1 g/L, was added to the suspension. The suspensions were then shaken at 32 °C for a total of 3 h. Samples (1 mL) were taken every hour, centrifuged, and 750 μL of supernatant was collected for HPLC analysis to monitor the production of either *tCA* or *p*-coumaric acid.

2.8. Assaying PADC activity in recombinant E. coli whole cells

The plasmids pTpad1Sc-fdc1Sc, pTpad1Sc, pTfdc1Sc, pTpdcLp, and pTpadcBs were each individually transformed into chemically competent *E. coli* BL21(DE3). A seed culture of each strain was then grown in LB broth overnight. Shake flasks (250 mL) containing 50 mL of LB were inoculated with 1 ml of seed culture. Cultures were grown at 32 °C until an OD_{600} of ~0.6, at which point they were induced by adding 0.2 mM IPTG and then incubated for an additional 6 h. Cells were then collected and re-suspended in 12 ml PBS buffer (as previously described) and the substrate (*tCA* or *p*-coumaric acid) was added at a final concentration of 1 g/L. Samples (1 mL) were removed from the culture at both the time of initiation as well as after 12 h of incubation at 32 °C and analyzed by HPLC using the methods described herein.

2.9. Co-expression of PAL and CADC isoenzymes in E. coli NST74 to convert L-phenylalanine to styrene

The L-phenylalanine over-producing strain *E. coli* NST74 (Table 1) was co-transformed with the plasmids pSpal2At and pTfdc1Sc and selected for on LB agar supplemented with 100 mg/L ampicillin and 34 mg/L chloramphenicol. The resultant transformant was then grown overnight at 32 °C in 5 mL LB broth. Shake flasks (250 mL) containing 50 mL LB were inoculated with 1 mL of seed culture. The culture was then grown at 32 °C until

Please cite this article as: McKenna, R., Nielsen, D.R., Styrene biosynthesis from glucose by engineered E. coli. Metab. Eng. (2011), doi:10.1016/j.ymben.2011.06.005

an OD_{600} of ~0.6, at which time it was induced with 0.2 mM IPTG and then incubated for an additional 8 h. Cells were then collected and re-suspended in 12 ml PBS buffer (as previously described) and L-phenylalanine added at a final concentration of either 400 or 900 mg/L. Samples were taken periodically and analyzed by HPLC to determine the content of L-phenylalanine, tCA, and styrene, according to the methods presented below.

2.10. Co-expression of PAL and CADC isoenzymes in E. coli NST74 to convert glucose to styrene in shake flask cultures

The L-phenylalanine over-producing strain *E. coli* NST74 was co-transformed with each of the following combinations of plasmids: pSpalAv and pTfdc1Sc, pSpalNp and pTfdc1Sc, pSpal1At and pTfdc1Sc, and pSpal2At and pTfdc1Sc.Single colonies were then selected from the resulting transformants and grown in 5 mL LB broth for 12 h at 32 °C while shaking at 250 rpm to prepare a seed culture. Each seed (1 mL) was then used to inoculate 50 mL MM1. These cultures were performed in 100 mL serum bottles outfitted with septa caps that were tightly sealed upon inoculation. A closed system was used to avoid volatile product (i.e., styrene) losses. A large headspace was used to preclude the depletion of oxygen from the bottle. Cultures were grown for 10 h prior to being induced by the addition of IPTG at a final concentration of 0.2 mM. Culturing continued for 29–48 h post-induction while 1 ml samples were periodically taken from each culture and analyzed for relevant metabolite contents via HPLC, using the methods described herein.

2.11. Metabolite analysis by HPLC

Samples were prepared by removing 1 mL of culture from a shake flask culture and pelleting the cells at 11,000 × *g* for 2 min. The supernatant (0.75 mL) was then transferred to a glass HPLC vial and sealed with a Teflon-lined cap. HPLC analysis was carried out using a Hewlett Packard 1100 series HPLC system equipped with an auto sampler, diode array (UV/vis) detector, and reverse-phase Hypersil Gold SBC18 column (4.6 mm × 150 mm; Thermo Fisher, USA). Samples (5 μL) were injected for analysis at a total constant flow rate of 1.0 ml/min and constant column temperature of 45 °C. The column was eluted with 'solvent A' (consisting of double-distilled water) and 'solvent B' (consisting of methanol (99.8% grade) plus 0.1% trifluoroacetic acid (TFA)). The eluent began as a mixture of 95% solvent A and 5% solvent B before a linear gradient was applied over 8 min to then reach a mixture of 20% solvent A and 80% solvent B. This eluent composition was then held constant for 2 min before a second linear gradient was then applied over the course of 4 min to achieve a final mixture of 95% solvent A and 5% solvent B. The eluent was monitored at each of 215 nm for L-phenylalanine and 258 nm for *tCA*, *p*-coumaric acid, hydroxystyrene, and styrene. Under these conditions L-phenylalanine, *p*-coumaric acid, *tCA*, *p*-hydroxystyrene, and styrene were eluted at 4.5, 6.7, 8.67, 8.78, and 10.4 min, respectively.

2.12. Confirmation of styrene biosynthesis by GC–MS

Culture supernatant (1 ml) was added to hexane (1 ml) and vortexed for 20 min at maximum speed. The biphasic mixture was centrifuged for 1 min at 11,000 × g to settle. 750 μl of the hexane layer was removed for analysis by GC–MS. GC–MS analysis was performed on a Hewlett Packard 5890 Series II gas chromatograph with a flame ionizing detector and Supelco MDN-5S (30 m × 0.25 mm id) fused-silica capillary column using helium as the carrier gas. The injector, column, and detector temperatures were initial set at 280, 240, and 220 °C, respectively. The column temperature was then increased from 40 to 320 °C at 14 °C/min.

3. Results and discussion

3.1. Assaying styrene toxicity

Under the studied conditions, maximal titers of L-phenylalanine produced by *E. coli* NST74 after 48 h of culture in MM1 (with 1.5% glucose) reached about 700–1000 mg/L (results not shown). Whereas the theoretical yield of L-phenylalanine on glucose is 0.55 g/g (Baez-Viveros et al., 2004), achievable yields by *E. coli* NST74 under the present conditions in shake flask cultures were only 0.052–0.074 g/g (or about 10% of theoretical). Assuming complete conversion of all endogenously produced L-phenylalanine to styrene were possible (for example, if our engineered pathway could achieve a particularly high flux), we would expect to be able to synthesize up to 440–630 mg/L of styrene under the same culture conditions (corresponding to a glucose yield of 0.051–0.072 g/g), barring any physiological limitations such as product toxicity. Through a preliminary screening study, however, we have found that the products *trans*-cinnamate ($\log K_{ow}=2.13$) and styrene ($\log K_{ow}=3.05$) can each inhibit *E. coli* growth when added to cultures at final concentrations above ~800 and ~300 mg/L, respectively, as demonstrated in Fig. 2. For reference, *p*-hydroxystyrene ($\log K_{ow}=2.28$) has previously been shown to inhibit *P. putida* at concentrations of ~540 mg/L (Verhoef et al., 2009). Solvent hydrophobicity is typically a good

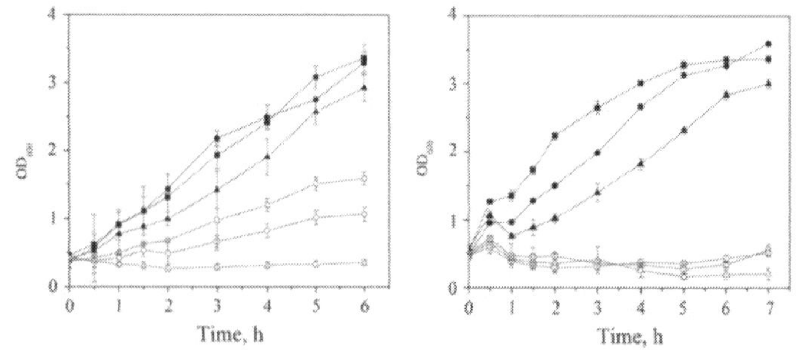

Fig. 2. Left: growth response of *E. coli* NST74 to tCA at concentrations of 0 mg/L (solid square), 200 mg/L (solid circle), 400 mg/L (solid triangle), 600 mg/L (open square), 800 mg/L (open circle), and 1000 mg/L (open triangle). Right: growth response of *E. coli* NST74 to styrene at concentrations of 0 mg/L (solid square), 100 mg/L (solid circle), 200 mg/L (solid triangle), 300 mg/L (open square), 400 mg/L (open circle), and 500 mg/L (open triangle). Error bars reported at one standard deviation from triplicate experiments.

Please cite this article as: McKenna, R., Nielsen, D.R., Styrene biosynthesis from glucose by engineered *E. coli*. Metab. Eng. (2011), doi:10.1016/j.ymben.2011.06.005

indicator of its toxicity towards microorganisms and may be quantified by log K_{ow}, the log of the octanol–water partition coefficient (Ramos et al., 2002). Even at low concentrations, compounds with a log K_{ow} between 1.0 to 5.0 tend to be very toxic (Heipieper et al., 1994). Accumulation of hydrophobic aromatics within the cytoplasmic membrane has been shown to disrupt its integrity, resulting in the leakage of ions, metabolites, lipids, and proteins, as well as affecting the cells ability to maintain its internal pH and an appropriate transmembrane proton gradient (Ramos et al., 2002; Weber et al., 1993). The measured toxicity thresholds were found to be consistent at both low (OD_{600} 0.6) and high (OD_{600} 2, data not shown) initial cell densities (reflecting early and late exponential growth, respectively), indicating growth inhibition was independent of cell density and growth stage. As a result of the inherent toxicity of styrene we anticipated that its biosynthesis using an NST74 host platform would be limited to below its maximum potential under our culture conditions. While no prior studies have specifically explored either the effects of styrene toxicity on *E. coli* or strategies to improve its tolerance, several *Pseudomonas* sp. have been shown to display enhanced styrene tolerance characteristics (Weber et al., 1993). Whereas the present study is solely focused on the prototyping of a novel pathway for styrene biosynthesis, it is clear from our preliminary assays that styrene toxicity must eventually be overcome or effectively circumvented if renewable styrene production is ever to become viable or sustainable.

Table 2
Specific activity of PAL isoenzymes from *A. variabilis*, *N. punctiforme*, and *A. thaliana* on L-phenylalanine and L-tyrosine when expressed in recombinant *E. coli* BL21(DE3). Errors reported at one standard deviation from triplicate experiments.

Strain	Substrate	Activity (U mg^{-1} protein)
BL21(DE3)	L-Phenylalanine	n.d.
	L-Tyrosine	n.d.
pSencPSm	L-Phenylalanine	n.d.
	L-Tyrosine	n.d.
pSpalAv	L-Phenylalanine	2.38 ± 0.64
	L-Tyrosine	n.d.
pSpalNp	L-Phenylalanine	0.91 ± 0.32
	L-Tyrosine	n.d.
pSpal1At	L-Phenylalanine	2.42 ± 1.07
	L-Tyrosine	n.d.
pSpal2At	L-Phenylalanine	4.08 ± 0.11
	L-Tyrosine	n.d.

n.d.—not detected.

3.2. Screening candidate PAL isoenzymes for activity in recombinant E. coli

The activities of recombinant PALs from various genetic sources were analyzed according to both *in vitro* (crude lysate) and *in vivo* (whole, resting cell) assays. All recombinant PALs, expressed from plasmids pSencPSm, pSpalAv, pSpalNp, pSpal1At, and pSpal2At (Table 1) in an *E. coli* BL21(DE3) background were recovered as crude lysates. According to our *in vitro* assay results, all of the tested PALs showed comparable levels of activity on L-phenylalanine as substrate (Table 2), with the exception of EncP from *S. maritimus* whose activity was non-measureable. It has previously been shown through kinetic studies on prokaryotic PALs that those derived from *N. puntiformes* and *A. variabilis* possessed 500–1000 times greater activity than EncP from *S. maritimus*; furthermore, the k_{cat}/K_m value of PAL from *A. variabilis* was found to be greater than that of *N. puntiformes* (72.2 and 43.8 $mM^{-1}s^{-1}$, respectively) (Moffitt et al., 2007b). Thus, it is possible that EncP was in fact functionally expressed in our study; however, its activity was too low to measure according to the protocols employed. The relative activities of candidate PALs were then further explored through the use of whole-cell assays in recombinant *E. coli*. Resting cells suspended in PBS buffer (pH 7) were supplemented with L-phenylalanine or L-tyrosine and product (tCA or *p*-coumaric acid, respectively) formation was monitored over the course of 3 h. Although the PALs from *A. thaliana* were found to possess the greatest specific activity in crude lysates (Table 2), assays of *in vivo* function provided a more stark contrast into the relative activities among all candidate PALs, as shown in Fig. 3. Under the studied conditions, the PALs of *A. thaliana* were found to display the greatest activity (being significantly higher that of either prokaryotic PAL) although being comparable for both isoenzymes. This result is more consistent with previous studies wherein k_{cat}/K_m values of PAL1 and PAL2 were estimated as 25,500 and 51,200 $M^{-1}s^{-1}$ (or six orders of magnitude greater than those reported for *N. puntiformes* and *A. variabilis*) when expressed in recombinant *E. coli* (Cochrane et al., 2004). The elevated relative activity of PAL2 is also consistent with the observation of higher initial rates of tCA accumulation when expressed from BL21(DE3) pSpal2At, as seen in Fig. 3.

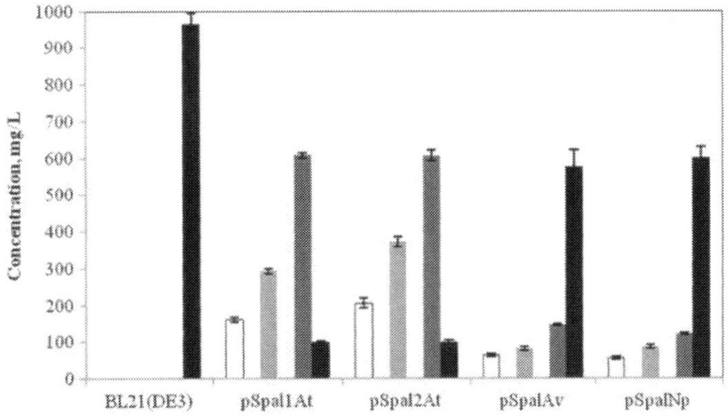

Fig. 3. Phenylalanine ammonia lyase activity from candidate genes cloned from *A. thaliana*, *A. variabilis*, and *N. punctiformes* in recombinant *E. coli* BL21(DE3) whole cells. 50 ml cultures were grown for 8 h (induced with 0.2 mM IPTG after 1.5 h), spun down and re-suspended in 10 ml PBS buffer. The conversion of 1 g/L L-phenylalanine (black) to *trans*-cinnamic acid after 1 h (no color), 2 h (light gray), and 3 h (dark gray). Error bars reported at one standard deviation from triplicate experiments.

Please cite this article as: McKenna, R., Nielsen, D.R., Styrene biosynthesis from glucose by engineered *E. coli*. Metab. Eng. (2011), doi:10.1016/j.ymben.2011.06.005

Again, however, none of the tested PAL isoenzymes were found to display any activity on L-tyrosine as the substrate (data not shown), which is consistent with the results of our *in vitro* assays (Table 2). Taken together, these analyses demonstrate that all of the studied PALs display rigid substrate specificity for L-phenylalanine. This important result contrasts numerous prior works on aromatic pathway development in recombinant *E. coli*, which have solely relied upon the use of bifunctional yeast PAL/TALs (Cui et al., 2008; Gilbert et al., 1985; Gilbert et al., 1983; Gilbert and Tully, 1982) and which was specifically used to synthesize *p*-hydroxystyrene (Qi et al., 2007). As these results illustrate that the first committed step in our pathway is highly specific for the intended substrate (L-phenylalanine) alone, we anticipate that this advantageous outcome will ultimately help to control product purity (specifically styrene over *p*-hydroxystyrene or a mixture of products) while also improving the activity and flux of our desired pathway. Meanwhile, the same cannot be assured had the first committed step of our pathway been catalyzed by PAL/TAL, as was the case for all engineered *p*-hydroxystyrene pathways reported to date (Qi et al., 2007; Verhoef et al., 2009).

3.3. Screening candidate PADC isoenzymes for tCA decarboxylase activity in recombinant E. coli

Candidate PADC isoenzymes from *L. plantarum*, *B. subtilis*, and *S. cerevisiae* were screened for their ability to decarboxylate tCA to produce styrene when expressed in *E. coli*. Plasmids harboring the candidate PADC-encoding genes (Table 1) were individually transformed into *E. coli* BL21(DE3), as described above. Whole, resting cells were prepared in PBS buffer supplemented with 1 g/L of tCA acid or *p*-coumaric acid. The production of styrene or *p*-hydroxystyrene, respectively, was then followed periodically, and the results after 12 h of culture are compared in Fig. 4. With the exception of the strain expressing *PAD1* from *S. cerevisiae* alone, all other strains displayed decarboxylase activity on *p*-coumaric acid, leading to *p*-hydroxystyrene biosynthesis. These results are consistent with previous reports on the functional expression of *pdc* and *padC* in recombinant *E. coli* (Qi et al., 2007) to support p-hydroxystyrene biosynthesis from glucose. However, our findings further and importantly show that the enzymes from *L. plantarum* and *B. subtilis* cannot catalyze the conversion of tCA to styrene (thus making them more specific for *p*-coumaric acid). In the present study it was instead found that the sole expression of *FDC1* from *S. cerevisiae* (which was previously characterized to encode a ferulic acid decarboxylase) was sufficient for achieving decarboxylase activity on either tCA or *p*-coumaric acid (with perhaps a slight preference towards tCA, based on overall conversion). Though it was previously reported that the co-expression of both *FDC1* and *PAD1* is necessary to achieve tCA decarboxylase activity in the native *S. cerevisiae* (Mukai et al., 2010), we now report that functional tCA decarboxylase activity in *E. coli* depends solely upon *FDC1* over-expression and is not dependent upon the co-expression of *PAD1*. Furthermore, as can be seen in Fig. 4, comparable styrene titers were achieved when either *FDC1* was expressed alone or together with *PAD1*, indicating that the co-expression of *PAD1* does not increase (or otherwise alter) tCA decarboxylase activity in *E. coli*, as has previously been suggested in the native *S. cerevisiae* (Larsson et al., 2001). It is, however, plausible that the expression of *ubix* (which has been shown to be 50% similar to *PAD1* (Mukai et al., 2010)) in our *E. coli* background could have served to compensate for the absence of *PAD1* expression, enabling tCA decarboxylase activity to be achieved in *E. coli* expressing *FDC1* alone. Future studies will explore the deletion of *ubix* from the background of *E. coli* to further investigate the recombinant function of *FDC1*. Most importantly, these results have demonstrated that *FDC1* over-expression in *E. coli* uniquely enables the decarboxylation of tCA to styrene.

Although the proposed styrene biosynthesis pathway is somewhat analogous to that which was previously developed for

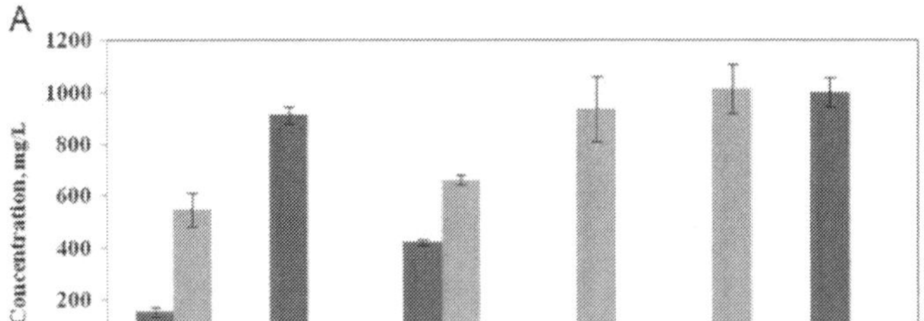

Fig. 4. Phenylacrylic acid decarboxylase activity from candidate genes cloned from *S. cerevisiae*, *L. plantarum*, and *B. subtilis* in recombinant *E. coli* BL21(DE3) whole cells. The conversion of (A) 1 g/L *p*-coumaric acid (dark gray) to *p*-hydroxystyrene (light gray) and (B) 1 g/L *trans*-cinnamic acid (dark gray) to styrene (light gray) after 12 h. Error bars reported at one standard deviation from triplicate experiments.

Please cite this article as: McKenna, R., Nielsen, D.R., Styrene biosynthesis from glucose by engineered *E. coli*. Metab. Eng. (2011), doi:10.1016/j.ymben.2011.06.005

p-hydroxystyrene biosynthesis (Qi et al., 2007; Verhoef et al., 2009), it is essential to realize that the specific enzyme 'parts' used in those two prior studies are wholly inadequate for achieving the present objective of styrene biosynthesis from glucose. In other words, said previously characterized enzymes were not simply transferrable among the two (styrene and p-hydroxystyrene) distinct pathways, but rather that styrene biosynthesis could only be supported through the identification and characterization of a specific tCA decarboxylase.

3.4. Biosynthesis of tCA and styrene from glucose in recombinant E. coli

The biosynthesis of tCA from glucose was first investigated using strains of *E. coli* NST74 that individually carried the plasmids pSpalAv, pSpalNp, pSpal1At, and pSpal2At. Cultures were grown in MM1 media with 1.5% glucose. 48 h after induction with 0.2 mM IPTG, the production of tCA was detected in each culture at final titers of 600, 473, 648, and 918 mg/L, respectively. Neither styrene nor p-coumaric acid were detected in any samples.

To test for styrene biosynthesis from glucose, *FDC1* from *S. cerevisiae* was then co-expressed with each of the PAL-encoding genes according to the following constructed strains: NST74 pSpalAv pTfdc1Sc, NST74 pSpalNp pTfdc1Sc, NST74 pSpal1At pTfdc1Sc, and NST74 pSpal2At pTfdc1Sc. Conversion of glucose to styrene was followed after IPTG induction and results for the strain NST74 pSpal2At pTfdc1Sc are shown in Table 3 (other results not shown as this dataset was representative of the results and trends observed with all other strains). Substantial L-phenylalanine accumulation was not observed until 17 h post-induction, after which time styrene titers then also rose considerably, reaching a final titer of 260 mg/L in the culture medium 29 h post-induction. The final styrene titers for the strains NST74 pSpalAv pTfdc1Sc, NST74 pSpalNp pTfdc1Sc, and NST74 pSpal1At pTfdc1Sc were found to be 210, 183, and 188 mg/L, respectively. As styrene-producing strains were cultured, a strong 'hydrocarbon' aroma was readily detected upon opening the sealed culture bottles. The high volatility of styrene necessitated the use of sealed jars for culturing. To ensure that sufficient oxygen remained available to the culture, a large headspace volume (200 mL headspace vs. 50 mL media) was used. Thus, cultures were maintained under aerobic conditions throughout the culture (as confirmed by the inclusion of the indicator dye resazurin). However, the use of such a large headspace volume also allows for the significant accumulation of styrene vapor, which should also be accounted for when assessing productivity. Under dilute and near-ambient conditions, the equilibrated headspace vapor composition can be estimated by application of Henry's Law. Using a dimensionless Henry's Law constant of 0.113 (Yang, 1992), it can be predicted that a headspace equilibrated with an aqueous phase containing 260 mg/L styrene will contain an additional 29 mg/L styrene, for a total of 18.9 mg in the 250 mL flask (note this would be equivalent to an aqueous titer of 376 mg/L if no volatilization had occurred). In an analogous manner, the final styrene production by strains NST74 pSpalAv pTfdc1Sc, NST74 pSpalNp pTfdc1Sc, and NST74 pSpal1At pTfdc1Sc would be estimated as 15.2, 13.4, and 13.6 mg, respectively.

Table 3
Biosynthesis of L-phenylalanine, *trans*-cinnamic acid, and styrene by *E. coli* NST74 pSpal2At pTfdc1Sc from glucose in MM1 media. Errors reported at one standard deviation from triplicate experiments.

Time (h)	L-Phenylalanine (mg/L)	*trans*-Cinnamic acid (mg/L)	Styrene (mg/L)
0	13.9 ± 0.03	0	0
13	56.4 ± 0.13	3.7 ± 0.01	7.3 ± 0.12
17	236.5 ± 0.57	6.3 ± 0.01	23 ± 0.38
21	167.4 ± 0.40	9.1 ± 0.02	205 ± 3.40
25	152.2 ± 0.36	13.5 ± 0.02	243 ± 4.03
29	179.5 ± 0.43	14.0 ± 0.03	260 ± 4.31

Following these fermentation studies, culture supernatants were extracted using hexane, and the extracts were analyzed by GC–MS to confirm that it was in fact styrene that was being synthesized by our cultures. Comparing the spectra of the dominant metabolite peak which was recovered from the extract to the NIST08 spectral database (Babushok et al., 2007), styrene was found to be the most probable compound. As can be seen from Fig. 5, the mass spectra of the extracted sample and the library reference provide an excellent match, providing confirmation that styrene was in fact synthesized by our engineered strains of *E. coli*.

Throughout the duration of culture, never was all of the L-phenylalanine observed to be fully assimilated into the styrene pathway (as seen in Table 3). Furthermore, tCA titers were observed to remain low throughout, indicating that almost all of the synthesized tCA could be quickly converted to styrene. Taken together, these observations suggest that low PAL activity presently remains as the flux limiting condition in the engineered styrene biosynthesis pathway. As the first committed step in the styrene pathway, high PAL activity is essential and must be

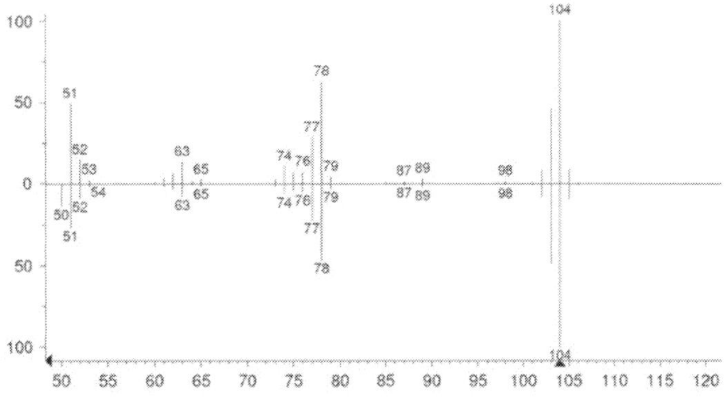

Fig. 5. Head to tail comparison of the standard mass spectra showing the relative abundance of the mass-to-charge ratio of styrene from the NIST08 library (lower) with that of the dominant metabolite peak obtained in hexane extractions of the culture broth (upper).

Please cite this article as: McKenna, R., Nielsen, D.R., Styrene biosynthesis from glucose by engineered E. coli. Metab. Eng. (2011), doi:10.1016/j.ymben.2011.06.005

improved in subsequent generations of our strains. To start, the expression of PAL can likely be improved through the use of codon optimized variants. This approach has shown to be useful, particularly in cases for which pathways involve the expression of plant-derived genes in *E. coli*. For example, codon optimization was applied to the amorphadiene oxidase from *Artemisia annua* (a plant species) as part as a synthetic pathway to produce artemisinic acid (Keasling, 2008). Whereas the native protein originally showed neither *in vivo* nor *in vitro* activity when expressed in *E. coli*, a codon optimized variant resulted in its functional expression in *E. coli*, contributing to successful pathway development.

Whereas improvements in PAL expression can lead to enhancements in the specific metabolite flux, the net flux can also be improved by increasing the pathway 'driving force'. That is, by promoting the increased availability of the pathway's immediate precursor, L-phenylalanine. This notion was tested in resting cells assays wherein *E. coli* NST74 pSpal2At pTfdc1Sc cell suspensions were supplemented with exogenous L-phenylalanine at initial concentrations of either 400 or 950 mg/L. As shown in Fig. 6, the conversion of 400 mg/L L-phenylalanine to styrene occurred rapidly and completely (i.e., no residual L-phenylalanine or tCA was detectable), yielding a final styrene titer of 250 mg/L after 30 h. Meanwhile, when 950 mg/L L-phenylalanine was added to the resting cell cultures, nearly 500 mg/L styrene could be produced as all of the L-phenylalanine was consumed. However, in this case, nearly an additional 250 mg/L tCA also remained in the culture. Noting that such a final styrene titer actually surpasses the toxicity threshold of growing cells, it is not altogether surprising that the reaction was unable to proceed to completion. It is speculated that the final styrene titer was only capable of surpassing the previously-determined toxicity threshold in resting cell studies because the biotransformation in this case was no longer growth dependent. That is to say that the stationary phase resting cells had previously assembled the necessary pathway enzymes and were not reliant upon endogenously synthesized L-phenylalanine as a precursor. Thus, the impediment of conversion observed at further elevated styrene titers most likely represents an inhibition imposed at the level of the pathway enzymes rather than an impairment of cell growth, as was observed in our previous characterization experiments. Further experiments will be required to characterize and understand this potential inhibitory mechanism. These results demonstrate that if the L-phenylalanine precursor pool can be enhanced the net production of styrene can also be increased. Clearly, however, the addition of exogenous L-phenylalanine is not a sustainable approach towards enhancing styrene biosynthesis, but rather endogenous L-phenylalanine production must be enhanced in our host platform. Subsequent studies will explore the incorporation of additional modifications into the genome of the *E. coli* NST74 host that have previously been shown by other groups to result in L-phenylalanine yield enhancements. For instance, it has been reported that the over-expression of endogenous transketolase I (encoded by *tktA*) supports elevated L-phenylalanine biosynthesis (Gosset et al., 1996). TktA over-expression enables the enhanced biosynthesis of erythrose 4-phosphate (E4P) which, when condensed with phosphoenolpyruvate (PEP), yields 3-deoxy-D-arabino-heptulosonate 7-phosphate (DAHP), the first committed intermediate in the aromatic amino acid biosynthesis pathway (Flores et al., 1996; Lütke-Eversloh and Stephanopoulos, 2008). Meanwhile, disruption of the global carbon storage regulatory system of *E. coli* through deletion of *csrA*, increases PEP biosynthesis, leading to up to a 2-fold enhancement in L-phenylalanine yield (Tatarko and Romeo, 2001).

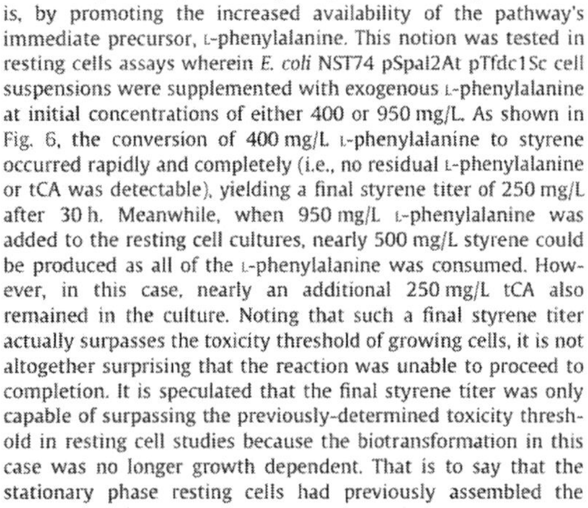

Fig. 6. Whole cell production of styrene (triangle) from *trans*-cinnamic acid (circle) after exogenous L-phenylalanine (square) addition by *E. coli* NST74 pSpal2At pTfdc1Sc. Open and filled shapes correspond to the initial addition of 400 and 950 mg/L L-phenylalanine.

For microbially derived, renewable styrene to become an economically viable and sustainable alternative to petroleum-derived predecessor, titers and productivity must ultimately be improved. Although the styrene titers achieved by our 1st generation strains were modest, in comparison to the toxicity assays performed above it can be seen that they are already approaching the inhibitory threshold. Thus, whereas increasing L-phenylalanine yields could eventually translate into elevated styrene production, product toxicity would soon become the subsequent limiting factor and must be addressed. As complex phenotypes such as solvent tolerance are simply not monogenic in nature (Alper et al., 2006), numerous specific mutations must applied in concert to achieve the specifically desired result. Since little is presently known of styrene toxicity or tolerance, rational approaches to engineering more tolerant strains may be less suitable than higher throughput, combinatorial strategies. Although commonly employed, combinatorial approaches towards enhancing desired phenotypes often rely upon the aggressive use of chemomutagenesis, such techniques are less desirable as they can also result an unforeseen and difficult to understand negative impacts on host fitness and/or productivity (Bonomo et al., 2006), in addition to requiring laborious screening and selection procedures. A more effective approach for the present application might involve the use of an alternative host platform which inherently boasts greater tolerance (Fischer et al., 2008), a strategy which has been applied for the production of biofuels, such as n-butanol (Nielsen et al., 2009). The bacterium *P. putida* S12 has been engineered, for example, as a solvent tolerant platform for the biosynthesis of both *p*-hydroxybenzoate and *p*-hydroxystyrene (Verhoef et al., 2007, 2009), and might also make excellent starting point for styrene production. Since our pathway is derived from the ubiquitous, proteinogenic amino acid L-phenylalanine, its transference to an alternative host platform remains wholly compatible. Alternatively, improved product tolerance can also be achieved through genome evolution, as accomplished through serial adaptations to increasing styrene concentrations. Such an approach has worked well for achieving tolerance to biofuels like n-butanol and could also be coupled with genomic library screens to provide the first comprehensive view of styrene inhibition and tolerance in *E. coli* (Reyes et al., 2011). In an interesting example, prior works have shown that *Pseudomonas* sp. may be adapted to styrene when also grown in the presence of acetate or similar carboxylic acids as the sole

Please cite this article as: McKenna, R., Nielsen, D.R., Styrene biosynthesis from glucose by engineered *E. coli*. Metab. Eng. (2011), doi:10.1016/j.ymben.2011.06.005

carbon source (Weber et al., 1993). It has been suggested that this evolved phenotype may be associated with genetic changes leading to reductions in membrane fluidity, as well as through the activation of genes believed to be specifically associated with enhanced tolerance to aromatic compounds. As an alternative strategy, combinatorial procedures such as the global Transcription Machinery Engineering (gTME) (Alper et al., 2006; Nicolaou et al., 2010) may be implemented to improve styrene tolerance in *E. coli* or alternative host platforms selected due to their elevated tolerance baseline. The principals of gTME have been successfully employed to enhance ethanol tolerance in both yeast (Alper et al., 2006) and *E. coli* (Alper and Stephanopoulos, 2007).

In addition to the development of styrene tolerant phenotypes, product toxicity can also be enhanced through the use of in situ product recovery (ISPR). Common ISPR approaches involve solvent extraction (Gyamerah and Glover, 1996; Malinowski, 2001; Weilnhammer and Blass, 1994), adsorption (Nielsen et al., 2010; Nielsen and Prather, 2009), gas and vacuum stripping (Loser et al., 2005), and membrane pervaporation (Vane, 2005). Each of these approaches has been successfully applied for the continuous recovery of biofuel compounds such as ethanol, for example, and even aromatics like L-phenylacetylcarbinol (Khan and Daugulis, 2010) and benzaldehyde (Jain et al., 2010). The latter three approaches may be particularly well-suited for styrene recovery given its volatile nature, as previously discussed.

4. Conclusion

For the first time, the present study has demonstrated the biosynthesis of styrene from renewable resources using an engineered microbial platform. Whereas low activity of pathway enzymes and product toxicity remain as challenges which limit the productivity of our styrene-producing strains, continued improvements will lead to the development of robust biocatalysts for the sustainable production of this important, large volume, commodity chemical.

Acknowledgments

We thank Prof. Bradley Moore (UCSD) and Prof. Kristala Prather (MIT) for their kind gifts. We also thank Tom Colella (ASU) for his skillful technical assistance in GC–MS analysis.

References

Alper, H., Moxley, J., Nevoigt, E., Fink, G.R., Stephanopoulos, G., 2006. Engineering yeast transcription machinery for improved ethanol tolerance and production. Science 314, 1565–1568.

Alper, H., Stephanopoulos, G., 2007. Global transcription machinery engineering: a new approach for improving cellular phenotype. Metabolic Engineering 9, 258–267.

Babushok, V.I., Linstrom, P.J., Reed, J.J., Zenkevich, I.G., Brown, R.L., Mallard, W.G., Stein, S.E., 2007. Development of a database of gas chromatographic retention properties of organic compounds. Journal of Chromatography A. 1157, 414–421.

Baez-Viveros, J.L., Osuna, J., Hernandez-Chavez, G., Soberon, X., Bolivar, F., Gosset, G., 2004. Metabolic engineering and protein directed evolution increase the yield of L-phenylalanine synthesized from glucose in *Escherichia coli*. Biotechnology and Bioengineering 87, 516–524.

Bonomo, J., Warnecke, T., Hume, P., Marizcurrena, A., Gill, R.T., 2006. A comparative study of metabolic engineering anti-metabolite tolerance in *Escherichia coli*. Metabolic Engineering. 8, 227–239.

Boudet, A.M., 2007. Evolution and current status of research in phenolic compounds. Phytochemistry 68, 2722–2735.

Clark, G., 1990. Phenethyl alcohol. Perfumer Flavor 15, 37–44.

Clausen, M., Lamb, C.J., Megnet, R., Doerner, P.W., 1994. PAD1 encodes phenylacrylic acid decarboxylase which confers resistance to cinnamic acid in *Saccharomyces cerevisiae*. Gene 142, 107–112.

Cochrane, F.C., Davin, L.B., Lewis, N.G., 2004. The Arabidopsis phenylalanine ammonia lyase gene family: kinetic characterization of the four PAL isoforms. Phytochemistry 65, 1557–1564.

Cui, J.D., Jia, S.R., Sun, A.Y., 2008. Influence of amino acids, organic solvents and surfactants for phenylalanine ammonia lyase activity in recombinant *Escherichia coli*. Letters in Applied Microbiology 46, 631–635.

DoE, Steam system opportunity assessment for the pulp and paper, chemical manufacturing, and petroleum refining industries. In: Energy, O. o. E. E. a. R. E. U. S. D. o. (Ed.), 2002.

Fernandez, X., Lizzani-Cuvelier, L., Loiseau, A.M., Perichet, C., Delbecque, C., Arnaudo, J.F., 2005. Chemical composition of the essential oils from Turkish and Honduras Styrax. Flavour and Fragrance Journal 20, 70–73.

Fischer, C.R., Klein-Marcuschamer, D., Stephanopoulos, G., 2008. Selection and optimization of microbial hosts for biofuels production. Metabolic Engineering 10, 295–304.

Flores, N., Xiao, J., Berry, A., Bolivar, F., Valle, F., 1996. Pathway engineering for the production of aromatic compounds in *Escherichia coli*. Nature Biotechnology 14, 620–623.

Gilbert, H.J., Clarke, I.N., Gibson, R.K., Stephenson, J.R., Tully, M., 1985. Molecular-cloning of the phenylalanine ammonia-lyase gene from rhodosporidium-toruloides in *Escherichia-coli* K-12. Journal of Bacteriology 161, 314–320.

Gilbert, H.J., Stephenson, J.R., Tully, M., 1983. Control of synthesis of functional messenger-rna coding for phenylalanine ammonia-lyase from rhodosporidium-toruloides. Journal of Bacteriology 153, 1147–1154.

Gilbert, H.J., Tully, M., 1982. Synthesis and degradation of phenylalanine ammonia-lyase of rhodosporidium-toruloides. Journal of Bacteriology 150, 498–505.

Gosset, G., Yong-Xiao, J., Berry, A., 1996. A direct comparison of approaches for increasing carbon flow to aromatic biosynthesis in *Escherichia coli*. Journal of Industrial Microbiology 17, 47–52.

Gyamerah, M., Glover, J., 1996. Production of ethanol by continuous fermentation and liquid-liquid extraction. Journal of Chemical Technology and Biotechnology 66, 145–152.

Heipieper, H.J., Weber, F.J., Sikkema, J., Keweloh, H., de Bont, J.A.M., 1994. Mechanisms of resistance of whole cells to toxic organic solvents. Trends in Biotechnology 12, 409–415.

Jain, A.N., Khan, T.R., Daugulis, A.J., 2010. Bioproduction of benzaldehyde in a solid-liquid two-phase partitioning bioreactor using *Pichia pastoris*. Biotechnology Letters 32, 1649–1654.

Jiang, H., Wood, K.V., Morgan, J.A., 2005. Metabolic engineering of the phenylpropanoid pathway in *Saccharomyces cerevisiae*. Applied and Environmental Microbiology 71, 2962–2969.

Jones, D.H., 1984. Phenylalanine ammonia-lyase: regulation of its induction, and its role in plant development. Phytochemistry 23, 1349–1359.

Keasling, J.D., 2008. Synthetic biology for synthetic chemistry. ACS Chemical Biology 3, 64–76.

Khan, T.R., Daugulis, A.J., 2010. Application of solid-liquid TPPBs to the production of L-phenylacetylcarbinol from benzaldehyde using *Candida utilis*. Biotechnology and Bioengineering 107, 633–641.

Larsson, S., Nilvebrant, N.O., Jonsson, L.J., 2001. Effect of overexpression of *Saccharomyces cerevisiae* Pad1p on the resistance to phenylacrylic acids and lignocellulose hydrolysates under aerobic and oxygen-limited conditions. Applied Microbiology and Biotechnology 57, 167–174.

Loser, C., Schroder, A., Deponte, S., Bley, T., 2005. Balancing the ethanol formation in continuous bioreactors with ethanol stripping. Engineering in Life Sciences 5, 325–332.

Lütke-Eversloh, T., Stephanopoulos, G., 2008. Combinatorial pathway analysis for improved L-tyrosine production in *Escherichia coli*: identification of enzymatic bottlenecks by systematic gene overexpression. Metabolic Engineering 10, 69–77.

Malinowski, J.J., 2001. Two-phase partitioning bioreactors in fermentation technology. Biotechnology Advances 19, 525–538.

Moffitt, M.C., Louie, G.V., Bowman, M.E., Pence, J., Noel, J.P., Moore, B.S., 2007a. Discovery of two cyanobacterial phenylalanine ammonia lyases: kinetic and structural characterization. Biochemistry 46, 1004–1012.

Moffitt, M.C., Louie, G.V., Bowman, M.E., Pence, J., Noel, J.P., Moore, B.S., 2007b. Discovery of two cyanobacterial phenylalanine ammonia lyases: kinetic and structural characterization. Biochemistry 46, 1004–1012.

Mukai, N., Masaki, K., Fujii, T., Kawamukai, M., Iefuji, H., 2010. PAD1 and FDC1 are essential for the decarboxylation of phenylacrylic acids in *Saccharomyces cerevisiae*. Journal of Bioscience and Bioengineering 109, 564–569.

Nicolaou, S.A., Gaida, S.M., Papoutsakis, E.T., 2010. A comparative view of metabolite and substrate stress and tolerance in microbial bioprocessing: from biofuels and chemicals, to biocatalysis and bioremediation. Metabolic Engineering 12, 307–331.

Nielsen, D.R., Amarasiriwardena, G.S., Prather, K.L.J., 2010. Predicting the adsorption of second generation biofuels by polymeric resins with applications for in situ product recovery (ISPR). Bioresource Technology 101, 2762–2769.

Nielsen, D.R., Leonard, E., Yoon, S.H., Tseng, H.C., Yuan, C., Prather, K.L., 2009. Engineering alternative butanol production platforms in heterologous bacteria. Metabolic Engineering.

Nielsen, D.R., Prather, K.J., 2009. In situ product recovery of n-butanol using polymeric resins. Biotechnology and Bioengineering 102, 811–821.

Pagot, Y., Belin, J.M., Husson, F., Spinnler, H.E., 2007. Metabolism of phenylalanine and biosynthesis of styrene in *Penicillium camemberti*. Journal of Dairy Research 74, 180–185.

Please cite this article as: McKenna, R., Nielsen, D.R., Styrene biosynthesis from glucose by engineered *E. coli*. Metab. Eng. (2011), doi:10.1016/j.ymben.2011.06.005

Qi, W.W., Vannelli, T., Breinig, S., Ben-Bassat, A., Gatenby, A.A., Haynie, S.L., Sariaslani, F.S., 2007. Functional expression of prokaryotic and eukaryotic genes in *Escherichia coli* for conversion of glucose to p-hydroxystyrene. Metabolic Engineering 9, 268–276.

Ramos, J.L., Duque, E., Gallegos, M.-T., Godoy, P., Ramos-González, M.I., Rojas, A., Terán, W., Segura, A., 2002. Mechanisms of solvent tolerance in gram-negative bacteria. Annual Review of Microbiology 56, 743–768.

Reyes, L.H., Almario, M.P., Kao, K.C., 2011. Genomic library screens for genes involved in n-butanol tolerance in *Escherichia coli*. PLoS ONE 6, e17678.

Santos, C.N.S., Koffas, M., Stephanopoulos, G., 2011. Optimization of a heterologous pathway for the production of flavonoids from glucose. Metabolic Engineering 13, 392–400.

SRI, 2010. Styrene. Access Intelligence LLC Inc.

Tatarko, M., Romeo, T., 2001. Disruption of a global regulatory gene to enhance central carbon flux into phenylalanine biosynthesis in *Escherichia coli*. Current Microbiology 43, 26–32.

Tran, N.P., Gury, J., Dartois, V., Nguyen, T.K.C., Seraut, H., Barthelmebs, L., Gervais, P., Cavin, J.-F., 2008. Phenolic acid-mediated regulation of the padC gene, encoding the phenolic acid decarboxylase of *Bacillus subtilis*. Journal of Bacteriology 190, 3213–3224.

Vane, L.M., 2005. A review of pervaporation for product recovery from biomass fermentation processes. Journal of Chemical Technology and Biotechnology 80, 603–629.

Vannelli, T., Wei, Qi, W., Sweigard, J., Gatenby, A.A., Sariaslani, F.S., 2007a. Production of p-hydroxycinnamic acid from glucose in *Saccharomyces cerevisiae* and *Escherichia coli* by expression of heterologous genes from plants and fungi. Metabolic Engineering 9, 142–151.

Vannelli, T., Xue, Z.X., Breinig, S., Qi, W.W., Sariaslani, F.S., 2007b. Functional expression in *Escherichia coli* of the tyrosine-inducible tyrosine ammonia-lyase enzyme from yeast Trichosporon cutaneum for production of p-hydroxycinnamic acid. Enzyme and Microbial Technology 41, 413–422.

Verhoef, S., Ruijssenaars, H.J., de Bont, J.A., Wery, J., 2007. Bioproduction of p-hydroxybenzoate from renewable feedstock by solvent-tolerant *Pseudomonas putida* S12. Journal of Biotechnology 132, 49–56.

Verhoef, S., Wierckx, N., Westerhof, R.G., de Winde, J.H., Ruijssenaars, H.J., 2009. Bioproduction of p-hydroxystyrene from glucose by the solvent-tolerant bacterium *Pseudomonas putida* S12 in a two-phase water-decanol fermentation. Applied and Environmental Microbiology 75, 931–936.

Weber, F.J., Ooijkaas, L.P., Schemen, R.M., Hartmans, S., de Bont, J.A., 1993. Adaptation of *Pseudomonas putida* S12 to high concentrations of styrene and other organic solvents. Applied and Environmental Microbiology 59, 3502–3504.

Weilnhammer, C., Blass, E., 1994. Continuous fermentation with product recovery by in-situ extraction. Chemical Engineering & Technology 17, 365–373.

Wierckx, N.J., Ballerstedt, H., de Bont, J.A., Wery, J., 2005. Engineering of solvent-tolerant *Pseudomonas putida* S12 for bioproduction of phenol from glucose. Applied and Environmental Microbiology 71, 8221–8227.

Wu, C., Koylinski, T., Bozik, J., 1981. Preparation of styrene from ethylbenzene US Patent 4,255,599.

Xiang, L.K., Moore, B.S., 2002. Inactivation, complementation, and heterologous expression of encP, a novel bacterial phenylalanine ammonia-lyase gene. Journal of Biological Chemistry 277, 32505–32509.

Xiang, L.K., Moore, B.S., 2005. Biochemical characterization of a prokaryotic phenylalanine ammonia lyase. Journal of Bacteriology 187, 4286–4289.

Xiang, L.K., Moore, B.S., 2006. Biochemical characterization of a prokaryotic phenylalanine ammonia lyase (187, pg4286, 2005). Journal of Bacteriology 188, 5331.

Yang, C.L.Y. a.H.C., 1992. Henry's Law Constant for Compound in Water. Gulf Publishing Company, Houston.

Young, M.R., Towers, G.H.N., Neish, A.C., 1966. Taxonomic distribution of ammonia-lyases for L-phenylalanine and L-tyrosine in relation to lignification. Canadian Journal of Botany 44, 341.

Please cite this article as: McKenna, R., Nielsen, D.R., Styrene biosynthesis from glucose by engineered *E. coli*. Metab. Eng. (2011), doi:10.1016/j.ymben.2011.06.005

(388d) Enhanced Xylitol Production Through Simultaneous Co-Consumption of Cellobiose and Xylose by An Engineereed Saccharomyces Cerevisiae Strain

Eun Joong Oh[1], Suk-Jin Ha[1], Soo Rin Kim[1], Jonathan M. Galazka[2], Jamie H.D. Cate[2] and Yong-Su Jin[1], (1)Department of Food Science and Human Nutrition, University of Illinois at Urbana-Champaign, Urbana, IL, (2)Department of Molecular and Cell Biology,, University of California at Berkeley, Berkeley, CA

Xylitol is a five-carbon sugar alcohol and has been used as a sugar substitute in the food industry because of its low caloric and anti-carcinogenic characteristics. In addition, xylitol is a building block for a variety of commodity chemicals. Biological routes for producing xylitol from cellulosic hydrolysates have been developed through metabolic engineering of *Saccharomyces cerevisiae*. *S. cerevisiae* cannot utilize xylose as a carbon source, but the expression of a xylose reductase (XR) gene from *Pichia stipitis* facilitates the production of xylitol from xylose with high yields. However, insufficient supply of NAD(P)H for XR, and inhibition of xylose transport by glucose, are major constraints for high productivity production of xylitol.

To overcome these problems, we engineered *S. cerevisiae* to convert xylose into xylitol through the simultaneous consumption of cellobiose and xylose. Specifically, xylose reductase from *P. stipitis* was chromosomally integrated into the *S. cerevisiae* genome, while a cellobiose transporter (*cdt-1*) and an intracellular β-glucosidase (*gh1-1*) from *Neurospora crassa* were introduced on a multi-copy plasmid. The resulting transformant was able to produce xylitol from a mixture of cellobiose and xylose. In contrast to glucose, cellobiose did not repress the transport of xylose, allowing the co-consumption of xylose and cellobiose. The co-consumption and metabolism of cellobiose regenerated NADPH for XR. As a result, an engineered *S. cerevisiae* strain (D-10-BT) exhibited 40% higher xylitol productivity when cellobiose was used as a co-substrate as compared to glucose. This result suggests that co-consumption of cellobiose and xylose allows efficient xylose uptake and cofactor regeneration without catabolic repression for the production of xylitol.

See more of this Session: Microbial Engineering for the Synthesis of Bulk and High-Value Products
See more of this Group/Topical: Sustainable Engineering Forum

(388e) α-Olefin Biosynthesis In the Cyanobacterium *Synechococcus Sp.* PCC7002

Daniel Mendez-Perez[1], Matthew B. Begemann[2] and Brian F. Pfleger[1], (1)Department of Chemical and BIOLOGICAL Engineering, UNIVERSITY OF WISCONSIN-MADISON, Madison, WI, (2) Microbiology Doctoral Training PROGRAM, UNIVERSITY OF WISCONSIN-MADISON, Madison, WI

Biological hydrocarbons are a promising alternative to petroleum-based liquid transportation fuels as they would be compatible with current engines and distribution systems, yet be sourced from renewable substrates. Cyanobacteria have received significant attention because the direct conversion of sunlight and carbon dioxide to lipids will by-pass recalcitrant intermediates found in terrestrial biomass. Although it has been known for decades that some species of cyanobacteria, including *Synechococcus sp.* PCC7002, can synthesize hydrocarbons of different lengths, very little is known about their biosynthesis. In this study, we analyzed the hydrocarbon profile of *Synechococcus sp.* PCC7002 and demonstrate the involvement of a gene in the biosynthesis of two C19 α-olefins. The gene encodes a large multi-domain protein with homology to Type I polyketide synthases, suggesting a new route for hydrocarbon biosynthesis from fatty acids via an elongation-decarboxylation mechanism. The combinatorial nature of polyketide synthesis raises the possibility of using this type of genes to produce a wide range of industrially relevant α-olefins and high value chemical building blocks.

See more of this Session: Microbial Engineering for the Synthesis of Bulk and High-Value Products
See more of this Group/Topical: Sustainable Engineering Forum

(388f) Building a Better Biofuel Production Platform: Engineering Metabolic Control Techniques to Control the Photosynthetic Cyanobacterium Synechocystis Sp. PCC 6803

Stevan Albers, Department of Cell and Molecular Biology, Colorado State University, Fort Collins, CO and Christie A.M. Peebles, Chemical Engineering, Colorodo State University, Fort Collins, CO

Building a Better Biofuel Production Platform:

Engineering metabolic control techniques to control the photosynthetic cyanobacterium *Synechocystis sp.* PCC 6803

Stevan C. Albers[1], Christie A.M. Peebles[1,2]

[1]Cell and Molecular Biology Graduate Program, [2]Department of Chemical and Biological Engineering

Colorado State University, Fort Collins, Colorado

Biofuel production in microorganisms has great potential to fulfill this nation's need for fungible "green" transportation fuels. Because of the simplicity of bacterial cells, photosynthetic cyanobacteria make good candidates as a platform for biofuel production. Many metabolic engineering techniques and procedures have been optimized for various heterotrophic bacteria, but few techniques have been designed with autotrophic cyanobacteria in mind. Designing rational control systems in cyanobacteria is the first step in controlling pathways that utilize carbon dioxide to produce long chain hydrocarbon like molecules. Understanding and controlling the metabolic pathways that produce these molecules is critical to demonstrate the true functionality of these photosynthetic organisms.

Synechocystis sp. PCC 6803 is a very versatile organism for a variety of reasons. It is a photoautotrophic bacterium for which it's haploid genomic DNA has been sequenced. Because of the simplicity of cyanobacterial cells, higher system efficiencies can be reached when compared to other photosynthetic algal cells. Manipulation techniques like natural uptake, electroporation, and homologous recombination are well annotated in this species. The organism is capable of rapid doubling times while harboring such metabolic pathways as flavonoid, clavulanic acid, and fatty acid biosynthesis while also harboring pathways capable of toluene, atrizine, DDT degradation. Our group plans to focus on understanding specific pathways endemic to *Synechocystis* and utilizing these pathways to further the molecular engineering toolset available for this versatile organism. Specifically, we will be discussing the design, construction, and expression of various promoter constructs capable of rational gene control in *Synechocystis sp.* PCC 6803. This talk will focus on our progress to date.

See more of this Session: Microbial Engineering for the Synthesis of Bulk and High-Value Products
See more of this Group/Topical: Sustainable Engineering Forum

AIChE Annual Meeting, October 16-21, 2011

Tie-Simplex Based Simulation of Complex Compositional Displacements: Gibbs Free Energy Minimization

Alireza Iranshahr, Denis Voskov, and Hamdi Tchelepi
Department of Energy Resources Engineering,
Stanford University, Stanford, CA

Introduction

Enhanced Oil Recovery (EOR) processes usually involve complex phase behaviors between the injected fluid (e.g., steam, hydrocarbon, CO_2, H_2S) and the in-situ rock-fluid system. For accurate and efficient simulation of these thermal-compositional displacements, several fundamental problems must be resolved.

In this work, we focus on the challenges associated with coupling the complex thermodynamic phase behavior with multiphase flow multi-component transport in porous media. Three or more phases can be formed as a consequent of the interaction between the injected fluid and the in-situ oil mixture. Moreover, miscibility can be developed given the conditions of the reservoir under study. In order to deal with thermal-compositional displacements of practical utility, we have developed a multiphase (three or more phases) compositional simulation framework where the complex thermodynamic interactions are modeled using an Equation of State (EoS).

An Overview of the Compositional Space Adaptive Tabulation Methodology

Our framework is based on the Compositional Space Adaptive Tabulation (CSAT) technique which was initially developed for two-phase displacements [7]. The methodology relies on tie-line tabulation of the compositional space. A table of tie-lines is (adaptively) constructed that parameterizes a given overall composition at different temperature and pressure levels. Tables of tie-lines are then used for phase-state identification of the mixtures that require stability analysis in the course of the simulation. For this purpose, interpolation in temperature and pressure is performed, which is possible because compositional space parameterization is continuous with respect to temperature, pressure, and composition [4]. From theoretical background and extensive numerical experiments, only a small number of tables parameterize a two-phase thermal-compositional displacement [9]. This approach is generalized to multi-contact miscible [8], and thermal [2] displacements.

Here, we briefly describe the development of CSAT for three-phase systems. For hydrocarbon systems studied in this work, a three-phase region (parameterized by tie-triangles), and three different two-phase regions exist inside the compositional space at low temperature and pressure conditions (see Figure 1). In these systems, the entire plane defined by the tie-triangle needs to be parameterized. Therefore, at each level of temperature/pressure in a three-phase table of tie-simplexes, one tie-triangle and three sets of tie-lines are stored. An extended negative-flash technique is used for the computations [3]. Upon interpolation in composition, as well as temperature and pressure, a plane of tie-simplexes is obtained which is used for the stability analysis of a gridblock in a three-phase compositional simulation. These techniques can be applied for conditions where the three- or two-phase regions are degenerated [4].

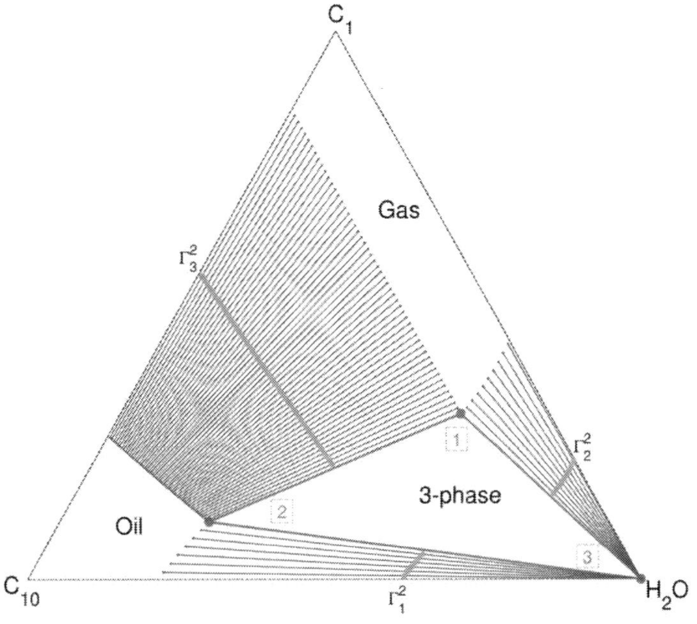

Figure 1: Complete parameterization of $\{C_1, C_{10}, H_2O\}$ at $T = 520\,\text{K}$, and $p = 75\,bar$ using one tie-triangle (in red) and three sets of tie-lines (in blue).

Extensive numerical examples have shown that the accuracy and efficiency of the numerical thermal-compositional simulation significantly improve using the Compositional Space Adaptive Tabulation technique. Here, we present an example to demonstrate the application of CSAT methodology for numerical simulation of complex thermal steam injection problems. We consider a 220×60 reservoir model (second layer of the SPE10 Problem [1]) with the permeability distribution shown in Figure (2). The injector and producer are placed at the opposite corners. We inject steam at $T = 610\,\text{K}, p =$

130 bar into oil made of $\{C_1(0.2), C_{10}(0.8)\}$ at $T = 420\,\text{K}, p = 110\,bar$ (see Figure 2). As shown, CSAT coupled with the flow simulator models the complex thermodynamic phase behavior of this displacement process. Figure (3) shows that the simulator is capable of taking very large CFL numbers ($\gg 1$) per time-step.

Global Minimization of the Gibbs Free Energy

Equality of the chemical potentials is not a sufficient condition for describing the thermodynamic phase behavior of a multi-component system; the Gibbs Free Energy, G, must be minimized globally at the solution. In this section, we use a numerical approach to demonstrate that CSAT yields the global minimum of the Gibbs Free Energy for a multiphase multi-component system. The following derivations follow the approach presented by Michelsen [5, 6].

Let N^C represent the number of phases that a given mixture, $\mathbf{z}$, splits into, as determined by CSAT. Suppose another (trial) distribution of multiple (N^T) phases exists for $\mathbf{z}$ at the same temperature and pressure conditions. Here, the superscripts C, and T stand for CSAT and trial, respectively. We want to derive the conditions for

$$G^C \leq G^T. \tag{1}$$

Note that N^C and N^T can take any value from 1 to N_p^{max}, and that they are not necessarily identical. We start by expressing G^C as

$$\frac{G^C}{N_{tot}} = \sum_i \sum_{j=1}^{N^C} \nu_j^C x_{i,j}^C \mu_{i,j}^C = \sum_i z_i \mu_i^C, \tag{2}$$

where

$$\mu_i^C = \frac{\sum_{j=1}^{N^C} \nu_j^C x_{i,j}^C \mu_{i,j}^C}{z_i}, \tag{3}$$

and N_{tot} represents the total number of moles in the system. Noting that CSAT yields equilibrium conditions (equal chemical potentials), $\mu_i^C = \mu_{i,j}^C$. From

$$z_i = \sum_{j=1}^{N^C} \nu_j^C x_{i,j}^C = \sum_{j=1}^{N^T} \nu_j^T x_{i,j}^T, \tag{4}$$

and

$$\frac{G^T}{N_{tot}} = \sum_i \sum_{j=1}^{N^T} \nu_j^T x_{i,j}^T \mu_{i,j}^T, \tag{5}$$

we obtain

$$\frac{\Delta G}{N_{tot}} = \frac{G^T - G^C}{N_{tot}} = \sum_{j=1}^{N^T} \nu_j^T \sum_i x_{i,j}^T (\mu_{i,j}^T - \mu_i^C). \tag{6}$$

Since $0 < \nu_j^T$, $G^C \leq G^T$ if and only if

$$F_j = \sum_i x_{i,j}^T (\mu_{i,j}^T - \mu_i^C) \geq 0. \tag{7}$$

We will expand this expression using

$$\mu_i(T,p,z) = \mu_i^\circ(T,p) + RT \ln \frac{\hat{f}_i(T,p,z)}{p} \tag{8}$$

written for some phase, where $\mu_i^\circ(T,p)$ represents the reference chemical potential for component i. We have

$$\mu_{i,j}^T = \mu_i^\circ(T,p) + RT \ln \frac{\hat{f}_{i,j}^T}{p}, \tag{9}$$

$$\mu_i^C = \mu_i^\circ(T,p) + RT \ln \frac{\hat{f}_i^C}{p}. \tag{10}$$

Therefore,

$$\mu_{i,j}^T - \mu_i^C = RT(\ln \hat{f}_{i,j}^T - \ln \hat{f}_i^C). \tag{11}$$

As a result, CSAT yields the global minimum of the Gibbs Free Energy for a multiphase multi-component mixture if and only if

$$F'_j(\mathbf{z}) = \sum_i z_i^T (\ln \hat{f}_{i,j}^T - \ln \hat{f}_i^C) \geq 0, \qquad \forall \mathbf{z}^T. \tag{12}$$

Figure (4) explains these developments for a three-component system. In this example, three liquid phases are at equilibrium. Here, we obtain the state of a (three-phase) test mixture using CSAT. Then, we plot $F'(\mathbf{z})$ (Eq. 12) to verify that the function is non-negative throughout the compositional space. Note that $F'(\mathbf{z}) = 0$ at the corners of the parameterizing tie-simplex.

References

[1] Christie M.A., Blunt M.J., "Tenth SPE Comparative Solution Project: A Comparison of Upscaling Techniques", *SPE Reservoir Evaluation & Engineering*, **4** (4), 308–317, 2001

[2] Iranshahr A., Voskov D., Tchelepi H., "Tie-Simplex Parameterization for EoS-Based Thermal Compositional Simulation", *Soc. Pet. Eng. J.*, **15** (2), 545–556, 2010

[3] Iranshahr A., Voskov D., Tchelepi H., "Generalized Negative-Flash Method for Multiphase Multicomponent Systems", *Fluid Phase Equilibria*, 2010

[4] Iranshahr A., Voskov D., Tchelepi H., "Multiphase Compositional Space Parameterization for General-Purpose Thermal-Compositional Simulation", The American Institute of Chemical Engineers Annual Meeting, Salt Lake City, UT, Nov. 7-12, 2010

[5] Michelsen M.L., The Isothermal Flash Problem. Part I. Stability Analysis. *Fluid Phase Equilibria*, 9:1-19 (1982a)

[6] Michelsen M.L. and Mollerup J.M., Thermodynamic Models: Fundamentals & Computational Aspects, Tie-Line Publications, Copenhagen, Denmark (2004)

[7] Voskov D., Tchelepi H., "Compositional Space Parameterization: Theoretical Background and Application for Immiscible Displacements", *SPE Journal*, **14** (3) 431-440 (2009)

[8] Voskov D., Tchelepi H., "Compositional Space Parameterization: Multi-Contact Miscible Displacements and Extension to Multiple Phases", *SPE Journal*, **14** (3) 441-449, 2009

[9] Voskov D., Tchelepi H., "Tie-Simplex Based Mathematical Framework for Thermodynamical Equilibrium Computation of Mixtures with an Arbitrary Number of Phases", *Fluid Phase Equilibria*, vol. 283(1-2) pp. 1-11, 2009.

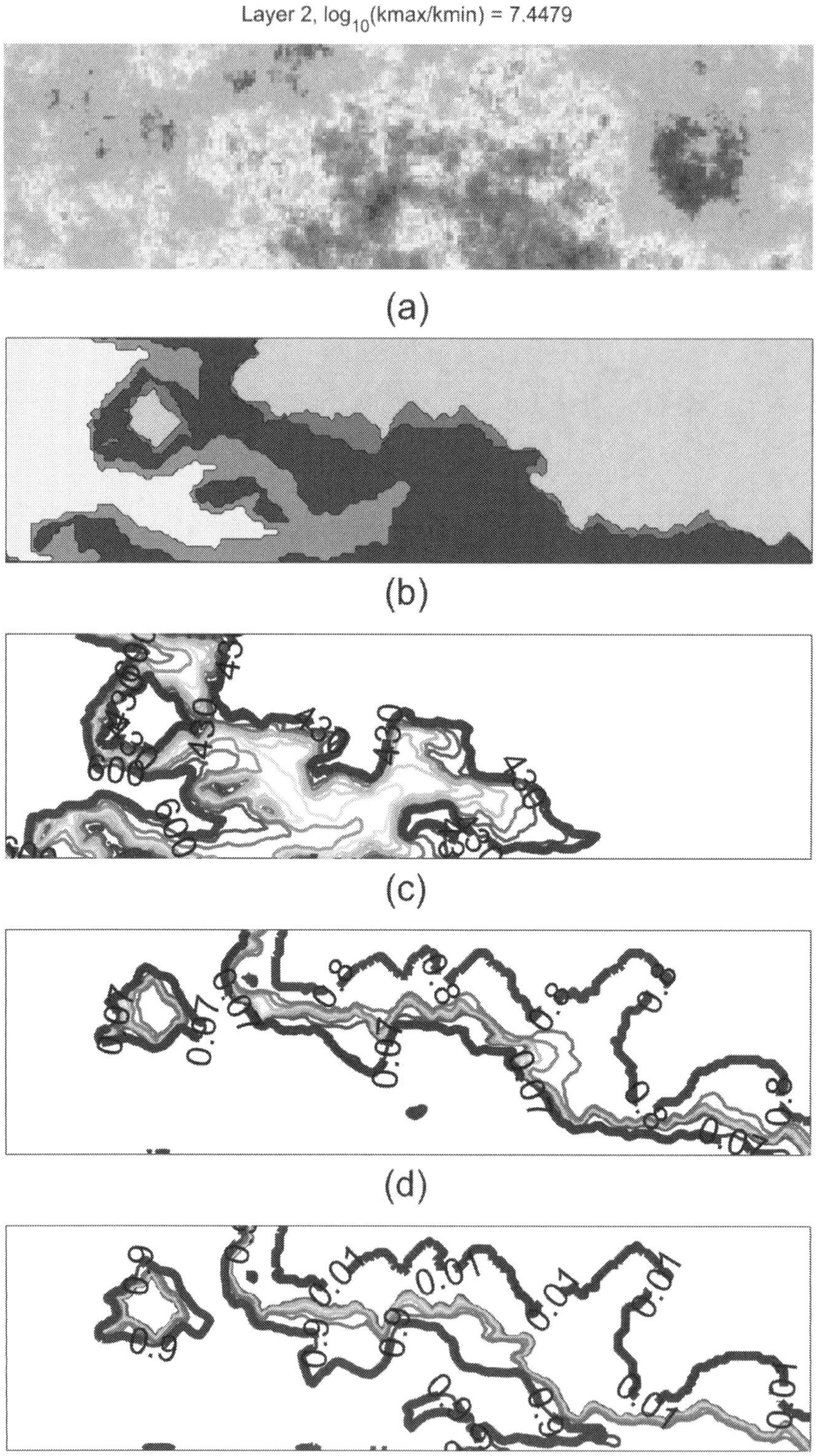

Figure 2: Solution profile at breakthrough: (a) Colored map of different phase-states: Gas, Gas-Water, Gas-Oil-Water, Oil-Water, Oil, (b) temperature, (c) $z_{C_{10}}$, (d) z_{H_2O}. The permeability distribution of the layer is shown on top.

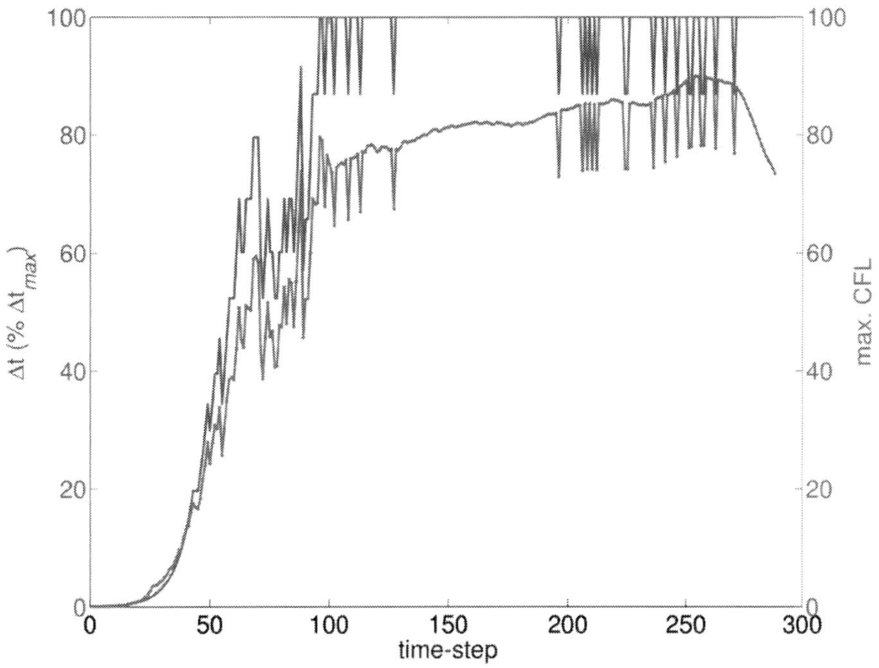

Figure 3: (left) Time-step size (as a fraction of the maximum allowed), and (right) CFL numbers (maximum of all gridblocks) at each time-step of the simulation

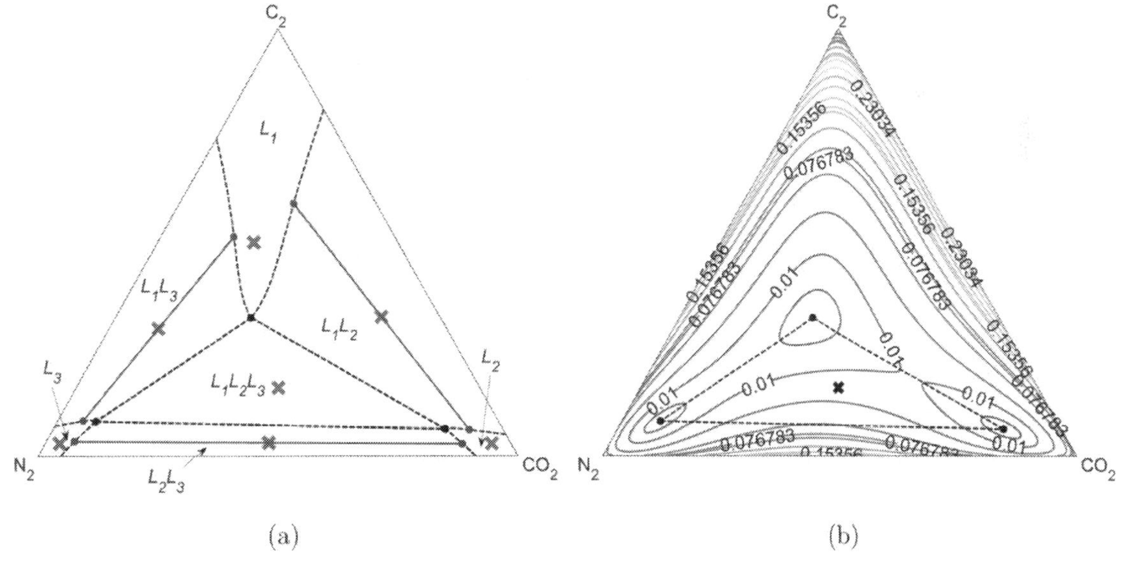

Figure 4: (a) Full parameterization of $\{C_2, N_2, CO_2\}$ at $T = 140$ K, and $p = 200\,bar$. (b) $F'(\mathbf{z})$ for a test mixture in the three-phase region.

A MODIFICATION TO THE APPLICATION OF ORTHOGONAL COLLOCATION METHOD FOR PETROLEUM RESERVOIR PRODUCTION

Mohammad Ranjbaran, M.Sc. candidate, Shiraz, Fars

Alireza Sadeghi, M.Sc. candidate, Shiraz, Fars

Mohammad Reza Rahimpour[1], College Dean, Shiraz, Fars

Alireza Shariati, Lecturer, Shiraz, Fars

1. Introduction

Although *diffusion equation* solution is already done, it could be used to test new modifications on available solution algorithms. This is what we are appealing to show.

Orthogonal collocation method (OCM) on modeling fluid motion in reservoir fields is new. Many authors solved *diffusion equations* using different mathematical methods to divine the behavior of reservoir. Arora et al. (2006) used orthogonal collocation on finite element (OCFE) for solving non-linear boundary value problems. He predicted that the results were more stable in the case of OCFE. Recently Al-Muftah (2009) used OCM for well testing oil reservoir. He showed that in many cases, it eventuated in solutions that are more practical than those resulted from numerical inversion of Laplace transform. He found that only 18 collocation points are sufficient for good accuracy and approach. Ghanaei and Rahimpour (2010) have modeled this equation for two simple cases. Their job was powerful for optimization part and attention to the effect of number of collocation points. However, their solution was very specific because of their assumptions. Furthermore they used the same dimensionless parameters as Al-Muftah (2009) did. By generalizing the problem, OCM could be developed in some more aspects without losing its simple structure.

In this paper a new challenge for reliability of OCM to deal with integral-differential equation is prepared. Lagrange interpolation function is used as the building block for integral part as long as the differential part. For the time aspect, three common initial value problem solvers are applied that are implicit, explicit Euler and Crank-Nicolson method.

2. Methodology

[1] The Corresponding Author, Professor of School of Chemical and Petroleum Engineering, Shiraz University, Shiraz, Iran, P.O. Box: 7134851154, Tel: +98-711-230-3071, Fax: +98-711-628-7294.

E-mail: Rahimpor@shirazu.ac.ir

To start the series of calculations and proof we put equations for each part on separated sections.

2.1. *One phase flow*

2.1.1. *Cylindrical geometry*

2.1.1.1. *Pseudo-steady state equations*

According to Ghanaei and Rahimpour (2010), the pseudo-steady state governing equation for oil production is as follows:

$$\frac{1}{r}\frac{d}{dr}\left(r\frac{dP}{dr}\right)=K_1,\ K_1=\frac{q_L B_o \mu_L}{\pi K_L h\left(r_w^2-r_e^2\right)} \quad [1]$$

And the boundary conditions are

$$@\, r=r_w\ P=P_w$$
$$@\, r=r_e\ \frac{dP}{dr}=0 \quad [2]$$

After some manipulations the applicable OCM based algebraic equations are obtained and given below.

For linear form

$$u_i=\sqrt{r_{Di}}\quad i\in\{1,2,\ldots,N\}\ N=\text{No. of collocation points} \quad [11]$$

$$4u_i\left(\sum_{j=1}^{N}B_{i,j}P_{D_j}+B_{i,N+1}\right)+2\frac{1-2(1-\delta)\sqrt{u_i}}{1-(1-\delta)\sqrt{u_i}}\left(\sum_{j=1}^{N}A_{i,j}P_{D_j}+A_{i,N+1}\right)=\frac{K_1(r_w-r_e)^2}{P_w} \quad [12]$$

For logarithmic form

$$B_{i,1}P_{D_1}+\sum_{j=2}^{N+1}B_{i,j}P_{D_j}+B_{i,N+2}=\frac{K_1 r_e^2(\ln\delta)^2\delta^{2u_i}}{P_w} \quad [13]$$

$$A_{1,1}P_{D_1}+\sum_{j=2}^{N+1}A_{1,j}P_{D_j}+A_{1,N+2}=0 \quad [14]$$

$A_{i,j}$ and $B_{i,j}$ are well-defined and calculated in Rice and Do (1995).

2.1.1.2. *Unsteady state equations*

The governing equation is well-known so that there is no need to derive any part.

$$\frac{1}{r}\frac{\partial}{\partial r}\left(r\frac{\rho_L K_L}{\mu_L}\frac{\partial P}{\partial r}\right)=\frac{\partial(\varphi\rho_L)}{\partial t} \quad [15]$$

And boundary plus the initial conditions are

$$-r\frac{\rho_L K_L}{\mu_L}\frac{\partial P}{\partial r}=0 \ @\ r=r_e$$

$$-r\frac{\rho_L K_L}{\mu_L}\frac{\partial P}{\partial r}=\frac{-\dot{m}_L}{2\pi h}=\frac{-q_L B_o \rho_L}{2\pi h} \ @\ r=r_w \qquad [16]$$

$$P=P_i \ @\ t=0$$

Since we are going to use OCM on the radius differential part, the first step is to normalize it. Moreover, because the logarithmic form of the pressure distribution in the space domain is very common in reservoir models, the following variable change is made and new parameters are defined.

$$\delta=\frac{r_w}{r_e}\rightarrow r=r_e\delta^u \rightarrow dr=(r_e\ln\delta)\delta^u du$$

Along with the new variable (u) substitution, it is also useful to make the equation dimensionless.

As long as the parameters like density, viscosity (not here) and effective permeability (not here) are nonlinear functions of Pressure and saturation (not here), it would be an approximation to expand the space differential part for these variables. Instead the equation can be once integrated to make the left hand site of the equation first order and the right hand side in the integral form. One of the boundary conditions can be put in the equation and the difficulty mentioned former is eliminated. The new format for the governing equation is as follows:

$$\frac{P_i}{(r_e\delta^u\ln\delta)^2}\frac{\partial}{\partial u}\left(\frac{\rho_L K_L}{\mu_L}\frac{\partial y}{\partial u}\right)=\frac{\partial(\varphi\rho_L)}{\partial t}\rightarrow\int_0^u\frac{\partial}{\partial\tilde{u}}\left(\frac{\rho_L K_L}{\mu_L}\frac{\partial y}{\partial u}\right)d\tilde{u}=\int_0^u\frac{(r_e\delta^{\tilde{u}}\ln\delta)^2}{P_i}\frac{\partial(\varphi\rho_L)}{\partial t}d\tilde{u}$$

$$\left.\frac{\rho_L K_L}{\mu_L}\frac{\partial y}{\partial u}\right|_u-0=\int_0^u\frac{(r_e\delta^{\tilde{u}}\ln\delta)^2}{P_i}\frac{\partial(\varphi\rho_L)}{\partial t}d\tilde{u}=\int_0^u F(\tilde{u},t)d\tilde{u} \qquad [19]$$

Now it is the time to start numerical operations. For the time derivative we use forward type. For the space derivative and integral part the related building blocks or weighting functions should be evaluated.

$$\frac{\partial(\varphi\rho_L)}{\partial t}=\frac{\Delta(\varphi\rho_L)}{\Delta t},\ \frac{\partial y_i}{\partial u}=\sum_{j=1}^{N+1}A_{i,j}y_j\ \&\ \int_o^{u_i}Fd\tilde{u}=\sum_{j=1}^{N+1}W_{j,i}F_j$$

For $A_{i,j}$ the related equation can be found in Rice and Do (1995), But for the weighting $W_{j,i}$ function the required relation was not found in literature and therefore derived. In OCM, Lagrange interpolation polynomials are used as building blocks for differentiation. So they could be used for integration.

$$W_{j,i}=\int_0^{u_i}L_j(x)dx \qquad [20]$$

$$L_j(x)=\prod_{k=1\ \&\ j\neq k}^{N+1}\frac{x-x_k}{x_j-x_k} \qquad [21]$$

The closed form of this integral function is hard to be evaluated. Thus, its value was obtained by recursive adaptive Simpson quadrature (*quadgk* function of MATLAB 2007b software) and used as an attached function to the main program.

Here the porosity (φ) and Liquid density (ρ) are only functions of pressure with a constant compressibility.

2.1.2. *spherical geometry*

2.1.2.1. *Pseudo-steady state equations*

The governing equation for this part is given below

$$\frac{1}{r^2}\frac{\partial}{\partial r}\left(r^2\frac{\partial P}{\partial r}\right)=K_1 \qquad [23]$$

$$K_1=\frac{3q_L B_o \mu_L}{4\pi\left(r_w^3-r_e^3\right)K_L} \qquad [24]$$

And the boundary conditions are

$$@\ r=r_w\ P=P_w$$

$$@\ r=r_e\ \frac{dP}{dr}=0 \qquad [25]$$

Finally, the discrete form of the governing equations based on OCM is given below

$$B_{i,1}P_{D_1}+\sum_{j=2}^{N+1}B_{i,j}P_{D_j}+B_{i,N+2}=\frac{K_1\left(\frac{1}{r_w}-\frac{1}{r_e}\right)^2}{P_w\left[\left(\frac{1}{r_w}-\frac{1}{r_e}\right)r_D+\frac{1}{r_e}\right]^4} \qquad [33]$$

$$A_{1,1}P_{D_1}+\sum_{j=2}^{N+1}A_{1,j}P_{D_j}+A_{1,N+2}=0 \qquad [34]$$

2.2. *Two phase flow*

The governing equations are given below:

Mass Balance + Darcy law

$$\frac{1}{r}\frac{\partial}{\partial r}\left(r\left(\frac{\rho_L k_L}{\mu_L}+\frac{\rho_g k_g}{\mu_g}\right)\frac{\partial P}{\partial r}\right)=\frac{\partial}{\partial t}\left(\varphi\left(S_L\rho_L+S_g\rho_g\right)\right) \qquad [35]$$

$$\frac{1}{r}\frac{\partial}{\partial r}\left(r\left(x_j\frac{\rho_L k_L}{\mu_L}+y_j\frac{\rho_g k_g}{\mu_g}\right)\frac{\partial P}{\partial r}\right)=\frac{\partial}{\partial t}\left(\varphi\left(x_j S_L\rho_L+y_j S_g\rho_g\right)\right) \qquad [36]$$

Equilibrium equations

$$y_j=K_j x_j\ j\in\{1,2,\ldots,n_c-1\}$$

$$\left(1-\sum_{j=1}^{n_c-1}y_j\right)=K_{n_c}\left(1-\sum_{j=1}^{n_c-1}x_j\right) \qquad [37]$$

The variable change is logarithmic as already discussed. After applying logarithmic variable change and integration, the new form of the governing equations are given below:

$$\left(\frac{\rho_L k_L}{\mu_L}+\frac{\rho_g k_g}{\mu_g}\right)\frac{\partial P}{\partial u}=\int_0^u\frac{\partial}{\partial t}\left(\varphi\left(S_L\rho_L+S_g\rho_g\right)\right)r_e^{\,2}\delta^{2\tilde{u}}(\ln\delta)^2\,d\tilde{u} \qquad [38]$$

$$\left(\frac{\rho_L k_L}{\mu_L} x_j + \frac{\rho_g k_g}{\mu_g} y_j\right)\frac{\partial P}{\partial u} = \int_0^u \frac{\partial}{\partial t}\left(\varphi\left(S_L \rho_L x_j + S_g \rho_g y_j\right)\right) r_e^{\,2} \delta^{2\tilde{u}} (\ln \delta)^2 \, d\tilde{u} \qquad [39]$$

The relative permeabilities are calculated according to Corey's model which is given below:

$$\frac{k_g}{K} = (1 - S_L)^3 \times (1 + S_L) \qquad [40]$$

$$\frac{k_L}{K} = S_L^{\,4} \qquad [41]$$

Both liquid and vapor viscosities are computed with corresponding states viscosity model modified by Pedersen et al, (1984), and the phase densities are obtained by SRK EoS with no binary interactions parameters.

In order to indicate effectiveness of the proposed variable change, orthogonal collocation on finite element method (OCFE) for the conventional variable change is also investigated. For further information about OCFE, Rice and Do (1995) is a good reference.

3. Result and discussion:

In this section the results are separated into two parts. First one is made for single phase flow and it contains the comparison between our model and the exact one to prove the correctness of the proposed idea. The next part shows the model applicability for two phase flow.

3.1. One phase flow

3.1.1. Pseudo-steady state model

3.1.1.1. *Cylindrical geometry(OCM)*

As noticed before, Ghanaei and Rahimpour (2010) used the conventional variable change in their job. Therefore, it is a good source to find out the effect of logarithmic variable change on their examples of oil production.

The data they extracted from literature are given in Tables 1 and 2.

Table 1: The data for reservoir#1 (Ahmad, 2006)

q_o STB/day	600	K_L md	120
μ_L cP	2.5	B_o RB/STB	1.25
P_w psia	1800	h ft	25
r_e ft	745	r_w ft	0.25

Table 2: The data for reservoir#2 (Slip Slider, 1983)

q_o STB/day	850	B_o RB/STB	1.1
μ_L cP	0.5	K_L×h md.ft	1830
P_w psia	3200	r_w ft	0.5
r_e ft	2000		

The comparison is given out of a typical percentage average absolute relative Error function (%AARD) as defined below

$$\%AARD = 100\frac{\int_{r_w}^{r_e}\left|\frac{P^{Exact}-P^{Calc.}}{P^{Exact}}\right|dr}{\int_{r_w}^{r_e}dr} \quad [42]$$

Error values and numbers of collocation points are given in Tables 3 & 4.

Table 1 : The results for linear Variable change for reservoir #1

	%AARD		%AARD
17	2.5863	18	2.7703
21	1.7511	22	1.8836
29	0.7925	30	0.8703
45	0.1516	46	0.1830
77	0.0031	78	0.0081
141	0.0003	142	0.0002

Table 2: The Err values for the use of Logarithmic variable change in reservoir #1

N	%AARD	N	%AARD
5	4.5105	6	2.0169
7	0.8277	8	0.3134
11	0.0109	12	0.0031
17	0.0000	18	0.0000

Not only better result is obtained from this variable change, but also faster changes in reduction of error values are seen for the same change of number of collocation points. Both of these results are interesting and good reasons for the use of logarithmic change of variable in the unsteady-state mode of reservoir modeling.

3.1.1.2. *Cylindrical geometry(OCFE)*

In this section the results of OCFE in cylindrical geometry is described. The results for reservoir 1 and 2 with conventional variable change are given in Tables 5 and 6.

Table 5: The result for reservoir#1 with Conventional variable change

W	M	N	%AARD
0.9990	5	25	22.48
0.9995	5	25	13.83
0.9999	5	25	33.25
0.9995	15	25	14.80
0.9995	25	25	15.00
0.9995	25	35	7.05
0.9995	25	45	3.60
0.9995	25	65	1.06
0.9995	35	65	1.05

Table 6: The result for reservoir#2 with Conventional variable change

W	M	N	%AARD
0.9990	5	25	56.1437
0.9995	5	25	12.6327
0.9999	5	25	11.9273
0.9999	15	25	12.0929
0.9999	25	25	12.1284
0.9999	25	35	6.3656
0.9999	25	45	3.7391
0.9999	25	65	1.5195
0.9999	35	65	1.5352

3.1.1.3. *Spherical geometry(OCM)*

In this part the results by OCM method in spherical geometry for reservoirs 1 and 2 with conventional variable change is donated in Tables 7 and 8 and with logarithmic variable change is presented in Tables 9 and 10.

Table 7: The result for reservoir#1 with Conventional variable change

N	%AARD	N	%AARD
17	82.2450	18	82.2630
25	82.2920	26	82.2984
45	82.3225	46	82.3234
109	82.3315	110	82.3315
155	82.3317	156	82.3317

Table 8: The result for reservoir#1 with logarithmic variable change

N	%AARD	N	%AARD
17	82.7581	18	82.6706
25	81.3493	26	81.0270
45	66.3237	46	65.1440
69	35.9207	70	34.7722
109	7.6587	110	7.3358
155	0.9423	156	0.8986

Table 9: The result for reservoir#2 with Conventional variable change

N	%AARD	N	%AARD
17	35.6314	18	35.6385
25	35.6470	26	35.6494
45	35.6547	46	35.6550
109	35.6558	110	35.6558
155	35.6558	156	35.6558

Table 10: The result for reservoir#1with logarithmic variable change

N	%AARD	N	%AARD
17	35.8601	18	35.8159
25	35.1751	26	35.0219
45	28.2849	46	27.7551
69	14.9276	70	14.4340
109	3.0474	110	2.9185
155	0.3577	156	0.3408

3.1.1.4. *Spherical geometry(OCFE)*

The upshot of OCFE method for reservoirs 1 and 2 in spherical geometry are given in Tables 11and 12.

Table 11: The result for reservoir#1 with Conventional variable change

W	M	N	%AARD
0.99	5	25	43.7742
0.995	5	25	17.1347
0.999	5	25	75.6822
0.995	5	45	20.9778
0.995	5	65	13.0635
0.995	5	151	0.819
0.995	10	151	70.809
0.995	105	151	19.0826

Table 12: The result for reservoir#2 with Conventional variable change

W	M	N	%AARD
0.9995	1	25	15.0104
0.9996	1	25	16.8858
0.9997	1	25	7.7489
0.9998	1	25	9.2094
0.9997	1	45	9.2078
0.9997	5	45	17.247
0.9997	5	85	17.2732
0.9997	105	151	23.6532

3.1.2. unsteady-state result

The needed data for a simple example is given in Table 13.

Table 13: Input data for a sample of Oil reservoirs (T. Ahmad 2006) [5]

φ_0	0.15	**q_0B_0 res. b/d**	375
ρ_{L0} lbm/ft^3	43.68	**K_L md**	60
μ_L cP	1.5	**c_t psi^{-1}**	1.2×10^{-5}
P_i psi	4000	**h ft**	15
r_e ft	2500	**δ**	0.0001

The following discussions are made in several Lines.

- Number of collocation points effect on the accuracy of the calculations
- The comparison between this variable change and the common variable change
- The difference in Implicit, explicit Euler and Crank-Nicolson method on the accuracy of the results for a suitable number of collocation points.
- Final comparison with accepted Log-method and the finite acting solution. (Slip Slider, 1983)

Part 1: Effect of number of collocation points

As can be seen in Figure 1 after N=12 the number of collocation points does not have any sensible effect on the accuracy of the solution for long times. This is because of the good distribution of the points on the space domain.

Part 2: The logarithmic variable change or the common one?

$$u=\frac{r-r_e}{r_w-r_e}$$ Linear form

$$u=\frac{\ln(r)-\ln(r_e)}{\ln(r_w)-\ln(r_e)}$$ Logarithmic form

The Logarithmic variable change presented a great improvement on the solution of a simplified pseudo steady state model. Although linear variable change was useful for those pseudo steady state problems mentioned by Ghanaei and Rahimpour (2010), it is not applicable here. Even for N=41 points no solution is obtained for implicit Euler method in time domain approximation. Probably only the symmetrical variable change was the reason of convergence for OCM in Ghanaei and Rahimpour (2010) which makes the points more concentrated near the wellbore.

Part 3: The difference in Implicit, explicit Euler method and Crank-Nicolson Method

The definition for these methods can be found easily in Rice and Do (1995). What is important here is to see which one is practical and if there is any significant difference in the results?

In Figure 2 for N=15 the bottom hole pressure vs. time is examined for these three methods for a same Δt set. There is not a meaningful difference in the three methods which shows that the Δt distribution made here is suitable and the error (if any) is for the space domain.

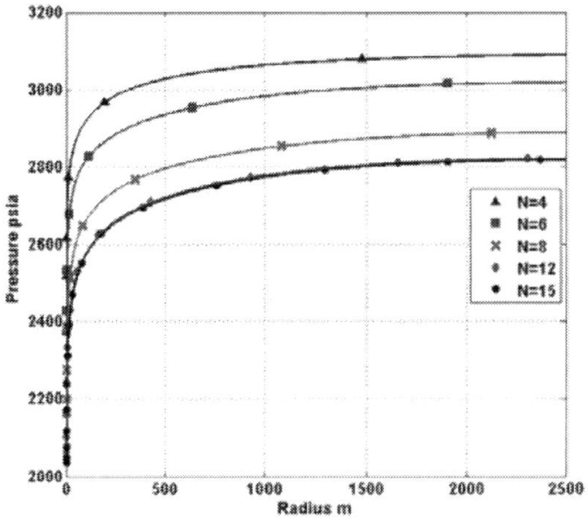

Figure 1: Effect N on the accuracy of the solution @ t ≈ 300 days

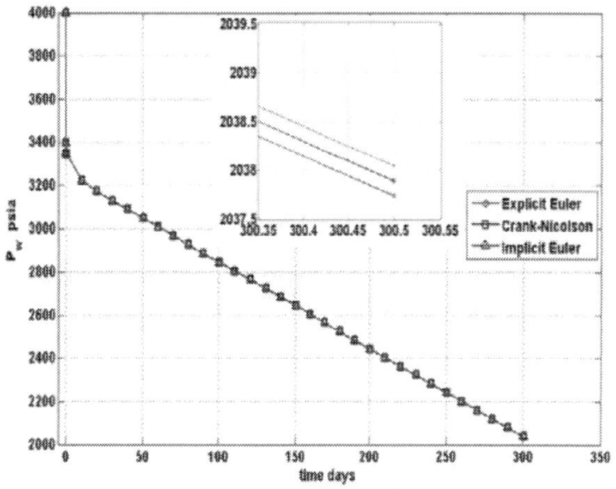

Figure 2: the effect of three different initial value numerical solutions for N=15 on bottom hole pressure

Part 4: comparison with exact solutions

The results of bottom hole well pressure calculation are assigned as the representative samples for the solution validity. The values of percentage average absolute relative deviations (%AARD) as defined below are given in Table 14.

$$100\times\frac{\sum_i \left|\frac{P_w^{calc}-P_w^{exact}}{P_w^{exact}}\right|_i}{n_t} \quad [43]$$

Table 14: Error values for one phase flow (time range= (0-300 days), Δt_1=0.1 day Δt_2 = 10 days)

N	% AARD	N	% AARD
5	12.66	6	10.41
9	2.41	10	0.82
13	1.03	14	1.01
17	0.73	18	0.51
23	0.50	24	0.50

According to Table 16, OCM works effectively on this type of problem. The more the number of collocation points are, the less the answers fluctuate. 17 or 18 points give satisfactory result.

The result of OCFE method solution reveals some weakness points

1) Increasing total number of collocation points does not elevate the accuracy.
2) Finding the best answer needs optimization.
3) The accuracy of result is very sensitive to the number of collocation points and the weight factor.

The above reasons confine OCFE usage and in this paper OCM method is so modified that those limitations are all eliminated.

3.2. Two phase unsteady-state flow (OCM)

The reservoir input data and reservoir fluid compositional information are given in Tables 15 and 16, respectively.

Table 15: Data for a simple GC Radial flow

φ	0.15	**h m**	10
m_L kg/s	0.835	**K md**	100
m_G kg/s	0.835	**δ**	0.0001
P_0 psi	2200	**T K**	360
r_e &r_w m	1000&0.1		

Table 16: Data for component

Components	z_i
C_1	0.8
nC_5	0.2

In Figure 3 and Figure 4, for N=21 the bottom hole pressure and liquid saturation versus radius are plotted.

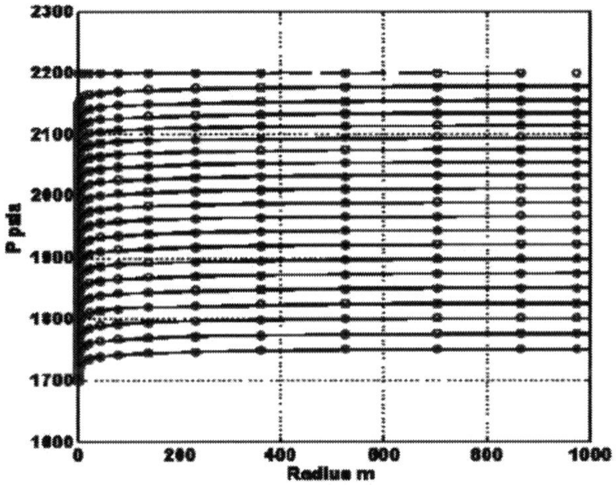

Figure 3: P vs. radius for 0-4000 days (each curve for 200 days, Δt=20 days), N=21

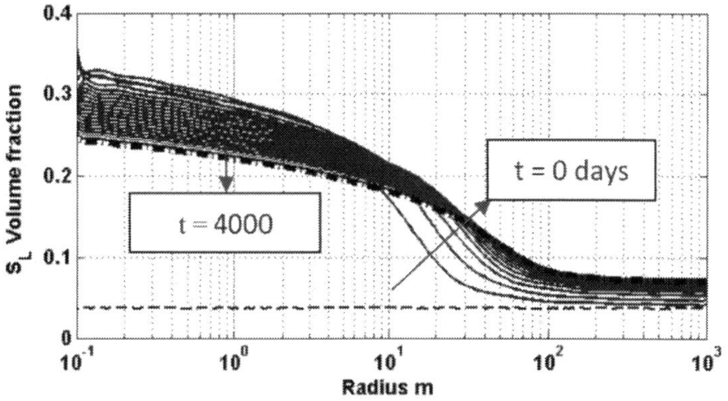

Figure 4: SL vs. radius for 0-4000 days (each curve for 200 days, Δt=20 days) On logarithmic scale

The Saturation profile seems reasonable for different times. It started to build up at wellbore, then enlarge to the whole reservoir and finally levels off because it reaches the end of the reservoir.

4. Conclusions

In the first step, logarithmic space domain variable change effect is investigated on a pseudo steady-state model and it shows a good combination with OCM method. By this idea there is no need for OCFE method.

After this proof, the variable change is used on an unsteady state one phase flow in porous medium that is effective and sufficient. Then, it is compared with accepted methods like log-method and it shows acceptable predictions.

Finally, the main challenge is on a differential algebraic system of equations made out of an unsteady state two phase flow model in porous medium and it is reasonable.

5. Acknowledgement

The authors are very grateful to thank Mr. Iranshahi, PhD candidate of Chemical Engineering in Shiraz University for his useful suggestions. They also wish to acknowledge the Southern Zagros Oil & Gas Production Company and Shiraz University for facilitating this collaboration.

6. Nomenclature

Symbol	Definitions
n_t	Number of time steps
q	Volumetric flow rate STB/day
P	Reservoir pressure, psia
μ	Viscosity, cP

K	Effective permeability, md
r_w	Well radius, m
r_e	Reservoir external radius, m
h	Reservoir thickness, m
ρ	Density, lbm/ft^3
S	Saturation
x & y	Liquid & vapor mole fraction
c	Compressibility factor
Greek letter	
φ	porosity
Subscript	
i	Initial
L	Liquid
g	Vapor
f	Formation
w	Well
Abbreviations	
SRK EoS	Soave/Redich/Kwong Equation of state

7. References

1. Ahmed, T. (2006), "Reservoir Engineering Handbook," 3rd ed. Gulf professional publishing, Houston.
2. AL-Muftah, A. E. (15- 18 March 2009), "Well test simulation using orthogonal," paper SPE 120668 was prepared for presentation at the 2009 *SPE Middle East Oil & Gas show and Conference* held in the Bahrain international exhibition centre, kingdom of Bahrain.
3. Arora, Sh., Dhaliwal, S.S., Kukreja, V.K. (2006a), "Application of orthogonal collocation on finite elements for solving non-linear boundary value problems," *Appl. Math. Comp.* 180, pp. 516–523.
4. Ghanaei, E. & Rahimpour, M. R. (2010), "Evaluation of orthogonal collocation and orthogonal collocation on finite element method using genetic algorithm in the pressure profile prediction in petroleum reservoirs," *Journal of Petroleum Science and Engineering*. 74, pp. 41–50.
5. Pedersen, K.S., Fredenlund, Aa., Christensen, P.L., and Thomassen, P. (1984), "Viscosity of crude oils," *Chemical Engineering Science.* 39, pp. 1011-1016.
6. Rice, R. G. & Do, D. D. (1995), "Applied Mathematics and Modeling for Chemical Engineers," John Wiley & Sons, Inc.
7. Slip Slider, H. C. (1983), "Worldwide Practical Petroleum Reservoir Engineering Methods," 1st edition. PennWell Books, Tulsa.

(445c) Improving Gas Condensate Reservoir Productivity

Adel Elsharkawy, Petroleum Engineering, Kuwait University, Kuwait, Kuwait

Gas condensate reservoirs experience great losses in productivity as the reservoir pressure drops below the dew point pressure due to formation and accumulation of liquid condensate around the well bore and subsequent reduction of gas relative permeability. Several methods have been proposed to minimize liquid condensate and or restore gas productivity of gas condensate well. Among these methods: pressure maintenance, gas re-injection, solvent injection for wettability alteration and near well bore massive hydraulic fracturing. Each of the previously mentioned method has its own advantages and shortcoming. One method which shows great success on laboratory scale and great potential in field application is wettability alteration to gas wetting which reduces condensate accumulation in the vicinity of well bore and hence maintain gas productivity. This paper evaluates the various techniques that have been used in the industry to maintain gas condensate productivity. Gas flow tests have been used to study the effectiveness of each method to restore or maintain gas production for gas condensate reservoir.

See more of this Session: Complex Subsurface Processes I
See more of this Group/Topical: Energy and Transport Processes

(445d) Role of Discrete Fracture Network In Tight Gas Reservoirs

Swapnil V. Thakur, Milind Deo and John McLennan, Chemical Engineering, University of Utah, Salt Lake City, UT

Role of Discrete fracture network in tight gas reservoirs.

Discrete fracture network play an important role by improving the connectivity and hence enhancing the productivity of gas in tight gas sands. This paper describes how the injecting fluid interacts with the discrete fracture network with geomechanical stresses acted upon them. Fluid is pumped into the reservoir domain having certain in-situ stress and with negligible matrix permeability and having significant fracture connectivity. In such reservoirs, influence of the geomechanical stress is substantial and fracture width controls the fluid flow. The fracture width is affected by the in-situ stresses which are acted upon them. A representative reservoir model is created and simulated using a weak coupling approach between a reservoir simulator (SMURF) and geomechanical simulator (3DEC). The two different simulators are weakly coupled via the width. With injection of the fluid the geomechanical stresses acting on reservoir domain are affected, this results in opening or closure of the fractures. This information is noted and is then transported to reservoir simulator to find out whether the geomechanical stresses facilitate the recovery of gas or hinder the recovery. A series of simulations are carried out with important parametric changes. Injection rates, Injecting fluids, and geo-mechanical stresses are varied to see how these affect the recovery of the gas from tight gas sands.

See more of this Session: Complex Subsurface Processes I
See more of this Group/Topical: Energy and Transport Processes

Morphology and Growth of Fractures in Unconventional Reservoirs

Nan Zhao, University of Utah, Salt Lake City, UT
Milind D. Deo, University of Utah, Salt Lake City, UT
John McLennan, University of Utah, Salt Lake City, UT

Abstract

Understanding the morphology and growth of hydraulic fractures is essential in the development of unconventional tight gas and shale gas resources. In this paper, we report the use of a computational tool consisting of geologic data representation, geomechanical modeling and multiphase flow simulations to predict reservoir performance. A finite element-based geomechanical module is interfaced with a control-volume finite element discrete-fracture reservoir simulator. The hydraulic fracture geometry is generated by following the fluid injection pathway in the existing fracture network and in the matrix. The permeabilities in the fractures and in the matrix are adjusted dynamically in the simulator based on coupling of the geomechanical and flow attributes. The approach described is applicable to tight gas and shale gas reservoirs. There are over 3700 active wells in the Greater Natural Buttes field in Uinta Basin, Utah with a cumulative production of 1.8 trillion cubic feet of gas. Most of this production comes from several tight gas formations (matrix permeability of less than 0.01 md). The importance of hydraulic fractures and their interaction with natural fractures in shale gas reservoirs characterized by nano- Darcy permeabilities is well known. This methodology provides a means of mapping a complex (non-planar) network of hydraulically-activated/induced fractures. The initial conductivity distribution in the fractures and the relative permeability of the matrix along with the stress tensor and mechanical properties of the matrix and the fractures control the geometry of the hydraulic fracture system and the ultimate well performance. Prediction of complex hydraulic fracture morphology and injected water balance are important in the development of tight gas and shale gas resources. Multiphase flow simulations with complex fracture networks will help delineate complex effects such as water blocks etc.

Introduction

Unconventional energy resources are becoming increasingly important all around the world. Gas production from tight sands and shale gas has increased significantly, and the unconventional portfolio accounts for more than half of the gas produced in the United States . These systems are characterized by low to ultra-low permeability. Hydraulic fracturing is critical in making these systems productive: and is further regulated by the mechanical properties of the reservoir and surroundings, the in-situ stress conditions and the fabric of the reservoir. Coupling these geomechanical parameters with transport-related reservoir properties is usually casually considered in conventional reservoir simulation; if at all. Especially when modeling fractured reservoirs using a discrete-fracture network representation, geomechanical effects are expected to have a significant impact on relevant reservoir characteristics. It has become more accepted that fracture growth, particularly in naturally fractured reservoirs with extremely low permeability, cannot be reliably represented by conventional planar simulations. Characterizing evolution of multiple, non-planar, interconnected and possibly non-vertical hydraulic fractures requires hydraulic and mechanical characterization of the matrix as well as existing, latent, or healed fracture networks. Recognizing these challenges, the Advanced Reactive Transport Simulator (ARTS) has been developed for generalized reservoir simulation at the University of Utah . ARTS is built on the Discrete Fracture Network (DFN) fracture representation method and can simulate diverse oil and gas recovery processes including waterflooding, steamflooding, Steam Assisted Gravity Drainage (SAGD, etc. A geomechanical model with functionality that allows different coupling schemes has been built into ARTS to simulate production from various unconventional oil and gas plays. The geomechanical model in ARTS is an independent module but the information from this module is shared with other modules in the simulator through a built-in data transport class. It is possible to use different constitutive relationships and coupling schemes in the geomechanical module. These attributes will be beneficial as the framework is further modularized and generalized.

Hydraulic fracturing is used extensively in this field to enable economic production. Several interesting phenomena have been observed during production-for example, water blocks and the so-called permeability jail. Water blockage is a phenomenon where capillarity and relative permeability effects prevent complete drainage of water from the hydraulic fractures, inhibiting or completely preventing subsequent gas production. A permeability jail occurs when all fluids in a multiphase setting have very low or near-zero mobilities over a broad saturation range. Representing the geometric complexity of fracture networks (natural fractures and hydraulic fractures) in these unconventional gas reservoirs also contributes to the difficulties of numerical modeling and production simulation. This called for a reservoir simulator with the capability of modeling flow through complex fracture networks and gas flow in the associated porous matrix; with a geomechanical model coupled to the reservoir simulation. ARTS was constructed to model explicitly fractures as discrete components, with built-in complex reservoir processes (multiphase black-oil, thermal-compositional, geochemical, etc.). A geomechanical module has recently been added. A workflow that has been adopted in evaluating this field has been described elsewhere. The specifics for the current analyses entailed inferring reservoir, fabric and geomechanical properties from field data, publications and other relevant sources. A DFN model was generated from these data and transferred to the reservoir model in ARTS. A series of reservoir simulations were then carried out to investigate some of the more critical aspects of multiphase phenomena in the unconventional reservoirs.

ARTS Frame Work and Governing Equations

The ARTS is a highly modularized simulation frame work, and it has several specific mofules with different features. The various combations of different modules enable the UFES to handle multiple applications. The major module of UFES is the reservoir module and the geomechanical module, under each module a variety of submodules is employed to handle specific problem.

Reservoir Model

The UFES maily employed two reservoir models, the black oil model[1] and the compositional k value thermal model. Both models are based on discreted frature network representation of reservoir and work with several discretization methods. The governing equation for a black oil model is:

$$\nabla \cdot (\frac{Kk_{rl}}{B_l \mu_l} \nabla \varphi_l) = \frac{\partial}{\partial t}(\phi \frac{S_l}{B_l}) + q_l \tag{1}$$

φ_l is the potenial of the species l:

$$\varphi_l = p + \gamma Z \tag{2}$$

K is the absolute preambility and k_{rl} is the relative preambility. B is the volumetric factor. ν is the viscousity.q is the source term of each species, in the reservoir simulation this term is generally refer to the injection or production well.

The governing equation for compositional model[2] is:

$$\frac{\partial(U_p + U_{rock})}{\partial t} = \nabla \cdot \bar{\bar{\mathbf{K}}}_{\mathbf{c}} \nabla T - \nabla \cdot (\sum_{p=1}^{Nf} \hat{H}_p \rho_p \vec{v}_p) + Q_e + Q_{loss} \tag{3}$$

where U_p and U_{rock} are

$$U_p = \sum_{p=1}^{Np} \phi S_p \rho_p \hat{U}_p \tag{4}$$

$$U_{rock} = (1 - \phi)\bar{U}_{rock} \tag{5}$$

Geomechanical Model

The geomechnical model is based on the assumption that the equilibrium state of mechanical is enforced in the reservoir. And the equation is:

$$\nabla \cdot \sigma + \rho \vec{b} = 0 \tag{6}$$

where σ is a tensor which represents stresses (symmetric):

$$\begin{bmatrix} \sigma_{x,x} & \sigma_{x,y} & \sigma_{x,z} \\ \sigma_{y,x} & \sigma_{y,y} & \sigma_{y,z} \\ \sigma_{z,x} & \sigma_{z,y} & \sigma_{z,z} \end{bmatrix}$$

Effective stress law is applied: $\sigma = \sigma' - \alpha \delta_{i,j} P$ where σ' is the effective stress,and P can be write as the saturation weighted pore pressure[3] ($P = \sum_{i=1}^{n} s_i p_i$).

$$\nabla \cdot (\sigma' - \alpha \delta_{i,j} P) + \rho \vec{b} = 0 \tag{7}$$

The density is defined as: $\rho = (1-\phi^*)\rho_s + \phi^* \rho_f$ Different models can be applied to the stress like linear elastic model, linear poro-elastic model and so on.In the simulation, a poroelastical model is applied:

$$\sigma_{i,j} = 2G\varepsilon_{i,j} + 2G\frac{\nu}{1-2\nu}\varepsilon_{k,k}\delta_{i,j} - \alpha p \delta_{i,j} \tag{8}$$

ε is the volumetric strain and defines as $\varepsilon_i = \frac{\partial u_i}{\partial x_i}$ and $\varepsilon_{k,k} = \sum_{i=1}^{i=3} \frac{\partial u_i}{\partial x_i}$. G is shear module, α is Biot-Willis coefficient and ν is Poission's ratio.

Coupling schemes

In general, there are two types of coupling schemes, soft coupling and hard coupling. Soft coupling means the geomechanical effect is not directly invoved in the reservoir simulation. Hard coupling method integrate geomechanical model in the reservoir simulator. Both of these two methods need some parameters to connection geomechanical model and reservoir model. One set of the parameter is porosity and permeabilty:

First the ture porosity of the reservoir is introduced[4]:

$$\phi^* = 1 - (1-\phi_0)e^{-\varepsilon_v}$$

or in linear form:

$$\phi = (\frac{1-\phi_0}{1-\phi^*})\phi^*$$

The equation above is come from the continuity of the rock:

$$\frac{\partial \rho_s (1-\phi)}{\partial t} + \nabla((1-\phi)v_s) = 0$$

where $\nabla v_s = \nabla \frac{du_s}{dt}$, $\nabla u_s = \varepsilon_v$

The permeability can be ralated to the geomechanic factor[4]:

$$\frac{k}{k_0} = \frac{1 + \frac{\varepsilon_v}{\phi_0}}{1 + \varepsilon_v}$$

where $\varepsilon_v = \frac{\delta V}{V} = \varepsilon_{xx} + \varepsilon_{yy} + \varepsilon_{zz}$ and $\varepsilon_{ij} = \frac{1}{2}(\frac{\partial u_i}{x_j} + \frac{\partial u_j}{x_i})$ Through these information, the impact of geomechanical effect is introduced in the reservoir model.

On the other hand, the geomechanical model is related to the reservoir model by the effective stress formulation. Also the geomechanical model is assumed to be at equilibrium all the time through the reservoir life. Based on the coupling parameter, soft and hard coupling can be applied.

Soft Coupling

To take into account of the geomechanical effect with out much change of the reservoir model. The "artifical" permeability and porosity can be introduced in the soft coupling scheme. In this way, the reservoir permeability and porosity is set to be a function of pressure (may plus saturation) based on the external information of this reservoir. This external information can be some real data and some pure mechanical simulation result of reservoir.

A soft coupling scheme is developed based on Itasca 3DEC simulation result for the multiphase flow simulation In the simulator the permeability and porosity is treaed from a look-up table generated from the geomechanical information of 3DEC. The pore pressure is using as a standard to maching these two simulation.

Hard Coupling

For the hard coupling approach, the geomechanical model is integrated in the reservoir simulator. The geomechanical equation set and flow equation set can be solved either "separately" or "together" depend on the real problem.

For the geomechanical model, a general virtual work formulation can be derived:(neglect the body force for convenience, acttually it just gravity in most case which only in z dirction:

$$\int_{\Omega} \delta\vec{\varepsilon}^T \Delta\sigma' d\Omega - \alpha \int_{\Omega} \delta\vec{\varepsilon}^T \Delta p d\Omega - \int_{\Gamma_\sigma} \delta\vec{u}^T \Delta\vec{t} d\Gamma = 0$$

Use the strain stress equation:

$$(\int_{\Omega} B^T D B d\Omega)(\Delta\vec{u}) - \alpha(\int_{\Omega} B^T \vec{m} N d\Omega)(\Delta p) - \int_{\Gamma_\sigma} \delta\vec{u}^T \Delta\vec{t} d\Gamma = 0$$

The coupling parameter is the volumetric strain.

References

1. Fu, Yao (2007), "Multiphase Control Volume Finite Element Simulation of Fracured Reservoirs", PhD thesis, University of Utah

2. Huang,Chung-Kan (2009), "Development of A General Thermal Oil Reservoir Simulator Under A Modularized Framework", PhD thesis, University of Utah

3. Zienkiewicz, O. C. and Shiomi, T(1984), "Dynamic behaviour of saturated porous media; the generalized biot formulation and its numerical solution", Internation Journal for Numerical and Analytical Method in Geomechanics, 87:71-96

4. L.K. Thomas and L.Y. Chin and R.G. Pierson and J.E. Sylte(2002), "Coupled Geomechanics and Reservoir Simulation" in SPE Annual Technical Conference and Exhibition

(445f) Gas Production Modeling From a Complex 3-D Description of Marine Gas Hydrate Sand Reservoir

Manohar Gaddipati, Chemical Engineering, National Energy Technology Laboratory/West Virginia University, Morgantown, WV and **Brian Anderson**, Chemical Engineering, West Virginia University, Morgantown, WV

In April and May of 2009, an industry research consortium led by Chevron in collaboration with the U. S. Department of Energy (DOE) National Energy Technology Laboratory (NETL), the U.S. Geological Survey (USGS), the U.S. Minerals Management Service, and others completed a marine hydrate drilling expedition in the Gulf of Mexico. During this 21-day logging-while-drilling (LWD) expedition targeted three drilling sites: Walker Ridge 313, Green Canyon 955, and Alaminos Canyon 21. Seven LWD holes were drilled and gas hydrate-bearing sediments with hydrate saturation greater than 50% were found in at least four wells. Following on this successful Joint Industry Project Leg II expedition, we have constructed a set of reservoir simulation descriptions based on the logging results and pre-drill seismic estimates of the extent and thickness of the gas hydrate deposits at the Walker Ridge 313 and Green Canyon 955 sites. By constraining the hydrate saturations and thicknesses using the resistivity logs at the drilling sites, reservoir models of complexity varying from two-dimensional homogeneous (cylindrical, *r-z*) to full three-dimensional, heterogeneous models of the deposits have been simulated. These models have been simulated using the CMG STARS simulator incorporating a non-orthogonal corner point grid to accommodate the geometry of the system. The full, heterogeneous three-dimensional model represents one of the most complete and complex reservoir models of a marine hydrate deposit to date. Results of the reservoir simulations indicate very high potential for producing methane from these marine hydrate deposits using depressurization due to their depth and *in situ* temperature.

See more of this Session: Complex Subsurface Processes I
See more of this Group/Topical: Energy and Transport Processes

(456a) Temperature Distribution In Activated Carbon Bed During Adsorption of Nitrogen (model gas for hydrogen); Experiment & Mathematical Model

Agnieszka Truszkowska[1], **Christopher Loeb**[1], Richard Chahine[2], Bruce Hardy[3] and Goran Jovanovic[4], (1)Chemical, Biological and Environmental Engineering, Oregon State University, Corvallis, OR, (2)Hydrogen Research Institute, Trois-Rivieres, QC, Canada, (3)Computational Sciences, Savannah River National Laboratory, Aiken, SC, (4)Oregon State University, Corvallis, OR

Temperaturedistribution in activated carbon bed during adsorption of nitrogen (model gasfor hydrogen); experiment & mathematical model

Agnieszka Truszkowska[1],Christopher Loeb[1], Richard Chahine[2], Bruce Hardy[3]and

Goran N. Jovanovic[1]

[1]Oregon State University,Department of Chemical, Biological, and Environmental Engineering, Corvallis, OR 97330, USA[2]Hydrogen Research Institute at University du Quebec a Trois-Rivires, PO BOX 500, Trois- Rivires, QC, Canada, G9A 5H7

3•
Savannah RiverNational Laboratory•
ProcessModeling and Computational Chemistry Group •
Building 999-2W, Rm. 122 •
Aiken, SC 29808, USA •
•

Emails: truszkoa@onid.orst.edu, loebc@onid.orst.edu, goran.jovanovic@oregonstate.edu, bruce.hardy@srnl.doe.gov, richard.chahine@uqtr.ca

Providinghigh volumetric and mass storage density of hydrogen gas is a major challengein the progress towards hydrogen-fueled automobiles. Of the various storage media,liquid chemical systems, and metal hydrides, activated carbon adsorbents havesubstantial advantages in cost, weight, ease of handling and safety.

In this paperwe present experimental data (temperature and pressure) obtained during adsorptionof nitrogen (model gas for hydrogen) in an activated carbon bed. We alsopresent the results of a numerical simulation of the same process.

Experimentalapparatus used in this study consists of a cylindrical activated carbon bed (2in diameter x 1.5 in high) placed into a sealed high-pressure cylindricalvessel. The carbon bed, consisting of densely packed activated carbon materialand binding agent, is instrumented with six thermocouples placed at suitablelocations throughout the bed. In addition, a thermocouple, and three pressuretransducers are placed into the high-pressure vessel outside the bed. Dynamic change of temperature andpressure during the charging process are recorded, corresponding to compressionand adsorption phenomena. The charging process starts at 1.0 bar and ends at50.0 bar pressure. A parametric study of filling time is performed for time intervalsbetween 10 and 30 seconds, to observe its effect on the charging process.

The mathematicalmodel is developed for the carbon bed pressure vessel system, which enablesnumerical simulation of the charging process. Initially, numerical simulationsare used to analyze experimental data and to explore characteristic parametersof the carbon bed. The same numerical tool will be used in the future todesign carbon bed with optimal characteristics imposed by requirements forhydrogen storage media in motor vehicles.

See more of this Session: Hydrogen Storage Systems and Materials: Modeling, Simulation, and Optimization:
See more of this Group/Topical: International Congress on Energy 2011

(456b) Optimization of Adsorbent Tanks for Hydrogen Storage for Automotive Applications

Stephen L. Garrison[1], David A. Tamburello[1], Matthew R. Kesterson[1], Claudio Corgnale[2] and Bruce Hardy[1], (1)Computational Sciences, Savannah River National Laboratory, Aiken, SC, (2)Hydrogen Technology Center, Savannah River National Laboratory, Aiken, SC

A key technical hurdle to an energy economy focused on clean-burning hydrogen over non-renewable petroleum is on-board storage of hydrogen for automotive vehicular applications. Adsorbents provide one possible solution to the need for improved gravimetric and volumetric capacities. However, while the heat of adsorption released upon sorption of hydrogen is relatively low, so it the thermal conductivity. The subsequent rise in storage temperature results in decreases in maximum hydrogen storage capacity and loading rate at the short fill times mandated by the Department of Energy (DOE) targets. Novel designs for integrated heat exchangers are needed to maximize the heat transfer out of system to maximize storage rate and capacity.

While many literature studies focus on maximum capacity assuming an (impossible) infinite fill time, full, detailed modeling of the coupled heat and mass transfer is needed. The focus of this presentation will be our work developing detailed numerical models that couple heat and mass transfer via the general purpose finite element solver COMSOL Multiphysics® for various adsorbent media and bed designs and the automatic optimization of the designs by linking the COMSOL models with routines in MATLAB®.

See more of this Session: Hydrogen Storage Systems and Materials: Modeling, Simulation, and Optimization:
See more of this Group/Topical: International Congress on Energy 2011

(456c) Molecular Simulation of Hydrogen Adsorption In Porous Material Made up with Silsesquioxane Units and Metal Catalytic Sites

Nethika S. Suraweera[1], Michael E. Peretich[2], Joshua Abbott[2], Craig E. Barnes[2] and David J. Keffer[3], (1)Chemical & Biomolecular Engineering, University of Tennessee, Knoxville, TN, (2)Chemistry, University of Tennessee, Knoxville, TN, (3)Chemical and Biomolecular Engineering, University of Tennessee, Knoxville, Knoxville, TN

The adsorptive hydrogen storage of molecular hydrogen on microporous materials is widely studied as an effort for efficient utilization of hydrogen as a pollution free renewable energy source. Porous structures made up with spherosilicate building blocks and consist of metal catalytic sites exhibit promising results for the ability of adsorption of gases. Experiments are carried out for investigating the hydrogen adsorption ability of these materials. In this study we performed computer simulations to analyze hydrogen adsorption in structures which consist of interconnected silsesquioxane units. This porous material has cubic silicate building blocks (Si_8O_{20}), which are cross-linked by $SiCl_2O_2$ bridges and decorated with $OTiCl_3$ catalytic sites and $SiMe_3$ ends.

The structure for the simulations was developed first using a coarse grain model, followed by an energy optimization and replacing the coarse grains with atoms. The physical properties of the structure such as density, surface area, accessible volume and pore size distribution were compared with experimental vales and they are in good agreement.

Adsorption of hydrogen was modeled by Path Integral Grand Canonical Monte Carlo (PI-GCMC) simulations using standard force fields. From the simulations, hydrogen adsorption isotherms at 300K and 77 K were generated for low pressures up to 10 bars. The energy of adsorption of hydrogen was calculated at each case. Density distribution was developed and analyzed. Classical MD simulations were performed to calculate the self-diffusivity of hydrogen in the porous structure and activation energy was calculated.

Keywords: spherosilicate building blocks, hydrogen adsorption, PI-GCMC simulation, silsesquioxane units, coarse grain model, energy optimization, adsorption isotherm, density distribution, MD simulations, self-diffusivity, activation energy

See more of this Session: Hydrogen Storage Systems and Materials: Modeling, Simulation, and Optimization:
See more of this Group/Topical: International Congress on Energy 2011

Density Functional Theory Study of the Effect of *3d* and *4d* Transition Metals on $NaMgH_3$ Complex Metal Hydride for Hydrogen Storage Applications

Authors: Fernando Soto[1], Daniela S. Mainardi[2]

[1]Institute for Micro-manufacturing, Louisiana Tech University
[2]Institute for Micro-manufacturing, Chemical Engineering, Louisiana Tech University

In order to achieve the Hydrogen Economy, it is necessary to use hydrogen as an energy carrier. Hydrogen as an energy carrier is a possible solution to neutralize the climate changes occurred due to carbon emission. It is also highly attractive to ensure a smooth and efficient provision of energy without depending on other nations. Vehicles powered by fuel cells are one of the applications in which hydrogen can be used as an energy carrier. However, the absence of a commercially viable hydrogen storage technology is delaying the large-scale use of fuel cell powered vehicles. In order to promote research in this area, the U.S. Department of Energy (DoE) has set targets for the year 2010 and 2015 for a system of on board hydrogen storage. Previous results suggest that the addition of transition metals enhances the desorption kinetics of complex metal hydride and lowers the hydrogen dissociation temperature and activation energies.

In the present work, study of catalytic additives with improved performance over Ti allows us to generalize the work on the procedure for 3d/4d doping in NaMgH3 and use it in systems with higher gravimetric densities and more favorable thermodynamics. For this purpose, plane wave density functional theory (PW-DFT) calculations helps us examine bulk and surface models of NaMgH3, Mg2FeH6 and KMgH3 complex metal hydride (CMHs) with impurities from the 3d element block. In this respect, first- principles calculations determined the cohesive energies of pure and doped bulk/surface CMHs as well as the adsorption/substitution energies. Furthermore, DFT coupled Molecular Dynamics (DFT-MD) calculations at elevated temperatures elucidate the effect of Ti placed on different sites on surface models. The results indicate that 5 out of 7 additives at the bulk model are thermodynamically stable relative to the pure model; therefore, possible. The most favorable elements in this regard came out to be; Ti, V, and Co in that order. Afterwards, from dynamics calculations performed at surface slab models, we conclude that Ti additives replacing sodium lattice sites and Ti additives placed at a four-fold vacant position are the stable arrangement in this system with potential formation of TiMgxHx complexes. Meanwhile, alloyed systems with 4d transition metals may be extremely stable for on-board hydrogen storage applications. Also, dynamics calculations performed at the surface slab models suggests that Ti additives replacing sodium lattice sites and Ti additives placed at a four-fold hollow site are the preferred, stable arrangements in NaMgH3 with possible formation of TiMgxHx complexes. This conclusion may serve as a guideline for the future design of nano-structured complex hydride for on-board hydrogen storage utilization.

Keywords: density functional theory, hydrogen storage.

1. Introduction

Finding a feasible solution to the problem of hydrogen storage for vehicular applications is essential for the transportation sector because it accounts for two-thirds of energy consumption in the U.S. [1]. Also, it is necessary to ensure an energy supply and avoid energy dependence from regions such as the Middle East which are politically unstable [2]. It is imperative to reduce carbon emissions in order to improve the air quality and avoid harsh climatic changes [3]. In order to promote research in this area, the U.S. Department of Energy (DoE) has set targets for the year 2010 and 2015 for a system of on board hydrogen storage. The use of hydrogen within H2/O2 fuel cell systems requires an adequate and accessible hydrogen storage medium. The production of liquid or highly compressed hydrogen requires an tremendous amount of energy and may be dangerous in certain situations. Therefore, solid-state hydrogen storage is currently the safest and most effective way of storing hydrogen. As explained by Grochala et al. an ideal hydrogen storage material (HSM) will have a low-molar weight, be inexpensive, have rapid kinetics for absorbing and desorbing H2 in the 25-120 °C temperature region, and store large quantities of hydrogen reversibly [4]. As noted by Grochala et al., at the moment, no single material fulfills all of these requirements. High stability and poor kinetics are significant challenges, which have to be solved in order for any material to be considered for on-board automobile applications [5].Therefore, the identification of material that can meet this criterion is a grand challenge that has attracted the interest of the global scientific community.

Light chemical elements are essential to establish a reversible hydrogen storage system [6]. In order to achieve hydrogen gravimetric and volumetric densities required for mobile applications, host elements in the hydrogen storage system should be lighter than Al metal. Typical elements considered among the light elements are: Li, Be, B, C, N, O, F, Na, Mg, and Al. As explained by Wu et al., metal hydrides are either too stable for efficient hydrogenation cycling or too unstable that desorption occurs at or below room temperatures but adsorption requires high pressures of hydrogen [7]. Nonetheless, CMH have been identified as promising candidates for on board hydrogen storage applications. CMHs may enhance kinetics and reversibility and inhibit toxic-gas side products. Since the pioneer work of Bogdanovic and Schwickardi in 1997, the focus has shifted to CMHs such as alanates, borohydrides, amides/imides, and destabilized CMH [8, 9].

Since the work of Bogdanovic and Schwickardi in 1997, it has been established that certain element addition can help CMH meet the criteria for HSM and are capable of showing hydrogen reversibility at moderate conditions [8, 10, 11]. Thus, enhanced CMH are a promising class of material for hydrogen storage that needs further study because high stability and poor kinetics are important challenges which have to be solved for the CMH to be considered for on-board automobile applications [12]. Recently several types of CMHs have attracted the attention as promising enhanced CMHs candidates for HSM. These CMHs are: Ternary hydrides composed of transition metals and different alkaline earth metals [13], alkaline transition-metal borohydrides [14], and ABH_3perovskite structures [15]. Ti-doped CMHs have shown to improve the kinetics of complex metal hydride decomposition and the thermodynamic reactions relative to the pristine material. Thus, the role of Ti and other *3d*/*4d* transition elements in different types of CMHs need further studying in order to achieve the goals established by the DoE [9].

$NaMgH_3$

Recently, Mg-based hydrides have attracted interest as a hydrogen storage medium due to their light weight and low cost [10].Several Mg-based hydrides show the perovskite structure expressed as ABH_3, where A is an alkali element and B is magnesium. $NaMgH_3$ is the only magnesium based ternary hydride with Na and it has attracted the attention as a possible candidate for hydrogen storage material because it has high gravimetric hydrogen density (6 %) and high volumetric hydrogen density (88 kg/m^3) [11]. This magnesanate has not been widely investigated as a candidate for hydrogen storage. The crystal structure of this material is known and it derives from cubic $KMgH_3$ by an orthorhombic distortion and has a perovskite structure similar to $GdFeO_3$ [10]. It belongs to the *Pnma* space group with orthorhombic symmetry in the temperature range of 4 to 370 K [11,12].The primitive unit cell contains four formula units for a total of 20 atoms.The $NaMgH_3$ unit cell is shown in figure 1.

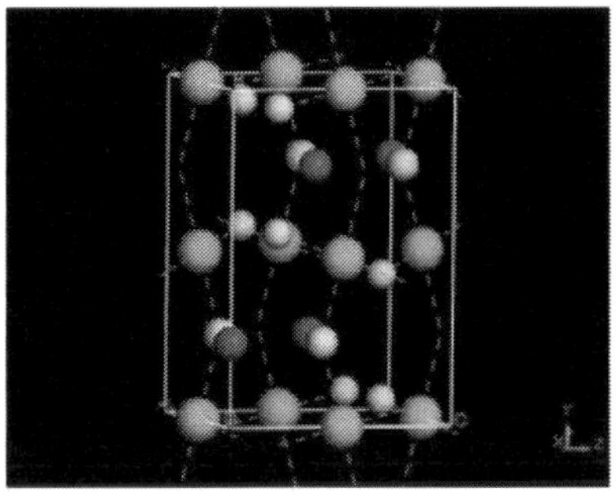

Figure 1. Unit Cell of Sodium Magnesium Trihydride.

Hydrogen disassociation in this material occurs at atmospheric pressure and approximately 670 K, which is well above the operable range for on board hydrogen storage applications. Thus, $NaMgH_3$ is more thermally stable than binary MgH_2, which has a hydrogen dissociation temperature ~550 K. The dehydrogenation reaction scheme takes place in two steps [12]:

$$NaMgH_3 \rightarrow NaH + Mg + H_2 \qquad (1)$$

$$NaH \rightarrow Na + \frac{1}{2}H_2 \qquad (2)$$

Numerous studies have been carried out to develop suitable Mg-based systems that show fast sorption kinetics at moderate temperatures. These studies achieved partial success in improving the kinetics by addition of dopants. Ma and co-workers demonstrated high affinity of transition metal cations toward hydrogen. Catalysts are usually mixed with the hydride by mechanical ball milling, which introduces defects and impurities. However, much research still needs to be done in order to identify appropriate catalysts, study the existence of stable intermediate phases and their crystal structure and understand the kinetics. Much insight can be gained from first principles simulations for the purpose of designing alloy catalysts.

For instance, Bouhadda et al. used ab initio calculations to report on the crystal, electronic and thermodynamic properties of $NaMgH_3$. Bouhadda et al. reported relaxed structures in agreement with experimental structures (Table 1).

Perovskite Structure	Experimental Lattice Constants (Å)	Calculated Lattice Constants (Å)
$NaMgH_3$	a=5.463	a=5.410
	b=7.703	b=7.628
	c=5.411	c=5.358

Table 1. Experimental and Calculated Cell Parameters

In addition, this author also reported that this structure is a wide band gap semiconductor with a calculated band gap of 3.4 eV. Bouhadda et al. explains that the interaction between Na and MgH_3 is ionic and that the interaction between Mg and H is both ionic and covalent.
To study the thermodynamic properties of the material, Bouhadda et al. calculated the standard enthalpy of formation H_f with the following reaction equation.

$$Na + Mg + 3/2\ H_2 \rightarrow NaMgH_3 \quad (3)$$

The result of 151.83 KJ/mol is well in agreement with the results obtained from Ikeda et al. (145.05 KJ/mol, 147 KJ/mol) [15]. Thus, this formation enthalpy is considerably larger than that of metal hydride NaH (-114 KJ/mol) [15]. Klaveness et al. also calculated the enthalpy of formation for this material obtaining similar results as those presented by Bouhadda et al. and Ikeda et al. Thus, the first principles approach tends to reproduce experimental results very well, and this is true also for ionic hydrides such as $NaMgH_3$ [16,17].

Furthermore, Shane et al. reported that the high rates of H motion in $NaMgH_3$ are due to regions with poorly organized crystal structure [17]. The author concludes that if the disorder could be sustained, it could provide a route towards improved reaction kinetics. Hao et al. presented a theory-guided computational study explaining the existence of intrinsic defects in $NaMgH_3$. The author reported that schottky defects are the dominant defects in this material and that hydrogen diffusion is determined by mobility of schottky defects. Hao et al. concluded that the diffusion induced by schottky defects can be tweaked by adding suitable dopants such as Cu and Co that can adjust the material's Fermi level [18]. Komiya et al. studied the substitution energies in $NaMgH_3$ by replacing the Na ion with an element from the same 1A group element from the periodic table (Li, K, Rb, Cs) and replacing the Mg ion with an element from the 2A group elements (Be, Ca, Sr, Ba) [19]. Komiya et al. reported positive substitution energies on the Na site, indicating that doping the structure makes the material unstable, except when doping with the K ion. Also, the addition of Be, Sr, and Ba destabilizes the structure and doping with Ca increases the stability. These results show that addition of impurities has a positive correlation on the destabilization of the material.

Moreover, ABH_3 structures such as $NaMgH_3$ may enhance kinetics and reversibility and inhibit toxic gas side products compared to the parents' borohydride, amide, and alanate compounds. Also, higher H mobility in this material due to their perovskite structure cause improved hydrogenation rates compared to pure MgH_2. It is clear from the literature review presented that further investigation is needed to investigate the positive correlation between hydrogen mobility, addition of dopants and hydrogenation/dehydrogenation reversibility. Therefore, the aim of this paper is to study the stability and dynamics of pure and doped $NaMgH_3$ for on-board hydrogen storage applications.

2. Computational Methods

2.1 Bulk Calculations

In this study, we use the first principles pseudopotential total energy method through the generalized gradient approximation (GGA) calculations with the Perdew and Wang (PW91) functional as implemented in the CASTEP® module [5] of the Materials Studio® software by Accelrys, Inc. [6] to study bulk and surface models of $NaMgH_3$. Spin-polarized calculations on Fe-, Ni-, and Co-doped systems show no significant effect to be created energetically, with respect to the non-spin polarized calculations. Hence, further conducted geometry optimization calculations were spin-unpolarized, and the cut-off energy was 400 eV with a fine k-point mesh in all cases.

We created hydride models alloyed with 3*d* and 4*d* block transition metals to investigate their effects on the energetics of bulk/surface of $NaMgH_3$ systems. The alloying elements substitute one Na atom at a center lattice location of each structure, resulting in a concentration between 5.65 and 7.57 at.wt. % in the case of 3*d* transition metals. Afterwards, bulk hydrides and surface slabs relaxed entirely in terms of all atomic degrees of freedom to obtain the ground-state energy system. After complete relaxation of the systems was obtained, single-point energies were calculated for the optimized structures. Since the main aim of this work is to illustrate the effect of transition metals on the prototype $NaMgH_3$ system, calculation of cohesive energies (ΔE_{coh})of the doped systems were doneaccording to equation (4).

$$\Delta E_{coh} = E_{TOT}(Na(X)MgH_3) - [NE(Na) + NE(Mg) + NE(H) + E(X)] \quad (4)$$

Where E_{TOT} is the total energy of the system, N is the number of atoms, E(Na) is the energy of a free neutral atom of Na, E(Mg) is the energy of a free neutral atom of Mg, E(H) is the energy of a single hydrogen atom and E (X) is the energy of a free neutral transition metal X.

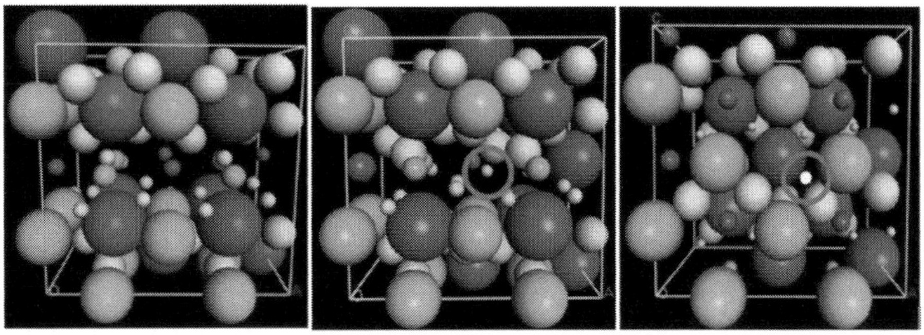

(a) (b) (c)

Figure 2. $NaMgH_3$ 2x1x2 Supercell consisting of five f.u. **(a)** pure model, **(b)**Ti_{sub}@Na in model, and **(c)**Vac_{sub}@Na in model.

2.2 Surface Calculations

Study of surface properties is vital to understand the catalytic behavior of impurities. An explanation for this is that the arrangement of atoms, distribution of electrons, and their response to external behavior is different at the surface when compared to the bulk.

Thus, besides the studies regarding bulk doped $NaMgH_3$ systems, we considered the adsorption/substitution behavior of Ti on the $NaMgH_3$ (001) surface. Accordingly, the calculations were performed for surface structures with a slab width of 2x1x2, and a gap of approximately 5 Å at the top of the $NaMgH_3$ surface. Moreover, k-point sampling is performed at the gamma point only. Doping on the surface structure proceeded in two distinct steps: first, one is the substitution adsorption, in which the dopant replaces a Na atom at the center lattice location of the uppermost layer of $NaMgH_3$ (001) surface. The second form of adsorption in the on-top of surface adsorption.

The substitutional/adsorption energy calculation for the (001) surface proceeds via the following equation:

THIS EQUATION WAS NOT AVAILABLE ON THE ORIGINAL E-MEDIA VERSION OF THIS PAPER (5)

Where N is the total number of atoms in the model and X is the transition metal added to the system.

3. Results and Discussion

The equilibrium cell parameters a, b and c of pure $NaMgH_3$ are 5.48, 7.68 and 5.40 Å, respectively. A brief comparison with experimental values showed that; these calculated values are of 0.20 % different than the initially set unit cell dimensions, which is a strong indication of the good agreement between the computational and experimental results. The calculated results for the lattice parameters are shown in Table 2.

Lattice Parameter (Å)	
experimental	**DFT (GGA/PW91)**
a = 5.463	5.47
b = 7.703	7.68
c = 5.411	5.39

Table 2. Experimental and Calculated Lattice Parameter

The calculated energy gap is 3.308 eV which is similar to the values found in literature [15]. Calculated band gap figure is shown in figure 3.

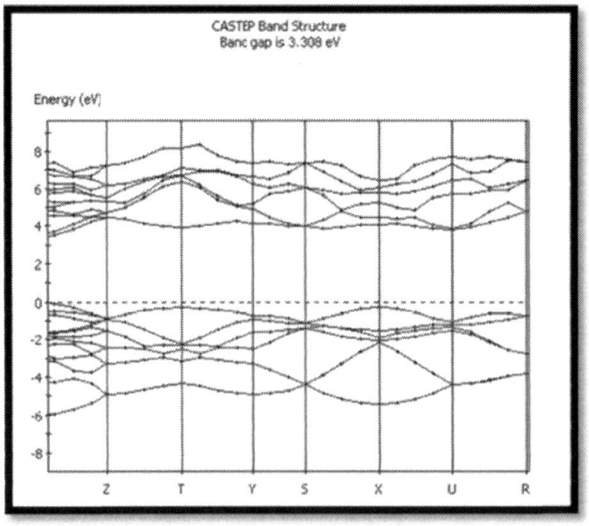

Figure 3. Calculated Band Gap as obtained from CASTEP®

Furthermore, electron density maps corroborate the results reported by Bouhadda et al., the interaction between Na and MgH_3 is ionic and that between Mg and H involves both ionic and covalent bonding schemes. In this respect, the theory level used is appropriate and the results encourage us to continue with the same theory level for subsequent ab initio calculations (Figure 4).

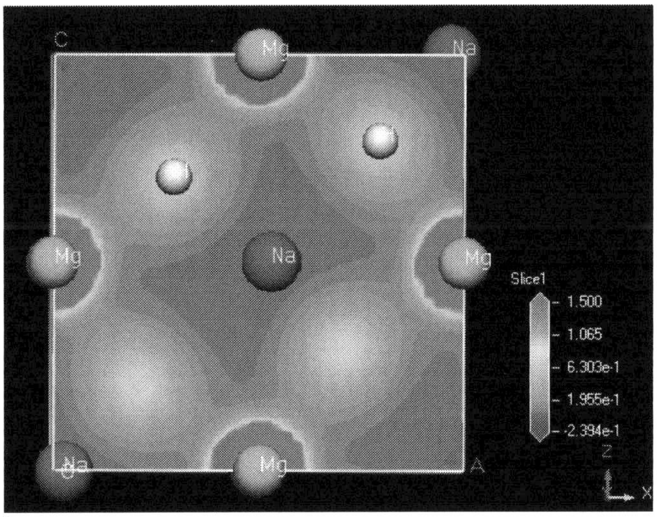

Figure 4. Total Charge Density of $NaMgH_3$ parallel to the x and z axis. Where the green balls are Mg atoms, purple balls are Na atoms and white balls are hydrogen atoms.

3.1.1 Bulk

Next, the primary goal in calculating the cohesive energy energies for each alloyed system was to investigate the effect of 3*d* and 4*d* block elements in the stability of the system. Modification in $NaMgH_3$ models took place via addition of selected impurity elements X (where X can be any element from 3*d* and 4*d* block: Sc, Ti, V, Nb, Y, Mo, etc) to the desired $NaMgH_3$ model. The cohesive energy of the pure model served as the reference value; thus, any energy above 0 eV may be considered as stable. It is interesting to note that several elements added as dopants in the $NaMgH_3$ structure are thermodynamically stable and; therefore, appealing for further consideration.

3*d* Transition Metals

In this respect, for the 3*d* block, Ti, V, Sc, Co, Ni and Cu in the decreasing order, are the most favorable elements with respect to the cohesive energy of the pure system for 3*d* transition metals (Figure 5). For instance, the cohesive energy due to the presence of the Ti dopant is 2.89 eV more favorable than the pure case where no dopants are available. Meanwhile, the removal of a Na ion from the pure model is 3.95 eV less preferred than the pure model. Also, V is 2.64 eV more energetically favorable, Sc has a value of +2.16 eV, and for Co and Cu those values are +1.62 eV and +0.62 eV respectively. On the other hand, Zn and Mn where found to be unfavorable in this system with cohesive energies of -2.58 eV and -0.68 eV, respectively. These results and their corresponding bond lengths with Mg and H are tabulated below.

Model	X in Bulk Site	System	*E_{coh} (eV)	E_{sub} (eV/atom)
Bulk	Na	$Na_{16}Mg_{16}H_{48}$	0.00	
	0	$Na_{15}Mg_{16}H_{48}$	-3.95	
	Sc	$(Sc)Na_{15}Mg_{16}H_{48}$	2.16	0.027
	Ti	$(Ti)Na_{15}Mg_{16}H_{48}$	2.89	0.036
	V	$(V)Na_{15}Mg_{16}H_{48}$	2.64	0.033
	Mn	$(Mn)Na_{15}Mg_{16}H_{48}$	-0.68	-0.009
	Co	$(Co)Na_{15}Mg_{16}H_{48}$	1.62	0.02
	Cu	$(Cu)Na_{15}Mg_{16}H_{48}$	0.62	0.008
	Zn	$(Zn)Na_{15}Mg_{16}H_{48}$	-2.58	-0.032

Table 3. Cohesive Energies of the different Bulks Models

X @Na site Bond Length (Å)	Na	Sc	Ti	V	Mn	Fe	Co	Ni	Cu	Zn
X-Mg	3.24	3.09	3.06	2.99	3.02	2.72	2.67	3.18	3.23	3.26
X-H	2.51	2.11	1.93	1.83	1.74	1.67	1.63	1.71	1.75	1.92
Mg-H	1.99	1.98	2.00	2.00	2.04	1.99	2.00	1.98	2.00	1.99

Table 4. Inter Bond Lengths of the different 3*d*Bulk Models

It should be pointed out that; that the cohesive energies of these 3*d* block elements in the bulk follow a strong correlation with the cohesive energies of their metallic elements counterpart. Also, the cohesive energies of the 3*d* block elements show a negative correlation with the ionization energies of the pure metallic elements and they also show a positive correlation with the melting point in Kelvin of the pure crystalline elements.

Thus, a high cohesive energy value of the metallic element helps to predict a high cohesive energy value in the bulk model, and transition elements with a high ionization potential energy appear to be responsible for the low thermal stability of the bulk model and the melting point of the metallic elements follow a strong positive correlation with the cohesive energy of the bulk suggesting that all these physical quantities have to be taken into account when destabilizing $NaMgH_3$ with partial substitution of the cation Na. Such prediction helps to identify additional elements that could improve on the effect of 3*d* block transition elements on CMHs.

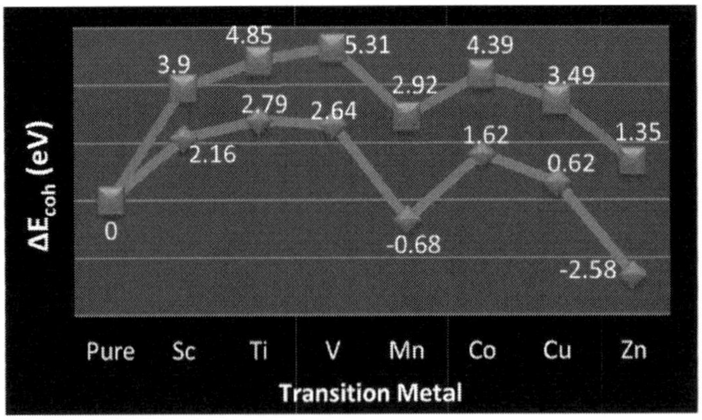

Figure 5. Cohesive Energy Results. Bulk substitution of transition metal in $NaMgH_3$(blue), pure crystalline transition metal (red).

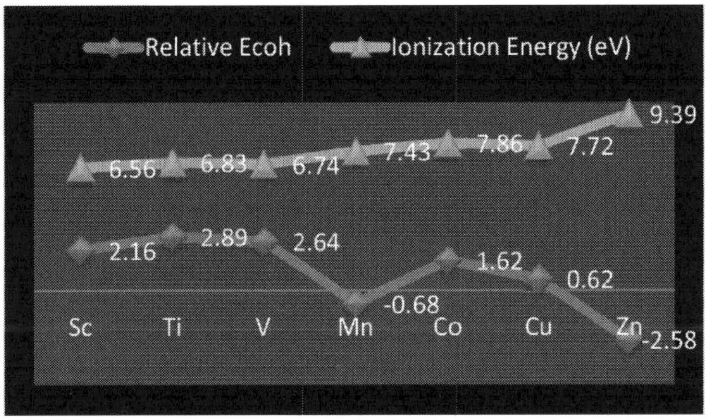

Figure 6. Cohesive Energy Results. Bulk substitution of transition metal in $NaMgH_3$ (blue), Ionization Energy (green)

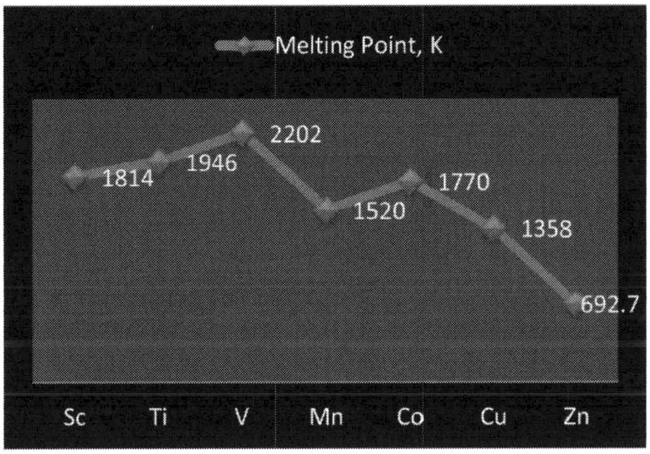

Figure 7.Melting point in Kelvin for transition metals in their pure crystalline state.

3*d* Total Charge Density

To further understand the effect of transition metals on bulk $NaMgH_3$ the total charge density of pure and Ti-doped models are shown below. The total charge density can be used to provide quantitative information at the atomic scale surrounding any atom or fragment of interest. The color corresponds to the total charge density, with the red regions corresponding to the highest density, which surrounds the atoms. The majority region of the model displays a very low charge density, colored blue. The plots suggest that the interaction between Na and Mg-H complex is mainly ionic in both models. However, the Ti-doped model shows an increased association of Ti with surrounding H and Mg atoms, as reported in table 4 above. Thus, a shortening of the X-H and X-Mg bond lengths implies an elongation and possible weakening of Mg-H bond lengths, making the desorption of H atoms more feasible.

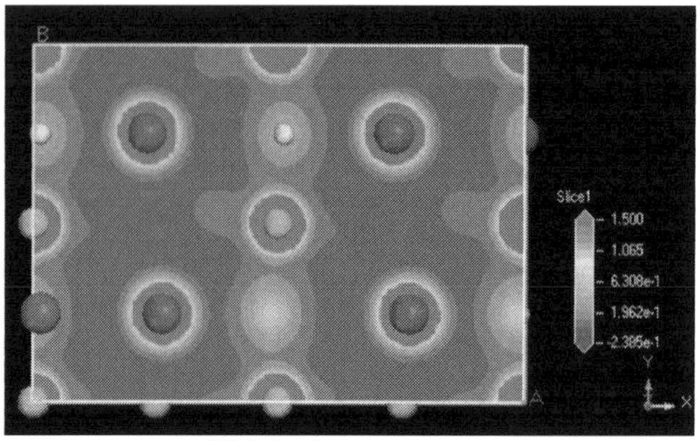

Figure 8. Total Charge Density of pure bulk $NaMgH_3$ parallel to the x and y axis. Where the green balls are Mg atoms, purple balls are Na atoms and white balls are hydrogen atoms.

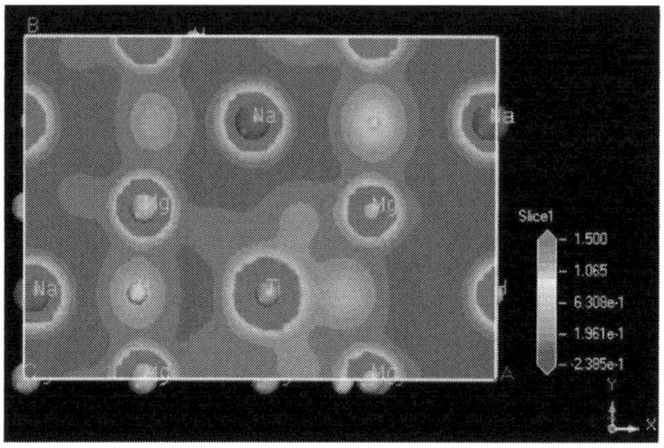

Figure 9. Total Charge Density of Ti-doped bulk $NaMgH_3$ parallel to the x and y axis. Where the green balls are Mg atoms, purple balls are Na atoms, white balls are hydrogen atoms and gray ball is Ti.

4*d* Transition Metals

In a similar fashion, cohesive energies for 4*d* block were calculated, Tc, Mo, Nb, Zr, and Y in the decreasing order, are the most favorable elements with respect to the cohesive energy of the pure . However, for the 4*d* block, all the models tested show a high positive energy with respect to the pure model. Thus, the alloyed systems are too stable for hydrogen storage applications. For instance, the cohesive energy due to the presence of the Tc dopant is 7.36 eV more favorable than the pure case where no dopants are available. This cohesive energy is 2.5 times greater than the cohesive energy of the system due to the presence of Ti. However, the cohesive energy of the system due to the presence of 4*d* block elements shows a strong positive correlation with the ionization energy of the crystalline element. Meanwhile, the correlation with the different melting points of the 4*d* element block and the cohesive energy of the crystalline elements is not as positive when compared to the 3*d* transition metal results.These results and their relationships are shown below.

Model	X in Bulk Site	System	*E_{coh} (eV)	E_{sub} (eV/atom)
Bulk	Na	$Na_{16}Mg_{16}H_{48}$	0.00	
	Y	$(Y)Na_{15}Mg_{16}H_{48}$	3.17	0.04
	Zr	$(Zr)Na_{15}Mg_{16}H_{48}$	5.15	0.064
	Nb	$(Nb)Na_{15}Mg_{16}H_{48}$	5.46	0.068
	Mo	$(Mo)Na_{15}Mg_{16}H_{48}$	6.21	0.077
	Tc	$(Tc)Na_{15}Mg_{16}H_{48}$	7.36	0.92

Table 5. Cohesive Energies of the different 4*d* Bulks Models

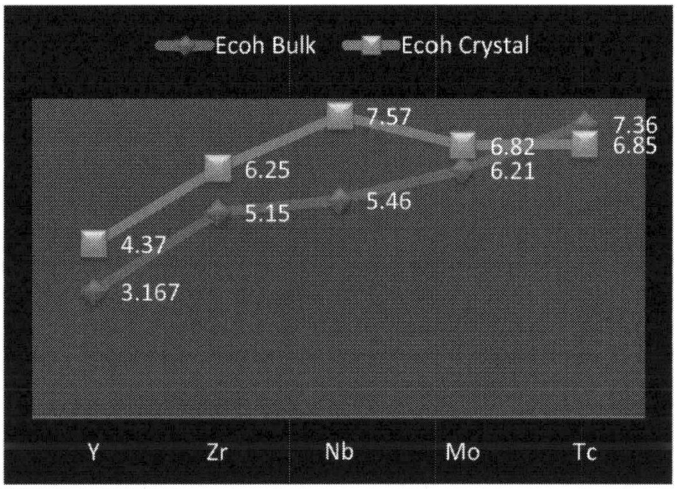

Figure 10. Cohesive Energy Results. Bulk substitution of 4*d* transition metal in $NaMgH_3$ (red), pure crystalline transition metal (blue).

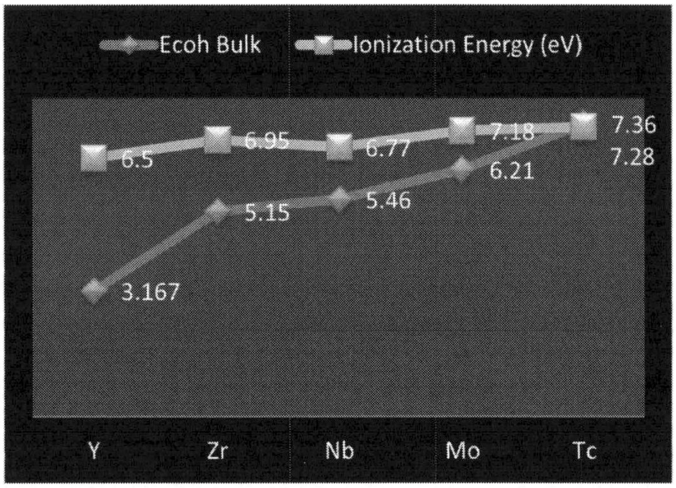

Figure 11. Cohesive Energy Results. Bulk substitution of 4*d* transition metal in NaMgH3 (blue), Ionization Energy (green)

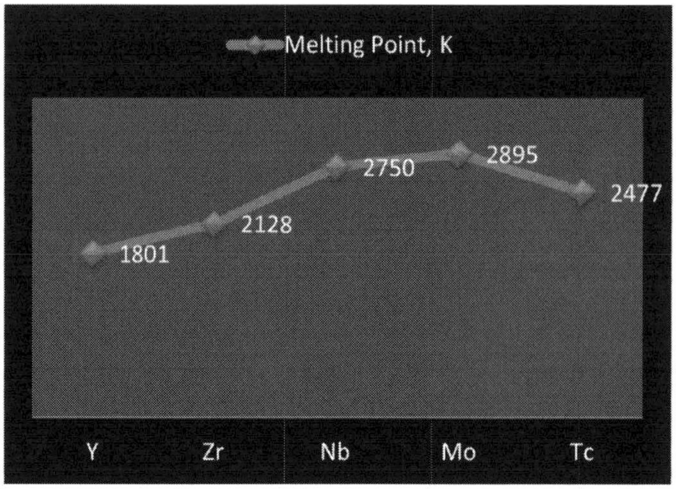

Figure 12. Melting point in Kelvins for 4*d* transition metals in their pure crystalline state.

4*d* Total Charge Density

The total density charge of the Nb-doped bulk model is taken as a sample from the 4*d* transition metal block. This plot also suggests that the addition of a dopant affects the local electron density of the model. This local modification of the 4*d* doped bulk model leads to an increase association of the dopants with the surrounding Mg and H atoms, thus; leading to a possible destabilization of the Mg-H complex as well.

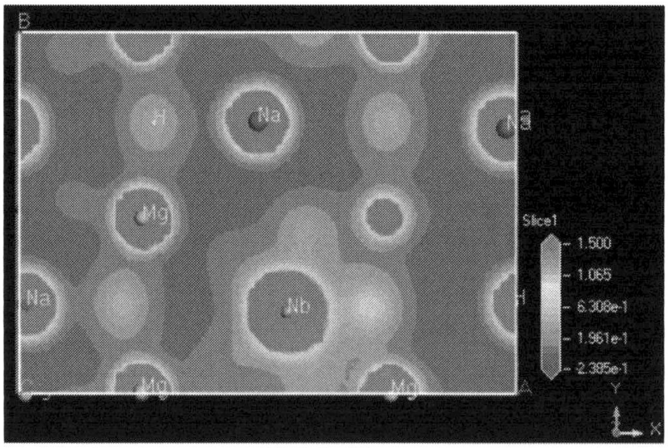

Figure 13. Total Charge Density of Nb-doped bulk $NaMgH_3$ parallel to the x and y axis. Where the green balls are Mg atoms, purple balls are Na atoms, white balls are hydrogen atoms and gray ball is Nb

3.2 Surfaces

Next, DFT calculations were conducted on five (001) surface models : Ti @ Na lattice site, Ti @ hollow site, Ti @ top of Na lattice site, Ti @ top of hollow site, and Ti @ Mg site. It is possible to observe the preferred Ti dopant site in terms of the substitutional/addition energy relative to the pure surface model. In this regard, possible formation of $TiMg_xH_x$ complexes was systematically investigated. The models used are showed below:

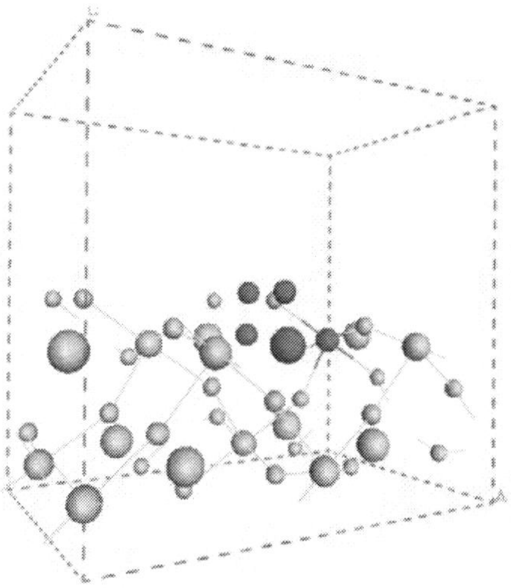

Figure 14 Model with Dopant (red) Sites

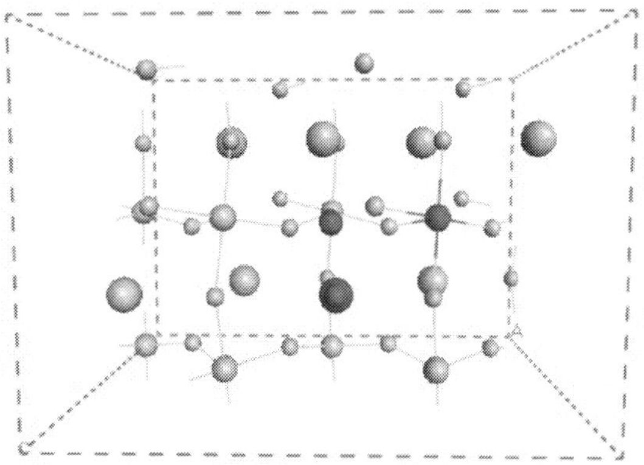

Figure 15 Model with Dopant (red) sites (Top View)

From our calculations, Ti @ Hollow site and Ti @ Na site are the preferred sites for the dopant in this system. Ti @ Hollow site has a cohesive energy of +5.47 eV and Ti @ Na site has a cohesive energy value of +4.60 eV (Table 5). Furthermore, Ti @ Mg shows a more favorable cohesive energy value than Ti @ Top sites with values of +4.23 eV and +3.22 eV, respectively. Simulation results also showed that it is more favorable to remove a hydrogen atom from a doped model than from the pure model, with values of +3.05 eV and –3.46 eV, respectively. Thus, for subsequent calculations only the most favorable models will be tested.

Site	**System**	$\mathbf{E_{coh}}$ **(eV)**	$\mathbf{\Delta E_{add}}$**(eV/ atom)**
Ti@ Na site	$TiNa_7Mg_8H_{24}$	4.60	N/A
Ti @ Hollow Site	$TiNa_8Mg_8H_{24}$	5.47	0.13
Ti@ Top Na site	$TiNa_8Mg_8H_{24}$	3.22	0.08
Ti@ Top Hollow Site	$TiNa_8Mg_8H_{24}$	3.22	0.08
Ti@ Mg Site	$TiNa_8Mg_7H_{24}$	4.23	0.11
H removed from Ti@ Na site	$TiNa_7Mg_8H_{23}$	3.05	N/A
H removed from Pure model	$Na_8Mg_8H_{23}$	-3.46	N/A

Table 5. Surface Energy Results

Formation of Ti-Mg-H complexes at the surface as shown in Figure 17 and Figure 18 may improve the reaction mechanism, in order to observe this phenomena surface models are investigated using DFT-MD at different temperatures (423 K and 448 K).

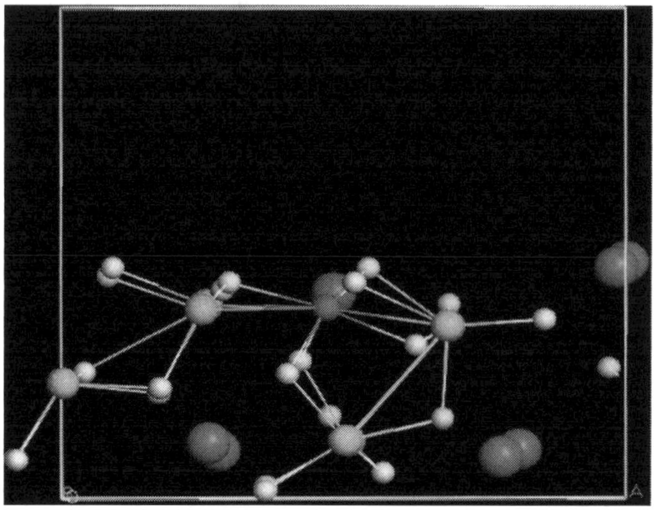

Figure 17 Optimized Ti@ Na site surface model

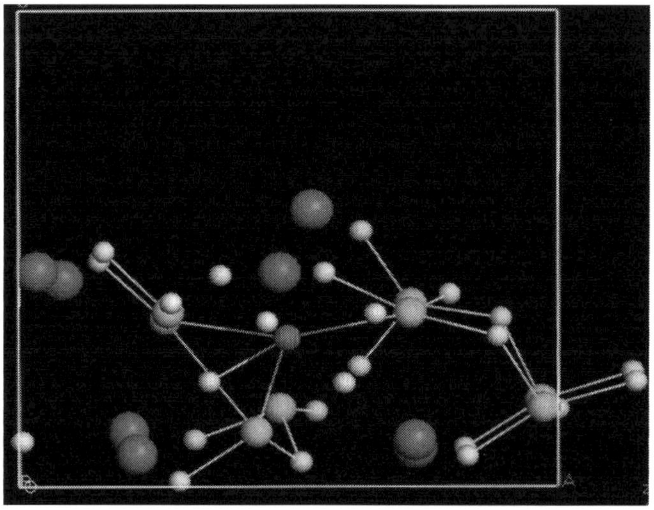

Figure 18 Optimized Ti@ Hollow site surface model

4. Dynamics

After completing the calculations regarding surface doped models and identifying the preferred sites of Ti dopants, the effect of high temperature on the Ti-Mg_x–H_x complexes is studied with ab initio molecular dynamics at 423K and 448 K in the canonical and micro-canonical regimes as mentioned earlier. First, via DFT coupled MD simulations, the effect of Ti impurity on surface models was studied to determine the stability of the complexes and to aid us in determining the role they play in hydrogen behavior at the surface. The calculations were spanned over an 8 ps time frame. The systems were considered to reach equilibrium once the drift in the constant of motion was calculated to be less than 8 meV/ atom. Also, dynamic equilibrium was analyzed through the use of radial distribution functions at the 1 ps interval. Electron density and electron density difference maps were generated to verify the distinguished effects of the Ti additive on the electronic structure of the system.

First, dynamic simulations were carried out for Ti@ Na site at 423 K for a total simulation time of 4 ps. At the end of the 4 ps the Ti-Mg_x–$H_x$was still present with a Ti-Mg_5 –H_7

formation. The calculated bond lengths as confirmed with radial distribution function plots were as follow: Ti-Mg 2.97 Å, Ti-H 1.87 Å. Then, the temperature is increased to 448 K to further study the temperature effect for a total simulation time of 8 ps. In a similar fashion, at the end of the 4 ps the Ti-Mg_x–$H_x$was still present with a Ti-Mg_4 –H_7 formation. The calculated bond lengths as were as follow: Ti-Mg 3.10 Å, Ti-H 1.93Å.

Next, the electron density difference in the surface model due to the Ti substitution at the Na lattice site is investigated. The electron density difference illustrates how the charge density changes during a chemical reaction or on docking of a molecule to a surface.

The electron density difference with respect to atoms can be represented in the following manner:

$$\Delta\rho = \rho Ti@ \text{ Na site} - \Sigma(\rho i) \qquad (6)$$

Where the subscript *i* runs over all atoms. This equation shows the changes in the electron distribution due to the formation of all the bonds. Furthermore, Figure 19 shows the formation of chemical bonds between Ti and the surrounding atoms by delocalization of the atomic electron density. In these plots, a loss of electron is indicated in blue, while regions rich in electrons are indicated in red. White colors indicate regions with small changes in the electron density. Based on this plots, it is clearly shown that Ti lost electrons; thus, confirming that electron affinity and ionization energy are important parameters in hydrogen storage research.

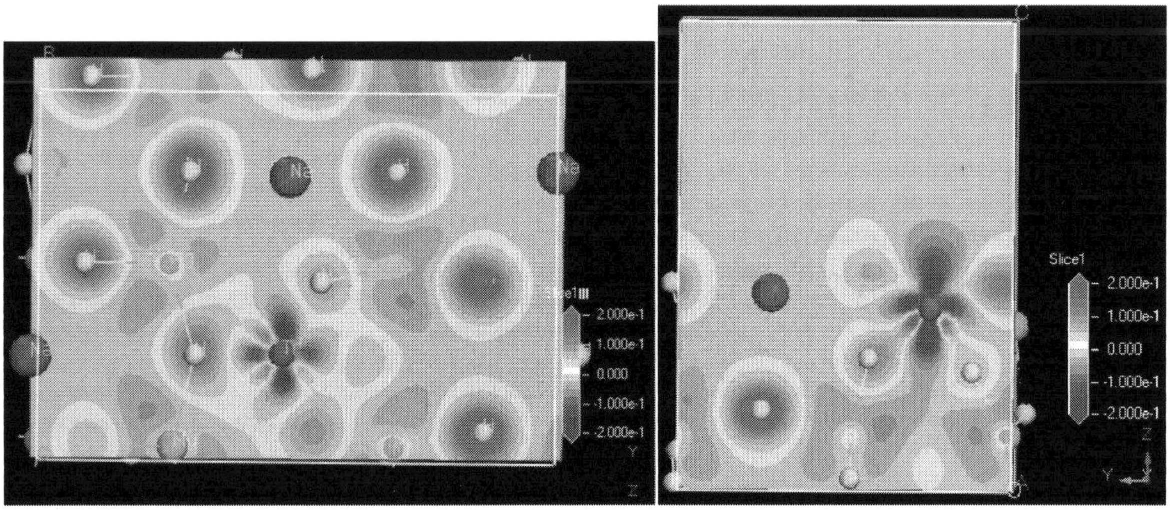

Figure 19 Electron Density Difference Maps of Ti@ Na lattice surface parallel to the x/y surface and y/z surface. Where red ball is Ti, green ball is Mg, purple ball is Na and white ball is H.

Then, following the same procedure DFT-MD simulations were carried out for Ti@ hollow site at 423 K with a canonical ensemble equilibrium run of 2 ps followed by a micro-canonical ensemble production run of 2 ps for a total simulation time of 4 ps. At the end of the 4 ps the Ti-Mg_x–$H_x$was still present with a Ti-Mg_6 –H_7 formation. The calculated bond lengths were as follow: Ti-Mg 2.92 Å, Ti-H 1.91 Å. Then, the temperature is increased to 448 K to further study the temperature effect for a total simulation time of 8 ps. Similarly, at the end of the 4 ps the Ti-Mg_x–H_x was still present with a Ti-Mg_5 –H_7 formation. The calculated bond lengths were as follow: Ti-Mg 2.94 Å, Ti-H 1.87Å.

Comparing the dynamic results of both models, it seems that the complex formed at Ti @ Hollow site shows increased association as predicted by cohesive energy values. However, it is apparently necessary to increase the number of accessible neighbors to the dopant to accurately elucidate the complex formation and exact role it plays in absorption/desorption mechanisms in CMHs.

In light of the MD simulation results, bulk/surface geometry optimization results several observations are reached. To summarize, Ti-Mg_x–$H_x$are observed after geometry optimization of doped surface models, DFT-MD show the existence of these complexes even at high temperatures, cohesive energy calculations at the surface models show that Ti@Na lattice site and Ti@ hollow site are the preferred Ti locations in this system, and the mechanism in which Ti @ hollow site ejects Na ion from lattice site requires further study with an extended slab model.This phenomenon will be further studied to determine the role that this plays in creation of nucleation sites or possible Na or Mg rich sites that ultimately leads to saturation and desorption of H from the system.

4. Conclusion and Future Work

The cohesive energies of the 3*d* block elements in the bulk follow a strong correlation with the cohesive energies of their metallic elements counterpart. For the 3*d* block elements, Ti, V, Sc, Co, Ni and Cu in the decreasing order, are the most favorable elements with respect to the cohesive energy of the pure system. On the other hand, Zn and Mn doped models are unstable in comparison to the pure model. Also, 4*d* block elements are too stable in comparison to the pure model to be used as viable dopants for CMHs. Furthermore, 4*d* block elements show a positive correlation with crystalline elements ionization energy, suggesting that dopants replacing Na ions should have high ionization energies to further weaken the Mg-H covalent bond, and ultimately reducing the thermodynamic stability of the system.

From the cohesive energy calculation on surface models, we conclude that dopants at the hollow and Na lattice sites are the most favorable sites for dopant locations in the $NaMgH_3$ system. Thus, future studies will have more emphasis on those models. Ti @ Na lattice site shows a possible Ti-Mg_4H_6 complex formation upon completion of geometry optimization. In the same manner, Ti @ Hollow site shows a possible $TiMg_5H_7$ complex formation; thus, showing an increased association of the bond lengths when compared to the Ti@ Na lattice site model. The increased association of the dopants in the bulk and surface model leads to elongation of Mg-H bond lengths, thus; possible weakening that facilitates hydrogen desorption is feasible.

It is worth mentioning that the Ti placed @ the hollow site of $NaMgH_3$ ejects a Na ion from its lattice site. This phenomenon will be further investigated to determine the role that it plays in desorption of H from the system. Thus, two destabilization methods can be rationalize based on bulk and surface models: 1) additives with a mix of low ionization energy (high ionization energy for 4*d* transition metals), low melting point and low crystalline cohesive energy should be considered to locally destabilize the CMHs by increasing the association with Mg-H complexes which ultimately results in regions rich in H and Mg atoms that initiate nucleation/growth sites that facilitate H kinetics and 2) additives with high ionization energy or low electron-negativity affinity should be considered to lower the overall thermal stability of the system which may lead to a decrease in H desorption temperature.

Furthermore, as a second step in the research program, the role of 3*d* transition metals such as Ti and Mn in the improved dehydrogenation kinetics of $NaMgH_3$ needs to be understood. Reactants and products will be optimized using DFT codes within the GGA approximation to determine the reaction barrier height for the first reaction. Reaction pathways will be determined through transition state search using LST/QST methods.

Acknowledgements

Financial support provided by NSF through the "Louisiana EPSCoR Research Infrastructure Improvement" grant (EPS 1003897) is acknowledged.

References

1) Hirst, E; Moyers, J.C.; "Efficiency of Energy Use in the United States." Science. 1973. 179. 1299-1304.

2) Salameh, M.G; "Quest for Middle East oil: the US versus the Asia-Pacific region." Energy Policy. 2003.31.1085-1091.

3) Kone, A.C; Buke, T.; "Forecasting of CO2 emissions from fuel combustion using trend analysis." Renewable and Sustainable Energy Reviews. 2010. 14. 2906-2915.

4) Grochala, W; Edwards, P. P.; "Thermal Decomposition of the Non-Interstitial Hydrides for the Storage and Production of Hydrogen." Chem. Rev. 2004. 104. 1283-1315.

5) Alapati, S. V; Johnson, K. J.; Scholl, D. S. "Identification of Destabilized Metal Hydrides for Hydrogen Storage Using First Principles Calculations." J. Phys. Chem. B. 2006. 110. 8769-8776.

6) Choudhury, P.; Bhetanabotla, V.R.; Stefanakos, E.; "Manganese Borohydride as a Hydrogen –Storage Candidate: First-Principles Crystal Structure and Thermodynamic Properties." J. Phys. Chem. C. 2009. 113. 13416-13424.

7) Wu, H; Zhou, W.; Udovic, T.J.; Rush, J.J.;Yildirim, T.; " Crystal Chemistry of Perovskite-Type Hydride NaMgH3 : Implications for Hydrogen Storage." Chem. Mater. 2008. 20. 2335-2342.

8) Bogdanovic, B; Schwickardi, M.; "Ti-doped alkali metal aluminum hydrides as potential novel reversible hydrogen storage materials." J. Alloys. Compd. 1997.253. 1-9.

9) Ozolins, V; Majzoub, E.H.; Wolverton, C.; "First-Principles Prediction of Thermodynamically Reversible Hydrogen Storage Reactions in the Li-Mg-Ca-B-H System." J. Am. Chem. Soc. 2009. 131. 230-237.

10) Lovvik, O.M; Swang, O.; "Crystal structures and electronic structures of alkali aluminohexahydrides from density functional calculations."J. Alloys. Compd. 2005. 404. 757-761.

11) Aguayo, A; Singh, D.J.; "Electronic Structure of the complex hydride NaAlH4."Physical Review B. 2004. 69.103-108.

12) Alapati, S. V; Johnson, K. J.; Scholl, D. S. "Identification of Destabilized Metal Hydrides for Hydrogen Storage Using First Principles Calculations." J . Phys. Chem. B. 2006. 110. 8769-8776.

13) Lovvik, O.M; Opalka, S. M.; "Density Functional calculations of Ti-enhanced NaAlH4." Physical Review B. 2005. 71. 1-9.

14) Yvon, K; Bertheville, B.; "Magnesium based ternary metal hydrides containing alkali and alkaline-earth metals." J. Alloys Compd. 2006. 38. 1-8.

15) Bouhadda, Y.; Boudouma, Y.; Fennineche, N.; Bentabet, A.; " Ab initio calculations study of the electronic, optical and thermodynamic properties of $NaMgH_3$, for hydrogen storage." J. Phys. Chem. Solids. 2010. 71. 1264-1268.

16) Klaveness, A.; Swang, O.; Fjellvag, H.; "Formation enthalpies of $NaMgH_3$ and $KMgH_3$: A computational study." Europhys. Lett. 2006. 76. 285-290.

17) Shane, D.T.; Corey, R.L.; Bowman, R.C.; Zidan, R.; Stowe, A. C.; Hwang, S.; Kim, C.; Conradi, M. S.; " NMR Studies of the Hydrogen Storage Compound $NaMgH_3$." J. Phys. Chem. C. 2009.113. 18414-18419.

18) Hao, S.; Sholl, D.S.; "Role of Schottky Defects in Hydrogen and Metal Diffusion in NaH, MgH_2, and $NaMgH_3$." J. Phys. Chem. Lett. 2010. 1. 2968-2973.

19) Komiya, K.; Morisaku, N.; Rong, R.; Takahashi, Y.; Shinzato, Y.; Yukawa, H.; Morinaga, M.; " Electronic Structure Calculations of Perovskite-type Hydrides for Hydrogen Storage." J. Alloys. Compd.2008. 453. 157-160.

20) Accelrys, "Materials Studio®," 4.0 ed. San Diego: Accelrys Software, Inc., 2006.

21) M. D. Segall, P. J. D. Lindan, M. J. Probert, C. J. Pickard, P. J. Hasnip, S. J. Clark, M. C. Payne;"First-principles simulation: ideas, illustrations and the CASTEP code." Journal of Physics: Condensed Matter. 2002.14.2717-2743

(456e) A Computational Study of Steam Hydrolysis of NaBH4

Ping Li, University of Pittsburgh, Pittsburgh, PA, Karl Johnson, Chemical and Petroleum Engineering, University of Pittsburgh, Pittsburgh, PA and Michael A. Matthews, University of South Carolina, Columbia, SC

Complex hydrides are promising hydrogen storage materials and have received significant attention due to their high hydrogen storage capacity. $NaBH_4$ has high gravimetric (10.7 wt%) and large volumetric (115 kg of H_2 m^{-3}) hydrogen content. The hydrolysis reaction of $NaBH_4$ releases hydrogen with both fast kinetics and high extent of reaction under technical conditions by using steam deliquescence of $NaBH_4$. This catalyst-free reaction has many advantages over traditional catalytic aqueous phase hydrolysis. However, the reaction mechanism and kinetic barriers associated with steam hydrolysis have not yet been elucidated. We are developing a fundamental molecular level understanding of the steam hydrolysis reaction of $NaBH_4$ using a combination of experiments and first-principles density functional theory (DFT) calculations. We have applied finite temperature *ab initio* molecular dynamics (AIMD) to investigate the reaction mechanism. AIMD provides clues to elementary reaction steps. We have studied the effect of $NaBH_4$ concentration on the initial reaction rate, and found that high $NaBH_4$ concentrations lead to high initial reaction rates. The reason for this is appears to be due to a cooperative interaction between two BH_4^- groups. We have computed the free energy profiles of possible reaction pathways observed in AIMD using a cluster approach with an implicit solvent model and explicit water molecules. Explicit water is critical because proton transfer is a key step in the hydrolysis reaction.

See more of this Session: Hydrogen Storage Systems and Materials: Modeling, Simulation, and Optimization:
See more of this Group/Topical: International Congress on Energy 2011

(470a) Silicon-Based Bulk Nanostructured Thermoelectric Generators

Akram Boukai, Materials Science and Engineering, University of Michigan, Ann Arbor, MI

State-of-the-art thermoelectric generators are typically based on exotic alloys and have low device efficiencies. Therefore, research targeting new discoveries in earth abundant thermoelectric materials that increase device efficiency are needed. This talk will present recent progress on bulk nanostructured thermoelectric materials that are silicon-based. These include functional silicon nanowire devices and silicide/silicon eutectics. These bulk nanostructured thermoelectric devices exploit dramatic reductions of thermal conductivity due to phonon scattering at disordered interfaces, which is expected to increase ZT. Thermoelectric properties, structural characterization, and device efficiencies will be presented.

See more of this Session: Nanomaterials for Thermal-to-Electric Conversion I
See more of this Group/Topical: Topical 5: Nanomaterials for Energy Applications

(470b) Advanced Nanostructures for Thermoelectric Energy Harvesting

Yue Wu, School of Chemical Engineering, Purdue University, West Lafayette, IN

Substantial efforts have been devoted to use nanostructured materials for thermoelectric energy harvesting and solid-state cooling in the past decade. However, the majority of the research is still limited in lab scale due to the incapability to mass-produce well-defined nanostructured materials with low yet industrial-compatible process. In addition, a lot of widely used thermoelectric materials contain toxic and expensive elements that prevent the large-scale deployment of the thermoelectric devices. In this presentation, we will discuss our research on the development of mass production of molecular scale nanostructures of chalcogenides and metal oxides, as well as their heterostructures, for the manufacture of thermoelectric generators operating at different temperature ranges. Particularly, we will talk about our approach to discover and investigate the non-toxic and abundant nanostructured materials to achieve an environmentally friendly process. Our preliminary research indicated that thermoelectric figure of merit (ZT) close to 2 can be achieved in the molecular scale nanowires of certain chalcogenide due to significantly enhanced power factor and reduced thermal conductivity as a result of quantum confinement.

See more of this Session: Nanomaterials for Thermal-to-Electric Conversion I
See more of this Group/Topical: Topical 5: Nanomaterials for Energy Applications

(470c) Molecular and Hybrid Solution Processible Thermoelectrics

Rachel A. Segalman[1], Shannon Yee[2], Boris Russ[2], Nelson Coates[2] and Jeffrey Urban[3], (1)Department of Chemical and Biomolecular Engineering, University of California, Berkeley, Berkeley, CA, (2)University of California at Berkeley, Berkeley, CA, (3)Lawrence Berkeley National Laboratories, Berkeley, CA

Thermoelectric materials for energy generation have severaladvantages over conventional power cycles including lack of moving parts,silent operation, miniaturizability, and CO2 freeconversion of heat to electricity. Excellentthermoelectric efficiency requires a combination of high thermopower(S, V/K), high electrical conductivity (σ, S/cm), and low thermalconductivity (κ, W/mK). To date the bestmaterials available have been inorganic compounds with relatively low earthabundance and highly complex, vacuum processing routes (and hence greaterexpense), such as Bi_2Te_3. Molecular materials andhybrid organic-inorganics bring the promise ofinexpensive, solution processible, mechanicallydurable devices. While highlyconductive polymers are now common place, theygenerally demonstrate low thermopower. Our work onmolecular scale junctions demonstrates that nanostructuringof organics allows them to act as thermionic filters between inorganicjunctions which can lead to enhanced thermoelectric properties. We have takeninspiration from this fundamental understanding to design material systems inwhich we combine a high electrical conductivity, low thermal conductivitypolymer with a nanoparticle that contributes high thermopower. Additionally, the work functions of the two materials are well-aligned which introduces the possibility of thermionicfiltering at the interface and an additional boost to the power factor. The combination of these effects results in a new hybrid,solution processible material with a thermoelectricfigure of merit within an order of magnitude of the Bi_2Te_3. In this talk, I will discuss both the use ofthermoelectric measurements to gain insight to molecular junctions and how thisinsight translates to design principles for polymer and hybrid thermoelectrics.

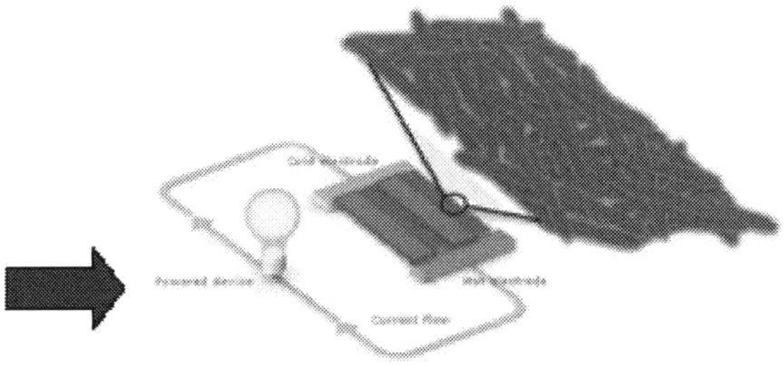

See more of this Session: Nanomaterials for Thermal-to-Electric Conversion I
See more of this Group/Topical: Topical 5: Nanomaterials for Energy Applications

(484a) Modeling of Long-Time Flux Decay In Cross-Flow Filtration

Philip P. Schonewill, Richard C. Daniel, Rick W. Shimskey, Justin M. Billing and Reid A. Peterson, Pacific Northwest National Laboratory, Richland, WA

Recent work at Pacific Northwest National Laboratory has been concerned with understanding how cross-flow filter flux decays with time. Understanding the cause(s) of flux decay is important for assessing long-term behavior of this separation technology, especially when used in challenging processing environments e.g. nuclear waste treatment. Fouling in the filter is a complex issue that is affected by many parameters: the chemistry of the solution being filtered, particle size and morphology, operating conditions, and cleaning methods. Based on filtration work performed at PNNL over the past several years, it is clear that a portion of the fouling can be "reversed" by back-pulsing the filter and another portion is irreversible (usually only rectified via cleaning protocols). However, the sheer number of parameters requiring consideration complicates effective modeling of both the long-time behavior of the cross-flow filter and its response to back-pulses.

To combat this, simple models were developed from first principles based on different suspected fouling mechanisms. The models were tested against long-term data (100-hour tests at a constant solids concentration) collected with PNNL's bench-scale filtration system. The flux was found to be well described by a cake mechanism model at intermediate times (starting at approximately 3 hours elapsed time); at longer times (approximately greater than 20 hours of filtration time) the flux behavior transitions to an intermediate mechanism behavior. The decay constant for these models can be predicted using a single dimensionless parameter, the ratio of the transmembrane pressure to the inertial pressure. Additional testing suggested that the filter could be conditioned by back-pulsing. These observations were used to build a comprehensive model to assist in determining a back-pulse strategy for cross-flow filters that will experience long-term decays in the filter flux.

See more of this Session: Theories and Applications of Cross Flow Filtration
See more of this Group/Topical: Separations Division

(484b) Control of Packed-Particle Flow Resistances During Cross-Flow Filtration In Microchannels

Nopphon Weeranoppanant, Levy Amar, **Michael Hill** and Edward F. Leonard, Chemical Engineering, Columbia University, New York, NY

Cross-flow filtration of fine suspensions through microsieves is a logical, sometimes necessary step in micro-processing. The observed limitation on steady-state filtrate flux accompanying this operation is generally attributed to particle accumulation upon the sieve. We consider two mechanisms: (1) blockage of filter pores by particles in contact with the filter surface, and (2) formation of a packed bed of particles that intrudes upon and is regulated by shear flow within the microchannel. This study examines the conditions under which a packed bed forms on a sieve, and provides a simple model for predicting its thickness when formed by hard non-interacting microspheres. Extensive experimental data are reported that corroborate the model. The good fit of these data lay a firm basis for the semi-quantitative analysis of the behavior of more complex suspensions.

See more of this Session: Theories and Applications of Cross Flow Filtration
See more of this Group/Topical: Separations Division

(484c) Cross-Flow Filtration of Plasma From Blood From a Microchannel

Xiangnan Ye[1], Levy Amar[1], Michael Hill[1], James Jones[2], Jacob Baggerman[3], Cees van Rijn[3] and **Edward F. Leonard**[1], (1)Chemical Engineering, Columbia University, New York, NY, (2)Department of Medicine -- St. Lukes'/Roosevelt Hospital, Columbia University, New York, NY, (3)Aquamarijn Micro Filtration, Zutphen, Netherlands

Steady cross-flow filtration of plasma through microsieves from blood in a microchannel is an important part of several biomedical applications under development, including a new artificial kidney. Erythrocytes, which comprise the largest volume fraction of all particulates in blood, are believed to create the controlling resistance at the sieve surface. Maximum steady-state filtration flux has been observed to be a function of wall shear rate, as predicted by any conventional cross-flow filtration theory, but to show weak dependence on erythrocyte concentration, contrary to theory based on convective diffusion. Maximum filtration flux has also been observed to be a function of sieve pore size and shape (in spite of their very low flow resistance) as well as sieve surface coating.

Recent experiments have correlated macroscopic measurements (filtration rates, transmembrane pressures) with direct observation of erythrocyte behavior at the filtering surface. At low filtration rates (low transmembrane pressures), erythrocytes roll across the filter surface, but at higher filtration rates (higher transmembrane pressures), erythrocytes are observed sticking to the sieve surface. Post-filtration SEM's, even those obtained at very low transmembrane pressures, reveal significant capture and deformation of erythrocytes in the filter pores. Careful balances among sieve design (which can be precisely defined by photolithography), surface treatments of sieves, and flow conditions, allow filtration fluxes at rates exceeding 0.1 cm^3/cm^2-min.

See more of this Session: Theories and Applications of Cross Flow Filtration
See more of this Group/Topical: Separations Division

An electrokinetic model for combined fouling during cross flow filtration of electrolyte solution and charged Colloids

Md Abdullaha Al Mamun, Reeshav Chatterjee, Subir Bhattacharjee,
Department of Mechanical Engineering, University of Alberta, Edmonton, AB, Canada T6G 2G8
Email: subir.b@ualberta.ca
Sirshendu De
Department of Chemical Engineering, Indian Institute of Technology, Kharagpur 721 302, India

Abstract

Colloidal fouling, formation of a cake due to deposition of colloids on to a membrane, is commonly encountered in pressure driven filtration processes, such as nanofiltration and reverse osmosis. The influence of these deposited charged colloids on the ion concentration polarization, ion rejection and the overall permeate flux is the focus of the present work. A common situation encountered during membrane filtration is the combined influence of charged colloids and the associated electrolyte on the flux decline and rejection. In this context, coupled model of concentration polarization (CP), cake layer formation by charged colloidal particles and ion transport through the interstitial spaces of the cake layer has been developed to predict the permeate flux decline and observed salt rejection during cross flow filtration of an electrolyte and colloidal mixture. The suspension structure of the cake layer is considered as a swarm of colloidal particles, represented using the Kuwabara cell model. For this system, the electroosmotic flow is calculated analytically using the Levine-Neale model. The analysis is then combined with the standard film theory of cross flow membrane filtration to predict the flux decline, observed rejection, cake layer thickness and resistance, and osmotic pressure difference. The model illustrates the combined influence of electrical and physical properties of colloids and electrolytes, mass transfer coefficient or crossflow condition and membrane resistance on the separation process. This approach provides a comprehensive insight into the mechanism of cake enhanced concentration polarization for charged colloids.

Keyword: Cake Enhanced Concentration Polarization, Kuwabara Cell Model, Levine-Neale Electrophoretic Model, Film Theory

1 Introduction

Membrane separation processes are increasingly being used in industries as a treatment process for separation of colloids, macromolecules, solutes, and ions. This is often achieved through processes like nanofiltration, whereby a single unit operation is used to remove multiple molecular weight and size species. The major operating costs associated with membranes include creating the pressure driving force for permeation, and mitigation of fouling. Both these operating costs are directly related to the overall energy consumption of membrane filtration. Several methods for fouling prevention and minimization has been suggested over the last few decades. But, most studies on fouling have typically ignored the concept of microscale manipulation of the fouling deposit itself to modify its permeability, or its ability to selectively adsorb ions or small organic contaminants. Therefore study of the fouling mechanism is important to develop a better understanding of the relation between the permeate flux, salt rejection, cake formation and to manipulate the fouled layer to enhance the preformece of the filtration process.

Hoek and Elimelech [1] conducted colloid fouling experiments during cross flow RO and NF, and reported cake enhanced osmotic pressure (CEOP) as the mechanism of flux decline. They proposed a model of CEOP, where hindered back-diffusion of ions within colloidal deposit layers was mentioned as the main mechanism of concentration polarization, resulting in an increase of osmotic pressure and decrease of pemeate flux.

Kim and co-workers [2] presented a detailed model of three component combined fouling to represent flux decline behavior of the system during dead end filtration. The model however introduces several parameters that need to be determined independently, such as, cake layer tortuosity, diffusive tortuosity, and hindered diffusivity of electrolyte through the cake layer. Additionally the model does not incorporate the effect of zeta potential of colloids and electrolyte solution conductivity in the flux decline behavior. However, the model presents a comprehensive mechanism for multi-component fouling from classical filtration concept.

Sim and co-workers [3] investigated the coupled effect of cake enhanced osmotic pressure and colloidal fouling in RO, using cross flow sampler-modified fouling index ultrafiltration (CFS-MFI$_{UF}$). They used CFS-MFI$_{UF}$ measurement to determine the deposition rate of colloidal mass and cake porosity, which then incorporates the CEOP effect in RO fouling. They also considered the hindered diffusivity of ions within the cake layer to model CEOP. Though, the technique provides good estimation of RO fouling profile, the model depends on the CFS-MFI$_{UF}$ or dead-end RO experimental data.

De and Bhattacharjee [4] proposed an electrokinetic model for flux decline during combined filtration of charged colloids and electrolyte. They assumed the cake layer as a bundle of charged cylindrical nano-pores similar to the membrane. Their model considered the electroviscous effect created by the charged nano-pores, which results in additional resistance of the cake layer. Though their model represents an improvement, the actual flow model should consider the flow of electrolytes around the charged colloids in a cake layer to be more realistic.

In this paper, a coupled model of concentration polarization, cake layer formation by charged colloidal particles and ion transport through the interstitial spaces of cake layer has been developed to predict the permeate flux decline and observed rejection due to fouling during cross flow nanofiltration of an electrolyte and colloidal suspension. The model cosiders the cake layer as a swarm of charged spherical particles instead of cylindrical nanopores, which is more realistic, and it can predict the flux decline and observed rejection without considering any correction parameter except the cake tortuosity. The model illustrates the combined influence of colloid mass transfer coefficient, electrical and physical properties of colloids and electrolytes, crossflow geometry and membrane resistance on the cake enhanced concentration polarization of the electrolyte.

2 Mathematical Model

During combined fouling of colloidal particles of radius a, surface potential ψ_c and electrolyte feed concentration of $C_{i,f}$ in a crossflow membrane filtration operation, the colloidal particles form a cake layer on the membrane and the electrolyte solution forms a concentration polarization layer on the cake surface as shown in Fig. 1. The dimensionless permeate flux through the cake layer and membrane can be represented by the following modified Darcy's equation [4]:

$$v_w^* = \frac{v_w}{v_w^0} = \frac{1-\Delta\pi_m^*}{1+R_g^*} \tag{1}$$

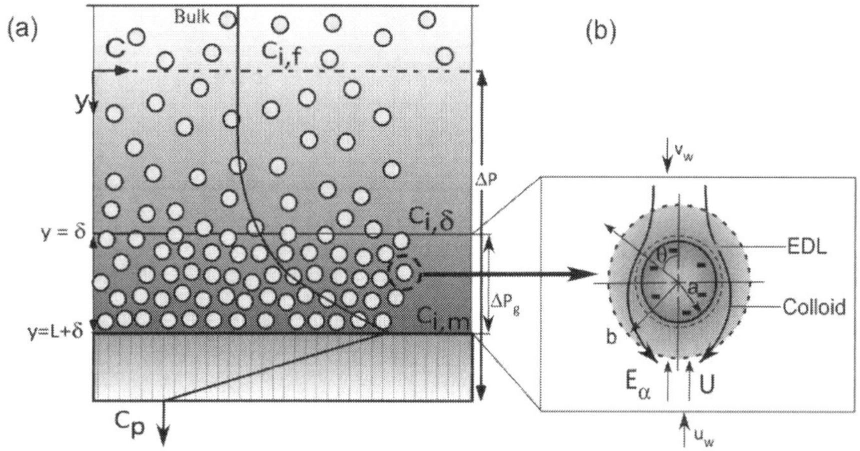

Figure 1: (a) Cake enhanced concentration polarization (b) Electrophoretic flow model around a charged colloid according to the Levine-Neale and Kuwabara cell model

Here, $v_w^o = \Delta P/\mu R_m$ is the pure water flux, $\Delta\pi_m^*$ is the dimensionless osmotic pressure across the membrane and R_g^* is the dimensionless resistance. The cake layer resistance R_g^* has been formulated based on the Kuwabara cell model and Levine-Neale electrophoretic flow model [5] of charged particles in a concentrated colloidal suspension as shown in Fig. 1b. The model considers the combined influence of hydrodynamic and electrokinetic resistance to the transport through the cake layer. According to the Levine-Neale model, electroosmotic flow (U) through the interstices of the cake layer due to induced streaming potential is given by [5]:

$$U = E_\infty\left(\frac{\varepsilon\psi_c}{\mu}\right)g_2(\kappa a,\varphi_p) \tag{2}$$

The expression of induced electric field (E_∞) in Eq. 2 at zero current condition, in accordance with Onsager principle of reciprocity for irreversible phenomena and for thin Debye layer, is [5]:

$$E_\infty = \frac{\Delta P_g}{L}\left(\frac{\varepsilon\psi_c}{\mu\sigma^\infty}\right) \tag{3}$$

Thus, the expression of electroosmotic flux (u_w) from the continuity assumption using Eqs. 2 and 3 is given as follows:

$$u_w = U(1-\varphi_p) = \frac{\Delta P_g}{L\sigma^\infty}\left(\frac{\varepsilon\psi_c}{\mu}\right)^2(1-\varphi_p)g_2(\kappa a,\varphi_p) \tag{4}$$

Therefore, the total flux across the cake layer at no current condition, due to effective pressure across the cake ΔP_g and reverse electroosmotic flow is given as follows:

$$\begin{aligned} v_w &= \frac{\Delta P_g}{\mu r_g L} - u_w \\ \text{or, } v_w &= \frac{\Delta P_g}{\mu r_g L}\left[1-9\varphi_p(1-\varphi_p)A_K\frac{\beta g_2(\kappa a,\varphi_p)}{2(\kappa a)^2}\right] \\ \text{or, } v_w^* &= \frac{\Delta P_g^*}{L^*/Mg} = \frac{\Delta P_g^*}{R_g^*} \end{aligned} \tag{5a}$$

$$r_g = \frac{9\varphi_p A_K}{2a^2}; \quad M = \frac{2a^2 R_m}{9\varphi_p A_K d_e} \quad \text{and} \quad g = \left[1-9\varphi_p(1-\varphi_p)A_K\frac{\beta g_2(\kappa a,\varphi_p)}{2(\kappa a)^2}\right] \tag{5b}$$

Here, r_g is the hydrodynamic resistance of the cake layer, $A_K = 1/(1-9\varphi_p^{1/3}/5+\varphi_p-\varphi_p^2/5)$ accounts for the presence of neighbouring particles, $\beta = (\varepsilon\psi_c\kappa)^2/(\mu\sigma^\infty)$ and $L^* = L/d_e$. The function g accounts for the electroosmotic effect in a swarm of colloidal particles, depends on dimensionless Debye length κa, surface potential of colloidal particles ψ_c, solution conductivity σ^∞ and volume fraction of the cake layer, φ_p.

The dimensionless osmotic pressure $\Delta\pi_m^*$, for simple electrolyte (1:1) can be written from the vant Hoff equation:

$$\Delta\pi_m^* = \frac{2R_{ug}T}{\Delta P}(C_{i,m}-C_{i,p}) = \alpha\bar{C}_{i,m}R_r \tag{6}$$

where, $\alpha = 2R_{ug}TC_{i,f}/\Delta P$ and $\bar{C}_{i,m}$ is dimensionless salt concentration at the membrane surface. The expression of $\bar{C}_{i,m}$, derived by integrating the mass balance of electrolytes within the concentration polarization and cake layer with appropriate boundary conditions is:

$$\bar{C}_{i,m} = \frac{C_{i,m}}{C_{i,f}} = \frac{\exp[v_w^*(R_i+L^*R)]}{R_r+(1-R_r)\exp[v_w^*(R_i+RL^*)]} \tag{7}$$

Here, $R_i = v_w^o/k_i$, $R = v_w^o d_e/D_i^*$. The observed rejection of the membrane can be written as:

$$R_o = 1-\frac{C_{i,p}}{C_{i,f}} = 1-\bar{C}_{i,m}(1-R_r) \tag{8}$$

The dimensionless governing equation for colloidal cake layer growth, obtained by integrating the mass balance equation of colloids in the cake layer with appropriate boundary conditions, is given by [6]:

$$\frac{dL^*}{d\tau} = v_w^*\frac{C_{c,m}^*-\exp(v_w^*R_c)}{1-\exp(v_w^*R_c)} \tag{9}$$

Here, $\tau = v_w^o C_{c,f} t/(d_e \rho_c)$, $R_c = v_w^o/k_c$, $C_{c,m}^* = C_{c,m}/C_{c,f}$. The dimensioless pressure drop with respect to operating pressure can be written as:

$$\Delta P_m{}^* + \Delta\pi_m^* + v_w^* R_g^* = 1 \tag{10}$$

Here, $\Delta P_m{}^*$ is dimensionless pressure drop across the membrane.

3 Numerical Methodology

Equations 1 and 9 are solved simultaneously in conjunction with the Eqs. 6 and 7, using a differential algebric equation solver DDASSL. For a 50 nm particle with surface potential -45 mV and cake volume fraction 0.64 for random-closed-packing, in 0.01 M NaCl solution, the value of g is 0.67. The value of M depends on particle size, cake volume fraction and hydrodynamic radius of crossflow channel and is less than unity for RO or NF. The volume fraction can be as high as 0.72 for hexagonal packing. The viscosity and relative permitvity of water is 0.89 mPa.s and 80.0, respectively. The electrolyte flow turbulence parameter R_i, is considered to be in the order of 1.0 becuase pure water flux and mass transfer coefficient of electrolyte both are of the order 10^{-5}. The corresponding value of R_c is considered to be of the order 10.0, because the diffusivity of colloidal particle is at least one order magnitude less than the electrolytes. For the silica colloidal particles, the bulk concentration is 0.2 kg/m^3, whereas the cake concentration is about 173 kg/m^3 [7]. Typical equivalent diameter of the cross flow channel is 10^{-3} m. For colloid feed concentration 0.2 kg/m^3, with particle density 2317 kg/m^3, τ= 0.01 represents 6.5 hr of operation in real time. The range of dimensionless parameters are listed in table 1.

Table 1: Range of dimensionless parameters used in simulations

Parameter	Range
κa	1 – 100
g_2	0 – 0.9
β	1.0 – 5.0
g	0.05 – 0.95
M	0.005 – 0.5
R_i	0.05 – 1.5
R_c	10.0 – 60.0
R	100 – 1000
α	0.09 – 0.9
$C_{c,m}^*$	800 – 900
R_r	0.8 – 0.9

4 Results and Discussion

Each of the dimensionless parameters listed in Table 1 affect the flux decline and observed rejection during cross flow filtration process. For the present parametric study, the effect of four parameters M,

g, R and R_c are considered and explained. The physical properties of the colloidal particle, cake layer and membrane are incorporated in the membrane and colloidal system interaction parameter M. The value of M is proportional to the square of particle size, i.e., for small particles, the value of M is small resulting in higher resistance since specific resistance is inversely proportional to M. As a result, the permeate flux decline is higher as shown in Fig. 4a. The observed rejection also increases when M is small, this is due to lower convection of salt compared to back diffusion across the cake layer. Therefore, concentration at the membrane surface is less significant for M=0.05 resulting in lower osmotic pressure as shown in Fig. 4a. Therefore, permeate flux and observed rejection can be changed by changing the size of the cake forming particles.

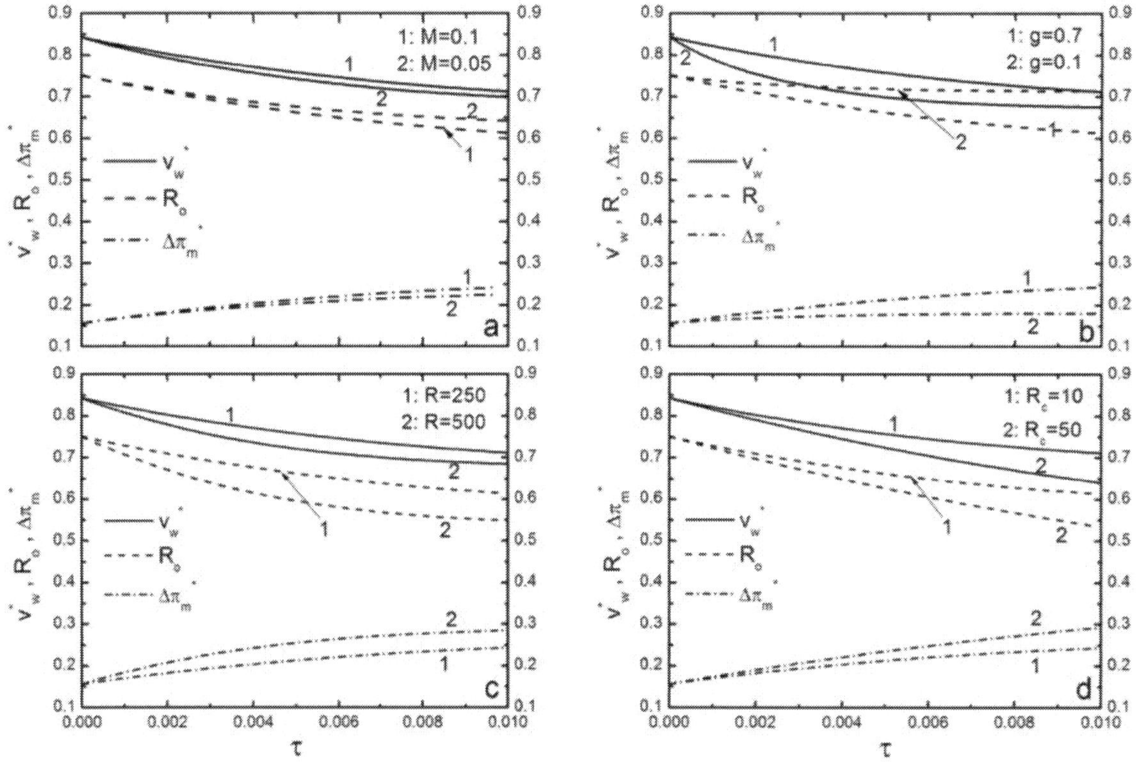

Figure 2: Variation of normalized permeate flux (v_w^*), observed rejection (R_o) and osmotic pressure $\Delta\pi_m^*$. (a) g=0.7, R=250.0, R_i=1.0, R_c=10.0, R_r=0.875, α=0.09 and $C_{c,m}^*$=850.0; (b) M=0.1, R=250.0, R_i=1.0, R_c=10.0, R_r=0.875, α=0.09 and $C_{c,m}^*$=850.0; (c) M=0.1, g=0.7, R_i=1.0, R_c=10.0, R_r=0.875, α=0.09 and $C_{c,m}^*$=850.0; (d) M=0.1, g=0.7, R=250.0, R_i=1.0, R_r=0.875, α=0.09 and $C_{c,m}^*$=850.0

The effect of electroosmotic parameter g, induced due to the presence of charged colloids and electrolyte solution, is shown in Fig. 4b. In absence of surface potential, the value of g goes to unity, implying absence of any induced electroosmotic flow. In presence of charged particles and salt solution the value of g is always less than 1. The parameter g is a strong function of κa and it decreases with increase of κa or with increase of electrolyte feed concentration. The parameter g is also a function of surface potential ψ_c and decreases with increase of ψ_c. Therefore, increase of surface potential of colloidal particles or increase of feed electrolyte concentration causes decrease of g. Since cake resistance is inversely proportional to g, increase of surface potential and feed concentration increases the electroviscous resistance of cake layer. As a result, permeate flux decline is more severe due to higher reverse electroosmotic flow induced from charged colloids. Due to the reverse electroosmotic flow of electrolyte, electrolyte concentration is less at the membrane surface, increasing the observed rejection and decreasing the transmembrane osmotic pressure as shown in Fig. 4b. Therefore, smaller value of g or higher value of κ and ψ_c, decreases the permeate flux and increases the salt rejection siginificantly. Thus, formation of cake layer with nano colloidal particles of high surface potential in concentrated salt solution can enhance the observed rejection of nanofiltration membranes or can enhance concentration polarization on cake surface significantly, which means the cake layer can be modeled as a secondary low salt rejection membrane of nano size pores with high surface potential.

The effect of parameter R on permeate flux and observed rejection is shown in Fig. 4c. The parameter R is defined as the ratio of pure water flux through the membrane to the diffusive flux through the cake layer. Higher value of R indicates that the convective flux is dominant compared to back diffusion through the cake layer which results in enhanced accumulation of electrolyte within the cake layer and increase in transmembrane osmotic pressure. As a result, the permeate flux decline is comparatively higher and observed rejection is lower for large value of R, according to Fig. 4c. Therefore, higher value of R enhances salt concentration at membrane surface and declines the permeate flux and observed rejection.

The parameter R_c is defined as the ratio of pure water flux to colloids mass transfer coefficient. The mass transfer coefficient solely depends on the geometry of cross flow channel and velocity. Since the model is for cross flow filtration lower value of R_c indicates higher mass transfer coefficient or, higher degree of turbulence and shear at the cake surface. Due to higher shear rate at cake surface, the cake layer can not grow much which leads to lower cake layer thickness and resistance, resulting in higher permeate flux as shown in Fig. 4d. Therefore, convective flux of electrolyte is higher for small value of R_c. Though the convective flux is higher, observed rejection is higher and transmembrane osmotic pressure is lower, due to smaller of ratio of convective flux to diffusive flux within the thinner cake layer for small R_c. Thus, higher mass transfer coefficient or small R_c enhances the permeate flux and observed rejection.

5 Concluding Remarks

The electrokinetic model presented in this paper is a simple coupled model for quantifying the combined effect of colloid-electrolyte system during crossflow nanofiltration. Due to reverse electroosmotic flow, the permeate flux and salt transport through the cake layer decrease. The salt concentration at the cake surface increases, which enhances the concentration polarization on cake surface. The cake enhanced concentration polarization will be severe for nano-colloids with higher surface potential. A detailed electrokinetic model of ion transport through the cake layer of nano-colloids and membrane will lead to a better understanding of the combined fouling process for high salt rejection membranes. Consideration of the electrokinetic interaction between the electrolyte and the colloid is important to develop a detailed model of the inter-relations between the permeate flux, salt rejection and cake formation. The study also includes that manipulation of the meso-scale structure of the cake will allow us to alter the porosity of cake layer, extent of adsorption of different species, and alter the salt rejection of the membranes.

Nomenclature

ΔP	Applied operating pressure, Pa
R_m	Membrane resistance, 1/m
R_r	Real rejection of membrane
d_e	Hydraulic diameter of cross flow channel, m
μ	Solvent viscosity, Pa.s
ε	Solvent electrical permitivity
R_{ug}	Universal gas constant, J/mole K
T	Absolute temperature, K
C_f	Solute concentration at feed (i.e., i= ions, c=colloids)
C_m	Solute concentration at membrane surface (i.e., i= ions, c=colloids)
ρ_c	Density of colloids, kg/m^3
L	Cake thickness, m
$D = \frac{k_B T}{6\pi\mu r}$	Brownian diffusion coefficient of solutes (i.e., i= ions, c=colloids), m^2/s
$D_i^* = \frac{(1-\varphi_p)*D_i}{[1-\ln(1-\varphi_p)^2]}$	Hindered diffusivity of ions within the cake layer, m^2/s
$k = 1.86\,\frac{D}{d_e}(\frac{u_c d_e^2}{DL_c})^{\frac{1}{3}}$	Mass transfer coefficient co-relation of solute (i.e., i= ions, c=colloids) for laminar flow in thin channel [8], m/s

References

[1] E.M.V. Hoek and M. Elimelech. Cake-enhanced concentration polarization: A new fouling mechanism for salt-rejecting membranes. *Environmental Science and Technology*, 37:5581–5588, 2003.

[2] A.S. Kim, A.E. Contreras, Q. Li, and R. Yuan. Fundamental mechanisms of three-component combined fouling with experimental verification. *Langmuir*, 25:7815–7827, 2009.

[3] L.N. Sim, Y. Ye, V. Chen, and A.G. Fane. Investigations of the coupled effect of cake-enhanced osmotic pressure and colloidal fouling in ro using crossflow sampler-modified fouling index ultrafiltration. *Desalination*, 273:184–196, 2011.

[4] S. De and S. Bhattacharjee. Flux decline during cross flow membrane filtration of electrolytic solution in presence of charged nano-colloids: A simple electrokinetic model. *Journal of Colloid and Interface Science*, 353:530–536, 2011.

[5] J. Masliyah and S. Bhattacharjee. *Electrokinetic and Colloid Transport Phenomena*. Wiley Interscience, 2006.

[6] S. De and P. K. Bhattacharya. Modeling of ultrafiltration process for a two-component aqueous solution of low and high (gel-forming) molecular weight solutes. *Journal of Membrane Science*, 136:57 – 69, 1997.

[7] S.K. Zaidi and A. Kumar. Experimental analysis of a gel layer in dead-end ultrafiltration of a silica suspension. *Desalination*, 172:107–117, 2005.

[8] S. Kim and E.M.V. Hoek. Modeling concentration polarization in reverse osmosis processes. *Desalination*, 186:111–128, 2005.

(484e) Comparison of Membrane Performance: Analysis with Fouling Models

Mingyan Zhou[1], James E. Kilduff[1] and **Georges Belfort**[2], (1)Department of Civil and Environmental Engineering, Rensselaer Polytechnic Institute, Troy, NY, (2)Department of Chemical and Biological Engineering, Rensselaer Polytechnic Institute, Troy, NY

Membrane research and engineering applications often involve optimizing flux and rejection for a particular application. This process frequently presents the need to compare the performance of membranes having different inherent permeabilities; an important performance criterion is resistance to fouling. To identify the most appropriate (least biased) approaches to compare membrane performance, an analysis using fouling models, with an emphasis on the combined pore blockage and cake filtration model, was performed. We illustrate how the models can be used to provide a general framework for assessing fouling and developing ways to compare membrane performance. Comparisons at constant transmembrane pressure using the same initial flux and the same initial pressure were emphasized. This analysis demonstrates that membrane resistance has a significant effect on flux decline - flux decline is smaller for membranes having greater resistance, especially when the feed solution and membrane-foulant interactions are similar. This effect cannot be normalized regardless of whether flux is plotted as a function of filtration time and volume throughput. A better approach is to compare the effects of fouling by evaluating fouling resistance as a function of volume throughput. Alternatively, comparing membranes based on total resistance provides some advantages because it can be related to permeate production and energy cost. When comparing membranes having different resistance and rejection, the effects of fouling should be compared by evaluating fouling resistance as a function of mass accumulation on the membrane surface. The effects of cake compressibility are also discussed.

See more of this Session: Theories and Applications of Cross Flow Filtration
See more of this Group/Topical: Separations Division

(498a) State-of-the-Art Techniques and Challenges in Characterization of New Hydrogen Storage Materials

Andre Levchenko, Setaram Inc, Pleasanton, CA

The new hydrogen storage materials discovered in the last few years present a number of challenges to the research community from a viewpoint of materials characterization. These include both evaluation of the performance in real-world applications of hydrogen storage materials and understanding the underlying fundamental mechanisms controlling their properties. This presentation gives an overview of state-of-the-art characterization techniques for hydrogen storage materials. It covers simultaneous TPD-gas sorption/mass-spectrometry measurements, volumetric PCT isotherm measurements on thin-film and extremely small metal-hydride samples, as well as direct van't Hoff and in-situ calorimetric measurements of enthalpy of formation of hydrides. Challenges in performing measurements on particular types of materials or under demanding conditions are discussed.

See more of this Session: Aspects of Hydrogen Storage Materials
See more of this Group/Topical: International Congress on Energy 2011

(498b) A New Metal-Decorated Carbon Catalyst for Dehydrogenation of Complex Hydride Hydrogen Storage Materials

Sean S.-Y. Lin[1], Jun Yang[2] and Harold H. Kung[1], (1)Chemical and Biological Engineering, Northwestern University, Evanston, IL, (2)Fuel Cell and Hybrid Electric Vehicle Department, Ford Motor Company, Dearborn, MI

In recent years, many new solid-state hydrides with a high hydrogen capacity that meet DOE requirement (> 7.5 wt.% H_2) have been synthesized as promising on-board hydrogen storage materials.

In these hydrides, however, only a small fraction of hydrogen can be released at a desired on-board operating temperature (<85°C) due to a slow dehydrogenation kinetics of the hydrides. It has been discovered that the kinetics can be improved by adding transition metal catalysts to form metal-hydride intermediates that facilitate hydrogen transportation and desorption. Nevertheless, identification of such catalyst for various hydrides is a challenging task because the catalytic effect is component-specific. Here we report a new form of catalyst for dehydrogenation of complex hydride, using a well-studied $NaAlH_4$ as a test case. Dehydrogenation of $NaAlH_4$ can be greatly facilitated by activated carbon catalysts. The catalytic function can be further enhanced by decorating the carbon with Co, Ni, or Cu nanoparticles. The decomposition temperature was lowered by as much as 100°C using a 3 wt.% Co or Ni-decorated activated carbon, comparable to a Ti-based catalyst, which were the most effective among the metals tested. The catalytic effect is likely due to a combination of hydrogen spillover effect, high contact area between carbon and the hydride, and confinement of the hydride as nano-sized domains in the pores of the carbon matrix. The catalysts were also effective in facilitating rehydrogenation of $NaAlH_4$ under moderate pressure (75.8 bar H_2) and low temperature (120°C), when no rehydrogenation would occur without the catalyst. The fact that this new catalyst system is not specific to any hydride offers many potential applications.

See more of this Session: Aspects of Hydrogen Storage Materials
See more of this Group/Topical: International Congress on Energy 2011

A Future Energy Supply Based on Liquid Organic Hydrogen Carriers (LOHC)

Daniel Teichmann, Wolfgang Arlt, Peter Wasserscheid
University of Erlangen-Nuremberg, Germany

Introduction

Our current energy system is mainly based on the availability of fossil fuels that can be extracted from the ground to a comparably cheap price and that are also excellent energy carriers in terms of storage density and manageability. The on-going transition towards renewable sources is a big challenge for our energy system, mainly due to the highly intermittent character of many renewable energies like wind or solar.

While hydrogen is often considered as a suitable energy vector in a future sustainable energy system [1,2], the weight and volume-efficient storage of sufficient amounts of hydrogen is a difficult task due to its low density. A comprehensive hydrogen infrastructure does not exist yet and makes high investments necessary.

Liquid Organic Hydrogen Carriers (LOHC)

This contribution describes a concept for the establishment of a competitive energy distribution network based on Liquid Organic Hydrogen Carrier (LOHC) compounds [3]. These compounds are characterized by the fact that they can be loaded and unloaded with considerable amounts of hydrogen in a cyclic process. In contrast to other hydrogen storage methods (like compressed gaseous or cryogenic hydrogen) LOHCs can be safely stored under ambient conditions in big quantities. As they have many physico-chemical similarities to Diesel, LOHCs can make use of the existing energy infrastructure (e.g. storage tanks or road tankers).

Many different substances are discussed in the literature as potential LOHC compounds [4,5]. In this contribution we concentrate on heterocyclic aromatic hydrocarbons such as N-ethylcarbazole (Fig. 1) which were first considered by the company AirProducts [6]. The fully hydrogenated substance perhydro-N-ethylcarbazole offers a hydrogen capacity of up to 5.8 wt%. The hydrogenation reaction for this particular LOHC system is typically carried out at 130–160 °C with 70 bars pressure using supported noble metal catalysts like ruthenium on aluminum oxide. The dehydrogenation reaction proceeds usually in the presence of supported Pt or Pd catalysts. The reaction is typically carried out at 200–230 °C and pressures slightly above ambient conditions [7-9].

Energy storage and mobility applications

In times of high loads from intermittent sources like wind or solar energies, excess energy can be used to locally produce hydrogen by water electrolysis, which is efficiently stored in the form of LOHC. The so-produced energy carrier can be stored at the spot of production and re-converted to electricity in times of high demand.

Alternatively it can be easily transported to be used in various off-site applications like mobility (as discussed in the following), heating, local hydrogen production, etc.

Compared to other energy storage technologies like pumped hydroelectric or compressed air energy storage [10], LOHCs show a high energy storage density, thus enabling a safe and loss-free storage of renewable energy with only limited land use. These reasons make LOHCs a promising technology especially for seasonal energy storage where large amounts of energy have to be stored over extended periods of time.

Another promising field of application for LOHCs is mobility [11]. To allow cruising ranges of 500km, about 5 kg of on-board hydrogen are necessary which correspond to about 100 liters of carrier compound. The hydrogen which is produced from the LOHC on-board in a catalytic converter can either be used in a combustion engine or a fuel cell (see Fig. 2). At the gas station the unloaded carrier compound is returned in exchange to fresh material and can be reloaded with hydrogen for further use.

Conclusion

The proposed concept links the technical challenge of storing temporary and local energy over-production from regenerative sources with the vision of a sustainable, hydrogen-based mobility. Due to their beneficial handling and storage properties we consider LOHCs as an attractive way to provide wind and solar energy for mobility applications.

Figures

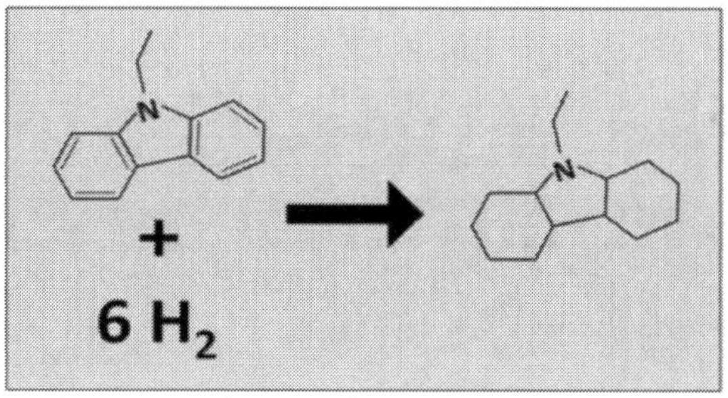

Figure 1: Chemical energy storage via hydrogenation of a LOHC compound [3]

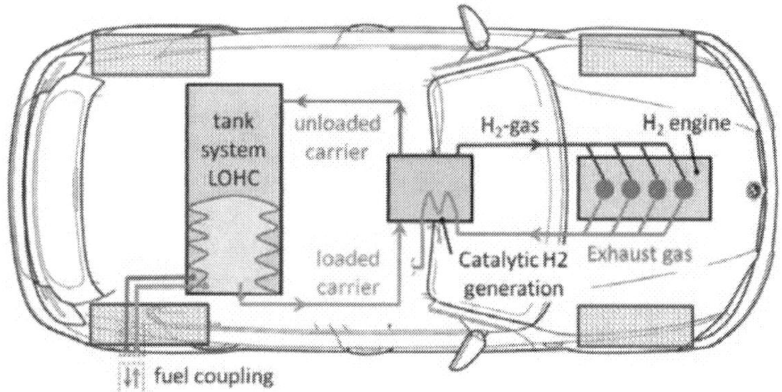

Figure 2: Concept of a hydrogen car with liquid organic hydrides [3]

References

1. Sartbaeva, V. L. Kuznetsov, S. A. Wells and P. P. Edwards, Energy Environ. Sci., 2008, 1, 79–85.
2. U.S. Department of Energy, Hydrogen Posture Plan, http://www.hydrogen.energy.gov.
3. D. Teichmann, W. Arlt, P. Wasserscheid, R. Freymann, Energy Environ. Sci., 2011, 4 (8), 2767 - 2773
4. G. W. H. Scherer, E. Newson, A. Wokaun, Int. J. Hydrogen Energy, 1999, 24, 1157–1169.
5. R. H. Crabtree, Energy Environ. Sci., 2008, 1, 134–138.
6. Pez et al., World Pat., WO 2005/000457A2, 2005.
7. M. Sobota et al., Chem.–Eur. J., DOI: 10.1002/chem.201101311.
8. F. Sotoodeh and K. J. Smith, Ind. Eng. Chem. Res., 2010, 49, 1018–1026.
9. K. M. Eblagon, D. Rentsch, O. Friedrichs, A. Remhof, A. Züttel, A. J. Ramirez-Cuesta and S. C. Tsang, Int. J. Hydrogen Energy, 2010, 35, 11609–11621.
10. Verband der Elektrotechnik, Elektronik, Informationstechnik (VDE), Energiespeicher in Stromversorgungssystemen mit hohem Anteil Erneuerbarer Energieträger, Frankfurt, 2009.
11. J. von Wild, R. Freymann and M. Zenner, Potentiale von alternativen Wasserstoffspeicherungstechnologien, VDI-Berichte: Innovative Fahrzeugantriebe, 2008, vol. 2030, pp. 273–298.

(498d) Accelerated Thermal Decomposition of Ammonia Borane by Carbon Dioxide

Junshe Zhang[1], Yu Zhao[2], Daniel L. Akins[2] and Jae W. Lee[1], (1)Department of Chemical Engineering, The City College of New York, New York, NY, (2)Department of Chemistry, The City College of New York, New York, NY

This talk presents the enhanced dehydrogenation of ammonia borane (AB) at 85 °C under CO2, an abundant and naturally occurring material. The presence of CO_2 significantly accelerates the thermal decomposition and it kinetics becomes faster as the CO_2 pressure increases from 1.70 to 3.67 bar. Around 4 bar of CO_2, 1.33 equivalents of hydrogen is released from AB in 1 hour. The accelerated AB thermolysis is attributed to CO_2 reacting with AB, or intermediates like diammoniate of diborane (DADB), or primary thermolysis products including linear dimers of aminoborane (LDAB), which accelerates AB thermolysis, and this reaction has been confirmed by ATR-FTIR, Raman, and ^{11}B and ^{13}C solid sate MAS-NMR measurements. The results open up a potential avenue to convert CO_2 to other useful chemicals under very mild conditions.

See more of this Session: Aspects of Hydrogen Storage Materials
See more of this Group/Topical: International Congress on Energy 2011

(498e) Hydrogen Production From Deliquescence of Ammonia Borane

Chang-Chen Chou[1], Cheng-Hong Liu[1], **Bing-Hung Chen**[1] and Duu-Jong Lee[2], (1)Department of Chemical Engineering, National Cheng Kung University, Tainan, Taiwan, (2)Department of Chemical Engineering, National Taiwan University of Science and Technology, Taipei, Taiwan

Chemical hydride is one of the most potential hydrogen storage materials which can effectively increase hydrogen storage capacity. To meet the new goal of U.S. department of energy (DoE), 4.5wt% and 5.5 wt% of gravimetric hydrogen density capacity by 2010 and 2015, respectively, ammonia borane (AB) is an attractive material due to high intrinsic hydrogen density (i.e. 19.6 wt%). Such a high intrinsic hydrogen density increases the possibility to apply to be the on-board power source of vehicle system, not only for 3C electric product.

In this study, hydrogen evolution from hydrolysis of ammonia borane in limiting water conditions is studied. The solid composite of ammonia borane hydrogen fuel is synthesized with pulverized catalyst through the high energy ball milling process. Various parameters such as the amount of water dosage, temperature, loading of catalysts, on hydrogen production characteristics will be discussed. Kinetic behavior of the hydrolysis reaction in such limiting water supply conditions will also be investigated. Based on experimental outcomes, optimal conditions on hydrogen production from this hydrogen storage system will be proposed.

See more of this Session: Aspects of Hydrogen Storage Materials
See more of this Group/Topical: International Congress on Energy 2011

Kinetics of Self-Hydrolysis of Concentrated Sodium Borohydride Solutions at High Temperatures

Rajasree Retnamma[1], Lin Yu[2], Carmen M. Rangel[3], Augusto Q. Novais[1], Karl Johnson[4], Michael A. Matthews[2]

[1]*Energy Systems Modeling and Optimization Unit, National Laboratory of Energy and Geology, Lisbon, Portugal*
[2]*Department of Chemical Engineering, University of South Carolina, USA*
[3]*Fuel Cells and Hydrogen Unit, National Laboratory of Energy and Geology, Lisbon, Portugal*
[4]*Department of Chemical and Petroleum Engineering, University of Pittsburgh, USA*

1. Introduction

Hydrolysis of sodium borohydride ($NaBH_4$) solution over metal catalyst is accepted as a potential technology for delivering H_2 to portable fuel cells. However, unresolved issues such as minimizing the amount of water and the nature of hydration of metaborate by-products limit its hydrogen storage capacities. An alternative method, steam/water vapor hydrolysis, can enhance the hydrogen storage capacities without catalyst, if operating conditions can be optimized [1]. In this approach, solid $NaBH_4$ absorbs water and deliquesces, forming a highly concentrated viscous solution at near boiling point of water, followed by hydrogen evolution from the concentrated viscous solution. It is also well understood that self-hydrolysis occurs even at room temperature when $NaBH_4$ is mixed with water and the rate becomes significant at high temperatures. Self-hydrolysis must be arrested to increase the shelf-life of the solution.

The knowledge of the kinetics of self-hydrolysis of concentrated solutions at elevated temperature is thus crucial for: (1) developing steam/water vapor hydrolysis technology; and (2) handling and storage of $NaBH_4$ solutions. Concentrated solutions have not been studied extensively so far and the reported studies, except two [2, 3], were for buffered dilute solutions of $NaBH_4$ (< 0.4 wt%) at low temperatures (< 25 ^{0}C). The present study reports the kinetics of self-hydrolysis of unbuffered, unstabilized concentrated $NaBH_4$ solutions (10-20 wt%) at high temperatures (25-80 ^{0}C) by a differential isothermal isoconversional method.

2. Theoretical

The $NaBH_4$ hydrolysis (self or catalytic) reaction occurs according to [4]:

$$NaBH_4 \text{ (aq or s)} + (2+x)\, H_2O \text{ (l)} \longrightarrow NaBO_2.\, x\, H_2O \text{ (aq or s)} + \text{heat} \qquad (1)$$

where x is the excess hydration factor.

The mechanism of this process is highly complex and too complicated to be described by a kinetic model within the practical operating conditions [5]. The isoconversional method is often used to describe such kinetics, where, assuming a single step kinetic approximation, the rate expression is [6].

$$\frac{d\alpha}{dt} = k'_{ap}(T, C_0)\, f(\alpha) \qquad (2)$$

$k'_{ap}(T, C_0)$ is the apparent rate constant, which depends on temperature T and initial reactant concentration C_0, α is the extent of conversion and $f(\alpha)$ the conversion function representing the mechanism of the process.

The $NaBH_4$ hydrolysis process exhibits maximum rate at the beginning of the reaction, which decreases continuously with the extent of conversion. The most appropriate form of $f(\alpha)$ for such decelerating type process is the 'reaction order model' [7]:

$$f(\alpha) = (1-\alpha)^m \qquad (3)$$

where m is the reaction order.

Combining Eqs. (2)- (3), and considering the definition of extent of reaction α gives:

$$\frac{d\alpha}{dt} = k_{ap}(T)\, C_0^{m-1}\, (1-\alpha)^m \qquad (4)$$

where,

$$k'_{ap}(T, C_0) = k_{ap}(T) * C_0^{m-1} \qquad (5)$$

For a fixed conversion (α), Eq. (4) reduces to

$$\left(\frac{d\alpha}{dt}\right)_\alpha = k''_{ap}(T,\alpha)\, C_0^{m-1} \qquad (6)$$

where,

$$k''_{ap}(T,\alpha) = k_{ap}(T) * (1-\alpha)^m \qquad (7)$$

Equation (6), after a logarithmic transformation:

$$\ln\left(\frac{d\alpha}{dt}\right)_\alpha = \ln k''_{ap}(T,\alpha) + (m-1)\ln C_0 \qquad (8)$$

For an isothermal process, the values of $\ln k''_{ap}(T,\alpha)$ and m are determined by Eq. (8) for each temperature and α, for different initial borohydride concentrations.

The temperature dependence of the rate constant $k_{ap}(T)$ is then expressed by Arrhenius relationship:

$$k_{ap}(T) = A_{ap} \exp\left(\frac{-E_{ap}}{RT}\right) \quad (9)$$

The apparent activation parameters, i.e. apparent activation energy (E_{ap}) and pre-exponential factor (A_{ap}) are estimated by linear regression of ln $k_{ap}(T)$ vs. 1/T data.

3. Results and discussion

3.1. Kinetic study

Effect of $NaBH_4$ conversion and temperature on reaction rate

The experimental kinetic data used were reported by Yu and Matthews [8], in which the evolution of metaborate formation and $NaBH_4$ consumption were generated *in-situ* by ^{11}B NMR technique for unbuffered, unstabilized 10, 15 and 20 wt% $NaBH_4$ solutions at 25, 60 and 80 ^{0}C. The estimated reaction rates from the experimental data for various $NaBH_4$ conversion, initial $NaBH_4$ concentration and temperature are given in Table 1. As evident, the reaction rate increases with both temperature and initial $NaBH_4$ concentration for a fixed $NaBH_4$ conversion and decreases with increase in $NaBH_4$ conversion for a fixed temperature and initial $NaBH_4$ concentration. Table 1 is then exploited to determine the reaction order and apparent activation parameters with respect to $NaBH_4$ conversion and temperature.

Table 1. Reaction rates for different $NaBH_4$ conversions at various temperatures and initial $NaBH_4$ concentrations.

$NaBH_4$ (wt %)	T(^{0}C)	Reaction rates (r, mol L^{-1} hr^{-1}) at various $NaBH_4$ conversions									
		5 %	10 %	15 %	20 %	25%	30 %	35 %	40 %	45%	50%
10		0.086	0.048	0.025	0.015						
15	25	0.136	0.080	0.037	0.027						
20		0.241	0.108	0.057	0.036						
10		0.637	0.559	0.485	0.415	0.345	0.273	0.210	0.158	0.110	0.100
15	60	0.740	0.654	0.569	0.483	0.398	0.317	0.242	0.181	0.144	0.130
20		0.897	0.797	0.697	0.598	0.499	0.402	0.310	0.227	0.160	0.143
10		2.209	2.089	1.970	1.848	1.723	1.597	1.468	1.340	1.201	1.062
15	80	2.384	2.241	2.130	1.981	1.862	1.720	1.687	1.470	1.352	1.232
20		2.698	2.555	2.410	2.264	2.115	1.964	1.810	1.653	1.493	1.326

Effect of $NaBH_4$ conversion and temperature on reaction order

Fig. 1 shows the reaction order as estimated by linear regression of rate versus $NaBH_4$ concentration for various $NaBH_4$ conversions and. As seen, the reaction order remains constant at a given temperature irrespective of $NaBH_4$ conversion, while it decreases with increase in temperature irrespective of $NaBH_4$ conversion. The order decreases from first- to 0.25 with increase in temperature from 25 to 80 ^{0}C. The estimated rate constant, $k_{ap}(T)$ (not given), decreases with increase in $NaBH_4$ conversion at constant temperature and increases with temperature at constant $NaBH_4$ conversion. It is noteworthy that the variation of reaction order with temperature has been reported by Vyazovkin and Sbirrazzuolui [9] for complex epoxy-anhydride cure kinetics.

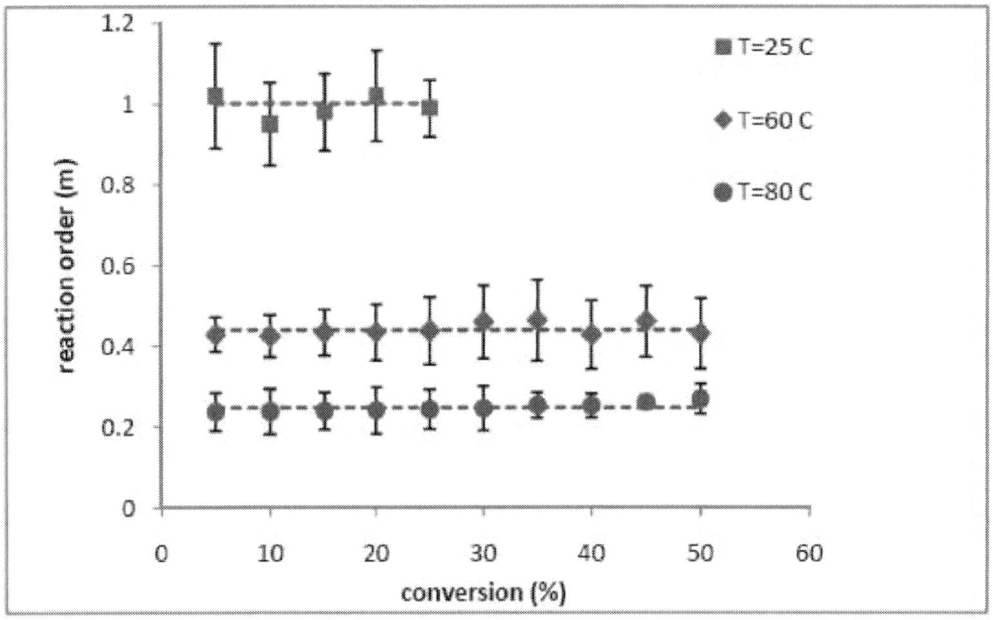

Figure 1. Effect of $NaBH_4$ conversion (%) on reaction order

It is well known that the solution viscosity increases during the course of hydrolysis reaction and this has been considered the reason for the observed decrease in rate with reaction time or with $NaBH_4$ conversion [5]. The effect of temperature on solution viscosity can be described in terms of Arrhenius-type relationships [10]:

$$\eta = \eta_0 \exp\left(\frac{E_{af}}{RT}\right) \qquad (10)$$

where η is the viscosity, η_0 the pre-exponential factor and E_{af} the activation energy of flow. This implies that increase in temperature decreases the solution viscosity. Viscosity is an indication of the solution resistance to flow, and reduced flow resistance at high temperature enhances the interaction between reactant molecules. This in turn lessens the relative contribution of the effect of the concentration on the reaction rate. The deviation from first-order kinetics at low temperatures to fractional-order kinetics at higher temperatures observed in the present study, is attributed to this difference in interaction between reactant molecules with respect to temperature due to change in solution viscosity. It is worthwhile noting that the earlier studies also reported first-order kinetics for self-hydrolysis of $NaBH_4$ for lower conversions and temperatures and for buffered solutions [11-15]

Effect of $NaBH_4$ conversion on activation energy

The classical kinetic theory defines reaction orders to be constant with temperature. In the present case the observed difference in the reaction order with temperature (Fig. 1) suggests a $NaBH_4$ conversion effect on the activation energy of the hydrolysis reaction (E_a), as given in Table 2. As seen, activation energy increases with $NaBH_4$ conversion/with formation of metaborate by-products.

It is well documented in literature that the formation of highly basic metaborate by-products in the course of the hydrolysis reaction results in highly viscous solutions. Investigation on the influence of soluble solids content on viscosity, by various researchers, showed that depending on the type of soluble solids content the activation energy of flow (E_{af}) might increase or decrease. The effects of soluble solid contents are described by the following expressions [10]:

$$\eta = \eta_1 C^{b_1} \qquad (11)$$

$$\eta = \eta_2 \exp(b_2 C) \qquad (12)$$

where η_i and b_i are constants and C is the concentration. Hence one may assume that the activation energy of the flow increases with an increase in $NaBH_4$ conversion due to metaborate by-product formation, which in turn increases the activation energy of the hydrolysis reaction.

Table 2. Apparent activation energy determined for various $NaBH_4$ conversions

$NaBH_4$ Conversion (%)	$^aE_{ap}$ (kJ mol^{-1})	R^2	Standard error	F	P-value
5	64. 43 ± 14.5	0.9944	1.14	3914.89	0.025
10	71.03 ±11.5	0.9998	0.91	6721.39	0.038
15	82.50 ± 13.5	0.9998	1.03	6725.19	0.008
20	89.62 ± 12.5	0.9994	1.96	2152.88	0.013

[a]with ± 95% confidence interval

3.2. Modeling

The dependence of apparent activation energy (E_{ap}) on α (Table 2) is described by [16]:

$$E_{ap} = \varepsilon_0 + \varepsilon_1 \ln(1-\alpha) \qquad (13)$$

with ε_0 = 54.2 ± 1.7 kJ mol^{-1} and ε_1 = -114±12.2 kJ mol^{-1} (within a 95% confidence interval)

The change in apparent reaction order with temperature (Fig. 1) can be evaluated if the parameters of the kinetic compensation effect (KCE) are known [16]. The present study shows that $NaBH_4$ hydrolysis follows the following KCE relationship:

$$\ln A_{ap} = \alpha^* + \beta^* E_{ap} \qquad (14)$$

with α^* =0.84 ± 1.2 (mol $L^{-1})^{1-(m+n)}$ hr^{-1} and β^*= 0.36 ± 0.04 kJ mol^{-1}(within a 95% confidence interval)

Combining Equations (4), (9), (13) and (14) results in:

$$\frac{d\alpha}{dt} = A_0(1-\alpha)^{(m+n)} \qquad (15)$$

where:

$$n = \beta^* \varepsilon_1 - \frac{\varepsilon_1}{RT} \qquad (16)$$

$$A_0 = \exp\left(\alpha^* + \beta^* \varepsilon_0 - \frac{\varepsilon_0}{RT}\right) C_0^{m+n-1} \qquad (17)$$

Equation (15) shows that the overall process is apparently described with a reaction order n+m as illustrated in Fig 1. If the values of n are estimated from equation (16) and n+m from Fig 1, the values for m obtained are given in Table 3. As expected, m tends to a constant value, i.e. 0.22 ± 0.01.

Table 3. Values of n, n+m and m at different temperatures

T(^{0}C)	$n^{(1)}$ (Eq. 16)	$n+m^{(2)}$ (Fig. 1)	$m^{(3,*)}$
25	0.820	1.04	0.214
60	0.222	0.44	0.218
80	0.024	0.245	0.220

[1]dependent on temperature, [2]apparent reaction order
[3] independent on temperature, [*] m = 0.22± 0.01 (mean ± standard deviation)

4. Conclusions

The complex $NaBH_4$ self-hydrolysis process is analyzed by a differential isothermal isoconversional method, based on a single-step kinetic approximation. The overall process is apparently described by a 'reaction order' model with temperature dependent reaction order in a wide practical operating window (10-20 wt% $NaBH_4$ solution and 25-80 ^{0}C).

The apparent activation parameters, i.e. apparent activation energy and pre-exponential factor are conversion dependent and interrelated through the kinetic compensation effect, which explains the observed effect of the temperature on the reaction order.

The apparent activation parameters obtained provide some insight into the theoretical kinetic mechanism, which however remains a matter of heated debates.

5. References

[1] Beaird AM, Davis TA, Matthews MA. Deliquescence in the hydrolysis of sodium borohydride by water vapor, Ind Eng Chem Res 2010;49(20):9596-9599.

[2] Gonçalves, A., Castro, P., Novais, A.Q., Fernandes, V., Rangel, C. and Matos, H., Dynamic modeling of hydrogen generation via hydrolysis of sodium borohydride, In Proceedings of PRES'07, 24-28 June, Ischia, Italy, (In Chemical Engineering Transactions, Vol. 12, 2007. Editor: Jiri Klemes, AIDIC, Italy, 243.)

[3] Andrieux J, Demirci UB, Hannauer J, Gervais C, Goutaudier C, Miele P. Spontaneous hydrolysis of sodium borohydride in harsh conditions. Int J Hydrogen Energy 2011;36:224-233.

[4] Cavaliere S, Hannauer J, Demirci UB, Akdim O, Miele P. Ex situ characterization of N_2H_4-, $NaBH_4$- and NH_3BH_3- reduced cobalt catalysts in $NaBH_4$ hydrolysis. Catal Today 2011;170:3-12.

[5] Retnamma R, Novais AQ, Rangel C.M. Kinetics of hydrolysis of sodium borohydride for hydrogen production in fuel cell applications: A review. Int J Hydrogen Energy 2011; 36: 9772-9790 (and references thereof).

[6] Simon P. Isoconversional methods: Fundamentals, meaning and application. J. Therm. Anal. Calorimetry 2004;76:123-132.

[7] Vyazovkin S, Burnham AK, Criado JM, Perez-Maqueda AA, Popescu C, Sbirrazzuoli N. ICTAC Kinetics committee recommendations for performing kinetic computations on thermal analysis data. Therm Chim Acta 2011;520:1-9.

[8] Yu Lin, Matthews MA. Hydrolysis of sodium borohydride in concentrated aqueous solution. Int J Hydrogen Energy 2011;36:7416-7422.

[9] Vyazovkin S, Sbirrazzuoli N. Kinetic methods to study isothermal and nonisothermal epoxy-anhydride cure. Macromol Chem Phys 1991;200:2294-2303.

[10] Ibarz A, Garvin A, Costa J. Rheological behavior of sole (prunus spinosa) fruit juices. J Food Engng 1996;27:423-430.

[11] Pecsok R. Polarographic studies on the oxidation and hydrolysis of sodium borohydride. J Ame Chem Soc 1953;75:2862-2864.

[12] Stockmayer WH, Miller RR, Zeto RJ. Kinetics of borohydride hydrolysis. J. Phys. Chem 1961;65:1076-1077.

[13] Davis RE, Swain CG. General acid catalysis of the hydrolysis of sodium borohydride. J Ame Chem Soc 1960;82(22):5949-5950.

[14] Davis RE. Bromels E, Kibby CL. Borohydrides III. Hydrolysis of sodium borohydride in aqueous solution. J Ame Chem Soc 1962;84:885-889.

[15] Mesmer RE, Jolly WL. The hydrolysis of aqueous hydroborate. Inorg Chem 1962;1:608-612.
[16] Budrugeac P, Popescu C, Segal E. Change of the apparent reaction order with temperature as a consequence of the reaction mechanism complexity. J. Therm. Anal. Calorimetry 2001;64:821-827.

Nano- and Micro-scale contact angle measurement under pressure by vertical shift interferometry

Hiroaki Kadono[1], Akira Ueda[2,1], Hisao Satoh[3], Yunfeng Liang[1],
Satoru Takahashi[4], Hiroshi Okabe[4], Yasuhiro Fukunaka[1] and Toshifumi Matsuoka[1]
[1]Kyoto University, Kyoto, Japan
[2]Toyama University, Toyama, Japan
[3]Mitsubishi Materials Corporation, Ibaraki, Japan
[4]Japan Oil, Gas and Metals National Corporation (JOGMEC), Chiba, Japan

Introduction

Economic growth of developing country like China and India dramatically enhances the worldwide demand of oil and the price of oil is rising along with it. However it is difficult to discover the new and big oil fields and we have to rely on the present oil fields. On the other hand various Enhanced Oil Recovery (EOR) methods like water injection, chemical method or carbon dioxide injection are proposed on the field, but the recovery rate is no more than 60%. Enhancing the recovery rate of these fields is comparable with discovering the new oil field.

It is of essential importance to better understand the wettability phenomenon between the crude oil, underground water and the reservoir pore surface. This is a key issue of EOR technology. Dependent on the geology condition, the reservoir pore diameter may be distributed from a few hundred micrometers to a few micrometers. The pore pressure and temperature under subsurface may also vary site by site in broad ranges. To improve our understanding on the wettability of crude oil and fluid on mineral surfaces, *in situ* observation is required down to micro- or even nano-meter scale under such severe environmental conditions. The interfacial phenomena are still quite limited in the petroleum engineering field, while many experimental and theoretical researches have been accumulated to discuss the wettability inside three dimensional porous structure encounterd in microelectronics and energy storage fields.

Wettability is frequently considered from various aspects like contact angle measurement, interfacial tension measurement and so on. Generally speaking, contact angle is frequently measured by sessile drop method [1]. In this method, contact angle is measured by photographing the sessile droplet illuminated from the side direction. Therefore, many measurements deal mainly with the macro-meter scale droplet. On the other hand, the techniques of measuring the contact angle of nano- to micro-scale droplet were made practical by applying AFM [2] as well as interferometry [3].

An experimental method for evaluating the contact angle of nano- to micro- scale droplet was proposed with a vertical shift interferometry (VSI) in order to acquisite a profound understanding of the interfacial phenomena encountered in petroleum engineering. A simple optical cell is also designed install in the interferometry arrangement. VSI has another merit which can observe not only the size but also simultaneously the height of many droplets within the image of 235 × 176 μm and thus make it possible to detect the contact angles of droplets only in one time measurement.

Experiments

Materials

The liquids used in this experiment were n-heptane (99.5+ %) and toluene (99.5+ %) purchased from Wako Pure Chemical Industries, Ltd. They were, generally speaking, fairly transparent and used without further purification. Crude oil was supplied from Japan Petroleum Exploration Co., Ltd. (JAPEX) for the present measurement. It was sampled from Sarukawa oil field, which is located in Akita prefecture and owned by JAPEX. The transparency or visibility of crude oil was rather poor.

It is well known that the measured data of contact angle are significantly influenced depending on the surface roughness of substrate [4]. Quartz was used as a model mineral in this experiment. It is simply because optically flat quartz surface can be easily obtained compared with other minerals. The quartz substrate was purchased from Sigma Koki Co., Ltd. It has a relatively flat surface whose surface accuracy level is of the order of visible light wave length λ. Surface accuracy is conventionally defined as the distance of the peak and valley of the substrate. Its unit λ is taken as the wavelength of He-Ne laser, 632.8nm.

Experimental apparatus

VSI is installed in the clean room maintained at a constant temperature of 22℃. Figure 1 schematically shows the optical arrangement of VSI. The shape of substrate surface is detected by the interference fringe shift between the reference light and the reflected light. Two kinds of fluid cells demonstrated in Figs. 2 & 3 were selected for contact angle measurement depending on the apparent visibility of organic phase of crude oil and heptane/toluene, Figure 2 simply shows two crude oil droplets in water attached to optically flat quartz surface facing downward under atmospheric pressure and temperature. This is the system to measure the contact angle of crude oil droplet in water. Another cell shown in Figure 3 can be used under high pressure to 20MPa and high temperature to 200℃. In the present experiment, the contact angle of water droplet in heptane or toluene was measured under pressure up to 10MPa. The light wave reflected on the surface of water droplet attached on the upward facing optically flat quartz surface makes the interference of reference beam.

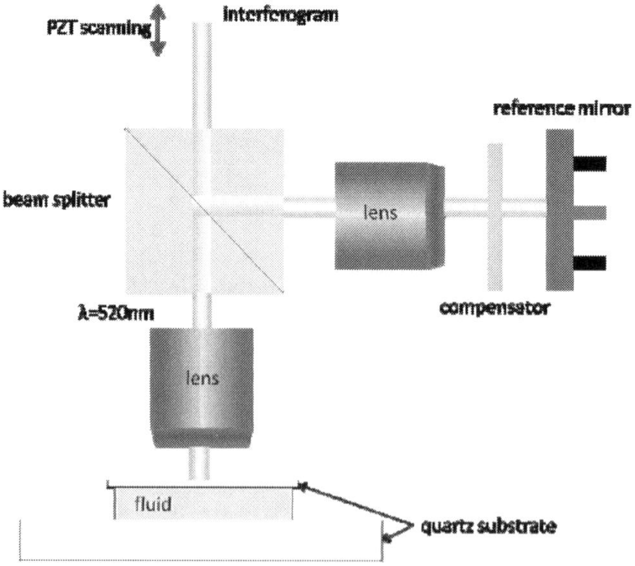

Figure 1. Optical system of VSI

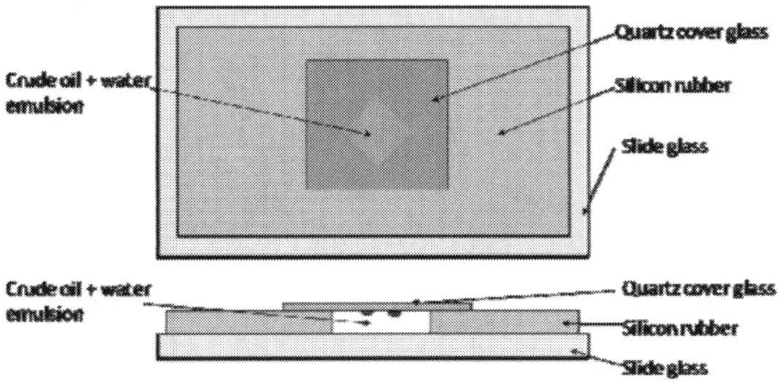

Figure 2. Fluid cell under atmosphere pressure

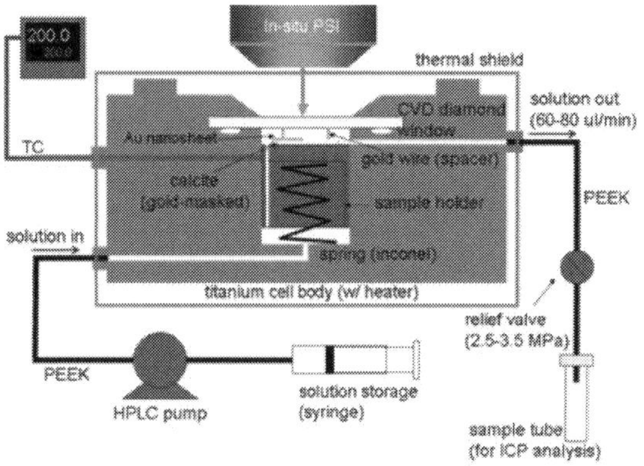

Figure 3. Fluid cell under high pressure

Analysis

The contact angle was calculated by theta/2 method defined by the following equation (1). Figure 4 shows the basic concept of the method.

$$\theta = 2\tan^{-1}\left(\frac{2h}{R}\right) \qquad (1)$$

Figure 5 shows the interference fringe of the water droplet being convex upward. A light-dark cycle means the difference in height which corresponds to integral multiples of half wavelength. By counting the number of interference fringe, the height of droplet is quantified. In addition, the phase shift method principally provides the measurement accuracy to achieve to nanometer scale. A dedicated soft VS-Viewer for measuring the width and height of droplet was employed. The crude oil emulsion droplets are not so transparent or visible that only the reflected light from the top of the droplet was detected as Figure 6 demonstrates. The left side of Figure 6 shows an image of layer profile and the right side does a strength of interference of crude oil droplet. On the other hand, a line to cross a droplet is clear to measure the width and height along the cross sectional line with VS-Viewer, when a water droplet in heptane or toluene is observed.

VSI measure the optical path length distance. Thus, the optical path distance have to be corrected with the refractive index in order to get the geometrical distance. Refractive index value was cited from the data of Handbook of Physics and Chemistry (CRC) [5] and previous literature [6]. The real distance d shown in Figure 7 is derived from the following equations (2) and (3).

$$n_m d + d_0 = n_h d \qquad (2)$$

$$d = d_0 / (n_h - n_m) \qquad (3)$$

The characters $n_m d$ and $n_h d$ correspond to the optical height of water and hydrocarbon respectively, d_0 is the measured height by VSI, and d is the real height of droplet.

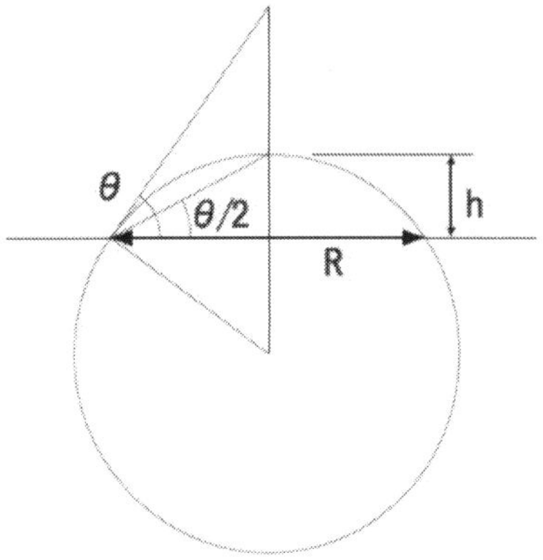

Figure 4. Basic concept of the theta/2 method

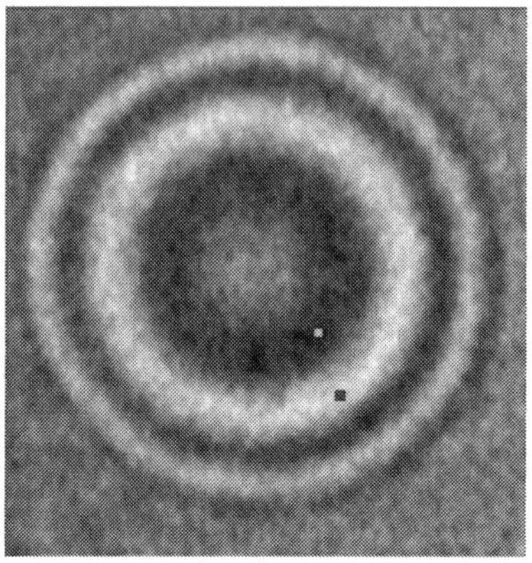

Figure 5. Droplet image which has an interference fringe

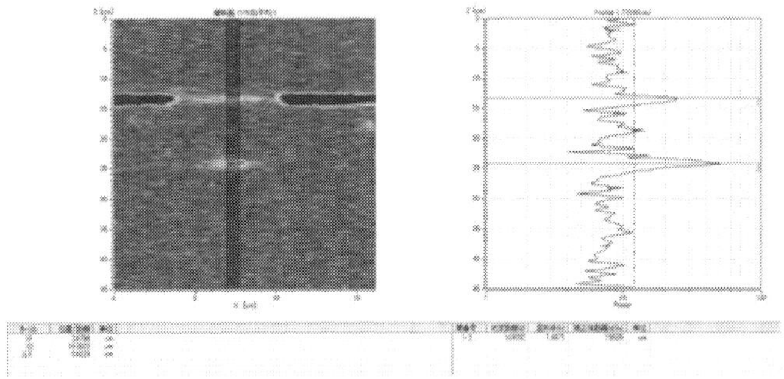

Figure 6. Image of analysis screen

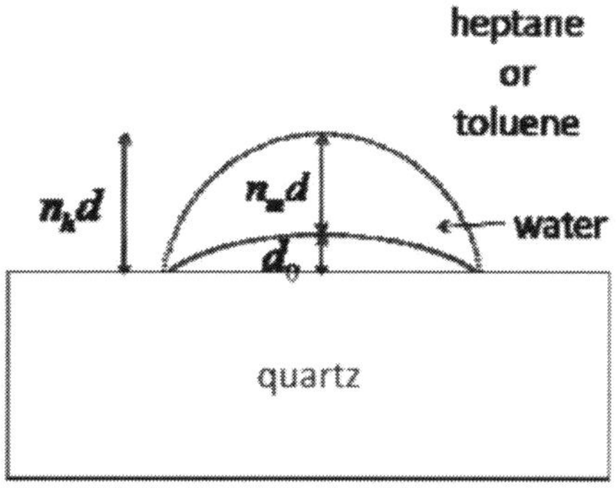

Figure 7. The concept of height correction

Results and Discussion

Contact angle of crude oil droplet in water

This section describes the measured wettability of crude oil droplet in water by means of contact angle. From the image of oil droplets attached on the surface of downward facing SiO2 substrate immersed in water, a size distribution of contact width of crude oil droplet was at first obtained by applying the soft Image Pro Plus. Based on this size distribution, 50 droplets with various size distribution were selected. Next, the height of particular droplet was individually determined from the interference fringe image. The contact angle was correspondingly determined by theta/2 method. It was assumed that there are no effects on where the droplet is located on downward facing quartz substrate. Scatter chart between contact angle and droplet width shows 2D normal distribution characteristics. That is, no apparent dependence of contact angle on droplet width was not detectable within the present experimental conditions. In this case of crude oil droplet in water, a supplementary angle was measured for convenient comparison with the other system of water droplet in organic phase of heptane or toluene. The experimental data are plotted in the plane ($\cos(\pi - \theta)$, $1/r$), as shown in Figure 8.

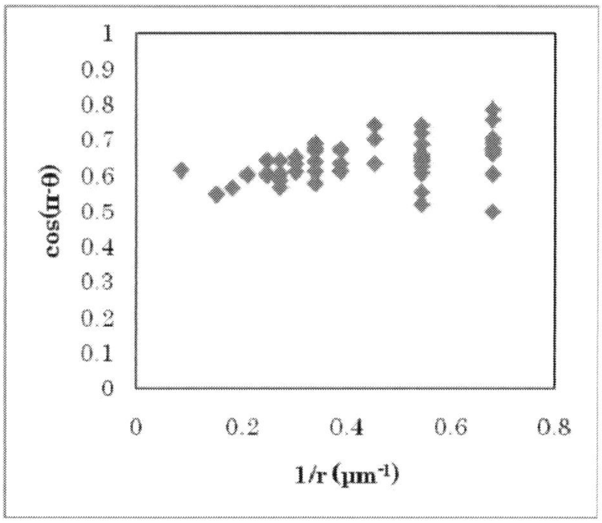

Figure 8. Cosine of the contact angle of water vs. inverse of the contact width of crude oil

Contact angle of water droplet in heptane or toluene

Different from the case of crude oil, the wettability behavior of water droplet on quartz surface immersed in hydrocarbon organic phase was reversed. In this case, the contact angle of water droplet in heptane or toluene was measured. The contact angles of water in heptane and toluene under the applied pressures of 0, 2, 4, 6, 8 and 10MPa were plotted in Figs. 9 (heptane) and 10 (toluene). From these figures, the difference of the effect of pressure could not be observed clearly. Considering only in the case of atmospheric pressure, the measured contact angle behaves similarly to the previous study [7]. They are converted into the relationship between cosine θ and 1/r as shown in Figs. 11 and 12 in order to compare the wettability behavior of crude oil.

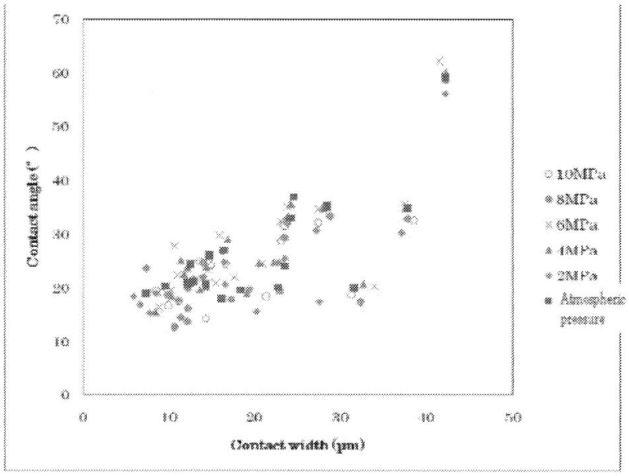

Figure 9. Contact angle of water in heptane under each pressure

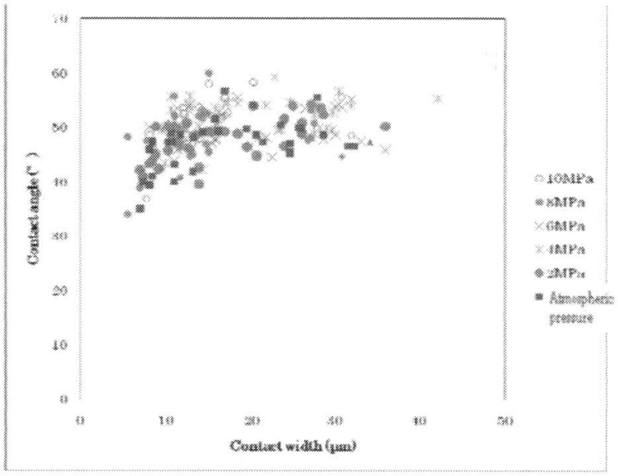

Figure 10. Contact angle of water in toluene under each pressure

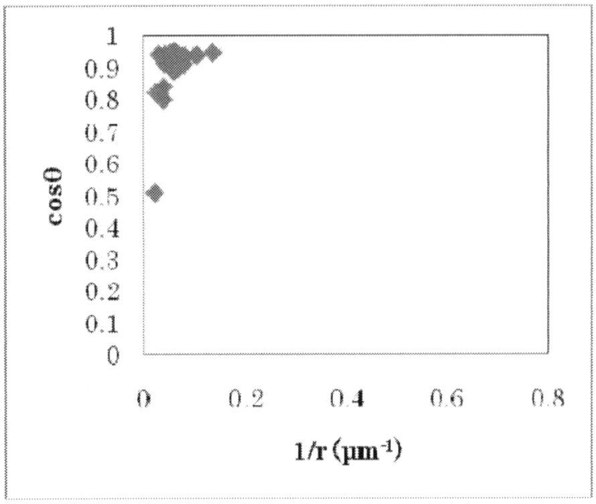

Figure 11. Cosine of the contact angle of water vs inverse of contact width of heptane

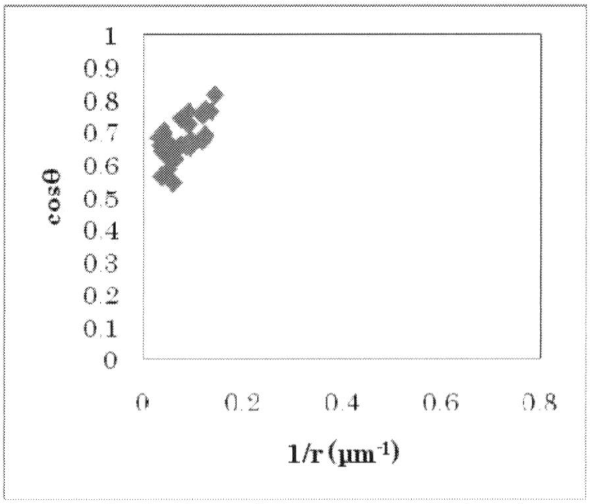

Figure 12. Cosine of the contact angle of water vs inverse of contact width of toluene

Comparison between the case of crude oil, heptane and toluene

From Figs. 8, 11 and 12, the cases of crude oil and toluene illustrate similar dependence of cosine θ around 0.6, while the cosine value of heptane reaches to above 0.9 and contact angle is at most 20 to 30° . It should be emphasized that the wettability of toluene on SiO2 surface behaves similar to that of crude oil, while that of heptane does significantly different.

Moreover it has been reported by our research group that aromatics like toluene tends to be accumulated around the oil-water interface from molecular dynamics calculation [8]. These facts suggest that interfacial phenomenon of crude oil is similar to that of toluene because contact angle consists of the balance of interfacial tension.

Conclusion

The technique to measure the contact angle of liquids on SiO2 surface immersed in fluids was reported under various pressures by using VSI in this study. The pressure was increased up to 10MPa, but the pressure dependence of contact angle was not detected under the present pressure variation range. On the other hand, it becomes evident that the contact angle behavior of crude oil droplet is similar to that of toluene. It may be caused by the accumulation of toluene on the oil-water interface as clearly understood in Figure 13. The interface of water and oil is the aromatics-rich region [8]. This schematic diagram is illustrated based on the calculated density profile across the water and oil system derived from MD computation in our research group in Kyoto University. These results may provide a similar theoretical background on three-phases interfacial phenomena between toluene and crude oil. In addition, contact angle of water droplet in heptane is considerably lower than the case of toluene. That is, the existence of heptanes on the oil-water interface makes more water-wet or hydrophilic characteristics on SiO2 surface. It conventionally provides a positive condition from the technological aspect of oil enhanced recovery process. On the other hand, the accumulation of toluene on the interface actually prevents from the mineral surface being water-wet. Thus, the proposal of getting rid of accumulation of toluene is still a remaining issue for enhanced oil recovery. More appropriate surfactant or aromatic component must be surely designed through MD computation in order to concentrate these molecules to the heterogeneous interface.

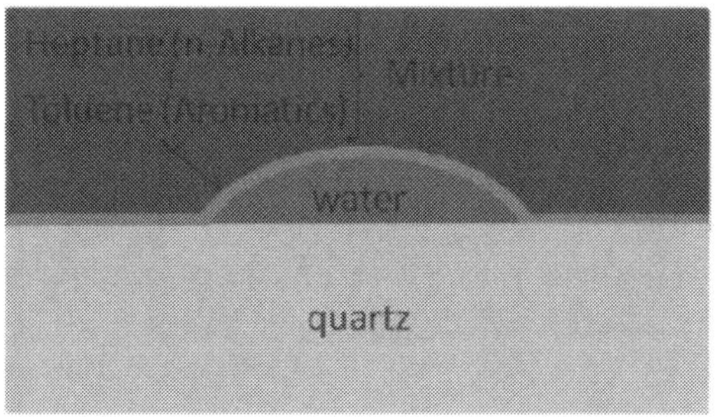

Figure 13. A model of the toluene accumulation

References

1. C. Bourges-Monnier and M. E. R. Shanahan (1995), "Influence of Evaporation on Contact Angle," Langmuir 1995, 11, 2820-2829
2. Antonio Checco (2006), "Nanoscale Wettability of Self-Assembled Monolayers Investigated by Noncontact Atomic Force Microscopy," Langmuir 2006, 22, 116-126
3. J. Y. Wang, S. Betelu and B. M. Law (2001), "Line tension approaching a first-order wetting transition: Experimental results from contact angle measurements," PHYSICAL REVIEW E, VOLUME 63, 031601
4. Didem Öner and Thomas J. McCarthy (2000), "Ultrahydrophobic Surfaces. Effects of Topography Length Scales on Wettability," Langmuir 2000, 16, 7777-7782
5. Handbook of Physics and Chemistry(CRC)
6. Jelenka Savkovic-Stevanovic and Dusan Simonovic (1976), "The Refractive Indicies and Densities of Ternary Mixtures of Benzene-Toluene-Xylene," Journal of Chemical and Engineering Data, Vol. 21, No. 4, 456-459
7. Bernard P Binks and John H. Clint (2002), "Solid Wettability from Surface Energy Components : Relevance to Pickering Emulsions," Langmuir 2002, 18, 1270-1273
8. Makoto Kunieda, et al. (2010), "Self-Accumulation of Aromatics at the Oil-Water Interface through Weak Hydrogen Bonding," J. AM. CHEM. SOC. Vol. 132, No.51 18281-18286

(516b) Comparison Between Equilibrium and Kinetic Models for Methane Hydrate Dissociation

Isaac K. Gamwo and Yong Liu, National Energy Technology Laboratory, U.S. Department of Energy, Pittsburgh, PA

Current literature generally agrees that the equilibrium and kinetic models for methane hydrate dissociation are almost indistinguishable. Here, we used the equilibrium and kinetic models with two kinds of thermal boundary conditions—adiabatic and constant temperature at the walls—to study the dissociation of methane hydrates. We have found significant deviation between the two models for hydrate saturation, temperature, and cumulative gas produced for constant temperature boundary conditions. Parametric study of the kinetic constants shows that the kinetic model is comparable to the thermodynamic model when the kinetic constant approaches idealistically high values. However, the difference between the two models is decreased for adiabatic boundary conditions. We have further shown deviations in the dissociation pattern between the equilibrium and kinetic models. For constant temperature boundary condition, the hydrate dissociates by shrinking in all directions for the equilibrium model, while for the kinetic model, it dissociates with no specific pattern throughout the entire reservoir.

See more of this Session: Complex Subsurface Processes II
See more of this Group/Topical: Energy and Transport Processes

(516c) Green Gas Hydrate Inhibitors

Nicolas von Solms and Lars Jensen, Chemical Engineering, Technical University of Denmark, Lyngby, Denmark

A combination of low temperature, high pressure and presence of water and natural gases provides the ideal conditions for formation of gas hydrates. Gas hydrates may precipitate in pipes and production gear during recovery and transportation of oil and gas resulting in production stops implying major economical losses. Presently large quantities of thermodynamic inhibitors, mainly methanol or glycol, are utilized to inhibit gas hydrate formation. Large amounts of thermodynamic inhibitor are required to prevent hydrate formation (up to 50% of the aqueous phase)

Alternatively it is possible to control formation of gas hydrates using small amounts of polymeric compounds; so-called kinetic inhibitors (KHIs), which instead of inhibiting the formation thermodynamically, work by slowing the formation rate of hydrates. Although using kinetic inhibition offers several advantages over thermodynamic inhibition, such as low chemical usage and savings in operational and capital expenditures, their implementation in field operations has been limited due especially to environmental restrictions as a result of their poor biodegradability.

We have tested the efficacy of antifreeze proteins, from fish and insects, as kinetic gas hydrate inhibitors. Especially antifreeze proteins from the Danish bark beetle have been shown to be particularly effective. These proteins have the greatest thermal hysteresis (difference between melting and freezing point of ice) of any known antifreeze protein, and initial results indicate they are astonishingly good hydrate inhibitors.

In this work we present results of kinetic inhibition tests of these proteins compared with other antifreeze proteins and commercially available kinetic inhibitors. Strategies for improvement of the inhibitors are presented in the framework of a larger project.

See more of this Session: Complex Subsurface Processes II
See more of this Group/Topical: Energy and Transport Processes

(516d) Nanoparticles Transport Through the Porous Media for Enhanced Oil Recovery

Devesh Kumar Agrawal and Ramanan Krishnamoorti, Chemical & Biomolecular Engineering, University of Houston, Houston, TX

Breakthrough in chemical injection based enhanced oil recovery technologies are limited by the need of better and superior materials which can withstand and function at harsh reservoir conditions. The application of nanoparticles is envisaged for EOR applications because of their unique transport & surface wettability properties and their relatively smaller sizes compared to the size of pores in porous rocks of an oil reservoir. Nanoparticles based viscosity and wettability modifiers are expected to produce nano-fluids with superior desired properties with the addition of small nanoparticles concentration. If the suitable functional groups are coated on the surface, these nanoparticles can perform the desired functions like viscosity improvement of the water flood or wettability change of the rock or both. Thus, a study of the transport behavior of such nanoparticles through porous media and establishing the fact that their interactions in the porous media do not cause plugging is essential. Though, factors like the size of nanoparticle as compared to those of the pore's and the pore throat's impact their transport without plugging but factors like nanoparticle-nanoparticle and rock-nanoparticle interactions can play dominant roles in governing their movement through porous media. In the current work, stable dispersion of the surface modified nanoparticles (<100 nm in size) are step input through the sandstones and carbonates cores to displace brine. The pressure profiles and breakthrough curves are obtained. The effects of nanoparticle-nanoparticle and rock-nanoparticle interactions on the transport of these nanoparticles are studied. The impact on enhanced oil recovery is also calculated. The modeling of the system is performed and the transport parameters determined suggest significant dependence of such parameters on the underlying molecular interactions.

See more of this Session: Complex Subsurface Processes II
See more of this Group/Topical: Energy and Transport Processes

(516e) Spent Shale Formations: Potential Source for CO2 Sequestration

Prashanth Mandalaparty, **Robert Krumm** and Milind D. Deo, Chemical Engineering, University of Utah, Salt Lake City, UT

Spent Shale Formations: Potential source for CO_2Sequestration

Prashanth Mandalaparty, RobertKrumm and Milind Deo

While producing oil fromshale in-situ, the shale is retorted in the geological formation without miningto the surface. The objective of the current study is to explore thepossibility of using the (spent) shale repository as a potential source forgeological sequestration of CO_2. For this research, the shalesamples used were retorted at different temperatures in order to address thefollowing problem: the response ofdifferent initial mineral matrices and unreactedorganic components to CO_2 injection.The experimental apparatus consists of isolated, high-pressure reactorsoperated at varying conditions with pure CO_2 as the feed gas. Initial XRD analysis of the raw and spentshale samples from Green River formation reveal carbonates, quartz, andfeldspars as the principal inorganic components. Calcite and dolomiteconstitute the carbonate sections in the shale and microcline (K-feldspar) andplagioclase feldspar make up the silicate section. Kaoliniteis present in significant portions and silica is mainly present as quartz. Illite forms the clay portion of the shale. SEM analysis ofthe reacted spent shale sample at 100^0C reveals precipitation ofcalcite, magnesite and possibly dolomite. Carbonationof feldspars was also observed, made evident by the precipitation of carbonateswith their respective principal ions. These changes in rock mineralogy arecomplemented by the changes in brine chemistry. Base case kinetic models weredeveloped using Geo-Chemists Work Bench (GWB). The results from these modelsindicate a rapid pace of sequestration reactions when compared to experiments.Thus this study provides useful laboratory data (with model comparisons) whenconsidering CO_2 sequestration in different spent shalerepositories.

See more of this Session: Complex Subsurface Processes II
See more of this Group/Topical: Energy and Transport Processes

(516f) Physicochemical and Hydrodynamics Aspects of Electrokinetics In Soil Remediation

Cameron A. Grey, Mario Oyanader and Pedro E. Arce, Chemical Engineering, Tennessee Technological University, Cookeville, TN

Heavy metal contaminants found in soil locations pose a tremendous problem to their removal and proper disposal. As to their removal, the extraction of these metals from the ground has been tested theoretically with the application of an electrical field in a soil matrix. This method has been reported to extract specific contaminants with little to no excavation through low permeability soils. In the presence of low permeability soils, these electric fields also theoretical prove to be very promising; however, there is still the need for furthering this technology in modeling as well as in its physical application. A major key in determining the driving forces of this system are the different components of the force (electroosmosis, electromigration, buoyancy, and advection) and how they contribute to the bulk flow of the fluid in the ground. Optimizing this force is one such discovery that will be unveiled as well as the different geometries and the analysis of their benefits. The Debye-Hückel approximation will be employed in order to help solve the differential equations of the electrostatic potential. As a result of this research, these contaminant sites will be readily cleaned with an efficient technology that is easily applied to the area; as well as further the advancement of the use of electric fields in cleaning applications.

See more of this Session: Complex Subsurface Processes II
See more of this Group/Topical: Energy and Transport Processes

Considerations for Research, Development and Commercialization/Scale-up of Biomass Pretreatment Processes

Dale Monceaux[1], *Richard C. Agar*[1] *and Charles E. Wyman*[2], [(1)]*AdvanceBio, LLC, Milford, OH,* [(2)]*College of Engineering-Center for Environmental Research and Technology, University of California, Riverside, Riverside, CA*

Global efforts to cost effectively convert biomass-based polysaccharides to sugars via enzymatic hydrolysis and subsequently to fuels and chemicals have been ongoing since the early discoveries by the US Army Natick Research & Development Command. These, and other early developments related to enzymatic hydrolysis of cellulose were followed by advancements in microbial genetic engineering. Genetically modified microorganisms capable of utilizing xylose-rich hemicellulose fractions of biomass feedstocks in addition to the cellulose derived glucose, promised improved ethanol yields and process economics. The past fifteen years of research has led to other major advancements in cellulase protein chemistry as well as an improved understanding of the microstructure of lignocellulose, moving the world ever closer to commercially viable biomass-based fuels and chemicals.

Early research ultimately revealed the sensitivity of the feedstock, pretreatment, enzyme and organism regime. This, in turn has exposed the need for consistency in methodology as research proceeds from concept to commercial-scale operations.

Much current activities focusing on the commercialization of biomass based fuels and chemicals hinge on the development of cost-effective biomass pretreatment processes. This "first step" in the process of converting biomass to end-products is critical to successful commercialization efforts for a number of reasons. First, it provides reactive cellulose that can be efficiently enzymatically hydrolyzed to oligomeric or monomeric glucose. Second, it efficiently solubilizes hemicellulose fractions of the biomass, producing oligomeric and or monomeric xylose, mannose, galactose, arabinose, etc. Third, it produces minimal reaction by-products that inhibit enzymatic hydrolysis and fermentation processes. Last, it is mechanically and economically scalable.

Many pretreatment technologies employ thermo-chemical reactions that reduce the recalcitrance of ligno-cellulosic matrices, solubilize fractions of the biomass and expose reactive sites to subsequent reactions. Various pretreatment technologies are under development, each with a range of design basis conditions. **Table 1** illustrates general conditions for six common pretreatment chemistries.

Pretreatment	Temp (ºC)	Residence Time (min)	Water Loading*	Catalyst	Catalyst Loading**	Catalyst Recyclable
AFEX	150	30	200	Anhydrous NH_3	152	Yes (95-95% of loaded)
Dilute Acid	140	40	891	1% H_2SO_4	9	No
Lime	120	240	1500	CaOH/O_2	100/100 psi	Maybe/Ne
Liquid Hot Water	200	10	663	Water	N/A	N/A
SAA	160	60	765	15% Aqueous NH_3	135	Yes, ecess
SO_2	180	10	900	5% SO_2/Steam	5/?	Yes, excess

* Water Loading = L/100 kg of dry biomass input to pretreatment

** Catalyist Loading = kg/100 kg of dry biomass input to pretreatment

Table 1

Typical Thermochemical Pretreatment Chemistries

Reaction kinetics varies significantly with feedstock type, mechanical pretreatment, catalyst type, catalyst loading, water loading, residence time and temperature. Tight control of key parameters is required to maximize conversion efficiency. Elevated catalyst concentration and residence times increase reaction rate but also xylose destruction as illustrated in **Table 2**. Consequently, optimum reaction conditions require mechanical designs that deliver controlled residence times under near-plug flow conditions.

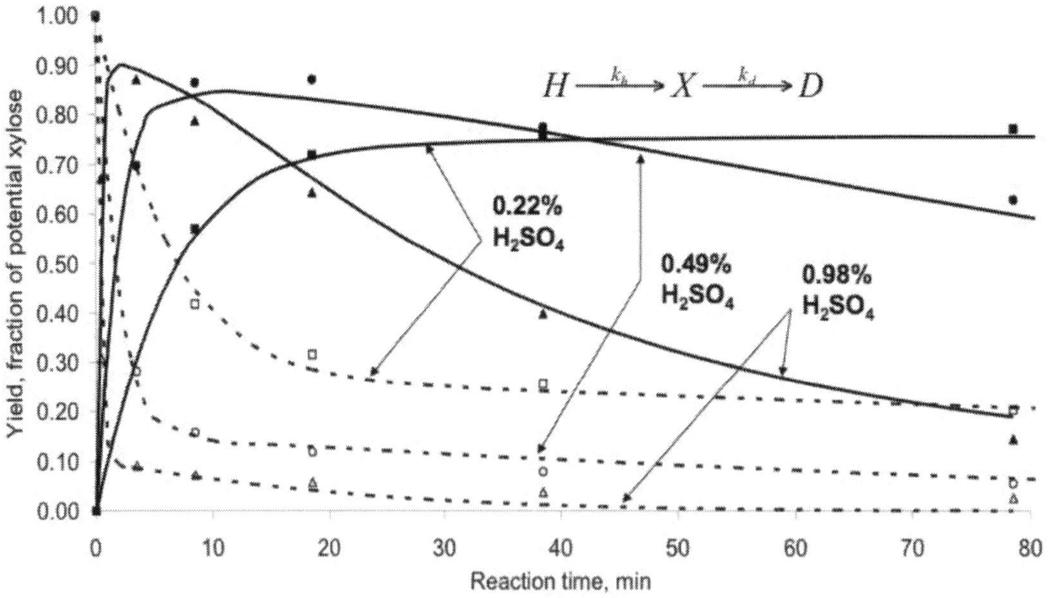

Table 2

Xylose Yield in Pretreatment vs Time at 160°C

Resource limitations, including feedstock supply, workspace, staffing and funding constrain the research and development efforts at each stage. Real-world feedstock properties such as foreign matter content, chemical composition variability, recalcitrance, bulk density, etc. introduce constraints that further complicate research and commercialization efforts. Feedstock selection dramatically impacts capital investment and production cost. Feedstock recalcitrance in conjunction with chemical composition effectively determines yield, and therefore the quantity of feedstock required per unit of fuel or chemical produced. Feedstock chemical composition and bulk density, in conjunction with reactor residence time establishes the size of the major components of the capital and energy intensive pretreatment reactor unit operation, as illustrated in **Table 3**.

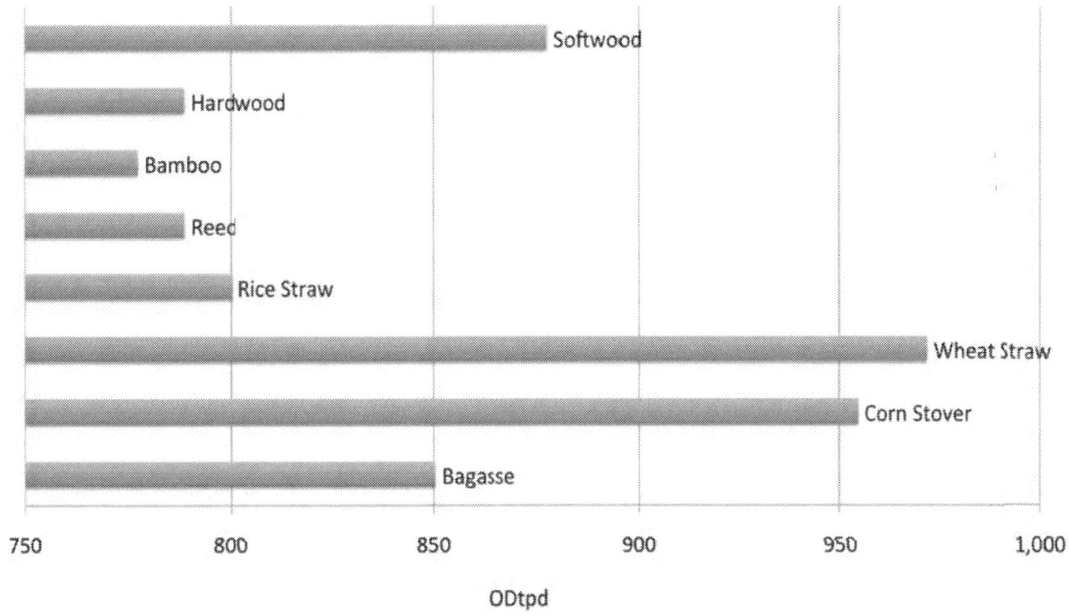

Table 3

Estimated Feedstock Tonnage per 100 Megaliters of Ethanol

High yields are critical to distribute both operating and capital costs. Low operating costs, including chemicals, energy and labor, are essential to providing adequate margins for return on capital. General operating cost categories must be lower for a new process than for established technologies making conventional products since low capital costs are essential to minimize financial exposure. The driving forces to contain capital investment include general technologies and processes that minimize equipment sizes, operate at lower pressures, operate at neutral or higher pHs as well as thermal processes that operate at lower temperatures and biochemical processes that operate at elevated temperatures.

Technical compromises frequently lead to approaches that ultimately expand the research and development program effort, schedule and budget. The impact of decisions made early in the technology development phase of a project may not fully materialize until late in the project after significant design and engineering expenditures have been made. The time period between research decision-making and engineering-based capital cost estimating delays the commercialization or leads to technology dead-ends. Visualization of the expected commercial scale process should occur very early in the technology research and development effort. A process "scale-down" approach to optimize research and technology development, scale-up and commercialization efforts has proven to be and effective strategy to accelerate project development and finance.

(522b) Improving Corn Stover Reactivity In Low Severity Pretreatment Through Deacetylation and Disc-Refining

Xiaowen Chen[1], Joseph Shekiro[1], mary Ann Franden[1], Min Zhang[1], Bonwook Koo[2], Sunkyu Park[2] and Melvin Tucker[1], (1)National Bioenergy Center, National Renewable Energy Laboratory, Golden, CO, (2) Department of Forest Biomaterials, North Carolina state university, Raleigh, NC

The current "state-of-art" dilute acid pretreatment to produce bio-ethanol from corn stover requires severe conditions to overcome biomass recalcitrance, which, in turn, offers 70~80% xylose monomer yield and 5~7% furfural yield. However, high acid loadings and high reaction temperatures harm the economic feasibility of the process. In addition, the amount of acetate and furfural produced during pretreatment is in a range that strongly inhibits cell growth and impedes ethanol fermentation. To decrease the impact of these issues, deacetylation, pretreatment with lower acid concentrations, lower temperatures, and integration of disc-refining was carried out in this work. The results show that deacetylation with dilute NaOH (0.1M) combined with low severitydilute-acid pretreatment increased corn stover reactivity by increasing the xylose monomer yield by 20% in pretreatment while degradation of xylan to furfural remained under 2%. Glucose yield increased by 10% and xylose yield by 20% during dilute solids enzymatic hydrolysis as a result of deacetylation of the feedstock prior to pretreatment. Particle size reduction by mechanical refining after pretreatment is shown to assist achieving> 90% cellulose yield in high solids (20 wt% total solids) enzymatic hydrolysis. Bioscreen C growth assays and mini fermentation at high solids (20wt%) showed less toxicity of deacetylated pretreated corn stover and enhanced xylose utilization by 20%.

See more of this Session: Developments In Biobased Alternative Fuels I
See more of this Group/Topical: Sustainable Engineering Forum

Reducing Process Complexity for Cellulosic Ethanol

Lonnie O. Ingram, Department of Microbiology and Cell Science
University of Florida, Box 110700, Gainesville, FL 32611 Ingram@UFL.edu

Lignocellulosic biomass represents a potential source of carbohydrates (cellulose and hemicellulose) for microbial fermentation into fuels and chemicals. Steam pretreatment with dilute mineral acids serves as an efficient approach to depolymerize hemicellulose and enhance cellulose digestion with enzymes. However, side products (furfural, 5-hydroxymethyl furfural, acetate, phenolics,and others) formed during pretreatment inhibit microbial growth and retard fermentation. Potential engineering solutions to this problem (separation of hemicellulose hydrolysate, counter-current-washing of fibers, and mitigation of toxins) add complexity and increase cost. The level of inhibitors produced during acid pretreatment is affected by the severity of conditions and the choice of acid. Weaker acids such as phosphoric acid produce lower levels of toxic side products and reduce the need for exotic metal alloys. Since mineral acids are not consumed by the pretreatment process, the higher cost of phosphoric acid as compared to sulfuric acid could be offset in part by recovery and reuse as a dilute fertilizer when crops are nearby. Recent progress has been made in the development of more robust *E. coli*-based biocatalyst with increased resistance to dilute acid hydrolysates. Using this biocatalyst and phosphoric acid pretreatment, we have been able to co-ferment sugarcane bagasse hemicellulose hydrolysate and enzyme-hydrolyzed cellulose in a single vessel using a variation of the simultaneous saccharification and co-fermentation process (SScF; Fig. 1). Hexose and pentose sugars from phosphoric acid pretreated sugarcane bagasse were fermented in a single vessel, eliminating process steps for solid–liquid separation and sugar cleanup. After pretreatment, a liquefaction step (L) with cellulase was included to improve mixing and saccharification (L + SScF), analogous to a corn ethanol process. Bagasse slurries containing 10% and 14% dry weight (fiber plus solubles) were tested using pretreatment temperatures of 160–190 C (1% phosphoric acid, 10 min). Enzymatic saccharification and inhibitor production both increased with pretreatment temperature. The highest titer (30 g/L ethanol) was obtained after incubation for 122 h using 14% dry weight slurries of pretreated bagasse (180 C). The highest yield (0.27 g ethanol/g dry bagasse) was 82 gal ethanol per dry ton of bagasse using this L+SScF process and 10% dry weight slurries.

Comparison of conversion processes

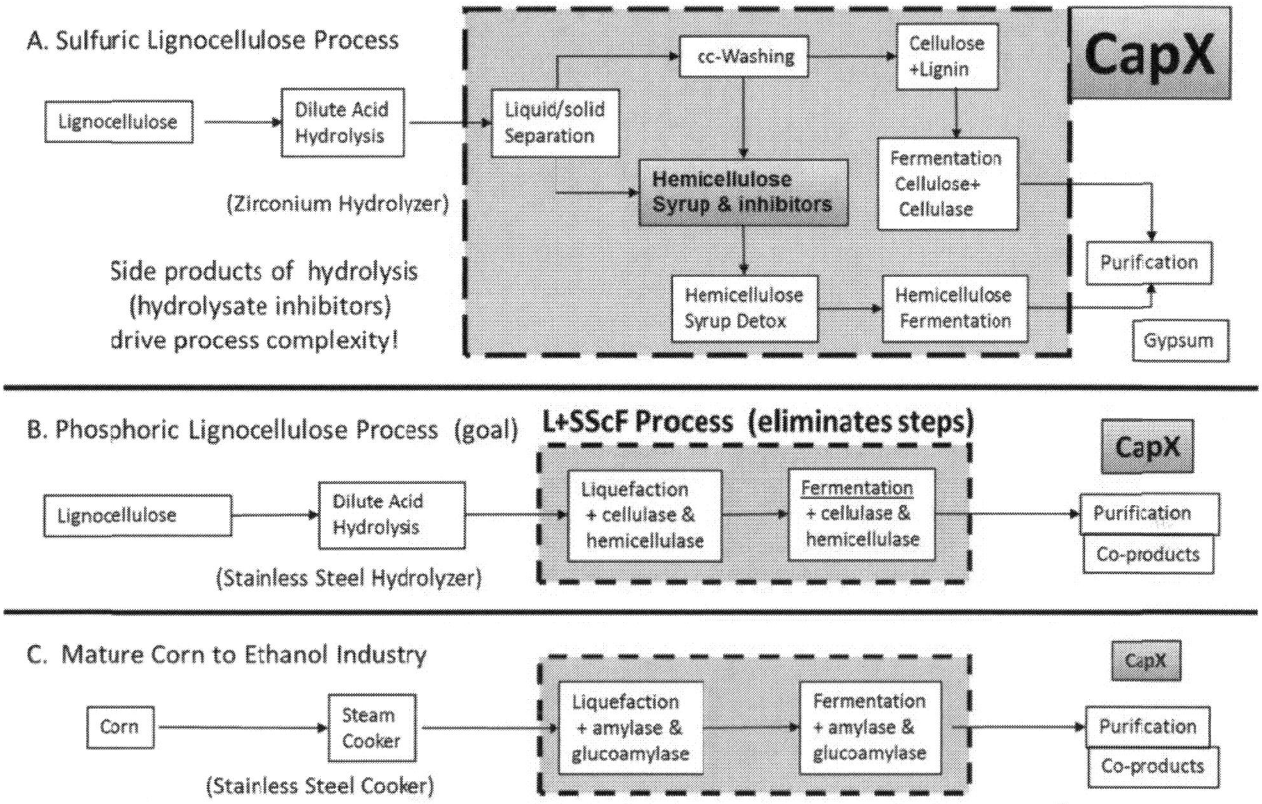

(522d) Co-Utilization of Methane In Steam-Biomass Gasification Using Concentrated Solar Energy

Aaron W. Palumbo, Erica L. Jorgensen and Alan W. Weimer, Chemical & Biological Engineering, University of Colorado at Boulder, Boulder, CO

Presented is a novel gasification application where biomass is steam-gasified and natural gas is steam-reformed simultaneously in a solar reactor. Solar-driven gasification is a proven technology, and can augment the biofuel industry as a domestic fuel source. Gasification performed in an indirectly heated gasifier can achieve temperatures greater than 1150 °C [1], mitigating or eliminating the production of CO_2, tar, char, and light hydrocarbon gases in the product gas. This reduces the need for gas clean-up technology and will not require a downstream water-gas shift reactor to adjust H_2:CO. Thermodynamics favor the formation of only hydrogen and carbon monoxide in the product synthesis gas. Therefore, a predictable H_2:CO can be controlled and tuned for downstream fuel conversion. Since the majority of biomass carbon is converted to CO, more than twice as much biomass is converted to a useable liquid fuel than in a traditional gasification process. Preliminary experiments show that reactor temperatures must exceed 1350 °C in order to reduce the effects of the water-gas shift reaction and reform practically all light hydrocarbon gases, including CH_4. This operating temperature is consistent with realizable concentrated solar energy systems.

1. Piatkowski, N., et al., *Solar-driven gasification of carbonaceous feedstock-a review.* Energy & Environmental Science, 2011. 4(1): p. 73-82.

See more of this Session: Developments In Biobased Alternative Fuels I
See more of this Group/Topical: Sustainable Engineering Forum

(522e) Poly(oxymethylene) Dimethyl Ethers As Components of Tailored Diesel Fuel - Properties, Synthesis and Purification Concepts

Jakob Burger, Laboratory of Engineering Thermodynamics, Technische Universitt Kaiserslautern, Kaiserslautern, Germany

Poly(oxymethylene) Dimethyl Ethers as components of tailored diesel fuel - properties, synthesis and purification concepts

Jakob Burger [1*], Markus Siegert [2], Eckhard Ströfer [2], Michael Nilles [2], Hans Hasse [1]

[1] Laboratory of Engineering Thermodynamics, University of Kaiserslautern, Germany;

[2] BASF SE Company, Ludwigshafen, Germany
*Corresponding author: Jakob Burger, Laboratory of Engineering Thermodynamics, University of Kaiserslautern, Germany. Email: jakob.burger@mv.uni-kl.de

Keywords: Fuel additive, Soot reduction, Process development

Poly(oxymethylene) Dimethyl Ethers (POMDMEs), oligomers of the general structure CH_3-O-(CH_2-O)$_n$-CH_3, open a new route for tailoring diesel fuels. POMDMEs belong to the group of oxygenates which reduce soot formation in the combustion when added to diesel fuels. They can be produced on a large scale based on Methanol which can be produced from renewable resources. This contribution presents the process chain, possible synthesis routes and some fuel related properties of POMDMEs. Comparisons to other oxygenates, e.g. Dimethyl Ether, are given.

A particularly favourable route for the large scale production of POMDMEs in which they are formed from Methylal and Trioxane is proposed. The chemistry behind this route is given in detail. Measurements of the reaction kinetics and the reaction equilibrium were carried out in a batch reactor using acidic ion exchange resin as heterogeneous catalyst. The parameters of a pseudo-homogeneous model were fitted to the results of these experiments. The formation of side-products was studied experimentally and concepts to minimize them are discussed. Based on the reaction model a process concept was developed. It consists of a reactor and a downstream distillation sequence to yield pure POMDMEs of the most favourable chain length n=3-5.

The use of distillation technology allows an easy scale-up to huge single train plants which can potentially fulfil the need for millions of tonnes of the innovative oxygen carrier POMDME needed for the supply of clean diesel fuel in the future.

See more of this Session: Developments In Biobased Alternative Fuels I
See more of this Group/Topical: Sustainable Engineering Forum

(522f) Synthesis of Green Diesel Fuel From Fatty Acid Feedstocks Via Electrochemical Hofer-Moest Decarboxylation

Cody Wagner, Charles Coronella and Rainer Busch, Chemical Engineering, University of Nevada, Reno, Reno, NV

Production of biodiesel by esterification or transesterification of long-chain fatty acids is a well known process, but the fuel product has characteristics making it not fully compatible with petroleum diesel engines. A better option is to convert fatty acid molecules to hydrocarbons by removing their carboxyl group. Decarboxylation of fatty acid feedstocks was performed by electrolysis using graphite electrodes, which selectively causes a Hofer-Moest (or non-Kolbe)-type reaction. Methanol was used as a solvent for a robust selection of fatty acid feedstocks.

The liquid product created by the electrochemical reaction was composed of mainly alkenes with at least one double bond, ethers, and methyl esters. This fuel product has advantageous cold-flow properties, no sulfur content, heating values close to those of diesel fuels, and is capable of running in current diesel engines as a drop-in replacement for petroleum diesel, unlike biodiesel. The temperature, pH, concentrations of ions, and type of base used were all varied to optimized the current density, which limits the production rate. A separation method was tested for removal of the fuel product from the electrolyte. An efficiency was then calculated for several varieties of fatty acid feedstocks, comparing the electrical energy input to the product generated from the electrolysis. Overall first-law thermodynamics and economics were also evaluated, examining production rates and the large-scale viability of the process.

See more of this Session: Developments In Biobased Alternative Fuels I
See more of this Group/Topical: Sustainable Engineering Forum

Quantitative Comparison of Chemical and Biological Cardboard's Lignin Degradation

E. Isaza
Department of Chemical Engineering, Universidad de Los Andes, Bogotá, Colombia
R. Sierra R.
Advisor, Universidad de Los Andes, Bogotá, Colombia

ABSTRACT: An efficient way to increase biomass digestibility, thereby, make it usable in the production of second generation biofuels, is pretreatment. However, it demands important energy inputs which make it one of the most expensive stages in biofuels processing. Comparisons between chemical (oxidative with lime) and biological (with commercial *Laccase* extract) pretreatments were performed both by experimental measurements and by simulations. All experiments were performed following a full factorial experimental design, with lignin and sugar yields as response variables. The influence of the type of pretreatment (i.e., pretreatment yields) on the process that leads to biogasoline and jet fuel production (MixAlco®) was simulated in AspenPlus®. This simulation showed higher productivity and energy efficiency (15 and 60% respectively) for the chemical pretreatment over the biological pretreatment.

1 INTRODUCTION

First generation biofuels have emerged as an alternative to fossil fuels, whose price and environmental impacts have been subject of intense debate over the last years. From all biofuels, biodiesel and bioethanol are the most popular. They are currently produced using high energy crops such as soy, sugar cane, oil palm and corn starch as feedstocks (U.S. State Department, 2010). These resources, however, are characterized by such high commercial and alimentary value that they become restrictive for biofuels and generate issues of alimentary sustainability (Lee, Biological conversion of lignocellulisic biomasss to ethanol, 1997).

The utilization of more inexpensive feedstock in fermentative process such as lignocellulosic residues is a good baseline for the optimization of this technology (Lee, Biological conversion of lignocellulisic biomasss to ethanol, 1997). Good feedstock resources are agricultural, industrial, and municipal wastes among others. The *MixAlco®* process, created by Dr. Mark Holtzapple in Texas A&M University and patented by Terrabon L.L.C, uses this kind of waste to produce ethanol and higher energy alcohols.

In this process, four main stages may be differentiated: pretreatment, fermentation, chemical dewatering, gasification and cogeneration, thermal conversion involving hydrogen, and final conversion into biofuels. Among these, pretreatment is the most expensive (Gousamme, 2011).

In biological and some chemical pretreatments (such as lime pretreatment), biomass compositional changes are sought. These changes, include but are not limited to delignification. As a result, an increased biomass digestibility is observed. Other effects of pretreatment are: disruption of the crystalline structure of cellulose, increased pore size to enhance enzyme access to carbohydrates, and some degree of separation of the main lignocellulose constituent polymers.

Lignin is a highly heterogeneous tridimensional aromatic polymer, formed by three different units of phenylpropylene: coumaric alcohol, coniferyl alcohol and sinaphyl alcohol; bounded randomly through C-C and C-O-C type bounds (Sylvia, Fuhrman, Hartel, & Zuberer, 1998), (Lee, Biological conversion of lignocellulisic biomasss to ethanol, 1997).

Widely extended delignification processes such as Kraft pulping, (Lachental, Mortha, Sevillano, & Zaroubine, 2004) are controversial due to the use of environmental harmful chemicals. Delignification in alkali with possible oxygen addition is preferred because comparable delignification levels are obtained with less toxic waste products. Pretreatments such as oxidative and non-oxidative lime (i.e., calcium hydroxide $(Ca(OH)_2)$, have been reported to promote lignin degradations close to 70% of the compound (Chang, Nagwani M, Kim, & Holtapple, 1999).

On the other hand wild occurring delignification processes are promoted by extracellular enzymes released by fungi, bacteria (Zhang, Kjonas, & Flurkey, 1999) or even plant cells; however, white rot fungi are the degraders by excellence (Lee, Biological conversion of lignocellulisic biomasss to ethanol, 1997). They generate extracellular enzymes such as *Laccase*, Manganese peroxidase and lignin peroxidase. Laccase (EC. 1.10, 32), benzendioloxigenoxidoreductase, is a copper containing enzyme capable of lignin degradation in the presence of oxygen. The only product of this transformation is water (Rodriguez Cuoto & Toca-Herrera, 2006). Because of this reason, this enzyme has been widely studied and used in industrial applications.

Since there are many advantages and drawbacks for both chemical and biological delignifications, it is desirable to formally compare these processes. However, a systematic comparison on the same basis (same feedstock, analytical procedures and general methods) was not found in the literature. This study aimed to make such comparison. To accomplish this goal, a full factorial design was implemented to both: oxidative lime delignification of cardboard (high in lignin content) and delignification of the same material using purified lacasse (provided by Sigma Aldrich)

The experimental data obtained from the best global conversion of carbohydrates was loaded to an existing preliminary simulation of the *MixAlco®* process (Gousamme, 2011) to determine which of the delignification processes is best on the basis of highest productivity and energy efficiency.

2 METHODOLOGY

2.1 *Chemicals and reagents*

2.1.1 Feedstock for pretreatments and substrate for enzymatic hydrolysis

Recycled cardboard is commonly used in Colombia to transport eggs. This biomass is characterized by high lignin content; a desirable feature for a delignification study. Untreated cardboard was the feedstock for pretreatment. The samples were obtained from *Kikes eggs*, Leader Company in the egg market at Colombia. Pretreated cardboard was used as substrate for enzymatic hydrolysis.

2.1.2 *Reagents for enzymatic delignification*

The selected enzyme was *Laccase Suberase* brand *Novartis*. It was activated in a phosphate buffer at pH 3, 4.5, 5, 5.5, 6, 6.5 and 7.

2.1.3 *Chemical pretreatment*

Hydrogen peroxide was obtained from *Panreac* at (%w/w) 0.5, 1, and 2. Calcium hydroxide $Ca(OH)_2$ also was obtained from Panreac.

2.1.4 *Enzymatic saccharification*

It was performed using cellulase 15 FPU obtained from *Sigma Aldrich*.

2.2 *Characterization of the selected biomass*

Quantification of biomass components (i.e., lignin, structural carbohydrates, and ash), was obtained following the laboratory analytical procedures (LAP) produced by the *National Renewable Energy Laboratory (NREL)* for lignocellulosic biomass (Sluiter A. , Hames, Ruiz, Scarlata, Sluiter, & Templeton, 2004). A summary of experimental protocols follows.

2.2.1 *Biomass conditioning*

In accordance to NREL Lap entitled "Preparation of Samples for Compositional Analysis" (Hames, Ruiz, Scarlata, Sluiter, Sluiter, & Templeton, 2008), all biomass was conditioned in a step previous to experimentation, by reducing: particle size (the fraction taken for experimentation was +40/-80 Thyler mesh) and the moisture content was reduced to less than 10%, by drying in a convection oven at 45°C for >24 h.

2.2.2 *Ash content determination*

The ash content of raw biomass was determined by exposing a known dry weight of biomass (7 replicates) to calcination during four hours in a muffle at 575°C. The weights of the resulting ashes were measured and related to the initial dry weight of each sample. (Sluiter A. , Hames, Ruiz, Scarlata, Sluiter, & Templeton, 2008).

2.2.3 *Structural sugars and lignin*

These components were measured in accordance to NREL laboratory analytical procedure entitled "Determination of Structural Carbohydrates and Lignin in Biomass" which involves two acid hydrolysis steps with H_2SO_4. The first stage is concentrated at 72% and the second diluted with 4%. This procedure results in solubilization of structural carbohydrates while lignin is precipitated (Scarlata, Slutier, Termpleton, Crocker,

Hames, & Ruiz, 2008). Sugar concentration is determined by the colorimetric method dinitrosalicilic acid or DNS, and lignin is separated and gravimetrically measured.

2.3 *Chemical pretreatment*

Cardboard (1 g), lime (0,5 g) and 45 mL of hydrogen peroxide were well mixed into a stainless steel reactor (Parr 500). Afterwards, the reactor was tightly closed, and submitted to a pretreatment at the temperature, hydrogen peroxide concentration and time of pretreatment that corresponds to the experiment according to the factorial design 2^3 summarized in Table 1.

Table 1. Factorial design of chemical pretreatment.

Factor	Level 1	Level 2
Hydrogen peroxide concentration%	0.5	2
Operating time (min)	30	120
Operating temperature (°C)	80	150

Pretreated samples were neutralized using 5 N HCl, washed with abundant distilled water, filtered through Whatman #1 filter paper and weighted. Afterwards, the harvested biomass was divided into three parts. The first part was kept frozen to be submitted to enzymatic hydrolysis later. Another part was put to dry in a 45°C convection oven, to be used in lignin and sugar quantification. The last part was used to determine moisture content.

A parameter was calculated to determine total solids yield after pretreatment (Y_T) as follows:

$$Y_T = \frac{\text{recovered biomass (g)}}{\text{biomass before pretreatment (g)}} \quad (1)$$

With this parameter, it is possible to determine structural carbohydrates and lignin yields (Y_s) (Y_L) as defined in equations (2) and (3),

$$Y_s = \frac{\text{recovered carbohydrates (g)}}{\text{carbohydrates before pretreatment (g)}} = Y_T \frac{C_0}{C_t} \quad (2)$$

$$Y_L = \frac{\text{recovered lignin (g)}}{\text{lignin before pretreatment(g)}} = Y_T \frac{L_0}{L_t} \quad (3)$$

Where, (L)=Lignin; (C)= Sugars; (0)=initial; and (t)=at the end of the pretreatment.

For lime pretreatment, kinetics of lignin and sugar degradation were modeled by standard power law (Eq. 4), accounting for two reacting moieties (fast and slow) as proposed by (Sierra, et al., 2011). Model parameters are obtained by minimization of the sum of squared residuals.

$$-r_i = -\frac{dY_i}{dt} = k_{if} M_{O_2}{}^{\beta if} Y_{if} + k_{is} M_{O_2}{}^{\beta is} Y_{is} \quad (4)$$

Where, (i)=Lignin or sugars; (j)=slow or fast reaction; k_{ij}= rate constant for substance; R=universal constant law for ideal gases (8.314×10^{-3} kJ/(mol K); M_{O_2}=oxygen loading (kg initial oxygen/kg dry biomass); and β_{ij}= reaction order for hydrogen peroxide.

2.4 *Enzymatic pretreatment*

2.4.1 *Activity determination*

Activity determination of the enzyme, *Laccase Suberase* brand *Novartis*, was made according to the *Sigma Aldrich* protocol (Aldrich, 1997). This protocol recommends the use of 0.216 mM Syringaldazine solution, water and *Laccase Suberase*, to determine the activity by a continuous spectrophotometric rate determination (Aldrich, 1997). Because experiments were run at several different pH (Table 2), one result of this experimentation was the recommended pH.

2.4.2 *Selectivity*

Enzymatic delignification is supposed to be highly (if not completely) selective. Nevertheless, a selectivity experiment was performed in order to ratify that the enzyme was free of cellulases. For this purpose, the enzyme (20 μL of *Laccase*) was put in contact with 0.1 g of cellulose at pH of 6, 6.5 y 7, and room temperature and 55°C for a day.

2.4.3 *Experimental Design*

A full factorial design 3^3 was made in order to determine enzyme loading, substrate concentration, residence time and temperature that would maximize enzyme performance at optimum pH of 6.5 and optimum temperature of 55°C. (Baldrian, Fungal Laccasses-ocurrance and properties, 2006). It is summarized in Table 2.

Table 2. Factorial design for enzymatic pretreatment

Factor	Level 1	Level 2	Level 3
Enzyme loading (mL/g substrate)	0.0 1	0.1	0.5
Substrate concentration (w/w)	1	5	10
Residence time (min)	60	360	1440

Immediately after pretreatment, enzyme pretreated samples were submitted to thermal shock to denature the enzyme and therefore stop the reactions. Subsequently, they were profusely washed with distilled water.

Based on the obtained results a decrease on the number levels of the factorial design is proposed. Reducing: residence time(0-1 h), enzymatic loading (20,150 μL) and substrate concentration (2 g). For a better compression of the degradation of lignin it is modeled by Michaelis Menten:

$$-r_s = \frac{V_{max} \times (S)}{K_m + (S)} \qquad (5)$$

Where (S)=substrate concentration; K_m=affinity constant; and V_{max}=maximum degradation rate of substrate.

2.5 *Enzymatic hydrolysis*

To determine the effect of pretreatment in biomass digestibility, an enzymatic hydrolysis is made quantifying the cellulose fraction that is hydrolyzed by cellulase at optimum conditions.

The procedure is the *NREL Lap* for *Enzymatic Saccharification of Lignocellulosic Biomass* (Brown & Torget, 1996). It consists on exposing pretreated biomass with an excess of cellulase (15 FPU/g biomass) at 50°C and pH 4.8 for 3 days. The protocol is modified adding 1g of biomass instead of 1 g cellulose equivalent in biomass.

The monomers of glucose and xylose are determined later by DNS, and digestibility is calculated with relation (5) and (6):

$$Y_{saccharification} = \frac{\text{digeted celulose (g)}}{\text{added celulose (g)}} \qquad (5)$$

$$Y_{saccharification} = \frac{\text{sugars}_e \times V_T}{\text{biomass(g)} \times \text{SST} \times \text{sugars}_p} \qquad (6)$$

Where, $sugars_e$=sugar content of saccharification extract (mg sugars/ml); V_T=total volume of saccharification (10.6 ml), biomass =pretrated biomass= 0.1 g; SST= total solid content in pretreated biomass; and $sugars_p$ =sugar content of pretreated biomass (g sugars/g pretreated biomass).

2.6 *Global conversion*

The combined effect of pretreatment and enzymatic hydrolysis a global recuperation factor is found:

(6)

Mentioned relation allows choosing the conditions of chemical and enzymatic pretreatment to load in the *MixAlco®* process simulation, where higher values of correspond to superior efficiency in pretreatment, therefore, preservation of sugars, breakage of crystalline structure of celluloses and hemicelluloses and higher lignin degradation.

2.7 *MixAlco® process simulation*

The simulation of the *MixAlco®* process, is modified from a existing one made by (Goussame,2011) where experimental data of both pretreatments are loaded. Accomplishing finally, the comparison of the two pretreatments, at a preliminary stage, because the simulation is an ideal modeling of the process.

The energy index was defined by (Gousamme, 2011) as,

(7)

3 RESULTS AND DISCUSSION

3.1 *Biomass characterization*

The obtained results of biomass characterization protocols are, that non pretreated biomass is composed in dry weight: 43% by lignin, 3% by ash, and a 44% of sugars, Illustration 1.

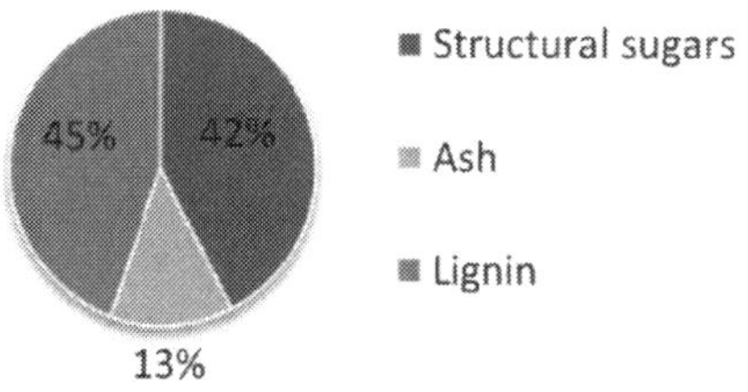

Ilustration1. Composition of cardboard in dry weight.

3.2 *Chemical pretreatment*

3.2.1 *Lignin degradation*

The average of the experimental results obtained for factorial design 2^3, for chemical pretreatment are presented in Illustration 2 with respective standard deviation.

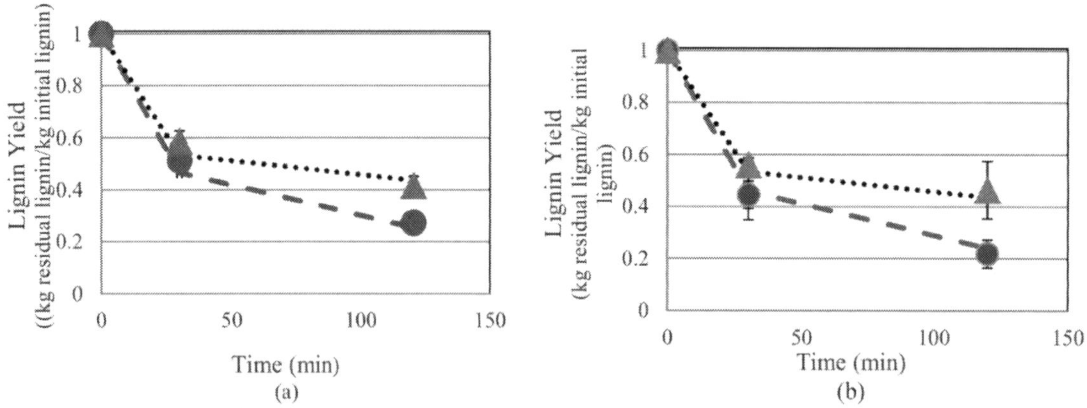

Illustration 2. Lignin yield for chemical pretreatment with lime Ca(OH)2 and respective kinetic model. (a) 80°C(b)150°C. (··▲··) 2% hydrogen peroxide, (—●—)0.5% hydrogen peroxide.

According to Illustration 2 at higher peroxide concentration, temperature and time, greater lignin degradation. This can be explained because higher hydrogen peroxide concentration generates more quantity of superoxide ions responsible of breaking C-C and C-O-C bounds.

As shown in Ilustration2, the time that superoxide ions take in breaking the lignin chain ends, where the degradation begins, is half hour. With higher residence time an increase of the lignin degradation is measurable, because the numerical relation between superoxide ions and attack sites become more efficient.

The highest delignification is at two hours, with lime ($Ca(OH)_2$) at 150°C and hydrogen peroxide at 2%.

Calculated lignin yields are presented as cutter lines in the operating rank of each pretreatment. Data is obtained by least squares minimization of the model presented by Sierra ET all, 2009, whose solution generates the parameters show in Table 3.

Table 3. Achieved values for parameters of lignin degradation with lime $Ca(OH)_2$.

Paramet er	Value	Units
Y_{ijf0}	0.545	g residual lignin/g initial lignin
a_{iijf}	0.0597	min^{-1}
B_f	0.0257	
E_{ijf}	0.0825	kJ/mol
a_{ifs}	4.96E-07	min^{-1}
B_s	2.10	
E_{ijs}	0.878	kJ/mol

3.2.2 *Sugar Yields*

Average of experimental data obtained for sugar yield in chemical pretreatment are presented as marks in Illustration 3.

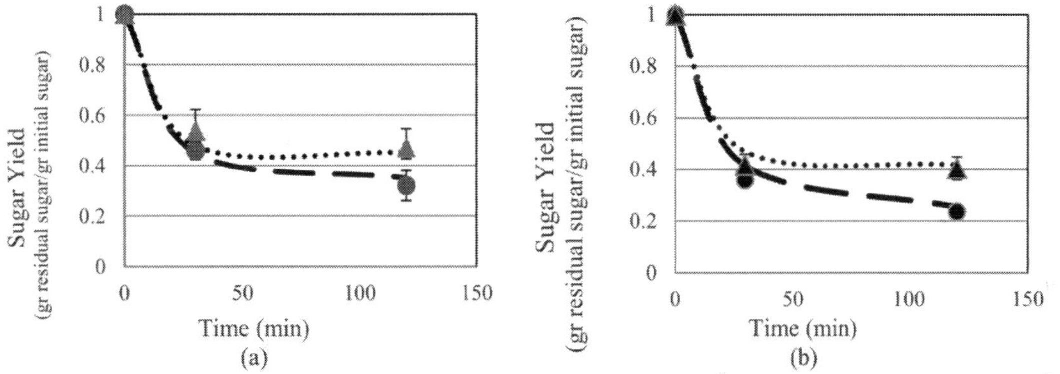

Illustration 3. Sugar yield for chemical pretreatment (a) 80°C, (b) 150°C. (▲) hydrogen peroxide at 2%, (●)hydrogen peroxide.0.5%

The behavior of sugar yield is similar to lignin yield, because at higher operation conditions lower sugar yield, this according to the presence of superoxide ions and harder thermal conditions.

Comparing part (a) and (b) of the Illustration 3, can be observed that sugars are sensitive to temperature degradation because at every pretreatment with equal time and peroxide concentration and higher temperature lower sugar yield is obtained.

The effect of time in sugar degradation is least at lower hydroxide concentration.

The kinetic modeling is made same as the lignin degradation, the obtained parameters are:

Table 4. Obtained values of kinetic parameters for sugar degradation with lime Ca(OH)2.

Parameter	Value	Units
Y_{ijf0}	0,489	g residual sugar/g initial sugar
a_{ijf}	0,00156	min^{-1}
B_f	1,02	
a_{ifs}	911×102	min^{-1}
B_s	4,315	
E_{ijf}	11,8	kJ/mol
E_{ijs}	0,869	kJ/mol

3.3 *Enzymatic pretreatment with Laccase*

3.3.1 *Activity determination and enzyme selectivity*

The higher activity is 864 U ml^{-1} at pH 6.5, Illustration 4.

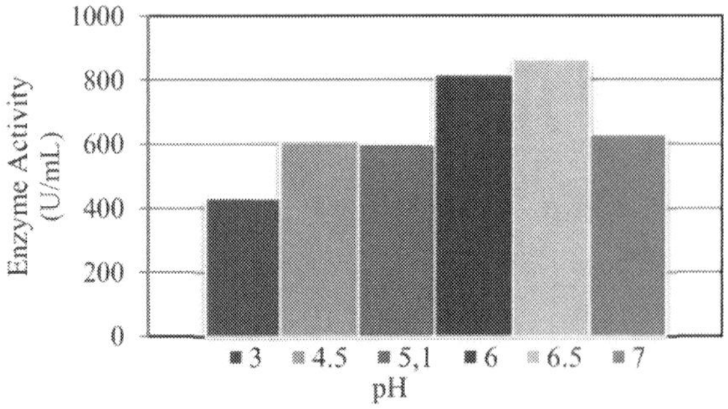

Illustration 4. Activity of the enzyme.

Obtained results for enzyme selectivity showed changes of 10^{-3} g, attributed to moisture content in the samples.

3.3.2 *Enzymatic lignin degradation*

Enzymatic experimental average results of the first substrate level (2 g) of the factorial design 3^3 are presented in Illustration 5. Only this level is presented because the behavior obtained for the other two levels of substrate, 1.5 and 1.0 g are very similar; (where at every enzymatic loading there is an increase of the lignin

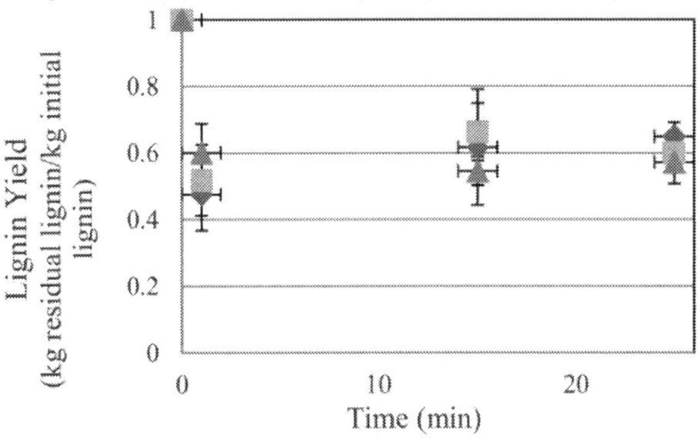

Illustration 5. Enzymatic delignification (◆) 20 μL, (■) 150 μL, (▲) 1000 μL.

For this reason the experimental conditions were cut to: a maximum of an hour, for residence time, and two levels of enzymatic loading, 20 and 150 μL, with a single substrate concentration, 2 g.

In Illustration 6, the results of the last Laccase experimentation are presented where the higher lignin degradation is 0.45(g residual lignin/g initial lignin), accomplish by 150 μL of enzyme at one hour. While the minimum yield is 0.75(g residual lignin/g initial lignin) obtain with 20 μL of Laccase at 30 min.

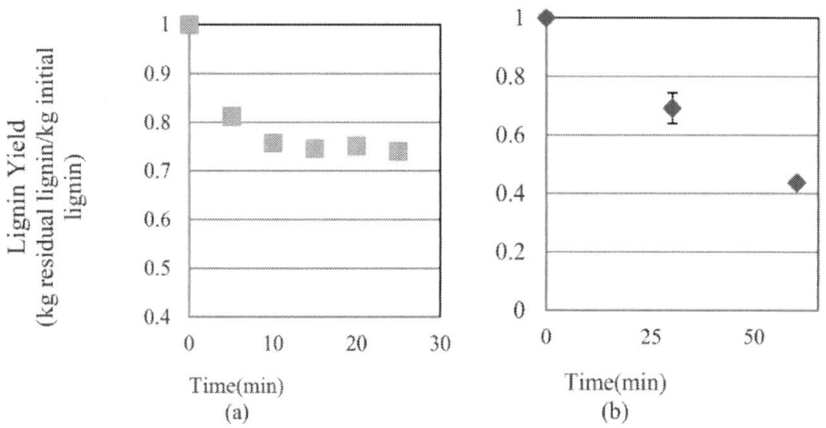

Illustration 6. Enzymatic delignification of 2 g of substrate (a) 20μL enzyme, (b) 150 μL enzyme.

The kinetic modeling for the enzymatic lignin degradation was made by the equation of Michaelis Menten as a first approximation.

The Lineweaver-Burk linearization was used to model the kinetic parameters, for 20 y 150 μl of enzyme data.

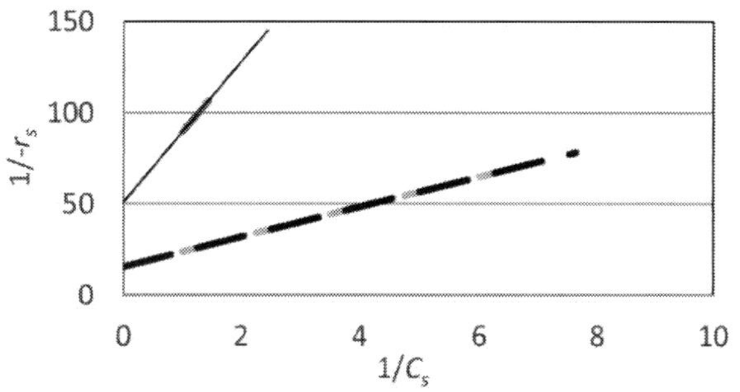

Illustration 7. Linewever-Burk for 2 g of biomass. (- -) 20 μL, ( ) 150 μL.

The values obtained for the kinetic parameters are presented at the Table 5.

Table 5.Obtained values for the enzymatic degradation of lignin.

Parameter	Level	Value	Units
V_{max}	20μl	0,0646	U mg^{-1}
K_m	20μl	0,529	g cm^{-3}
V_{max}	150μl	0,0195	U mg^{-1}
K_m	150μl	0,752	g cm^{-3}

3.3.3 *Sugar Yield*

As shown at Illustration 8, it can be identified that the higher sugar yield is obtained with enzymatic pretreatment at ten minutes and 20 μL of enzyme, while the lowest is also obtained with 20μL at 20 minutes.Thereby time is an important factor at this enzymatic loading level.

At 150 μL of enzyme, sugar yield keeps near 0.6 (g recover sugar/g initial sugar) in the selected time range.

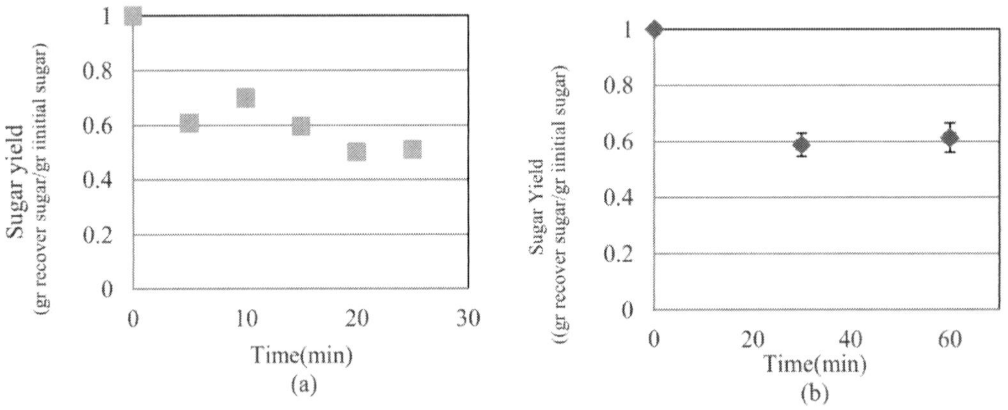

Ilustration8. Sugar yield for 2 g of substrate (a) 20 μL of enzyme (b) 150 μL of enzyme.

3.4 *Enzymatic saccarification*

Higher digestibility values were achieved for chemical pretreatment with the higher level of: time (2 h) and hydrogen peroxide concentration(2%), but since results are similar with different temperatures, Illustration 9, it is necessary to calculate global sugar conversion to select the best chemical pretreatment conditions.

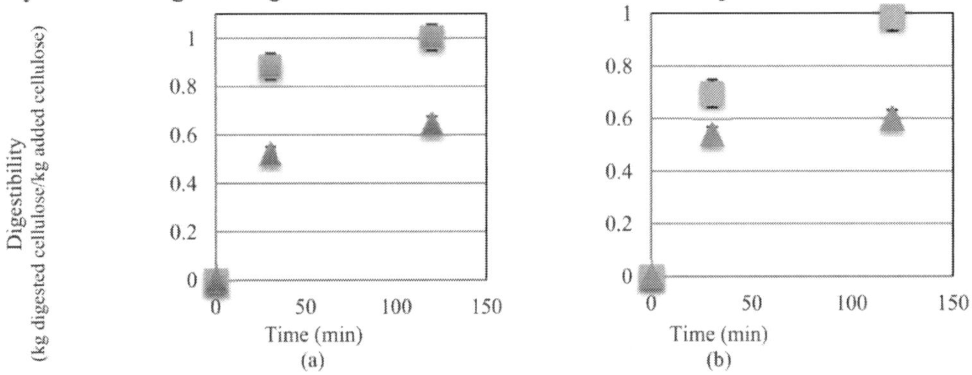

Illustration 9. Digestibility of chemical pretreated biomass.(a) 80°C, (b) 150°C. (▲) 0.5% peroxide,(■)2% peroxide.

Highest digestibility for enzymatic pretreatment is at one hour of pretreatment with *Laccase,* with 150 μL of enzyme loading, Illustration 10.

The biomass, pretreated with 20 μL of *Laccase* obtained the highest digestibility at 10 minutes of pretreatment, which can be justified with the fact that low enzyme concentration triggers quicker enzyme action because there is lower enzyme competition, but also is less efficient compared to 150 μl of enzyme where are more competition but higher lignin degradation.

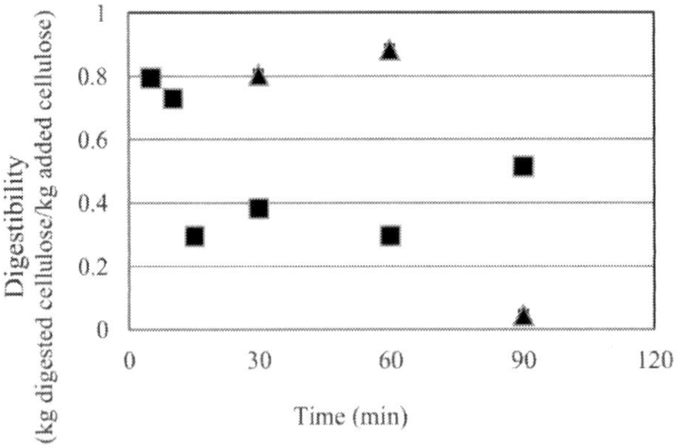

Illustration 10. Digestibility enzymatic pretreated biomass (■) 20μL and (▲)150 μL de enzyme.

The calculation of global conversion is necessary to determine which chemical and enzymatic pretreatment is the proper one to implement at Aspen® simulation of the *MixAlco®* process.

The global conversion results are presented in a contour graph, made using *MiniTab Statistical Software©*.

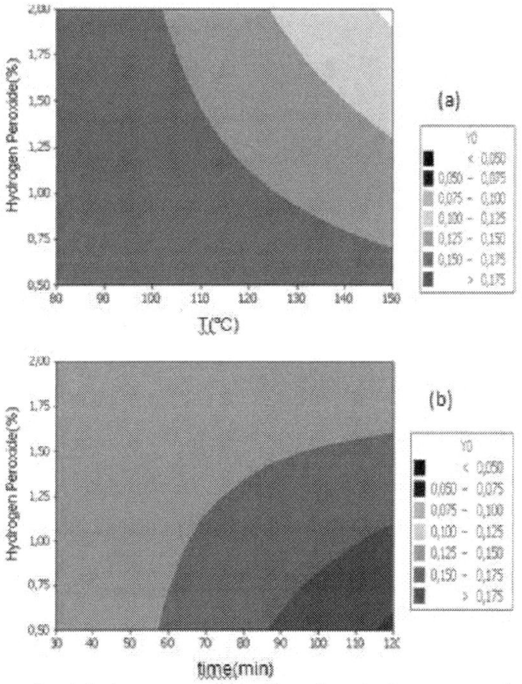

Illustration 11.Contour graph of global conversion for chemical pretreated biomass. (a) %Peroxide vs. temperature(°C),(b) %peroxide vs. time(min).

The global sugar conversion, Y_0, of chemical pretreatment samples are shown in Illustration 11, where at less peroxide concentration, lower temperature y higher time results in the best sugar conversion which can be explained by sensibility of sugar degradation to temperature and lower superoxide ions concentration.

The global sugar conversion for enzymatic pretreated biomass is presented in Illustration 12, where the zone with higher sugar conversion correspond to one hour, 150 μL of enzyme loading which can be explained due to enzyme selectivity to lignin degradation and the inverse relation between enzyme competivity and residence time.

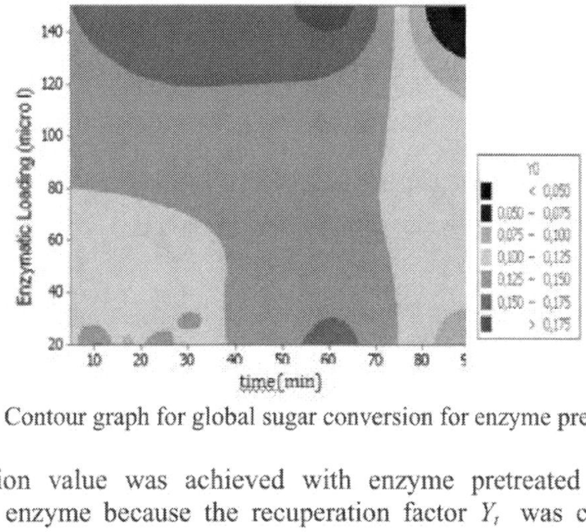

Illustration 12. Contour graph for global sugar conversion for enzyme pretreated biomass.

The best global conversion value was achieved with enzyme pretreated biomass at one hour of pretreatment and 150 μL of enzyme because the recuperation factor Y_t was close to one and the sugar degradation was low compare to chemical degradation recuperation factor close to ,5 and higher sugar degradation.

3.5 *Aspen Plus® Modeling*

The pretreatment conditions chosen to be loaded at the *MixAlco®* process simulation, design by (Gousamme, 2011) are the corresponding to chemical pretreatment 0.5% peroxide, 80°C y 120 minutes and the corresponding to enzymatic pretreatment with 150 μL and one hour of residence time.

The higher biofuel production was achieved with chemical pretreatment with an energy index of 4,477 vs. 4,189 (kJ out/kJ in) of the enzymatic pretreatment.

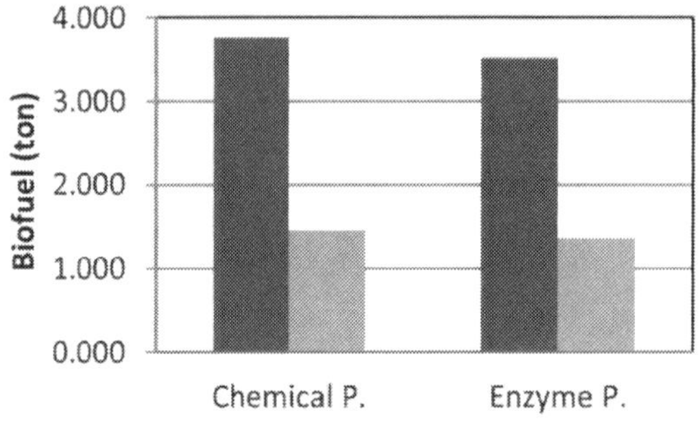

Illustration 13. Bar graph for biofuel production with chemical and enzyme pretreatment. ( ) Jet biofuel, ( ) gasoline.

4 CONCLUSIONS

Comparing chemical with enzymatic pretreatment, the enzymatic is more sensitive to time and pH, which complicates the implementation of this kind of pretreatment in the industry.

Enzymatic pretreatment also achieved the best values for digestibility and global sugar conversion.

The simulation, final step of the comparison between chemical and biological pretreatment, gives better results for the chemical pretreatment in biofuel generation and energy index.

5 BIBLIOGRAPHY

1. Aldrich, S. (1997 11-03). *Enzymatic Assay of LACCASE.* Retrieved 2011 12-03 from http://www.sigmaaldrich.com/etc/medialib/docs/Sigma/General_Information/2/laccase.Par.0001.File.dat/laccase.pdf
2. Baldrian, P. (2006). Fungal Laccasses-ocurrance and properties. FEMS Microbial Review 30 , 215-242.
3. Brown, L., & Torget, R. (1996). *Enzymatic Saccharification of lignocellulosic Biomass.* NREL.
4. Carvajal Montoya, S., & Giraldo Garcia, D. (2010). Simulación del proceso MixAlco en Aspen Plus y análisis de sensibilidad de los parámetros involucrados en las principales etapas del proceso. Bogota D.C: Proyecto de grado, Universidad de los Andes.
5. Chang, V., Nagwani M, M., Kim, C., & Holtapple, M. (1999). *Oxidative Lime Pretreatment of High-Lignin Biomass.* College Station: Texas A&M.
6. David, C., Salvador, S., Dirion, J., & Quintard, M. (2003). Determination of reaction scheme for cardboard thermal degradation using thermal gravimetricanalysis. *Journal of Analitycal an d Applied Pyroysis* , 307-232.
7. Fogler, S. H. (2008). Elementos de ingeniería de las reacciónes químicas. Naucalpan de Juarez, mexico: Pearson Prentice Hall.
8. Gandi, J., Holtzapple, M. T., Ferrer, A., Byers, M. F., Turner, N. D., Nagwani, M., et al. (1997). Lime treatment of agricultural residues to improve rumen digestibility. *Animal Feed and Science Technology* , 195.
9. Gousame, P. (2011). Simulation of the MixAlco Process. *Special Project* .
10. Granda, C. B., & Holtzapple, M. T. (2008). Part 1: Comparison of Three Biomass Conversion Platforms. In *Carboxylate Platform: The MixAlco Process.* Texas: Department of Chemical Engineering Texas A&M University.
11. Hames, B., Ruiz, R., Scarlata, C., Sluiter, A., Sluiter, J., & Templeton, D. (2008). *Preparation of Samples for Compositional Analysis.* Cole Boulevard, Golden, Colorado : National Renewable Energy Laboratory.
12. Hayatsu, R., Winans, R., McBeth, R., Scott, R., Moore, L., & Stuider, M. (1979). Lignin-like polymers in coals. *Nature* , 278.

13. Holtzapple, M., Davison, R. R., Ross, M. K., Aldrett-Lee, S., Nagwani, M., Lee, C.-M., et al. (1999). Biomass Conersion to Mixed Alcohol Fuels Using the MixAlco Process. *Applied Biochementry and Biotechnology* , 609-630.
14. Kaar, W. E., & Holtzapple, M. T. (2000). Using lime pretreatment to facilitate the enzymic hydrolysis. *Biomass and Bioenergy* , 189.
15. Lachental, D., Mortha, G., Sevillano, R.-M., & Zaroubine, M. (2004). Isolation of residual lignin from softwood kraft pulp. Advantages of the acetic acid acidolysis method. *Compites Rendus Biologies* , 911-916.
16. Lee, J. (1997). Biological conversion of lignocellulisic biomasss to ethanol. *Journal of Biotechnoly* , 3-4.
17. Lee, J. (1997). Biological conversion of lignocellulosic biomass to ethanol. *Journal of Biotechnology* , 4.
18. Mandels, M., Hontz, L., & Nystrom, J. (1974). Enzymatic Hydrolysis of Waste Cellulose. *Biotechnology and Bioengineering* , 1471-1493.
19. Morozova, O., Shumakovich, G., Gorbacheva, M., Shleev, S., & Yaropolov, A. (2007). "Blue" Laccases. *Biochemestry (Moscow) , 73* (10), 1396-1412.
20. Mosier, N., Wyman, C., Dale, B., Elander, R., Lee, Y., Holtappple, M., et al. (2005). Features o promising technologies for pretreatment of lignocellulosic biomass. *Bioresource technology* , 673.
21. Pham, V., Holtazpple, M., & El-Halwagi, M. (2010). Techno-economic analysis of biomass to fuel ocnversion. *Journal of Microbiology and Biotechnology* .
22. Phillips, T. (2006 йил June). Structural Formula of Ethanoic Acid reacting with Propanol to form Propyl-Ethanoate.
23. Robert J. Wooley, V. P. (1996 йил Abril). Development of an ASPEN PLUS Physical Property Database for Biofuels Components. Golden, Colorado, Estados Unidos.
24. Rodriguez Cuoto, S., & Toca-Herrera, J. L. (2006). Lacasses in the textile industry. *Biotechnology and Molecular Biology review* , 115-120.
25. Scarlata, C., Slutier, J., Termpleton, a. D., Crocker, D. A., Hames, B., & Ruiz, R. (2008). *Determination of Structural Carbohydrates and Lignin in Biomass LAP.* Cole Boulevard, Golden, Colorado: National Renewable Energy Laboratory.
26. SE HOON, K. (2004). *LIME PRETREATMENT AND ENZYMATIC HYDROLYSIS.* Texas, USA: Texas A&M University Dissertation.
27. Sierra, R., & Holtzapple, M. T. (2009). Selectivity and Delignification Kinetics for Oxidative and Non-Oxidative Lime Pretreatment of Poplar Wood - Part III: Long-therm. *(Under Review)* .
28. Sierra, R., & Holtzapple, M. T. (2009). Selectivity and Delignification Kinetics for Oxidative Short-therm Lime Pretreatment of Poplar Wood - Part I: Constanr Pressure. *(Under Review)* .
29. Sierra, R., Smith, A., Granda, C., & Holtzapple, M. (2008). Producing fuels and chemicals from lignocellulosic biomass. *Chemical Engineering Progress* , 510-518.
30. Sluiter, A., Hames, B., Ruiz, R., Scarlata, C., Sluiter, J., & Templeton, D. (2008). *Determination of Ash in Biomass (LAP).* Cole Boulevard, Golden, Colorado: National Renewable Energy Laboratory.
31. Sluiter, A., Hames, B., Ruiz, R., Scarlata, C., Sluiter, J., & Templeton, D. (2004). *Determination of Total Solids in Biomass LAP.* Cole Boulevard, Golden, Colorado: National Renewable Energy Laboratory.
32. Smufit Kappa. (2011). *Preguntas frecuentes.* Retrieved 2011 6-Mayo from http://www.smurfitkappa.com.co/DropdownMenu/Contact/FAQ/Recycling/
33. Sylvia, D., Fuhrman, J., Hartel, P., & Zuberer, D. (1998). Decomposition of Naturraly Occuring Organic Materials. In *Principles and applications of soil microbiology.* (pp. 228-234). USA: Prentice Hall.
34. Tchobanoglous, G., Theissen, H., & Eliassen, R. (1982). Producción de desechos sólidos. In *DESECHOS SÓLIDOS PRINCIPIOS DE INGENIERÍA Y ADMINISTRACIÓN* (pp. 67-82). Ambiente y los Recursos Naturales Renovables.
35. Terrabon, L. (2008). *Terrabon Renewable Energy Solutions.* Retrieved 2009 30-Octubre from MixAlco Process: http://www.terrabon.com/mixalco_overview.html
36. Tisma, M., Znidarsic-Plazl, P., Plazl, I., Vasic-Racki, D., & Zelic, B. (2009). *Oxidation of Coniferyl Alcohol Catalized by catalized from Trametes versicolor.* Croacia, Eslovenia.
37. Tuomela, M., Vikman, M., Hatakka, A., & Itavaara, M. (2000). Biodegradation of lignin in a compst enviroment: a review. *Bioresource Technology* , 169-183.
38. U.S. State Department. (2010 йил 15-March). *Raw materials of biofuels.* Retrieved 2011 йил 11-March from http://www.america.gov/st/develop-spanish/2010/March/20100409143620CMretroP3.654116e-02.html
39. UNL. (2004). *Capítulo 2. Aspectos Generales.* From bibliotecavirtual.unl.edu.ar:8180/tesis/.../1/.../07-Capítulo%202.pdf
40. Van Soest, P. J. (1973). Lignin. In *Nutritional ecology of the ruminant* (p. 193). Cornell, USA: Cornell University Press.
41. Yamauchi, K., & Fukushima, K. (2004). The regulation from guaicyl to syringyl lingnin in the differenciating xylem of Robinia pseudoacacia. *Comptes Rendus Biologies* , 791-797.

42. Zhang, J., Kjonas, R., & Flurkey, W. (1999). Does N-hydroxyglycine inhibit plant and fungal laccases? *Phytochemestry* , 775-783.

(523b) Isoconversional Methods Applied to Kinetic Modeling Oxidative Alkaline Pretreatment of Sugarcane Bagasse

Rocio Sierra, Cuellar Gerardo, Simbaqueba Andres, Auza Nicolas and Copete Diego, Chemical Engineering, Universidad de los Andes, Bogota, Colombia

Delignification (decrease of lignin content) has shown to be a valid method to significantly increase lignocellulosic feedstocks digestibility. This operation (widely known as pretreatment) makes these abundantly available and inexpensive feedstocks usable in fermentation processes whose main goal is to produce biofuels and other chemicals. Because pretreatment is one of the most expensive processes in biofuels production plants, it has been extensively studied; however, a throughout literature search shows that development of kinetic models to better understand, optimize and control pretreatment, are very scarce if available at all.

This study uses sugarcane bagasse which is a residue from the sugar industry abundantly produced particularly near and in the Equator zone. This lignocellulosic resource is traditionally used in combustion processes to supply energy to the production plant; however life cycle analysis have shown that if lignocellulose is used as substrate in fermentation processes, the energy yield would be much closer to the theoretically available energy.

In this work, sugarcane bagasse is submitted to an alkaline pretreatment that uses an oxidative agent, systematically applied at widely varied ranges of temperature and time using a full factorial experimental design

Model fitting methods do not achieve a clean separation between the temperature dependent Arrhenius constant and the reaction model. On the contrary, the isoconversional methods evaluate the activation energy values at progressive degree of conversion without modelistic assumptions. This work applies this methodology to calculate activation energy for alkaline pretreatment of sugarcane bagasse and compares with previous work where the traditional curve fitting is employed.

See more of this Session: Developments In the Pretreatment of Lignocellulosics for Bioconversion II
See more of this Group/Topical: Sustainable Engineering Forum

(523c) Alkaline Deacylation of Corn Stover to Reduce Toxicity of Prehydrolyzates and Improve Enzymatic Hydrolysis

Urvi D. Kothari and Y. Y. Lee, Chemical Engineering, Auburn University, Auburn, AL

Hydrolysates from dilute acid pretreatment of biomass contain various biomass degradation products which are inhibitory to enzymes and toxic to microorganisms. One of the main inhibitory compounds is acetic acid which is released from the lignin-hemicellulose matrix in the biomass. Acetic acid at the level of 1 g/L inhibits cellulase activity by 10%. In most pretreatment liquors its concentration is in the range of 3-10 g/L and inhibition of cellulase activity by acetic acid at these concentrations is 20-60%. In this study, sodium carbonate was investigated as a deacylation reagent that can selectively remove part of the acetyl components in corn stover and reduce the acetyl concentration in the pretreatment hydrolysate. Sodium carbonate was chosen since it is a mild alkali that can preserve carbohydrates under the deacylation process conditions. It is inexpensive and can easily be recovered through established Kraft recovery process. Sodium carbonate of 5-10 wt% was used over 30-60°C for deacylation. A dilute-acid pretreatment was then applied under the conditions of 0.5-1% H_2SO_4 at 60-110°C in order to assess the effects of deacylation. The two-step process operated under optimized conditions led to reduction of the final acetyl concentration by 50-80%, and minimal loss of hemicellulose at 2-12%. The combined process further improved the glucan and xylan hydrolysis yields of the pretreated corn stover over those of the dilute-acid pretreatment alone.

See more of this Session: Developments In the Pretreatment of Lignocellulosics for Bioconversion II
See more of this Group/Topical: Sustainable Engineering Forum

(523d) Deconstruction of Wheat Straw Lignin by Streptomyces Viridosporus As Insight Into Biological Degradation Mechanism

Jijiao Zeng[1], Department of Biological Systems Engineering, Washington State University, Pullman, WA, Deepak Singh, Edeniq Inc, Visalia, CA, Dhrubojyoti D. Laskar, Washington State University, Pullman, WA and Shulin Chen, Biological Systems Engineering, Washington State University, Pullman, WA

Deconstruction of wheat straw lignin by *Streptomyces viridosporus* asinsight into biological degradation mechanism

JijiaoZeng[1], Deepak Singh[2],Dhrubojyoti Dey Laskar[1]and Shulin Chen[1*]

1. Department of Biological Systems Engineering, Washington State University,Pullman, Washington 99164-6120, USA

2. Edeniq Inc, 2505 N Shirk Road, Visalia, 93291 CA

*Corresponding author. Tel: +1 509 335 3743; fax: +1 509 335 2722.

E-mail address: chens@wsu.edu

Abstract:

Lignins are secondary natural organic compositions of plant cell wall besides cellulose and hemicelluse. Structurally lignins are linked with hemicelluloses and cellulose to form recalcitrant complex that resist the environmental stress. Biological systems created by nature over millions of years of evolution can be capitalized for targeted deconstruction of plant cell walls through selective cleavage and structural modification of lignin. Such modification causes changes in cell wall structure for facilitating enzymatic attack to cellulose for sugar release. The study of biological degradation mechanism on lignocellulosic materials could provide the new knowledge to convert plant biomass to biofuel and high value co-products. In this study, we have investigated the effect of wheat straw lignin structural changes by *Streptomyces viridosporus* on downstream enzymatic hydrolysis. ^{13}C CPMAS NMR revealed that the rising of carbonhydrate/lignin ratio accompany with enhanced intensity of methoxyl group in bio-treated wheat straw. The degraded lignin were isolated and analyzed by Fourier Transmission Infrared (FTIR), Pyrolysis Gas Chromatography/Mass Spectrometry (Py-GC/MS) and two dimenssional heteronuclear multiple quantum coherence nuclear magnatic resonance (2D HMQC NMR) to elucidate their structural change. The results indicate that the biological pretreated lignin were associated with critical change and/or modification on their structure implying selective removal of guaiacyl (G) lignin unit which is more condense than syringyl (S) unit. Moreover, the treated lignin also dramatically reduced the inhibition of cellulose hydrolysis. This information provides new insights for understanding the biological degradation mechanism on lignin for cellulose utilization and exploring a possible route to convert lignin derived co-products.

See more of this Session: Developments In the Pretreatment of Lignocellulosics for Bioconversion II
See more of this Group/Topical: Sustainable Engineering Forum

Catchlight Energy LLC (CLE)

A 50/50 Chevron and Weyerhaeuser joint venture.

Our Mission: To commercialize the large-scale production of liquid transportation fuels from sustainable forest-based resources

Weyerhaeuser | Catchlight Energy | Chevron

Feedstocks at scale → Conversion Technology → High quality fuels to customers

A Unique End-to-End Value Chain Solution
Leveraging the strengths of two natural resource leaders

Catchlight Energy (CLE) brings strengths not found elsewhere - feedstocks development and procurement, fuels testing and refining, operations, conversion engineering.

CLE drivers aligned with national interests to develop renewable energy platform, increase energy independence & security and promote rural development.

October 15, 2011

Softwood – High Availability/High Recalcitrance

Catchlight Energy

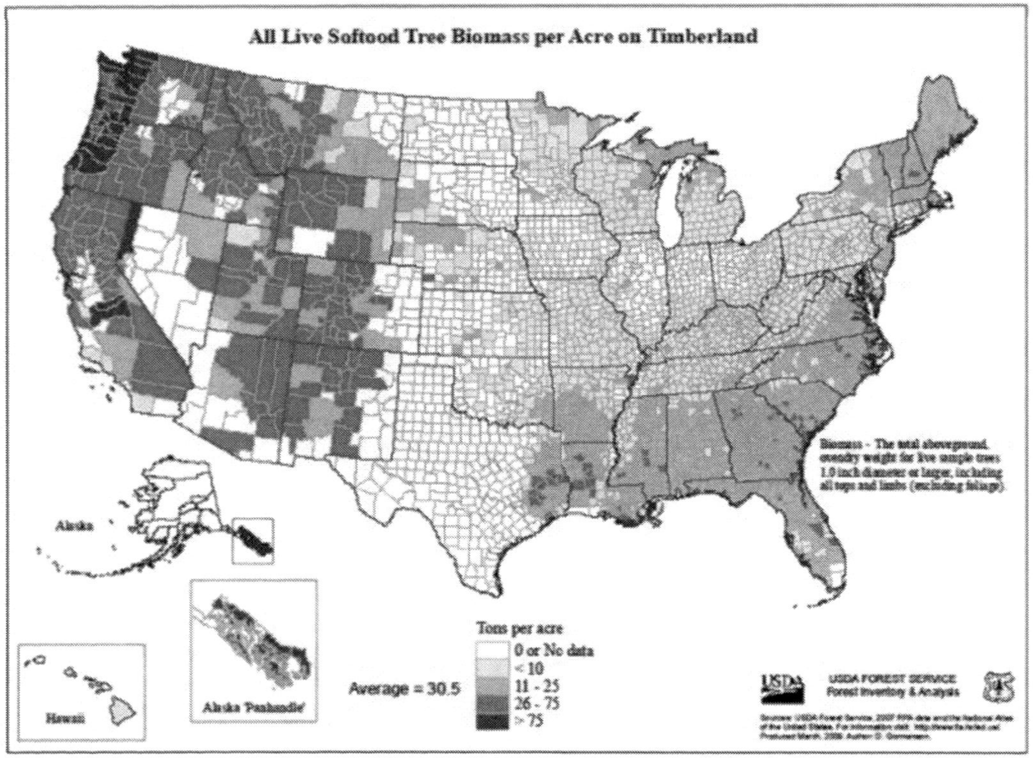

The use of softwood is of high critical value to the biofuels industry, but it remains a difficult feedstock for most pure bio-conversion processes.

October 15, 2011

Pretreatment Issues for Softwood and other Recalcitrant Biomass

Catchlight Energy

- **Dilute acid and base systems ineffective at reasonable levels.**
- **Known and proposed "Fractionation" pretreatments are excessively expensive, and not commercial scale.**
- **Pulping processes effective because of removal of lignin, but result in high and unsupportable costs.**
  - High capex (unless repurposed mill that is obtained at very low cost).
  - High opex (chemical costs or chemical recovery costs, wastewater treatment).
  - Fuels value can't compete with pulp value.

October 15, 2011

CLE Pretreatment

Catchlight Energy

- **Pulping, especially sulfite pulping, is an excellent but expensive pretreatment**
- CLE Pretreatment has
  - Higher throughput
  - Lower chemical consumption
  - Lower inhibitors
  - Excellent enzymatic hydrolysis results
  - Higher sugar yield

October 15, 2011

CLE Pretreatment Characteristics

Catchlight Energy

- Adaptable to existing infrastructure
- Scalable from pilot operations
- Feedstock flexible

October 15, 2011

CLE Pretreatment and Process to Sugar

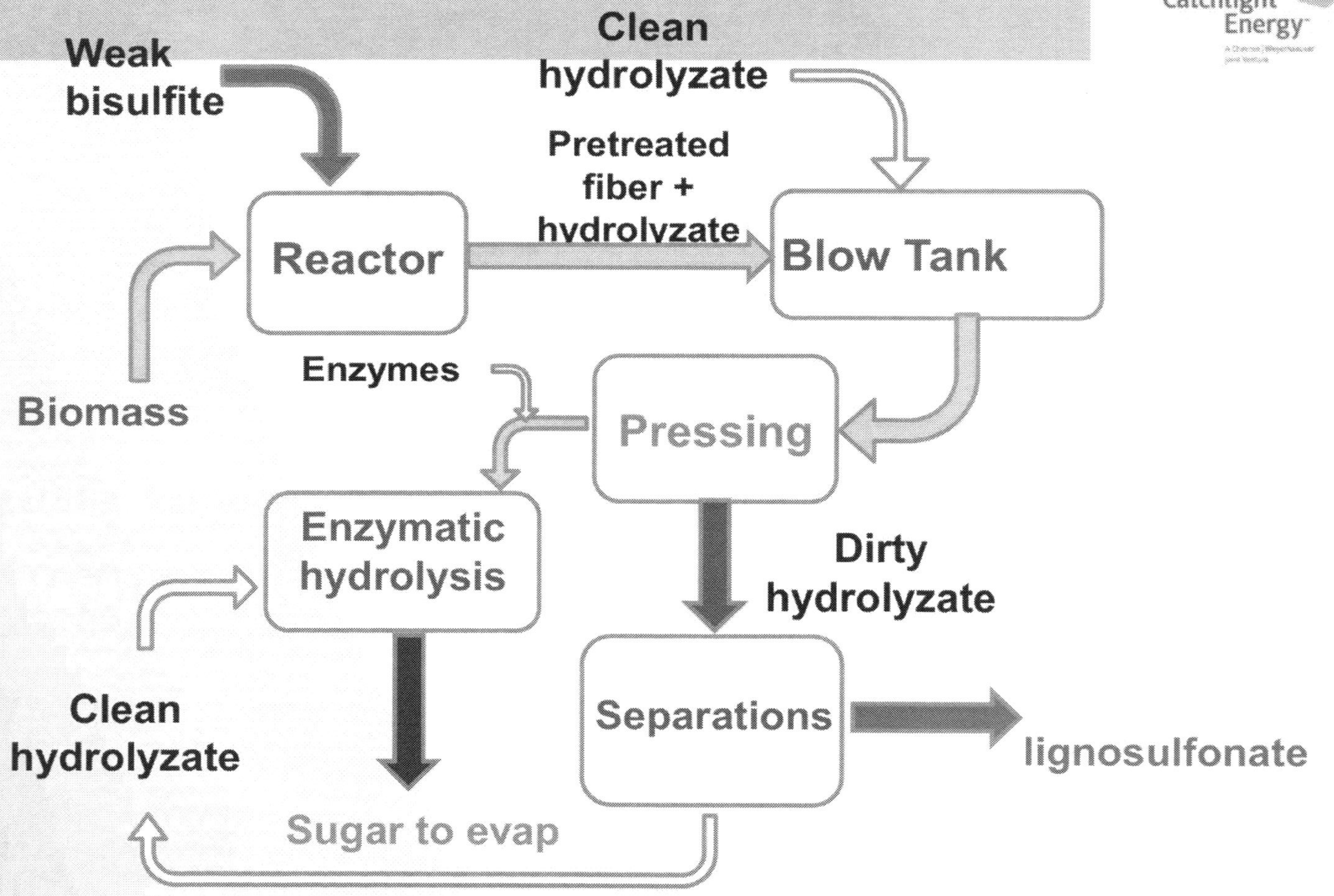

October 15, 2011

Softwood Sugar

Pretreated Softwood–Mill Residuals

Catchlight Energy

Softwood Residuals

Softwood Residual Sugar Composition:

Arabinan	1.01%
Galactan	2.09%
Glucan	39.5%
Xylan	6.05%
Mannan	10.5%
Total	59.2%

→ **Calcium Bisulfite Pretreatment** →

Pretreated Refined Pulp

Pretreated Residual Sugar Composition:

Arabinan	0.21%
Galactan	0.55%
Glucan	49.9%
Xylan	3.42%
Mannan	3.63%
Total	57.7%

Pretreatment Liquor

Liquor Sugar Composition:

Arabinose	0.07%
Galactose	0.17%
Glucose	0.36%
Xylose	0.48%
Mannose	1.10%
Total	2.18%

October 15, 2011

Softwood Mill Residual Chips

Pilot Pretreatment Results

Catchlight Energy

Pilot Pretreatment Size: 3 kg OD chips/1 cu ft digester	
Overall Pretreatment Sugar Yield on Wood Sugar	88-95%
Furan Formation on Wood Sugar	<1.5%
Acetic Acid on Wood	~1.0%
Initial Cook Liquor pH	2.0-4.0
Pulp Yield on Wood	65%-80%

CLE pretreatment produces high yield of pulp and ultimately sugar.

October 15, 2011

Enzymatic Hydrolysis

Catchlight Energy

- Near 80% conversion of glucan up to 17% solids in the hydrolysis step using low dose of enzymes.
- Overall sugar yields of 87% attained on clean softwood with economic enzyme loading.
- Overall sugar yields of 80% attained on softwood forest residuals.
- Switchgrass and hardwood behave similarly although have higher C5 sugar in the product.

October 15, 2011

CLE Pretreated Hydrolysis

Glucose Yield vs. CTec2 Enzyme Dosage

Catchlight Energy

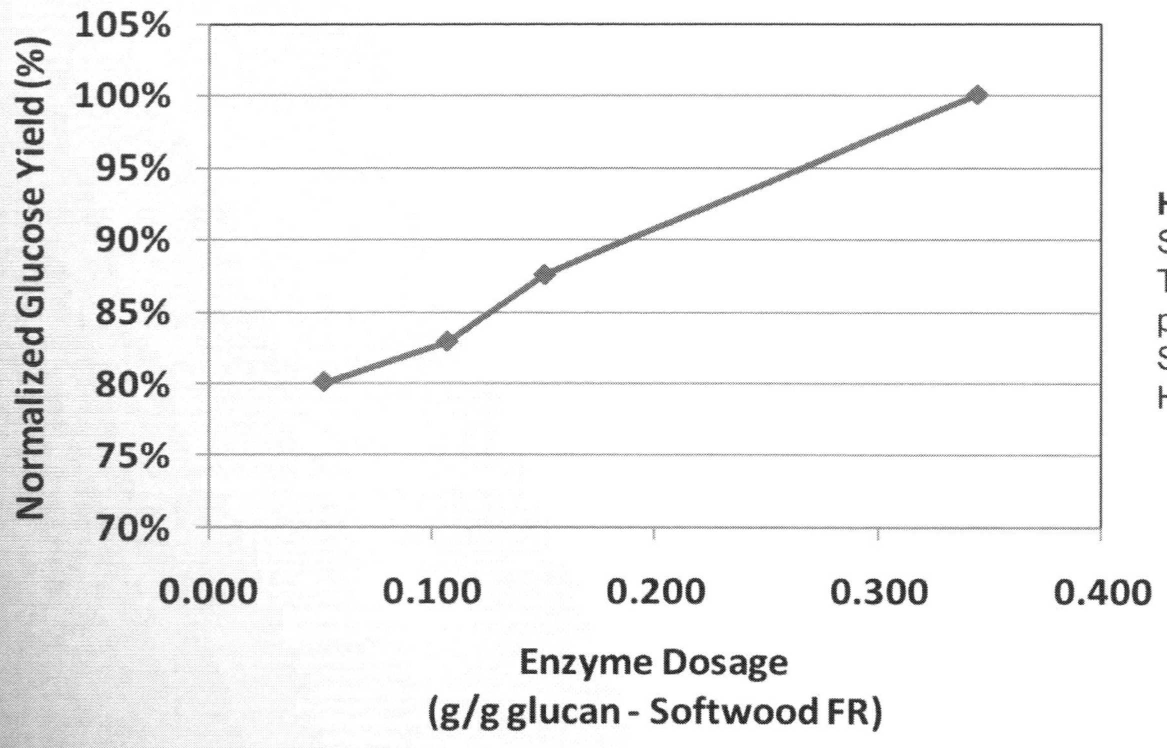

Hydrolysis Conditions:
Solid: 17%, unwashed
Temperature: 50°C
pH: 4.8-5.3
Shaking Speed: 200 rpm
Hydrolysis Time: 96 hrs

October 15, 2011

CLE Pretreatment Works on Multiple Feedstocks

Hydrolysis results at 4 days

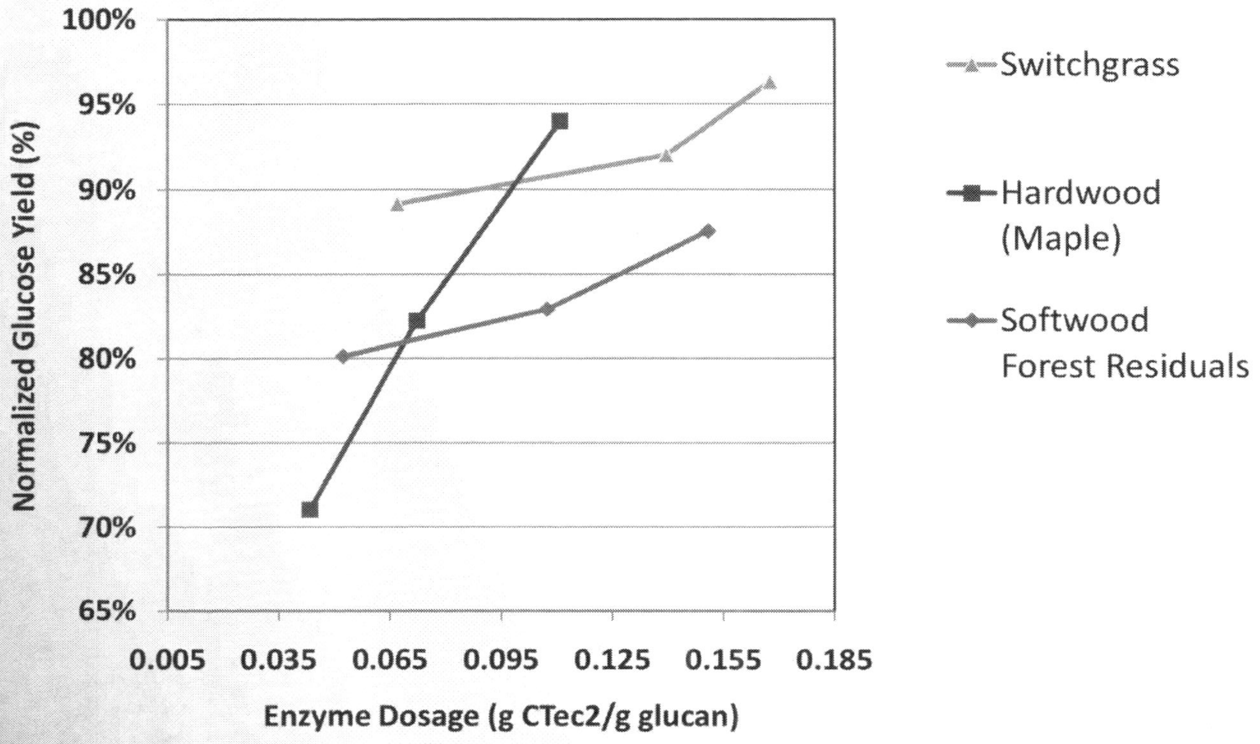

October 15, 2011

Commercial Data

CTec2 Dosage on Pretreated Corn Stover (Dilute Acid)

Catchlight Energy

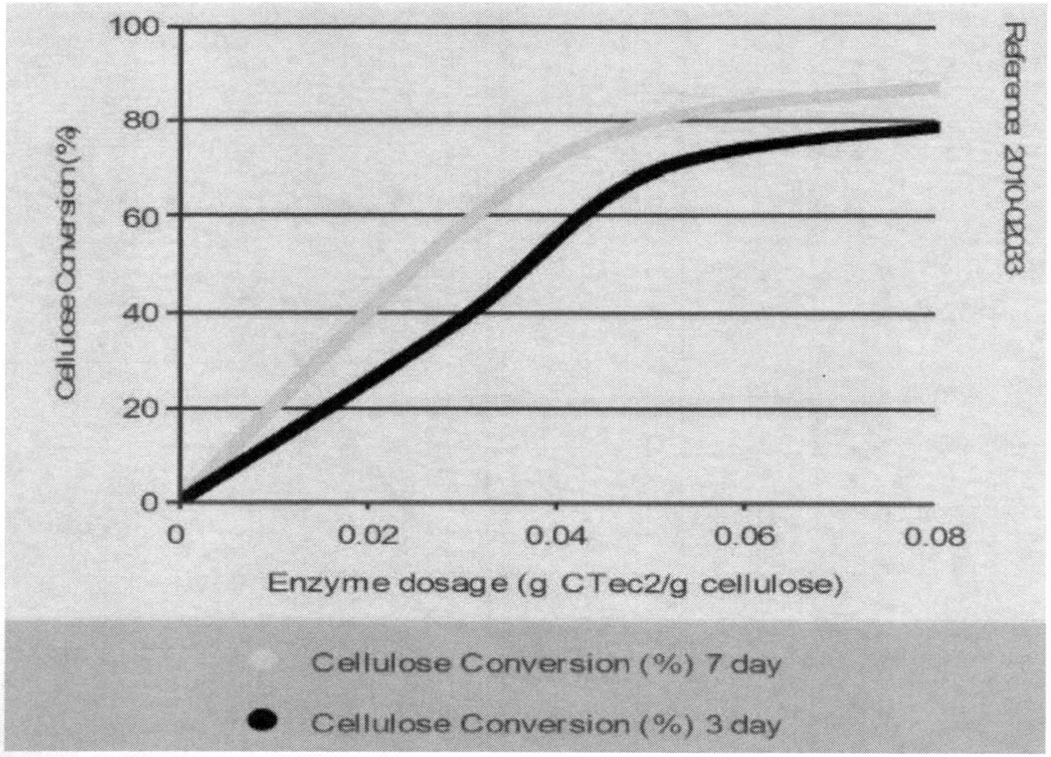

Reference: Corn stover data from CTec2 and HTec2 Application Sheet , www.novozymes.com
Used with permission

October 15, 2011

CLE Pretreatment Makes Softwood Hydrolyze Like Acid-Pretreated Corn Stover

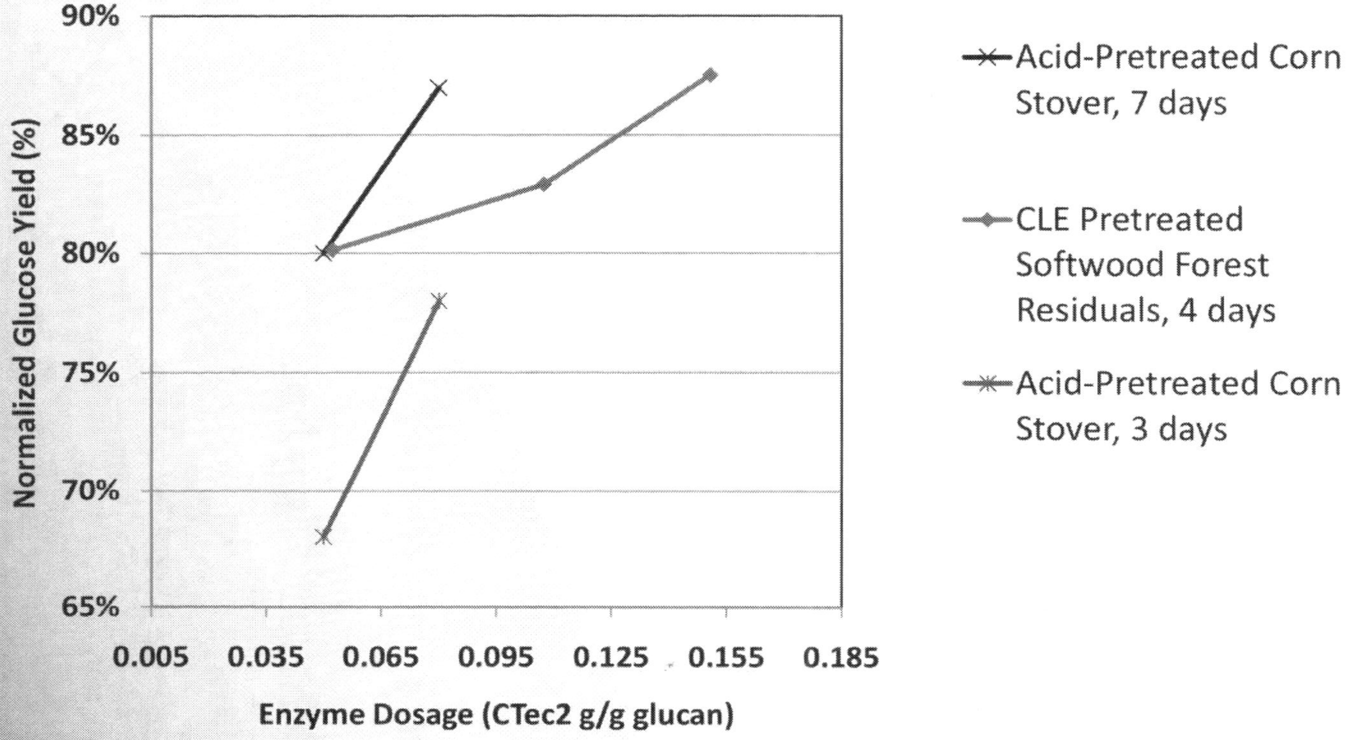

October 15, 2011

CLE Sugar: Relative Cost Components

Catchlight Energy

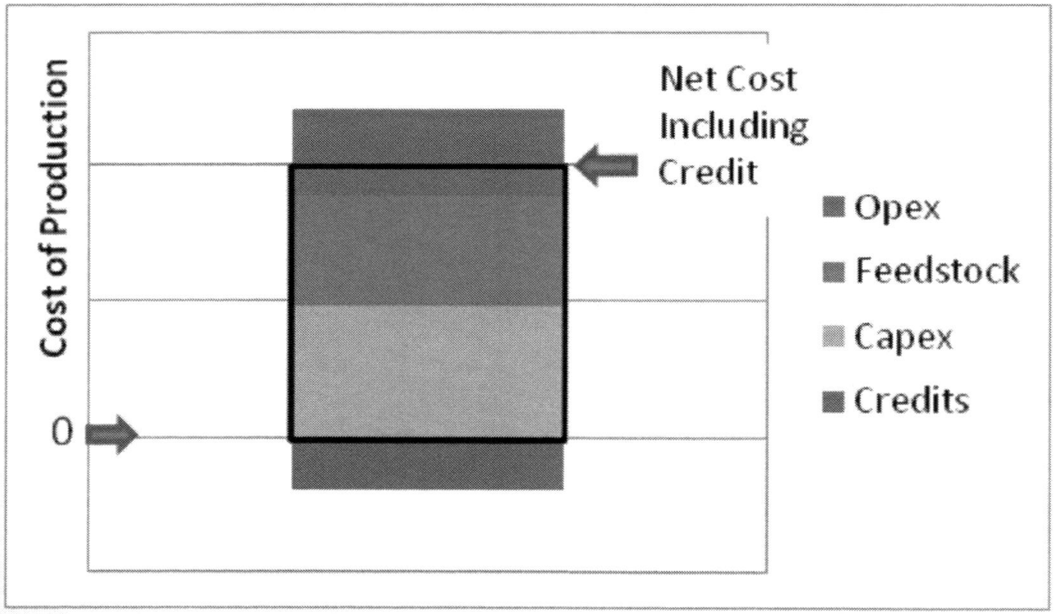

October 15, 2011

Conclusions

Catchlight Energy

- **CLE's pretreatment technology is advantageous for difficult to hydrolyze feedstocks, such as softwood residuals which are available today at commercial scale.**
- **The technology is proven, scalable, and cost effective.**
  - An engineering design has been developed for a large scale greenfield biofuels with downstream processing and separations for sugar, lignosulfonate, and energy production.
- **CLE is actively working with select partners and welcomes interest in the production of sugar, ethanol, and drop-in biofuels.**

October 15, 2011

(523f) Economics of Pretreatment for Biological Processing

Ling Tao, Andy Aden and Richard T. Elander, National Bioenergy Center, National Renewable Energy Laboratory, Golden, CO

Economics of Pretreatment for Biological Processing

Ling Tao[1], Andy Aden[1] and Richard T. Elander[1]

[1] NationalRenewable Energy Laboratory, 1617 Cole Blvd., Golden, CO 80401, USA

Abstract:

A broad understanding of pretreatment economics within anoverall conversion context will be presented. The importance of pretreatmentin the overall economics of the biorefinery is addressed along with the keyparameters that influence the pretreatment economics. A large array ofpretreatment technology options exist and are currently being researched anddeveloped in this study. There appear to be many performances and costtradeoffs for these varying technologies. Only through continued R&D willthese tradeoffs be more fully understood. Process economics, especially thepromising pretreatment technologies studied in the Consortium for AppliedFundamentals and Innovation (CAFI) projects, are analyzed and compared indetails in this work. The six pretreatment technologies vary greatly in termsof their process design and projected total capital investment. Pretreatmentarea capital costs are based on past research, existing databases andengineering judgments, and include not only pretreatment reactor costs, butalso pretreatment chemical recovery and recycle equipment costs for theprocesses that use high pretreatment chemical loadings. More accurate costestimates for each pretreatment process will require specific vendor quotationsbased on pretreatment process conditions and any pretreatment chemical recoveryrequirements. Overall ethanol yield, which is largely based on achieved overallsugar yield achieved in pretreatment and subsequent enzymatic hydrolysis steps,is one of the single-most important factors in determining projected ethanolcosts for each process. Many key variables and parameters can have significantimpact on the pretreatment economics. While many are discussed within this talk, continuing R&D will generate data that further guide the development of thebiomass conversion industry.

See more of this Session: Developments In the Pretreatment of Lignocellulosics for Bioconversion II
See more of this Group/Topical: Sustainable Engineering Forum

(542a) Formation and Control of Thermoelectric Composites At the Nanoscale

G. Jeffrey Snyder, Caltech, Pasadena, CA

Many new high efficiency thermoelectric materials are multi-phase composites at the nanometer scale. These include materials such as PbTe alloys, and recently Zn4Sb3, which contain small ~10nm nanoparticles that can be difficult to control. Typically the goal is to reduce the lattice thermal conductivity of the matrix by scattering phonons at the interfaces of the particles. Thus it is not the small size of the particles but the spacing between particles that matter. We have found that in PbTe with large (~50nm – 200nm) nanoparticles of Ag_2Te high $zT \sim 1.5$ can be obtained through the independent control of thermal conductivity reducing Ag_2Te large nanoparticles and electronic doping with La.

In general, because electrons are also scattered, a net improvement in thermoelectric efficiency will depend strongly on the spacing, size, shape, orientation and interfacial properties of the particles. At Caltech we have extensively studied the formation and control of composite microstructure in PbTe-Sb_2Te_3, identifying three distinct mechanisms for forming composites and controlling the characteristic length scales down to the nanometer scale. The coherent interfaces of these composites may help retain high carrier mobility.

[1] G. J. Snyder, E. S. Toberer. *Nature Materials* **7**, p 105 - 114 (2008).

[2] D.L. Medlin and G.J. Snyder "Interfaces in Bulk Thermoelectric Materials" *Current Opinion in Colloid & Interface Science* (2009)

[3] Yanzhong Pei, Jessica Lensch-Falk, Eric S. Toberer, Douglas L. Medlin, and G. Jeffrey Snyder "High Thermoelectric Performance in PbTe due to Large Nanoscale Ag_2Te Precipitates and La Doping" *Advanced Functional Materials* **21**, 241 (2011)

See more of this Session: Nanomaterials for Thermal-to-Electric Conversion II: Thermoelectrics, Thermophotovoltaics and Thermopower
See more of this Group/Topical: Topical 5: Nanomaterials for Energy Applications

(542b) New Concepts In Energy Transport within Carbon Nanotubes: Thermopower Waves

Michael Strano, Chemical Engineering, Massachusetts Institute of Technology, Cambridge, MA

Our laboratory has been interested in how carbon nanotubes (CNTs) can be utilized to illustrate new concepts in molecular and energy transfer. We predict and demonstrate the concept of thermopower waves for energy generation. Coupling an exothermic chemical reaction with a thermally conductive CNT creates a self-propagating reaction wave driven along its length. We realize such waves in multi-walled CNTs and show that they produce concomitant electrical pulses of high specific power-- greater than 7 kW/kg. Such waves of high power density may find uses as unique energy sources.

Choi W, Hong S, Abrahamson JT, Han JH, Song C, Nair N, Baik S, Strano MS: Chemically driven carbon-nanotube-guided thermopower waves. NATURE MATERIALS, 9 (2010) 423-429.

See more of this Session: Nanomaterials for Thermal-to-Electric Conversion II: Thermoelectrics, Thermophotovoltaics and Thermopower
See more of this Group/Topical: Topical 5: Nanomaterials for Energy Applications

(542c) Thermal Plasmonics As a Route to Photovoltaics?

David J. Norris, Chemical Engineering and Materials Science, University of Minnesota, Minneapolis, MN

Patterned metallic films allow the generation and manipulation of special electromagnetic waves known as surface plasmons that propagate along a metal interface. Because these waves allow the concentration of light in nanometer-scale volumes, they have implications for fundamental phenomena as well as applications such as imaging, sensing, and solar cells. Thus, the field of plasmonics has arisen to study and utilize surface plasmons. While light is typically used to create these waves, in some cases it would be more convenient to have a non-optical source for surface plasmons. One route is thermal excitation (*i. e.*, heat). Here, we will explore hot plasmonic structures for obtaining new optical behavior. For example, we will examine metallic films patterned with a series of circular concentric grooves (a bull's eye pattern). We show that, when heated, these films can emit light that is amazingly narrow, both in terms of its spectrum and its angular divergence. Thus, a simple metallic foil can generate a highly directional beam of monochromatic light by a thermal process. This effect has implications for creating efficient thermophotovoltaic devices, which convert heat into electricity. During these studies, we also developed a simple, high-throughput method for obtaining ultrasmooth patterned metal films. Previously, roughness in such films has hindered the field of plasmonics. Therefore, our approach solves a critical problem and allows many high-quality plasmonic structures to be available for the first time.

See more of this Session: Nanomaterials for Thermal-to-Electric Conversion II: Thermoelectrics, Thermophotovoltaics and Thermopower
See more of this Group/Topical: Topical 5: Nanomaterials for Energy Applications

(542d) New Absorber and Emitter Designs and High Temperature Materials for Thermophotovoltaics

Paul V. Braun, Department of Materials Science and Engineering, University of Illinois at Urbana-Champaign, Urbana, IL

Thermophotovoltaic (TPV) solar cells, where broad-band solar radiation is absorbed by an intermediate, which then emits narrow-band thermal radiation towards a solar cell have been proposed for some time as a route for high efficiency solar energy harvesting. However, due to limitations in both design of the active components, as well as the high temperature stability of the materials making up these systems, the theoretical maximum efficiency of 85% for such devices has not been realized. We are currently exploiting emerging opportunities in the design and fabrication of metallic photonic crystals for TPV applications. We have now demonstrated tungsten based 3D photonic crystals which retain micron-scale structure even at temperatures of 1300 K, and are exploring other materials systems that may have even higher temperature stability. Through simulation and modeling, we have designed a number of absorber/emitter structures that match both the broad-band solar spectrum and the desired narrow band emission required for efficient single-junction solar cells. The self and directed assembly methods used to fabricate the TPV structures can produce complex large-area structures at low cost.

See more of this Session: Nanomaterials for Thermal-to-Electric Conversion II: Thermoelectrics, Thermophotovoltaics and Thermopower
See more of this Group/Topical: Topical 5: Nanomaterials for Energy Applications

(542e) Hybrid Nanowire/Polymer Composite for Thermoelectric Energy Harvest

Yue Wu, School of Chemical Engineering, Purdue University, West Lafayette, IN

Successfully decreasing the thermal conductivity while maintaining or increasing the power factor is the key to improve the efficiency for thermoelectric energy harvesting. Up to date, most research efforts in the development of thermoelectric materials have been devoted to use the rigid inorganic materials and little progress has been achieved to use the hybrid composite. In this presentation, we will discuss our research on the hybrid nanocomposite between inorganic thermoelectric nanowires and conductive polymers and our results indicated that significantly lowered thermal conductivity using the organic polymer matrix and reasonably good electrical conductivity and Seebeck coefficient could be achieved depending on the mixing ratio between nanowires and conductive polymer, which demonstrates a great potential to form hybrid thermoelectric devices with high flexibility as portable energy source.

See more of this Session: Nanomaterials for Thermal-to-Electric Conversion II: Thermoelectrics, Thermophotovoltaics and Thermopower
See more of this Group/Topical: Topical 5: Nanomaterials for Energy Applications

(560a) Evaluation of Modified Feedstocks for Deconstrctionability

Seema Singh, Joint BioEnergy Institute, Emeryville, CA

Today, carbon-rich fossil fuels, primarily oil, coal and natural gas, provide 85 percent of the energy consumed in the United States. Fossil fuel use increases CO2 emissions, increasing the concentration of greenhouse gases and raising the risk of global warming. The high energy content of liquid hydrocarbon fuels makes them the preferred energy source for all modes of transportation. In the US alone, transportation consumes around 13.8 million barrels of oil per day and generates over 0.5 gigatons of carbon per year. This has spurred research into alternative, non-fossil energy sources. Among the options (nuclear, concentrated solar thermal, geothermal, hydroelectric, wind, solar and biomass), only biomass has the potential to provide a high-energy-content transportation fuel. Biomass is renewable resource that is carbon-neutral.

Currently, biofuels such as ethanol are produced largely from grains, but there is a large, untapped resource (estimated at more than a billion tons per year) of plant biomass that could be utilized as a renewable, domestic source of liquid fuels. Well-established processes convert the starch content of the grain into sugars that can be fermented to ethanol. Plant-derived biomass contains cellulose, which is more difficult to convert to sugars. The development of cost-effective and energy-efficient processes to transform cellulose in biomass into fuels is hampered by significant roadblocks, including the lack of specifically developed energy crops, the difficulty in separating biomass components, low activity of enzymes used to deconstruct biomass, and the inhibitory effect of fuels and processing byproducts on organisms responsible for producing fuels from biomass monomers.

The Joint BioEnergy Institute (JBEI) is one of three US Department of Energy Bioenergy Research Centers that will address these roadblocks in biofuels production. JBEI draws on the expertise and capabilities of three national laboratories (Lawrence Berkeley National Laboratory (LBNL), Sandia National Laboratories (SNL), and Lawrence Livermore National Laboratory (LLNL)), two leading US universities (University of California campuses at Berkeley (UCB) and Davis (UCD)), and a foundation (Carnegie Institute of Washington at Stanford University) to develop the scientific and technological base needed to convert the energy stored in cellulose into transportation fuels and commodity chemicals. Established scientists from the participating organizations are leading teams of researchers to solve the key scientific problems and develop the tools and infrastructure that will enable other researchers and companies to rapidly develop new biofuels and scale production to meet US transportation needs, and to develop and rapidly transition new technologies to the commercial sector. This talk will present a summary of the organization and efforts at JBEI, highlight the efforts on the evaluation of modified feedstock for deconstructionability and development of novel biomass pretreatment methods that enable the efficient conversion of biomass into next-generation biofuels.

(560b) Discovery and Characterization of Cellulolytic Enzymes From the Wood Wasp Symbiont Streptomyces Sp. ActE

Taichi E. Takasuka, Adam J Brook, Cameron R. Currie and Brian G. Fox, Great Lakes Bioenergy Center, University of Wisconsin, Madison, WI

One of the major hurdles for producing cost effective biofuels is the enzymatic conversion of lignocellulosic biomass into fermentable sugars. The US Department of Energy has issued the challenge of discovering and designing enzymes with novel biomass-deconstruction capabilities. This study focuses on the discovery of better enzymes and accessory proteins for biomass deconstruction from a novel bacterial species. *Streptomyces* sp. ActE, an isolate from wood-boring wasps of the genus *Sirex*. Its genomic sequence has been determined and initial physiological studies suggested high cellulolytic potential. We analyzed the genome-wide gene expression of ActE grown on insoluble substrates such as cellulose, xylan, chitin and others as the sole carbon source in the culture medium, and compared these expression profiles with glucose-grown control ActE cells. The secretomes obtained from growth on different substrates were analyzed by shotgun LC-MS/MS and enzymatic assays. From the combination of the genomic, transcriptomic, and secretomic analyses, we discovered ~200 genes that have carbon source-specific regulation of gene expression or protein secretion. This work provides the first extensive systems biology characterization of the capabilities of the genus *Streptomyces* for wide variety of polysaccharide utilization.

This work was funded by the DOE Great Lakes Bioenergy Research Center (DOE BER Office of Science DE-FC02-07ER64494).

See more of this Session: Advances In Biofuels: DOE Bioenergy Research Centers II
See more of this Group/Topical: Sustainable Engineering Forum

(560c) Dissecting the Cellulosome Enzyme Complex: Assembly and Function

Yannick J. Bomble[1], Mark Nimlos[2], Michael E. Himmel[1] and Michael F. Crowley[1], (1)Biosciences Center, National Renewable Energy Laboratory, Golden, CO, (2)National Bioenergy Center, National Renewable Energy Laboratory, Golden, CO

Most bacteria and fungi use free enzymes to degrade plant cell walls. However, a small number of bacteria have developed a fundamentally different approach, where enzymes are tethered to a large protein scaffold forming a multi-domain complex known as, the cellulosome. The study of these large protein assemblies is an ongoing research topic that has already yielded numerous breakthroughs. However, the mechanism of assembly of the enzyme subunits onto the natural scaffoldin; as well as the modes of action of the cellulosome and its enzymatic components, are currently not well understood.

In this study, we focused on the cellulosome-integrating protein (CipA) of C. thermocellum; as well as cellulosomal enzymes from families 5, 9 and 48. These three enzymes are representative of the diversity of enzymes secreted by C. thermocellum. This work not only aims at understanding the mechanisms involved in the sequential binding of the cellulosomal enzymes to the CipA scaffold of C. thermocellum, but also the binding of CipA to secondary scaffolds, such as OlpB. We focused on the physical properties of the binding of component enzymes to the scaffoldin. The modularity of the enzymes was found to be one of the main influences on the cellulosome assembly process.

Additionally, we examined several modules from the multi-modular enzyme complex, CbhA. Several of these modules have undefined functions, such as the X1 modules found closely linked to the catalytic domain. We hypothesize that the X modules could serve as tethers between several modules in CbhA to fine-tune the positioning of this enzyme during cellulose deconstruction. We conducted steered molecular dynamics simulations (directly comparable to AFM pulling experiments) to evaluate the energy and the profile for unfolding these modules. We found that the energy required to partially unfold this domain is accessible by the cellulosome and the unfolding pathways are similar in the multiple simulations conducted.

See more of this Session: Advances In Biofuels: DOE Bioenergy Research Centers II
See more of this Group/Topical: Sustainable Engineering Forum

(560d) Construction of Efficient Gram-Negative Biocatalysts for Lignocellulose Conversion

David Keating, Michael Schwalbach, Jeffrey Gardner, Mary Tremaine, Patricia Kiley and Robert Landick, Great Lakes Bioenergy Center, University of Wisconsin, Madison, WI

Research within the Great Lakes Bioenergy Research Center aims to better understand the bottlenecks associated with bioconversion of lignocellulose to liquid transportation fuels, knowledge which will then be used to improve the efficiency of this process. Much of these studies are currently focused on the Gram-negative bacterium *Escherichia coli,* due to its sophisticated genetics, well-understood physiology, and widespread industrial use.

To better understand the limitations associated with microbial biocatalysis of lignocellulosic biofuels, we have engineered *E. coli* strains capable of high-level ethanol production and tolerance. To develop efficient *E. coli* ethanologens, we have constructed mutations that reduce alternative pathways of election flow, and introduced the pyruvate decarboxylase and alcohol dehydrogenase genes from *Zymomonas mobilis*. In addition, we have carried out extensive directed evolution of *E. coli*, which has resulted in strains capable of growth in the presence of 10% ethanol. These strains are currently being subjected to a systems-level microbial analysis during lignocellulosic bioconversion. In collaboration with the Dale laboratory, we have generated enzymatic hydrolysates of AFEX-pretreated corn stover at high solid loadings. Furthermore, we have developed methodology to measure extracellular amino acids, metabolic endproducts, intracellular metabolites, and global transcription. We have used these techniques to characterize the physiological state of our ethanologenic strains during fermentation in lignocellulosic hydrolysates. Our initial analysis has already generated several surprising findings. First, lignotoxins within the hydrolysate appear to strongly limit the production of biomass, but have a more modest effect the conversion of glucose to ethanol. However, these inhibitors appear to strongly limit the conversion of xylose to ethanol. Second, microbial growth in lignocellulosic hydrolysates is a dynamic process, in which global transcription and overall physiology change dramatically during the course of fermentation. Our systems-level analysis of these strains has identified transcriptional and metabolic changes that provide insight into the mechanism of action of the lignotoxins. We are using this information, as well as data from similar experiments with lignotoxin-tolerant mutants, to generate new ethanologens with improved productivities under these conditions. Furthermore, we expect these studies will produce an improved molecular understanding of the current limitations of ethanologenesis in these strains, which we expect will be applicable to diverse ethanologenic microorganisms.

See more of this Session: Advances In Biofuels: DOE Bioenergy Research Centers II
See more of this Group/Topical: Sustainable Engineering Forum

(560e) One-Step Production of Lactate and Isobutanol From Pretreated Biomass by Recombinant Cellulolytic Bacillus Subtilis

Xiaozhou Zhang, Biological Systems Engineering, Virginia Polytechnic Institute and State University, Blacksburg, VA, Chun You, Biological Systems Eng, Virginia Polytechnic Institute and State University, Blacksburg, VA and **Y.-H. Percival Zhang**, Biological Systems Engineering Department, Virginia Polytechnic Institute and State University, Blacksburg, VA

Consolidated bioprocessing (CBP) is low-cost cellulosic biomass bioconversion by integrating cellulase production, cellulose hydrolysis, and sugar fermentation into a single step. Although significant efforts had been made to create recombinant cellulolytic microorganisms, real recombinant cellulolytic microorganism that can produce sufficient secretory active cellulases, hydrolyze cellulose, and utilize soluble sugars for supporting cell growth and cellulase synthesis without any other organic nutrient is not available yet. We demonstrated that over-expression of *Bacillus subtilis* endoglucanase BsCel5 by more than 30-fold enables non-cellulose-utilizing *Bacillus subtilis* to grow on amorphous cellulose and pretreated lignocellulosic biomass as the sole carbon source without addition of costly organic nutrients. After directed evolution and screening on amorphous cellulose, both the expression/secretion level and specific activity of Cel5 and Cel9 had been increased successfully. An advanced recombinant cellulolytic *Bacillus subtilis* strain has been constructed with engineered cellulase and the genes knock out/knock in. Besides these, we also achieved the co-expression and secretion of family 5, 9 and 48 cellulases in a single *B. subtilis* strain. The one-step production of lactate and isobutanol from cellulose and pretreated biomass in the minimal medium without adding any organic nutrient had been demonstrated. We are reconstituting *in vivo* mini-cellulosome on the cell surface of *B. subtilis*. The recombinant cellulolytic B. subtilis would be an ultra-low-cost platform for producing biofuels and other value-added products from non-food biomass.

See more of this Session: Advances In Biofuels: DOE Bioenergy Research Centers II
See more of this Group/Topical: Sustainable Engineering Forum

(560f) Development of Alkene Biofuels

Harry R. Beller, Joint BioEnergy Institute (JBEI), Emeryville, CA and Ee-Been Goh, Fuels Synthesis, Joint BioEnergy Institute (JBEI), Emeryville, CA

Aliphatic hydrocarbons are appealingtargets for advanced cellulosic biofuels, as they arepredominant components of petroleum-based gasoline and diesel fuels and thuswould be compatible with existing engines and fuel distribution systems. Long-chainalkenes are useful as feedstocks that can be crackedto form diesel- or gasoline-range hydrocarbons. We have studied alkenebiosynthesis in the bacterium *Micrococcus luteus*,a close relative of which was reported in the 1960s to biosynthesize *iso*- and *anteiso*-branched,long-chain alkenes. The underlyingbiochemistry and genetics of alkene biosynthesis werenot elucidated in those studies. In fact, until very recently, there were no known sequences forprokaryotic enzymes catalyzing the synthesis of non-isoprenoidaliphatic hydrocarbons (> C_2). We have shown that heterologousexpression of a three-gene cluster from *M. luteus*(Mlut_13230-13250) in a fatty-acid overproducing *E. coli* strain resultedin production of long-chain alkenes, predominantly 27:3 and 29:3 (no. carbon atoms: no. C=C bonds). Heterologous expression ofMlut_13230 (*oleA*) alone produced no long-chainalkenes but unsaturated aliphatic monoketones,predominantly 27:2, and *in vitro* studies with the purified Mlut_13230protein and tetradecanoyl-CoA produced the same C_{27}monoketone. Gas chromatography-time of flight (GC-TOF) mass spectrometry confirmedthe elemental composition of all detected long-chain alkenes and monoketones (putative intermediates of alkenebiosynthesis). Negative controlsdemonstrated that the *M. luteus* genes wereresponsible for production of these metabolites. We have proposed a metabolic pathway for alkene biosynthesisstarting with acyl-CoA (or -ACP) thioestersand involving decarboxylative Claisencondensation as a key step, which we believe is catalyzed by OleA. Suchactivity is consistent with our data and with the homology of Mlut_13230 (OleA) to FabH, which catalyzes decarboxylative Claisencondensation during fatty acid biosynthesis. We will discuss how understanding of the alkenebiosynthetic pathway can lead to enhanced production.

See more of this Session: Advances In Biofuels: DOE Bioenergy Research Centers II
See more of this Group/Topical: Sustainable Engineering Forum

(583a) Production of Concentrated Monosaccharride Solutions From Biomass Using CO2-H2O Mixtures At Varying Temperatures

Jeremy S. Luterbacher[1], Jefferson W. Tester[1] and Larry P. Walker[2], (1)Department of Chemical and Biomolecular Engineering, Cornell University, Ithaca, NY, (2)Department of Biological and Environmental Engineering, Cornell University, Ithaca, NY

Sustainably producing concentrated solutions of monosaccharrides from biomass is a key bottleneck in the conversion of lignocellulosic biomass to biofuels or bioproducts. Most pretreatment and enzymatic hydrolysis processes are run at low-solid concentration (<10 wt%) and use chemical catalysts, while high-solids enzymatic hydrolysis experiments are almost always performed with air-dried pretreatment mixtures.

This study uses a stirred 1 L reactor to study high solids (40 wt%) and high-pressure (200 bar) H_2O-CO_2 pretreatment of mixed hardwood and switchgrass. In addition, two-temperature stage pretreatment was introduced and optimized to take advantage of the temperature dependent biomass depolymerization reaction sequence (a short high-temperature stage at 200-210°C for 0 to 24 min was followed by a long low-temperature stage at 160-170°C for 60 min). Optimal glucan to glucose yields around 80% were obtained after 72 hrs of dilute hydrolysis (1 wt% glucan) for both substrates at similar pretreatment conditions (210°C, 16 min for hardwood and 210°C, 1 min for switchgrass, both followed by 60 min at 160°C). These optimally pretreated substrates were used as feedstock in high-solids enzymatic hydrolysis reactions, which gave glucose yields above 80% for both hardwood and switchgrass after 48 hrs of hydrolysis. Therefore, without additional chemical catalysts or any drying, two-temperature stage H_2O-CO_2 pretreatment coupled with high-solids enzymatic hydrolysis can produce monosaccharide solutions of 185 gr/L (170 gr/L of glucose and cellobiose 15gr/L of hemicellulose sugars) for mixed hardwood and 148 gr/L (115 gr/L of glucose and cellobiose 33gr/L of hemicellulose sugars) for switchgrass. This suggests that H_2O-CO_2 pretreatment is an attractive alternative to chemically catalyzed processes such as dilute acid pretreatment.

See more of this Session: Developments In the Pretreatment of Lignocellulosics for Bioconversion III
See more of this Group/Topical: Sustainable Engineering Forum

(583b) Structural Characterization of AHP-Pretreated Biomass

Muyang Li, Chemical Engineering and Material Science, Michigan State University, East Lansing, MI and David Hodge, Department of Chemical Engineering and Material Science, Michigan State University, East Lansing, MI

Alkaline hydrogen peroxide (AHP) pretreatment is exceptionally well-suited to grasses, yielding high digestibilities at low enzyme loadings, while generating relatively few fermentation innhibitors. For AHP pretreatment, the question of how structural and chemical compositional changes within the plant cell wall correlate to pretreatment effectiveness has not been effectively resolved, while knowledge of how selective modification of lignin during pretreatment may improve digestibility may yield insights into both improving pretreatments and tailoring plant cell walls for deconstruction. This study presents a comprehensive chemical and structural characterization of the changes in plant cell walls associated with switchgrass (*Panicum virgatum* cv. Cave-In-Rock), corn stovers (a commercial hybrid and inbred brown midrib lines bm1 and bm3), sugar maple (*Acer saccharum*) and hybrid aspen (*Populus tremula x tremuloides*) that were AHP pretreated at varied severities. Both the remaining solids and solubilized biomass hydrolyzates were subjected to a number of characterizations including total polysaccharide composition (monosaccharides, acetate, uronic acids), lignin content as both Klason and acetyl bromide lignin, and the ratio of H/G/S monolignols as determined by thioacidolysis GC/MS. Solid-state ^{13}C CP/MAS NMR and HSQC 2D NMR were applied to determine changes associated with intra-lignin, carbohydrate-carbohydrate, carbohydrate-lignin linkages and functional groups. Apparent molecular weight distributions of soluble polymeric lignins and hemicellulose aggregates were determined by HP-SEC. Py-GC/MS was utilized to characterize changes in the volatilized compounds before and after pretreatment implying pretreatment-induced alterations in pyrolysis-labile monolignol linkages.

See more of this Session: Developments In the Pretreatment of Lignocellulosics for Bioconversion III
See more of this Group/Topical: Sustainable Engineering Forum

(583c) Pretreatment of Corn Stover for Bioethanol Production Using Low-Moisture Anhydrous Ammonia (LMAA); Ammoniation and Pretreatment

Chang Geun Yoo, Agricultural and Biosystems Engineering, Iowa State University, Ames, IA, Nhuan-John Nghiem, Crop Conversion Science and Engineering, USDA-ARS-ERRC, Wyndmoor, PA, Kevin B. Hicks, Eastern Regional Research Center, Agricultural Research Service, Wyndmoor, PA and **Tae Hyun Kim**, Department of Agricultural and Biosystems Engineering, Iowa State University, Ames, IA

In the past several decades, various pretreatment methods have been suggested to enhance the enzymatic digestibility and fermentability of lignocellulosic biomass. Although a few of them may be effective, several cost barriers which prohibit scale-up exist including high chemical input and excessive water use. In order to solve these problems, a simpler pretreatment method using anhydrous ammonia was developed. In this method, corn stover with 30~70% moisture was used, which was contacted with anhydrous ammonia in a packed bed reactor. After the ammoniation step, biomass was treated at different reaction temperatures (40-120°C) and reaction times (24-96 h). Pretreated biomass was directly saccharified and fermented "with no additional washing step." Therefore, it is anticipated that this method can reduce energy cost substantially due to the extremely low liquid loading and moderate conditions applied. With the treated solids, we have obtained significantly improved ethanol fermentability (>85%) using *E.coli* KO11 strain.

In this investigation, it is speculated that ammonia swells cellulose structure effectively and ammoniation prevents methoxyl groups on lignin from absorbing cellulase enzymes, thus it enhances the enzyme hydrolysis rate and ethanol fermentation yield of lignocellulosic biomass. Ammoniation pretreatment can also supply assimilable nitrogen (up to 1.25 wt.% of the weight of dry biomass) for microorganism growth in the fermentor.

In this paper, various pretreatment conditions such as moisture levels, residual ammonia concentrations, reaction time, and reaction temperature were explored to reduce the pretreatment cost and its pertinent costs. Enzyme digestibility and ethanol fermentation test results will also be presented.

See more of this Session: Developments In the Pretreatment of Lignocellulosics for Bioconversion III
See more of this Group/Topical: Sustainable Engineering Forum

(583d) Feedstock Mixture Effects On Sugar Monomer Recovery Following Dilute Acid Pretreatment and Enzymatic Hydrolysis

Michael J. Brodeur-Campbell, Jordan Klinger and David Shonnard, Chemical Engineering, Michigan Technological University, Houghton, MI

Feedstock MixtureEffects on Sugar Monomer Recovery Following Dilute Acid Pretreatment andEnzymatic Hydrolysis

MichaelBrodeur-Campbell[1], Jordan Klinger[1], and DavidShonnard[1,2]

[1] Department of Chemical Engineering,Michigan Technological University

[2] Director Sustainable Futures Institute,Michigan Technological University

Increaseduse of lignocellulosic biomass for bioenergyproduction holds promise for improved domestic energy security, reducedgreenhouse gas (GHG) emissions, rural economic development, and developing amore sustainable energy infrastructure [Chum and Overend2001; Perlack, Wright, et al. 2005; U.S. DOE 2006]. Transportation fuels are especially difficultto replace as their replacement has to have a high energy density, be compatiblewith existing fuel distribution infrastructure, and be compatible with existingautomobile technology. In 2009, 97% oftransportation fuels were derived from fossil sources [EIA 2008], andtransportation accounted for one-third of carbon dioxide emissions in 2008 [EIA2009]. Liquid fuels derived from biomassare particularly attractive as a replacement for fossil-derived transportationfuels. Cellulose is the most abundantbiopolymer on Earth, and lignocellulosic feedstocks represent one of the fewnatural resources available in sufficient quantity to replace a significantfraction of liquid transportation fuel.

In acommercial biorefinery feedstocks are likely to come from a variety of sourcesand include many different materials, such as softwoods, hardwoods, agriculturalresidues, and herbaceous or woody energy crops.Much work has been done to understand and optimize the pretreatment andenzymatic hydrolysis conditions for pure feedstocks, but little is knownabout mixture effects on enzymatic hydrolysis.A primary concern in processing woody biomass to produce a range ofbiofuels is to maximize the production of desired intermediates, in this case,determining the quantity of monomer sugars that are recoverable. The biochemical platform as described by theNational Renewable Energy Laboratory (NREL) was the process studied. Briefly, lignocellulosic biomass which iscomposed primarily of hemicellulose, cellulose, and lignin is treated withdilute sulfuric acid to hydrolyze the hemicellulose into soluble sugars. The solid residue remaining after dilute acidpretreatment is then treated with cellulose-degrading enzymes to hydrolyze thecellulose into soluble sugars. Theremaining lignin may be utilized as a high energy co-product stream to provideheat and power for the cellulosic ethanol conversion processes.

This studyseeks to clarify the effects of processing mixtures of feedstocks on overall sugarrecovery. We wish to test the hypothesisthat sugar yields from mixed feedstocks can be predicted by a simple linearinterpolation model based on the behavior of the pure species. Experiments were performed on pure species(aspen, balsam, and switchgrass), and on 50/50 blends of aspen/balsam andaspen/switchgrass. Aspen was chosen as amodel species due to its high suitability for processing in a commercialbiorefinery. Balsam was chosen to mixwith aspen to study the effects of high lignin content on mixed feedstockprocessing, and switchgrass was chosen to mix with aspen to investigate theeffects of a high ash content energy crop.

One-halfgram (0.5 g) of dry sample was weighed and placed into one of a set of 9stainless steel Swagelok cylindrical reactor tubes, and 4.5 ml of dilutesulfuric acid (0.5% w/v) was added to achieve 10% (wt) solids loading in thereactor. Reactors were rapidly heated to160°C in a silicon oil bath (Dow Corning 550 fluid), and three tubes wereremoved from the oil bath at previously determined time intervals. From previous research involving pure samplesof aspen, balsam, and switchgrass (Jensen et al. 2010), the time to achievemaximum yields of monomer and oligomer sugars from combined dilute acidhydrolysis and enzymatic hydrolysis was determined; the "optimum" time. Threereactors were removed at one-half the optimal time, three reactors were removedat the optimal time point, and the last three were removed at two times theoptimal time. Reactor tubes removed fromthe oil bath were immediately submerged in an ice-water bath to quench thereaction. One milliliter of liquid fromeach reactor was removed, filtered, neutralized, and analyzed on an Agilent1200 series HPLC for glucose and xylose concentration. One milliliter of liquid from each reactorwas filtered and then further acidified to 4% (w/v) sulfuric acid, and treatedat 121°C for one hour tohydrolyze any soluble oligomers into monomer form. After neutralization these samples were analyzedby HPLC as above. Results from these twoliquid samples were combined into total soluble (monomer plus oligomer) sugarrecovery from dilute acid pretreatment.Solids were washed with excess distilled water on a glass-fiber filter(VWR) and carried on to enzymatic hydrolysis.

Solidresidues containing cellulose and lignin were subjected to enzymatic hydrolysisusing Accelerase 1500 (Genencore). Each of the three reactors from the samepretreatment time point were loaded with a different enzyme concentration 0.1, 0.25, and 0.5 ml Accelerase 1500 per gram ofcarbohydrate in the feedstock in order to study the interaction betweenpretreatment severity and enzyme loading on sugar yields. Reaction conditions were 50mM citrate buffer,pH 4.8, 50°C,250 RPM orbital shaking. Liquid sampleswere removed at 0, 1, 2, and 3 days, filtered, and analyzed for sugar concentrationsby HPLC as above.

Maximumyields of xylose from pretreatment were seen to vary with both time andfeedstock composition in the expected manner. Maximum glucose yields following enzymatichydrolysis were seen to be strongly affected by both pretreatment severity andenzyme loading. Results are summarizedin Table 1 below.

Table 1: Experimental sugar recoveries for Dilute AcidHydrolysis (DAH), Enzymatic Hydrolysis (EH), and experimental and predictedrecoveries for Total Sugar (TS) yield as percent of theoretical.

	DAH		EH		TS		TS (Predicted)	
	Xylose	**Glucose**	**Xylose**	**Glucose**	**Xylose**	**Glucose**	**Xylose**	**Glucose**
Balsam	72%	7%	0%	2%	72%	9%		
Aspen/Balsam	82%	5%	8%	27%	90%	32%	85%[1]	29%
Aspen (1)	83%	3%	4%	47%	87%	50%		
Aspen/Switchgrass	79%	5%	4%	53%	83%	58%	86%	60%
Switchgrass	69%	12%	5%	59%	74%	71%		
Aspen (2) w/ β-Glucosidase	93%	4%	6%	61%	100%	65%		

[1]Predicted value based on Aspen (2) pretreatment results.

Xyloseyields from pretreatment varied from 69-93% of the theoretical maximum. Pure switchgrass gave the lowest yield andpure aspen the highest. Each purespecies and mixture showed a peak hydrolysis yield near the expected optimaltime. Earlier time points showedincomplete hydrolysis with greater quantities of soluble oligomericsugars in the hydrolyzate. Later timepoints yielded sugar decomposition products such as furfural and hydroxymethylfurfural (HMF) in the hydrolyzate. Longer pretreatment reaction times alwaysresulted in higher glucose yields, but due to the degradation of xylose atlater time points total sugar yields were generally lower for extendedpretreatment times.

Glucoseyields following enzymatic hydrolysis varied from 2-61% with pure balsam givingthe lowest yield and pure aspen with supplemental β-glucosidase giving thehighest. The first aspen experiment wasrepeated due to incomplete pretreatment (indicated by lower than expected sugarrecoveries), and subsequently treated with the highest Acceleraseconcentration plus supplemental β-glucosidase to study the effect ofadditional enzyme loading on yield.

Using asimple linear interpolation model to predict sugar recoveries for mixedfeedstock streams based on the known yields obtainable for pure species gives aresult that is accurate to within about 2% of the experimental data for bothglucose and xylose recovery following dilute acid pretreatment, and to within3% for glucose recovery following enzymatic hydrolysis. Xylose recovery following enzymatichydrolysis for blends containing balsam could not be predicted accurately fromthese data, but since that is only a very small amount of the total sugar yieldthe discrepancy did not affect the model results for total sugar yield. Total sugar yield was predicted to within 5%of experimental results.

References:

Chum, Helena L. and Ralph P Overend (2001). *Biomass and renewable fuels*, FuelProcessing Technology 71, 187-195.

Perlack,R. D., L. L. Wright, et al. (2005). Biomass as Feedstock for a Bioenergy and Bioproducts Industry: The Technical Feasibility of aBillion-Ton Annual Supply. U. S. D. o. A. (USDA) and U. S. D. o. E. (DOE).

U.S. DOE (2006). *Breakingthe Biological Barriers to Cellulosic Ethanol: A Joint Research Agenda*,DOE/SC-0095, U.S. Department of Energy Office of Science and Office of EnergyEfficiency and Renewable Energy (www.doegenomestolife.org/biofuels/).

EIA (2008). *Emissionsof Greenhouse Gases in the United States*. U.S. Energy Information Administration Office of Integrated Analysis andForecasting (http://www.eia.doe.gov/oiaf/1605/ggrpt/carbon.html).

EIA (2009). *Annual Energy Review 2009*. U.S.Energy Information Administration, (http://www.eia.doe.gov/aer/pecss_diagram.html).

Jensen, Jill R., Juan E. Morinelly, Kelsey R. Gossen,Michael J. Brodeur-Campbell, and David R. Shonnard (2010). Effectsof dilute acid pretreatment conditions on enzymatic hydrolysis monomer andoligomer sugar yields for aspen, balsam, and switchgrass. BioresourceTechnology 101, 2317-2325.

See more of this Session: Developments In the Pretreatment of Lignocellulosics for Bioconversion III
See more of this Group/Topical: Sustainable Engineering Forum

(583e) Hydrolysate De-Acidification Using Resin-Wafer Electrodeionization

Yupo J. Lin, Saurav Datta, Michael Henry, Anthony Fracaro and Seth Snyder, Argonne National Laboratory, Argonne, IL

Dilute acid pretreatment of lignocellulosic feedstock offers an economic route for the extraction of fermentable sugars from non-food biomass for biofuel production by bioconversion. However, removal of acids from the hydrolysate is a critical step to ensure optimal conditions for the subsequent bioprocessing/conversion steps, i.e., enzyme hydrolysis of the cellulose to release C6 sugars and for the final fermentation to produce biofuel. The removal (conditioning step) of inorganic and organic acids (released together with C5 sugar molecules from the hemicellulose) after the acid treatment stage poses a great economic challenge to the overall biofuel production cost. To address this challenge, we used a proprietary resin wafer electrodeionization (RW-EDI) technology to remove both the inorganic and organic acids in a single step. Preliminary process economic analysis indicates that the easily scalable and small foot-print unit operation of RW-EDI can offer an economic and viable means of addressing the hydrolysate conditioning challenge. Selective removal of inorganic acids and organic acids was demonstrated when appropriate operation conditions were employed for the RW-EDI operation. This unique feature makes possible the separate recovery of both inorganic acid (e.g., sulfuric acid) and organic acids, which act as fermentation inhibitors, so that the inorganic acid can be recycled to the dilute acid pretreatment stage. The RW-EDI also provides an option to remove other fermentation inhibitor such the furfural and hydroxy methyl furfural (HMF) during the hydrolysate conditioning. In this presentation, process design and performance of hydrolysate de-acidification using RW-EDI will be discussed. Pilot-scale demonstration of technology viability using synthetic hydrolysate will be also presented along with preliminary assessments of economic and process benefits.

See more of this Session: Developments In the Pretreatment of Lignocellulosics for Bioconversion III
See more of this Group/Topical: Sustainable Engineering Forum

(583f) Biomimicry of the Fungal Consolidation of Biomass Pretreatment with Saccharification

Justin T. Kaffenberger and Jonathan S Schilling, Bioproducts and Biosystems Engineering, University of Minnesota, St Paul, MN

Efficient breakdown of ligno-cellulosic material prior to fermentation for bio-fuel production is a key step in optimizing bio-fuel production. Current industrial practice involves chemical pretreatment and acid or enzymatic saccharification to prepare ligno-cellulosic biomass for fermentation. The commercial viability of bio-fuels hinges in large part on production cost. One approach to reducing production cost is process consolidation. Consolidation not only eliminates the capital costs associated with the additional processing steps, but can also trim operational costs by reducing total reaction time and down time between steps.

In nature, the brown-rot fungus *Postia placenta* successfully performs both chemical pretreatment and enzymatic saccharification of biomass in a consolidated manner. As evidenced by its recently sequenced genome, *P. placenta* produces a limited suite of cellulases, lacking genes for exoglucanases. Despite this limitation, it is still capable of rapidly degrading wood. Brown-rot fungi quickly reduce the degree of polymerization of wood with little weight loss through a reduction-oxidation pathway that yields highly reactive hydroxyl radicals through a chelator-mediated fenton reaction. While hydroxyl radicals would readily damage its cellulases, the fungus manages to conduct both reaction types, enzymatic and oxidative, in the same general location. Examination of the spatial and temporal relationship of these two reaction systems as brown-rot fungal degradation of wood progresses will be presented.

See more of this Session: Developments In the Pretreatment of Lignocellulosics for Bioconversion III
See more of this Group/Topical: Sustainable Engineering Forum

(588a) Flexible Energetically Self-Sufficient Co-Production of Hydrogen and Formic Acid

Jorge Pena Lopez, Chemical Engineering Department, University of California, Los Angeles, Los Angeles, CA and Vasilios Manousiouthakis, Chemical Engineering Department, University of California at Los Angeles, Los Angeles, CA

The use of fossil fuels (gasoline, Diesel, etc.) for light vehicle based transportation has resulted in two major environmental quandaries: poor air quality in largely populated cities and increased carbon dioxide emissions into the Earth's atmosphere. The use of hydrogen fueled vehicles has been proposed as an avenue to address the aforementioned dilemmas. However, hydrogen production technologies have yet to achieve a generation system competitive with, for example, the production of gasoline. Several options exist for hydrogen production; one of the most explored has been the reforming of natural gas. However, conventional natural gas reformers release carbon dioxide into the atmosphere. A solution to this issue is redirecting carbon dioxide into formic acid, thus preventing its emission into the atmosphere.

In this work we propose a novel method to synthesize energetically self-sufficient and environmentally friendly process flowsheets, and we show hydrogen production as a case study. To this end we make use of commercially available technologies for the production of formic acid [1]. The resulting process flowsheet goes through heat and power integration techniques that allow the maximum utilization of energy in favor of hydrogen production, maintaining a self-powered plant and the carbon dioxide emissions at zero. Using the simulation tool UniSim Design, we show that there exists an optimal themodynamical and operational point for the generated process flowsheet.

[1] Reutemann W, Kieczka H. Ullmann's Encyclopedia of Industrial Chemistry, "Formic Acid". John Wiley & Sons, Ltd. 2000

See more of this Session: Fundamentals of Hydrogen Production and Utilization
See more of this Group/Topical: Environmental Division

The Attractiveness of Ammonia Borane to Generate Hydrogen for PEM Fuel Cell Vehicles

Ahmad Al-Kukhun, Hyun Tae Hwang, and Arvind Varma

School of Chemical Engineering, Purdue University,
480 Stadium Mall Drive, West Lafayette, IN 47907

Current promising methods to release hydrogen from Ammonia Borane (NH_3BH_3, AB; 19.6 wt% H_2) including neat thermolysis, thermolysis in ionic liquid bmimCl with or without proton sponge, thermolysis with nano-BN and hydrothermolysis, were investigated for hydrogen yield and ammonia formation. It was found that even trace moisture influences AB dehydrogenation significantly. The hydrothermolysis (13.5 wt% H_2, 1 mol% NH_3) at 85°C and thermolysis in bmimCl with 3 wt% moisture at 110°C (13 wt% H_2, 0.2 mol% NH_3) methods were found to be the most promising. This study demonstrates that AB dehydrogenation is an attractive approach to generate hydrogen for PEM fuel cell vehicle applications.

(588c) Solar Thermochemical Hydrolysis of Metal Nitrides for H2 Production and Integrated Storage In Form of Ammonia

Ronald Michalsky and Peter Pfromm, Department of Chemical Engineering, Kansas State University, Manhattan, KS

Solar thermochemical hydrolysis of metal nitrides forH_2 production and integrated storage in form of ammonia

RonaldMichalsky and Peter H. Pfromm

Department of Chemical Engineering, Kansas State University, Manhattan, Kansas, USA

Conversion ofabundant but intermittent solar energy at high temperatures to high valuechemical energy stored in the products of a solar thermochemical reaction cycle,such as H_2 or syngas has been studied widely. Ammonia (NH_3)has been proposed more recently as hydrogen carrier for sustainabletransportation fuel. Easily liquefied and transported, NH_3 exceedsthe H_2 storage requirements of the Department of Energy. Modifieddiesel engines can combust NH_3 releasing mainly H_2O and N_2. H_2 can also be recovered catalytically on board a vehicle from $NH_3$for subsequent combustion.

Solarthermochemical NH_3 synthesis from steam and nitrogen produces asustainable solar fuel and avoids the storage problem for H_2.Production of metallic nitrides by reduction of their metal oxides is a high-temperatureand energy-intensive process that may take advantage of concentrated solarradiation as inexpensive and sustainable source for process heat. This workpresents a solar thermochemical NH_3 synthesis process sequence ofnitride hydrolysis splitting H_2O and absorbing protons released inthe formation of NH_3 at ambient pressure, and endothermic metaloxide reduction and nitridation driven by concentrated solar energy.

A thermodynamicrationale is presented. Various characteristic metals are selected andexperimentally explored studying yield and kinetics of breaking the N_2triple bond during nitridation and liberating NH_3 when hydrolyzingthe nitride. Experimental data will focus transition metal interactions on N_2uptake from the gas phase and nitrogen distribution in the solid bulk materialstudied using various solid-state analytical techniques. Protonation of thenitrogen liberated during steam hydrolysis of various ionic, covalent,intermediate and interstitial nitrides is examined establishing a mass balanceon the nitrogen atom. The yield of NH_3 will be correlated with thenitride ionicity and with the activation energy of nitride hydrolysis to thusdiscuss the trade-off between desirable high yields of metal nitridation andnitride hydrolysis on one side and undesirable strong bonds between the metalliccomponent and oxygen formed during hydrolysis.

See more of this Session: Fundamentals of Hydrogen Production and Utilization
See more of this Group/Topical: Environmental Division

(588d) Quantification of [Fe-Fe]-Hydrogenase, MrcA and FTHFS Gene In Mixed Anaerobic Cultures During Dark Hydrogen Fermentation

Subba Rao Chaganti[1], Daniel D. Heath[1] and Jerald A. Lalman[2], (1)Great Lakes Institute for Environmental Research, University of Windsor, Windsor, ON, Canada, (2)Department of Civil and Environmental Engineering, University of Windsor, Windsor, ON, Canada

Energy security, climatic change and decreasing fossil fuel inventories are major issues driving researchers to develop technologies to produce cheap and dependable energy supplies. Hydrogen (H2) is a promising potential energy carrier for meeting future energy needs because it is eco-friendly, it has a high heating value (2.75 times higher energy content than hydrocarbon fuels) and it can be converted into electricity via fuel cells (Benemann, 1996; Momirlan and Vezrioglu, 2002). Currently, H2 is mainly produced from fossil fuel-based feedstock (Oil/naptha reforming, coal gasfication, methane steam reforming etc.,) chemicals because many commercial petrochemical processes are cost effective (Ewan and Allen, 2005). However, the dark fermentation process is emerging as a practical route for producing H2 because of fast reaction rates, technical simplicity and less area (1-5%) requirements compared to the direct photolysis or indirect photolysis or photo fermentation (Levin et al., 2004; Li and Fang 2007).

The successful operation of the dark fermentation process requires a good understanding of how mixed microbial cultures behave under a variety of operating conditions. Identifying microbial communities can provide descriptions of the interaction between different groups; however, this approach may not provide sufficient information about the function of the communities. Noticeable changes in the microbial community composition within the H2 fermenting reactor is expected during operation. Quantifying the gene expression and monitoring fluctuations within the community structure will provide an understanding of the changes linked to the operation of bioreactors. This approach can possibly serve as a rapid tool for bioprocess monitoring.

In the present study, genes responsible for H2 production and H2 consumption were monitored as a means of establishing changes within the microbial community structure. The Fe-Fe hydrogenase gene is monitored because it is related to H2 production. Methanogens and acetogens are major H2 consumers and monitoring includes examination of the McrA gene and formyltetrahydrofolate synthetase (FTHFS) gene, respectively. A monitoring protocol for three different sources of cultures (A, B and C) at pH 5.5 and 37oC was developed using qRT-PCR. Terminal Restriction Fragment Length Polymorphism (T-RFLP) analysis was also performed to characterize changes in the community. The results indicated that the expression levels of [Fe-Fe]-hydrogenase, MrcA and FTHFS gene varied with the culture source. The [Fe-Fe]-hydrogenase gene expression was shown to correlate with H2 production. Increasing MrcA expression as well as increasing FTHFS and [Fe-Fe]-hydrogenase activity was correlated with low H2 yields of approximately 1.65±0.25 mol.mol-1 glucose in culture B. In comparison, cultures A and C with lower FTFHS and higher [Fe-Fe]-hydrogenase activities had higher H2 yields of 2.4±0.12 and 2.5±0.23 mol.mol-1 glucose, respectively. This data clearly suggest that monitoring the [Fe-Fe]-hydrogenase gene alone is insufficient to gain an understanding of how mixed cultures function and genes related to H2-consumers (MrcA and FTHFS) should be monitored and down regulated to achieve higher H2 yields.

References

Bennemann, J. 1996. Hydrogen biotechnology: progress and prospects. Nat. Biotechnol. 14, 1101-1103.

Ewan, B.C.R., Allen, R.W.K. 2005. A figure of merit assessment of the routes to hydrogen. Int. J. Hydrogen Energ. 30 (8), 809-819.

Levin, D.B., Pitt, L., Love, M. 2004. Biohydrogen production: prospects and limitations to practical application. Int. J. Hydrogen Energ. 29 (2), 173-185.

Li. C., Fang, H.H.P. 2007. Fermentative hydrogen production from wastewater and solid wastes by mixed cultures. Crit. Rev. Env. Sci. Tec. 37 (1), 1-39.

Momirlan, M., Veziroglu, T.N. 2002. Current status of hydrogen energy. Renew. Sust. Energ. Rev. 6 (1-2), 141-179.

See more of this Session: Fundamentals of Hydrogen Production and Utilization
See more of this Group/Topical: Environmental Division

(588e) Hydrogen Car Fill-up: Thermodynamic Modeling Using the Generic Cubic Equation of State

Fernando Olmos, Chemical and Biomolecular Engineering, UCLA, Los Angeles, CA and Vasilios Manousiouthakis, Chemical Engineering Department, University of California at Los Angeles, Los Angeles, CA

In order to provide hydrogen fueling stations with required information about hydrogen's pressure, temperature, and mass flowrate, hydrogen was modeled as a non-ideal gas employing the generic cubic equation of state (GC-EOS), along with its special four cases (VdW, RK, SRK, PR). With the purpose of validating the GC-EOS hydrogen thermodynamic property predictions, comparisons were carried out with NIST hydrogen's data, real-life hydrogen fueling station data, and with a mass balance equation in integrable form. Both the results of the GC-EOS and the mass balance model matched the NIST data and the hydrogen fueling station data describing the time evolution of the molar volume inside of the hydrogen vehicle tank. In addition, based on the concept of residual properties, an explicit expression for constant-pressure and constant-volume heat capacity was derived based on the GC-EOS as a function of molar volume and temperature. The results from the heat capacities, based on GC-EOS, matched NIST hydrogen's data at low pressures, and at pressures ranging from 100 to 1000 bar there is deviation less than 2%. Finally, the fueling station was modeled employing a total mass and energy balance, along with the GC-EOS, establishing that, in the absence of heat transfer, the maximum temperature achieved by hydrogen inside of the vehicle's tank does not depend on how high or low the mass flowrate is, but rather it is a function of the amount of mass accumulated in the tank.

See more of this Session: Fundamentals of Hydrogen Production and Utilization
See more of this Group/Topical: Environmental Division

(609a) Exciton Engineering with Carbon Nanotubes and Graphene for Solar Energy Conversion: From Exciton Antennae to Nano-Heterojunctions

Michael Strano[1], Geraldine LC Paulus[2], Moon-Ho Ham[2], Chang Young Lee[2], Changsik Song[2], Kourosh Kalantar-Zadeh[3], Wonjoon Choi[4], Jae-Hee Han[1], Ryuichiro Maruyama[5], Esther S. Jeng[6], Daniel A. Heller[2], Woo-Jae Kim[1], Paul W. Barone[7] and Cristiano Fantini[8], (1)Chemical Engineering, Massachusetts Institute of Technology, Cambridge, MA, (2)Department of Chemical Engineering, Massachusetts Institute of Technology, Cambridge, MA, (3)School of Electrical and Computer Engineering, RMIT University, Melbourne, Australia, (4)Department of Mechanical Engineering, Massachusetts Institute of Technology, Cambridge, MA, (5)Materials Laboratories, Sony Corporation, Kanagawa, Japan, (6)Department of Chemistry, University of Illinois at Urbana-Champaign, Urbana, IL, (7)Chemical Engineereing, Massachusetts Institute of Technology, Cambridge, MA, (8)Departamento de Fisica, Universidade Federal de Minas Gerais, Belo Horizonte-MG, Brazil

Exciton engineering with carbon nanotubes and graphene for solar energy conversion: from exciton antennae to nano-heterojunctions

Michael S. Strano

Charles and Hilda Roddey Associate Professor of Chemical Engineering

66-566 Department of Chemical Engineering

77 Massachusetts Avenue

Cambridge, MA 02139-4307

Email: strano@MIT.EDU

phone: (617) 324-4323

fax: (617) 258-8224

http://web.mit.edu/stranogroup/

Our laboratory has been interested in how low dimensional materials, such as single walled carbon nanotubes or graphene, can augment and enhance solar conversion efficiencies and demonstrate new photonic concepts. We will discuss two recent examples of our work in this space.

In the first example, there is significant interest in combining carbon nanotubes with semiconducting polymers for photovoltaic applications because of potential advantages from smaller exciton transport lengths and enhanced charge separation. However, to date, bulk heterojunction (BHJ) devices have demonstrated relatively poor efficiencies, and little is understood about the polymer/nanotube junction. To investigate this interface, we fabricate a planar nano-heterojunction comprising well-isolated millimeter-long single-walled carbon nanotubes underneath a poly(3-hexylthiophene) (P3HT) layer (1). The resulting junctions display photovoltaic efficiencies per nanotube ranging from 3% to 3.82%, which exceed those of polymer/nanotube BHJs by a factor of 50-100. The increase is attributed to the absence of aggregate formation in this planar device geometry. It is shown that the polymer/nanotube interface itself is responsible for exciton dissociation. Typical open-circuit voltages are near 0.5 V with fill factors of 0.25-0.3, which are largely invariant with the number of nanotubes per device and P3HT thickness. A maximum efficiency is obtained for a 60 nm-thick P3HT layer, which is predicted by a Monte Carlo simulation that takes into account exciton generation, transport, recombination, and dissociation. We combine for the first time both optical T-matrix and kinetic Monte Carlo models to investigate the photocurrent generation in two state-of-the-art PHJ photovoltaics. The combined model takes into account the rates of exciton generation, transport, recombination and dissociation using literature values. By including the optical, electronic and structural properties of the different materials, we are able to predict the short-circuit current of recently reported P3HT/SWNT PHJ and also P3HT/PCBM PHJ solar cells from the literature. The experimental data for each of these devices show a maximum photocurrent output at a P3HT thickness of 60-65 nm, in contradiction to the expected value equal to the diffusion length of excitons in P3HT (8.5nm). The model demonstrates how a bulk exciton sink can explain this shifted maximum in the P3HT/SWNT case, whereas the maximum is mainly determined by PCBM interdiffusing in P3HT in the P3HT/PCBM case. This platform is promising for further understanding the potential role of polymer/nanotube interfaces for photovoltaic applications.

In the second example, there has been renewed interest in solar concentrators and optical antennas for improvements in photovoltaic energy harvesting and new optoelectronic devices. We dielectrophoretically assemble single-walled carbon nanotubes (SWNTs) of homogeneous composition into aligned filaments that can exchange excitation energy, concentrating it to the centre of core–shell structures with radial gradients in the optical bandgap (2). We find an unusually sharp, reversible decay in photoemission that occurs as such filaments are cycled from ambient temperature to only 357 K, attributed to the strongly temperature-dependent second-orderAuger process. Core–shell structures consisting of annular shells of mostly (6, 5) SWNTs (E_g = 1.21 eV) and cores with bandgaps smaller than those of the shell (E_g = 1.17 eV (7,5)–0.98 eV (8,7)) demonstrate the concentration concept: broadband absorption in the ultraviolet–near-infrared wavelength regime provides quasi-singular photoemission at the (8, 7) SWNTs. This approach demonstrates the potential of specifically designed collections of nanotubes to manipulate and concentrate excitons in unique ways.

1. Ham MH, Paulus GLC, Lee CY, Song C, Kalantar-zadeh K, Choi W, Han JH and **Strano MS**: Evidence for High-Efficiency Exciton Dissociation at Polymer/Single-Walled Carbon Nanotube Interfaces in Planar Nano-heterojunction Photovoltaics. ACS NANO, 4 (2010) 6251-6259

2. Han JH, Paulus GLC, Maruyama R, Jeng ES, Heller DA, Kim WJ, Barone PW, Lee CY, Choi JH, Ham MH, Song C, Fantini C, **Strano MS**: Exciton Antennas and Concentrators from Core-Shell and Corrugated Carbon Nanotube Filaments of Homogeneous Composition. NATURE MATERIALS, 9, 833 - 839 (2010).

See more of this Session: Semiconductor Nanowires and Nanotubes for Next Generation Photovoltaics From Fundamentals to Devices
See more of this Group/Topical: Topical 5: Nanomaterials for Energy Applications

(609b) Semiconductor Nanowire Fabric As a New Photovoltaic Material

Chet Steinhagen[1], Vahid Akhavan[1], Vince Holmberg[1] and **Brian A. Korgel**[2], (1)Chemical Engineering, University of Texas at Austin, Austin, TX, (2)Department of Chemical Engineering, University of Texas at Austin, Austin, TX

Significant quantities of semiconductor nanowires can be synthesized using solution-based methods, such as supercritical fluid-liquid-solid (SFLS) and solution-liquid-solid (SLS) growth. These nanowires are mechanically flexible because of their narrow (<50 nm) diameter and high aspect ratio, with lengths greater than millimeters in some cases. Because of the absence of crystallographic defects like grain boundaries, the nanowires are extremely strong, exhibiting nearly ideal mechanical strength. The nanowires can be formed into non-woven fabric, or paper, that is flexible and mechanically robust. We are in the process of trying to develop these nanowire fabrics as a new class of flexible photovoltaic material. This presentation will highlight the nanowire fabric itself--its properties and how it is made--as well as data from prototype photovoltaic silicon (Si) and CuInSe2 (CIS) nanowire fabric.

See more of this Session: Semiconductor Nanowires and Nanotubes for Next Generation Photovoltaics From Fundamentals to Devices
See more of this Group/Topical: Topical 5: Nanomaterials for Energy Applications

(609c) Solar Cells Based On Sensitized Nanowire Arrays

Jason B. Baxter, Department of Chemical and Biological Engineering, Drexel University, Philadelphia, PA

Solar cells can provide an abundant, clean, and sustainable source of electricity, but high costs have limited their implementation. The use of sensitized nanostructured architectures may enable both low-cost processing and high efficiency by decoupling the functions of light harvesting and charge transport into different materials. Sensitized solar cells consist of a bicontinuous interpenetrating network of electron- and hole-transporting materials with an interfacial absorber layer. In this architecture, interfacial recombination is the dominant loss process, so controlling the interfacial chemistry during deposition is critical. I will present our group's work on the use of ZnO nanowire arrays sensitized by thin CdSe coatings to produce extremely thin absorber (ETA) solar cells. Nanowire arrays can provide direct pathways for electron transport as well as sufficient surface area for sensitization. Progress in sensitized nanowire-based solar cells will be presented, as well as some of the challenges for improving their efficiencies. Overcoming these challenges will require detailed fundamental understanding and control of charge transfer in the materials and interfaces within the cells.

See more of this Session: Semiconductor Nanowires and Nanotubes for Next Generation Photovoltaics - From Fundamentals to Devices
See more of this Group/Topical: Topical 5: Nanomaterials for Energy Applications

(609d) Photoelectrochemical Complexes for Solar Energy Conversion That Dynamically, Reversibly and Autonomously Self-Assemble

Ardemis A. Boghossian[1], Jong Hyun Choi[2], Moon-Ho Ham[1], Esther S. Jeng[3], Rachel A. Graff[4], Daniel A. Heller[1], Alice C. Chang[3], Aidas Mattis[5], Timothy H. Bayburt[6], Yelena V. Grinkova[6], Adam Scott Zeiger[7], Krystyn J. Van Vliet[8], Erik K. Hobbie[9], Stephen G. Sligar[6], Colin A. Wraight[6] and Michael Strano[3], (1)Department of Chemical Engineering, Massachusetts Institute of Technology, Cambridge, MA, (2)Mechanical Engineering, Purdue University, West Lafayette, IN, (3)Chemical Engineering, Massachusetts Institute of Technology, Cambridge, MA, (4)Chemical and Biomolecular Engineering, University of Illinois at Urbana-Champaign, Urbana, IL, (5)Department of Biochemistry, University of Illinois at Urbana-Champaign, Urbana, IL, (6)Biochemistry, University of Illinois Urbana-Champaign, Urbana, IL, (7)Materials Science and Engineering, Massachusetts Institute of Technology, Cambridge, MA, (8)Materials Science and Engineering, MIT, Cambridge, MA, (9)Polymers Division, National Institute of Standards and Technology, Gaithersburg, MD

Photoelectrochemical complexes for solar energy conversion that dynamically, reversibly and autonomously self-assemble

Naturally occurring photosynthetic systems use elaborate pathways of self-repair to limit the impact of photo-damage. Here, we demonstrate a complex consisting of an aqueous solution containing photosynthetic reaction centers (RCs), membrane scaffold proteins (MSPs), phospholipids, and single-walled carbon nanotubes (SWCNTs) solubilized with the surfactant sodium cholate (SC) that mimics this process.(1) The components reversibly self-assemble into a highly ordered structure upon dialysis of the surfactant. The resulting assembled structure is photoelectrochemically active and consists of 4-nm-thick lipid bilayer disks (nanodisks, NDs) arranged parallel to the surface of the SWCNT with the RC housed within the bilayer such that its hole injecting site faces the nanotube surface. The structure can be assembled and disassembled autonomously with the addition or removal of surfactant. We model the kinetic and thermodynamic forces that drive the dynamics of this reversible self-assembly process.(2) The assembly is monitored using spectrofluorimetry during dialysis and subsequent surfactant addition and used to fit a kinetic model to determine the forward and reverse rate constants of ND and ND-SWCNT formation. The calculated ND and ND-SWCNT forward rate constants are 79 mM^{-1} s^{-1} and 5.4×10^2 mM^{-1} s^{-1}, respectively, and the reverse rate constants are negligible over the dialysis time scale. We find that the reaction is not diffusion-controlled since the ND-SWCNT reaction, which consists of entities with smaller diffusion coefficients, has a larger reaction rate constant. Using these rate parameters, we were able to develop a kinetic phase diagram for the formation of ND-SWCNT complexes, which indicates an optimal dialysis rate of approximately 8×10^{-4} s^{-1}.

The assembly is thermodynamically metastable and can only transition reversibly if the rate of surfactant removal exceeds a threshold value. Only in the assembled state do the complexes exhibit photoelectrochemical activity. We experimentally demonstrate a regeneration cycle that uses surfactant to switch between assembled and disassembled states, resulting in an increased photoconversion efficiency of more than 300% over 168 hours and an indefinite extension of the system lifetime. The system can disassemble and reassemble over an indefinite number of cycles. We also fit the model to these cyclic ND-SWCNT assembly and disassembly experiments and hence mimic the thermodynamic forces used in regeneration processes detailed previously. Such forces may form the basis of both synthetic and natural photoelectrochemical complexes capable of dynamic component replacement and repair.

1. Ham MH, Choi JH, Boghossian AA, Jeng ES, Graff R.A, Heller DA, Chang AC, Mattis A, Bayburt TH, Grinkova YV, Zeiger AS, Van Vliet KJ, Hobbie EK, Sligar SG, Wraight CA and **Strano MS**: Photoelectrochemical complexes for solar energy conversion that chemically and autonomously regenerate. NATURE CHEMISTRY, 2 (2010) 929-936.

2. Boghossian AA, Choi JH, Ham MH and **Strano MS**: Dynamical and Reversible Self-Assembly of Photoelectrochemical Complexes Based on Lipid Bilayer Disks, Photosynthetic Reaction Centers, and Single-Walled Carbon Nanotubes. Langmuir, 27 (2011) 1599-160

See more of this Session: Semiconductor Nanowires and Nanotubes for Next Generation Photovoltaics From Fundamentals to Devices

See more of this Group/Topical: Topical 5: Nanomaterials for Energy Applications

(609e) A Fundamental Investigation of Electron Transport and Recombination Characteristics of Titania Nanowire/Nanoparticle Hybrid Structures

Venkat Kalyan Vendra[1], Delaina A. Amos[1], Mahendra K. Sunkara[1] and Thad Druffel[2], (1)Chemical Engineering, University of Louisville, Louisville, KY, (2)Conn Center for Renewable Energy Research, Louisville

The electron transport and recombination of long titania nanowires and titania nanoparticle/nanowire composites are investigated. Understanding charge transport and recombination in 1D nanostructures of titania is critical for improving the performance of DSSC employing these 1D nanostructures as photoanodes in dye sensitized solar cells. 1D titania nanostructures are expected to have short transport times and high electron lifetimes. However, most of the 1D titania nanostructures have not shown these desirable properties. For instance, the electron diffusion coefficients and transport time constants in titania nanotubes were found to be very similar to that of titania nanoparticles.[1] But, the recombination time constants for titania nanotubes were 10 times higher than then nanoparticle based DSSCs. In contrast, DSSCs fabricated using titania nanowires showed both shorter transport time scales and higher electron lifetimes.[2] In a seminal paper, Richter and Schmuttenmaer showed that the lower the electron mobility in TiO2 nanotubes was due to the existence of exciton like trap states formed from Ti^{+3} donor states resulting from oxygen deficiencies, fluoride impurities and excess titanium introduced during synthesis.[3] Although thermal annealing is expected to remove the oxygen deficiencies, the structural integrity of the nanotubes would be compromised and hence a new synthesis technique is needed to prepare defect free TiO_2 nanotubes for enhanced DSSC performance.

Several approaches such as electrochemical anodization, solvothermal techniques, hydrothermal methods, thermal evaporation and template based approaches have been widely explored. However, most the methods produce short titania nanotubes (~ 2 μm in length) which is undesirable considering the low dye loading and slow electron transport in these architectures. In addition, all these synthesis techniques discussed above are very time consuming with the reaction time scales ranging from several hours to even days. To our knowledge there have not been any studies which have focused on the fast synthesis of single crystalline titania nanowires. Here we present a novel, scalable approach for making single crystalline titania nanowires arrays for dye sensitized solar cells.[4] Electron transport and recombination in these single crystalline nanowires and in single crystalline titania nanowires coated with titania nanoparticles are discussed.

[1] Zhu, K., Neale, N.R., Miedaner, A., Frank, A.J. Enhanced charge collection efficiencies and light scattering in dye sensitized solar cells using oriented TiO2 nanotubes. Nanoletters 2007, 7, 69-74.

[2] Tetreault, N., Horvath, E., Moehl, T., Brillet, J., Smajda, R., Bungener, S., Cai, N., Peng, W., Zakeeruddin, S.M., Forro, L., Magrez, A., Gratzel, M. High-Efficiency Solid-State Dye- Sensitized Solar Cells: Fast Charge Extraction through Self-Assembled 3D Fibrous Network of Crystalline TiO2 Nanowires. ACS Nano 2010, 4, 7644-7650.

[3] Richter, C., Schmuttenmaer, C.A., Exciton like trap states limit electron mobility in TiO2 nanotubes, Nature Nanotechnology 2010, 5, 769–772

[4] Kumar, V., Kim, J.H. Jasinski, J.B., Clark, E.L., Sunkara, M. Alkali metal assisted gas phase bulk production of titania nanowires. Journal of Crystal Growth and Design 2011, Article Accepted.

See more of this Session: Semiconductor Nanowires and Nanotubes for Next Generation Photovoltaics - From Fundamentals to Devices
See more of this Group/Topical: Topical 5: Nanomaterials for Energy Applications

(609f) Vertically Aligned Silicon Radial p-n Junction Micropillar Array Solar Cells

Kane J. Miller, Chemical Engineering, University of Louisville, Louisville, KY, Kevin M. Walsh, Department of Electrical and Computer Engineering, University of Lousiville, Lousiville, KY and Xiao-an Fu, Chemical Engineering, University of Lousiville, Lousiville, KY

Solar cell applications using silicon have been extensively studied in the past. The theoretical limit of the efficiency of single-crystalline silicon solar cells is well known. Several methods for optimizing silicon solar cells and obtaining the maximum possible efficiency exist in the literature. Among these methods are changes in the geometry and configuration of the solar cells, such as introducing radial p-n junction solar cells.

Micro- and nanoscale radial p-n junction solar cells, in theory, result in increased solar cell efficiency compared to planar p-n junctions by reducing carrier collection length. The geometry of arrays of these radial p-n junctions also introduces internal reflection and light trapping effects, ultimately improving the photon absorption characteristics.

We have fabricated a first-pass vertically-aligned micropillar p-n junction solar cell array with a micropillar diameter of 2 μm, length of approximately 10 μm, and distance of 2 μm between adjacent pillars. These arrays were fabricated using conventional photolithography and deep reactive ion etching on a (100) n-type silicon wafer, followed by boron diffusion using a spin-on dopant. Initial measurements demonstrated an open-circuit voltage of 0.36 V, short-circuit current density of 24 mA/cm^2, a fill factor of 0.6, and efficiency of 5.3%.

Nanopillar p-n junction solar cell arrays of varying pillar diameters and lengths, fabricated using the combination of electron beam lithography and deep reactive ion etching, will be discussed. The test results will be compared to planar silicon solar cells and to theoretical calculations. With various techniques, we expect to increase the efficiencies of our solar cells beyond 10%.

See more of this Session: Semiconductor Nanowires and Nanotubes for Next Generation Photovoltaics - From Fundamentals to Devices
See more of this Group/Topical: Topical 5: Nanomaterials for Energy Applications

(641a) Direct Conversion of Lignocellulosic Biomass to Drop-In Hydrocarbon Fuel

Xuejun Pan and Li Shuai, Biological Systems Engineering, University of Wisconisn, Madison, Madison, WI

This research reports an innovative process (HDA) for producing fuel-grade hydrocarbon from lignocellulosic biomass. The process directly converts polysaccharides in lignocellulosic biomass to furan-based precursors for producing fuel-grade hydrocarbons in one step under mild conditions without prior pretreatment and saccharification of the feedstock biomass. Specifically, in the presence of acetone, halide salt, and small amount of mineral acid, the lignocellulosic biomass undergoes hydrolysis of polysaccharides (both cellulose and hemicellulose) to monosaccharides, dehydration of the monosaccharides to furfural (from pentoses) and hydroxymethylfurfural (HMF, from hexoses), and aldol-condensation of HMF and furfural with acetone, and ends up with the formation of furan-derived oxygenates (furfural- and HMF-acetone adducts) with chain length of C5–C15. The resultant furan-based precursors can be subsequently converted to fuel-grade hydrocarbons via hydrogenation. The process integrates the three reactions (Hydrolysis-Dehydration-Aldol condensation) in one step, and is abbreviated as HDA. At the conditions of 120-140°C and 1-2hour, cellulose and hemicellulose in lignocellulosic material (e.g., ground or chipped wood) were completely hydrolyzed into sugars, and more than 80% of the sugars were selectively converted to the furan-acetone adducts. Meanwhile, lignin in the biomass is extensively depolymerized and extracted. Preliminary evaluation indicated the HDA lignin has high purity, low molecular weight with narrow distribution, and good reactivity. The lignin is expected to have great potential in co-products development.

See more of this Session: Chemical and Catalytic Conversions and Processes for Renewable Feedstocks
See more of this Group/Topical: Sustainable Engineering Forum

(641b) A Method for Rapid Screening of Catalysts for Hydrodeoxygenation of Biomass Intermediates

David K. Johnson, Biosciences Center, National Renewable Energy Laboratory, Golden, CO

The focus of biofuels development is shifting from alcohol fuels to advanced biofuels that would be drop-in replacements for petroleum-derived hydrocarbons. There is a wide range of intermediates that can be produced from biomass by chemical or biological routes, but virtually all require further deoxygenation before they can be used as fuel components. Potential routes for decreasing the oxygen content of biomass components include dehydration, hydrodeoxygenation and decarboxylation (possibly after controlled oxidation). Deoxygenation by HDO is preferred over dehydration and decarboxylation reactions, as it is likely to result in higher yields through conservation of carbon, retention of hydrogen, and the production of more energy dense products. Hydrodeoxygenation (HDO) is a hydrogenolysis process that removes oxygen by converting it to water, however, HDO is often performed at high temperatures and pressures that significantly increase the cost of the transformation. We need to develop HDO catalysts that more efficiently convert biomass-derived components into fuel products that are compatible with the existing fuel distribution infrastructure. The feedstocks for the chemical transformations can include sugars, lignins, bio-oils, and chemicals obtained by biological transformations of sugars and biomass. It is clear from previous research that in order to design commercially viable catalysts for HDO upgrading of biomass-derived intermediates, the catalyst system must (1) exhibit high activity under mild conditions (<200°C), (2) contain low-cost materials, (3) demonstrate high stability in an aqueous environment, and (4) have high selectivity for the desired products. To accomplish this, a microwave heated (to 250°C), pressurizable (to 250 psi), automated reaction system has been assembled to rapidly screen catalyst systems using model compounds. To allow automated operation, hydrogen is provided in-situ from a hydrogen donor material, such as formic acid, *i*-propanol or decalin. The rapid screening not only allows the rapid testing of catalysts, but also of reaction conditions, including temperature, reaction time, solvent, and hydrogen donor. The results from tests with model compounds such as acetovanillone and hydroxymethyl furfural, representing typical lignin and carbohydrate derived intermediates, with heterogeneous catalysts, will be described.

See more of this Session: Chemical and Catalytic Conversions and Processes for Renewable Feedstocks
See more of this Group/Topical: Sustainable Engineering Forum

A Novel Reactive Distillation Approach for Producing 1,3-Dihydroxyacetone from Glycerol

Xi Hong, Carl T. Lira, and Dennis J. Miller
Department of Chemical Engineering and Materials Science, Michigan State University
East Lansing, Michigan 48824

Introduction

Glycerol is a viable renewable feedstock that has emerged as a consequence of triglyceride conversion to biodiesel. In addition to glycerol conversion to commodity chemicals such as propylene glycol and epichlorohydrin, formation of higher-value specialty chemical products such as 1,3-dihydroxyacetone (DHA) will provide income for the biorefinery.

DHA is the active ingredient in sunless tanning products. Currently, DHA is industrially produced by fermentation of glycerol, but this microbial approach is challenging because of low production rates and high recovery costs [1]. As a result, the high selling price of DHA limits access of sunless tanning products in the general population.

Recently, scientists have attempted conventional chemical oxidation processes to make DHA from glycerol using metal catalysts with air or molecular oxygen. The goal is to achieve direct oxidization of the secondary hydroxyl group in glycerol to a carbonyl bond. Unfortunately, this reaction is always accompanied with primary alcohol oxidation and a variety of products (i.e. glyceroaldehyde, glyceric acid, etc.) are produced. The highest yield obtained from this method is in the range of 40-60% [2-4].

We present here a novel chemical process to convert glycerol to DHA that greatly improves the efficiency and reduces cost. The essential feature of this process is the selective formation of a 6-membered acetal of glycerol via reactive distillation, followed by selective oxidation of the secondary alcohol to its corresponding ketone product using ruthenium metal and a chemical oxidant. The ketone product is then hydrolyzed to obtain DHA.

A continuous process of making DHA from glycerol has been designed and will be presented. Reactions involved in the process are rapid and can be carried out under mild conditions. Yields of individual process steps are all in excess of 90%, suggesting that this process has good potential to reduce DHA production costs.

Experiments

Materials

Anhydrous glycerol (99.5%), 1-propanol (99.9%) and 1-butanol (99.9%) were obtained from J.T. Baker. Ethanol (99.5%), 1,1-diethoxyethane (99%) and sodium hypochlorite (15% aqueous solution) were purchased from Pharmco and Aldrich, respectively. Sodium carbonate (> 99%), sodium bromate (> 99%), potassium bromate (> 99.8%), calcium hypochlorite (> 98%), sodium periodate (> 99%), sodium iodide (> 99%) were purchased from Sigma-Aldrich. Amberlyst-15 cationic exchange resin (16-50 mesh), purchased from Sigma-Aldrich, was investigated as the catalyst. Before being used in reactions, it was pretreated at 90°C in a vacuum oven for 2 hours. Ruthenium dioxide hydrate (99.9% trace metals basis, 28% water content) and ruthenium trichloride hydrate (reagent plus, 46% Ru content) were obtained from Aldrich. Hydrogen (99.999%) and helium (99.999%) were used for GC analysis; both were purchased from Linde Gas, Inc.

Analysis

Gas Chromatography Analyses

1-Butanol and 1-propanol were used as internal standard and solvent, respectively, for this analysis. To prepare samples for analysis, 0.16 ml 1-butanol and 0.2 ml of reaction sample were added to a volumetric flask and the mixture was diluted with 1-propanol to 10 ml of solution. A Perkin-Elmer Sigma 2000 gas chromatograph with FID and a DB-Wax capillary column (15m × 0.53mm × 1.0 μm) was used for analysis. Ultra high purity helium was used as the carrier gas; the column temperature was ramped from 40°C (3min) at a rate of 30°C/min to 220°C and held for 1 min. The total program time was 10 min. On column injection was applied to analyze the samples. Injector and detector chamber temperatures were set at 60°C and 300°C, respectively.

GC-MS analysis

The structure of some products was identified using GC-MS analysis (GC-Agilent 6890, MS-Agilent 5973). A 1μL sample was separated in a DB-wax capillary column (30m × 0.25mm × 0.25 μm). The initial column temperature was 40 °C (for 5min), the temperature was then increased at 20 °C/min to 250 °C and held for 10 min. Total program time was 24.5 min. Injector temperature was set at 240 °C. Electron impact (EI) mass spectra were monitored from 20 to 300 m/z. The MS source was set at 230 °C.

Results and discussion

Synthesis of glycerol acetal

GA isomers were synthesized at 45 °C by mixing glycerol (3.1 kg, 33.9 mol), DEE (4.0 kg, 33.9 mol), and Amberlyst-15 cationic exchange resin (120 g) and stirring for 4 h. Following reaction and removal of volatile species (DEE and ethanol) by rotary evaporation, GC-MS analysis of the product mixture showed the presence of four different structures corresponding [5, 6] to the four GA isomers displayed in Table 1.

Table 1. Structure and IUPAC name of GA isomers

	Structure	IUPAC Name	Designation
a)	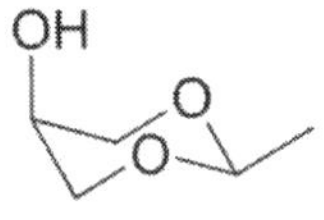	cis-5-hydroxy-2-methyl-1,3-dioxane	Cis-6
b)	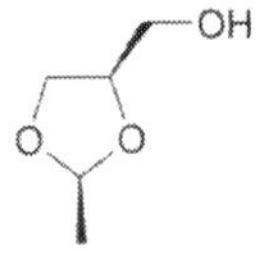	cis-4-hydroxymethyl-2-methyl-1,3-dioxolane	Cis-5
c)	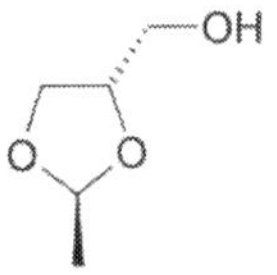	trans-4-hydroxymethyl-2-methyl-1,3-dioxolane	Trans-5
d)		trans-5-hydroxy-2-methyl-1,3-dioxane	Trans-6

GA isomerization equilibrium

In the presence of acid catalyst [7, 8], the four GA isomers interconvert and reach an equilibrated composition. To demonstrate this equilibration, three samples obtained from the distillation experiments below with widely differing isomer compositions (cis-6 from 0% to 84%, cis-5 from 2% to 54%, trans-5 from 2% to 41%, and trans-6 from 0% to 63%) were stirred for 5 hr at 25 °C in the presence of 5 wt% (relative to GA) Amberlyst-15 cationic exchange resin. Identical isomer compositions were observed in the products from the three samples: cis-6, cis-5, trans-5 and trans-6 were 37±1%, 17±1%, 13±1%, and 33±1%, respectively. The molar percentages of 6-membered isomers (cis-6 + trans-6) and 5-membered isomers (cis-6 + trans-5) are approximately 70% and 30%, respectively, at equilibrium.

Rapid isomerization requires vigorous agitation (to ensure mass transfer between solid catalyst and solution) and elevated temperature. For example, Sample 1 took 12 h to reach equilibrium at 0 °C, but only one hour at 80 °C. Increasing temperature also shifts the equilibrium composition. For Sample 1, increasing temperature from 0 °C to 80 °C decreased the sum of cis-6 + trans-6 isomers from 78% to 54%. However, the total quantity of the cis- form isomers (cis-6 + cis-5) and trans- form isomers (trans-6 + trans-5) isomers remained essentially constant over all temperatures at 55±2% and 45±2%, respectively.

Production of the cis-6 GA isomer

A glass vacuum distillation column was constructed to separate the four GA isomers. The column (height = 3.6 m, i.d. = 50 mm) was packed with 0.6 cm 316 Stainless Steel PRO-PAK distillation packing (Scientific Development Co.). The column reboiler, consisting of a two-liter round-bottom flask, was heated with a 750 W electric heating mantle. Vapor temperature was measured at the top of the column.

In the initial distillation, 896 g (800 ml) of a GA mixture containing 35% cis-6, 18% cis-5, 15% trans-5, and 32% trans-6 isomers was placed in the reboiler. To avoid isomer interconversion, the solution was maintained at a weakly basic state by adding 50 g of sodium carbonate to the mixture. This experiment was carried out at a pressure of 20.5 ± 1.0 mm Hg (except for Sample 1, where the pressure was temporarily reduced to 9.0 mmHg).

Twelve 60-mL distillate samples were collected over the course of the distillation and analyzed by GC. Ninety percent of the initial charge to the still pot was recovered as distillate in these samples (Table 2). The material remaining in the still pot consisted primarily of the trans-6 isomer and glycerol; inclusion of this material gave essentially complete (100±2%) closure of the material balance for the experiment.

Table 2. Summary of GA isomers recovered in the initial distillation

	Initial GA charge		Product recovery in distillate		Percent recovered in distillate (%)
	Mass (g)	Composit ion (wt%)	Mass (g)	Percent of total	
Cis-6	311	35	288	36	93
Cis-5	166	18	152	19	92
Trans -5	132	15	120	15	91
Trans -6	287	32	246	30	86
Total	896	100	806	100	90

Figure 1 gives the reboiler liquid temperature, vapor temperature at the top of the column, and composition (via GC) of the distillate samples taken during the distillation. The results clearly show the order of decreasing volatility of the GA isomers as cis-6, cis-5, trans-5, and finally trans-6. As the more volatile isomers are removed from the still pot, liquid and vapor temperature gradually increase.

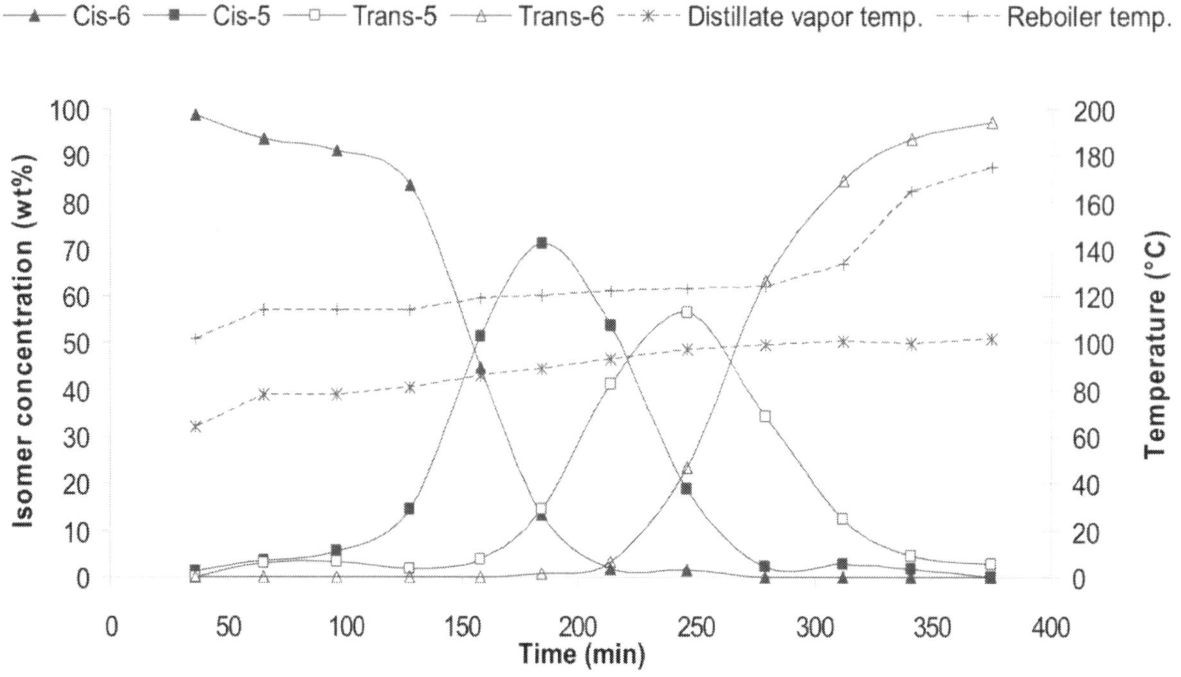

Figure 1. Initial distillation of GA isomer. Column pressure was 20.5±1.0 mmHg (except for the first sample, where the pressure was 9.0 mmHg).

Based on the result of the initial distillation and the fact that GA isomers can interconvert to each other and reach an equilibrated isomer composition in the presence of acid catalyst, a continuous reactive distillation experiment was then carried out to produce the cis-6 isomer as the sole product. In this run, 504 g (450 mL) of a mixture of 36 wt% cis-6, 16 wt% cis-5, 14 wt% trans-5, and 34 wt% trans-6 GA isomers was charged to the reboiler pot of the glass column along with 50 g of Amberlyst-15 cationic exchange resin. For this experiment, the column pressure was maintained at 2.9±0.2 mmHg. Once the distillate product started to condense at the top of the column, a continuous feed of the same GA isomer mixture was introduced to the column reboiler to maintain its liquid level constant. The feed rate was adjusted to match the distillate rate from the column, which was approximately 1.8 mL/min. During the distillation, 16 distillate samples were collected with an average volume of 59 mL.

As shown in Figure 2, this experiment produced samples with high concentration of cis-6 isomer (from 90% to 96%) continuously during operation of the column. The reboiler temperature and distillate vapor temperature were essentially constant during operation, indicating that the reboiler composition was constant and that the column was operating at steady state. The total quantity of distillate obtained was 1028 g, more than twice the amount of the initial charge to the reboiler, demonstrating that continuous production of the cis-6 isomer from the GA acetal mixture is feasible. Based on these results, it is fully expected that an appropriately designed commercial-scale column with sufficient separation capacity would provide essentially pure cis-6 isomer.

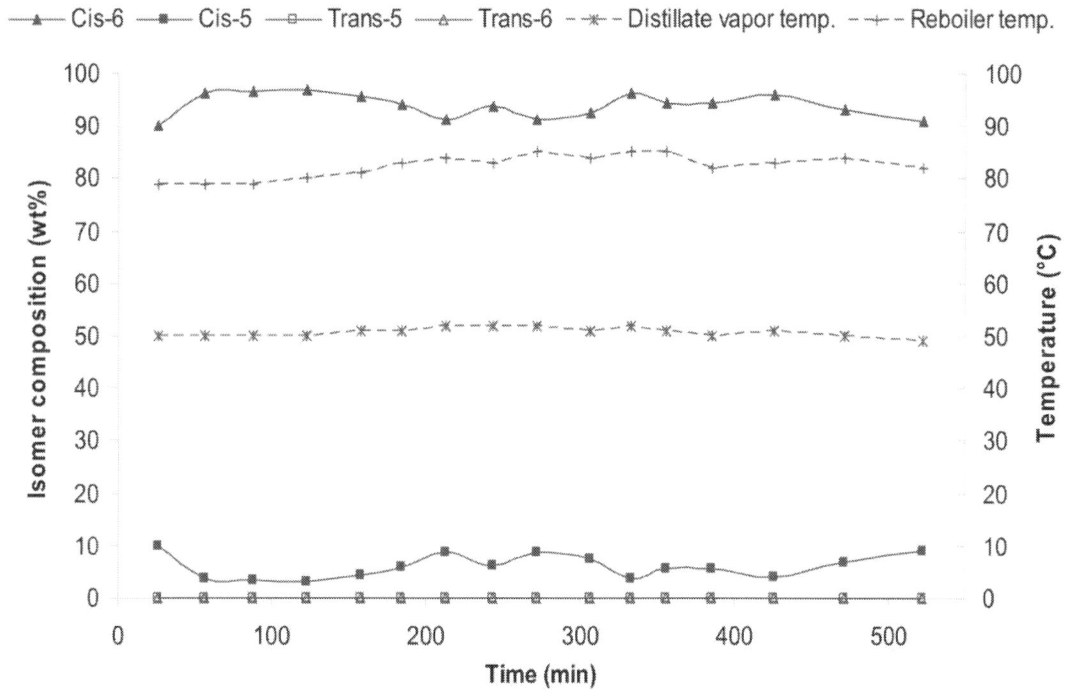

Figure 2. Continuous reactive distillation of GA isomers. Column pressure = 2.9±0.2 mm Hg.

Oxidation of cis-6 to the precursor of DHA

With the availability of cis-6 isomer, a new synthesis pathway can be developed to obtain valuable chemicals. For instance, by oxidizing the remaining free hydroxyl group in cis-6 and hydrolyzing the corresponding ketone product (2-methyl-1,3-dioxan-5-one, G-tone), 1,3-dihydroxyacetone (DHA) can be produced [7].

For this purpose, a variety of oxidation reagents including pyridinium chlorochromate (PCC), ruthenium tetroxide, hydrogen peroxide, molecular oxygen, and ozone were tested to convert cis-6 into its ketone product.

The most effective method is the ruthenium tetroxide-catalyzed oxidation. The reagents used in this reaction involves a ruthenium species (i.e. $RuCl_3$ and RuO_2) and another oxidizing salt (i.e. NaClO, $NaBrO_3$, $NaIO_4$, $Ca(ClO)_2$, $KBrO_3$, etc.) During reaction, the ruthenium species is first oxidized to RuO_4, and then RuO_4 oxidizes cis-6 to the corresponding ketone product. The co-product is a black precipitate of RuO_2, which can be re-oxidized to RuO_4 by the oxidizing salt in the aqueous solution. The regenerated RuO_4 then enters the organic phase again for another cycle of oxidation. In this way, the alcohol substrate can be continuously oxidized to the desired product with catalytic amounts of RuO_4 while the secondary oxidant is consumed. When the reaction is finished, the residue RuO_4 species in the organic phase can be easily recovered by adding isopropanol to make it RuO_2 precipitate [9].

Table 3 presents the results of cis-6 oxidation experiments catalyzed by ruthenium-based substances. It is clear that this type of reaction very efficient. In some reactions over 90% yield of G-tone was achieved within 30 min.

Table 3. Ruthenium tetroxide catalyzed oxidation of cis-6 at 25°C, 1.0 atm.

Reaction	Cis-6 (g)	Water (g)	Catalyst ($RuO_2 \cdot xH_2O$) loading (g)	Secondary oxidant solution/mass (g)/concentration	Time (min)	G-tone yield (%)
1	0.41	10	0	NaClO/4/10%	30	0
2	0.42	10	0	$KBrO_3$/10/saturated	1440	0
3	4.0	50	0.40	NaClO/100/10%	30	80
4	0.39	11	0.31	NaClO/4/10%	30	76
				NaClO/5/10%	60	95
5	0.45	10	0.05	$Ca(ClO)_2$/10/ saturated	20	87
6	0.40	10	0.30[a]	$NaIO_4$/10/saturated	75	88
7	0.42	10	0.26	$KBrO_3$/10/ saturated	45	>99
8	0.43	10	0.32[a]	$KBrO_3$/10/saturated	75	95
9	4.1	10	0.40	$KBrO_3$/100/saturated	20	97
10	0.30	10	0.41	$NaBrO_3$/10/10%	20	96
11	10.0	10	0.51	$NaBrO_3$/100/20%	40	90
12	3.0	10	0.10	$NaBrO_3$/12.5/20%	20	95

a. $RuCl_3 \cdot xH_2O$ was used as catalyst.

Hydrolysis of G-tone to DHA

Hydrolysis of G-tone to DHA is very fast and can be carried out at mild conditions (Scheme 1). In one experiment, 5 wt% (weight to substrate) of Amberlyst-15 cationic exchange resin was added to the final mixture of Reaction 12 in Table 3. When the mixture was heated to 30°C, complete hydrolysis of G-tone was achieved in 10 min.

Scheme 1. Processes of catalytic conversion of glycerol into useful chemicals.

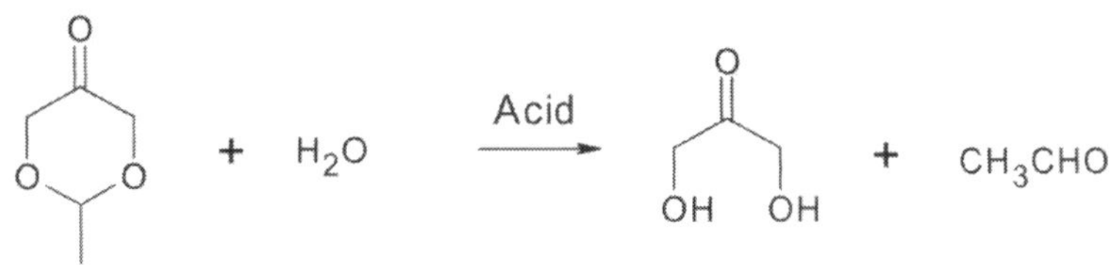

A novel process of making DHA from glycerol

Using cis-6 as an intermediate provides a relatively quick, environmentally safe and economical route to 1,3-dihydroxyacetone (DHA) from glycerol. Based on the experimental results obtained, a novel process of making DHA from glycerol is developed and described here.

Figure 3 presents the overall scheme of a novel DHA production process based on acetaldehyde or 1,1-diethoxyethane (DEE) as the glycerol acetal forming reagent. The process may also work with other aldehydes or ketones. Selection of the acetal forming reagent is dependent on price, availability and process feasibility.

In Figure 3, glycerol and acetaldehyde are introduced to Reactor 1 to form cyclic glycerol acetals. The reaction mixture, which contains four isomers of glycerol acetal, is then evaporated to get rid of water and unreacted acetaldehyde for recycling. After that, the mixture of glycerol acetal isomers is introduced to a vacuum distillation column, in which the cis-6 GA isomer is formed, distilled, and thus obtained from the top of the column. At the same time, other isomers, which have higher boiling points, react in the reboiler to re-establish isomerization equilibrium. The continuous recycling of the less-volatile glycerol acetal isomers through the equilibration-distillation procedure can transform all of the products into the desired 6-membered structure.

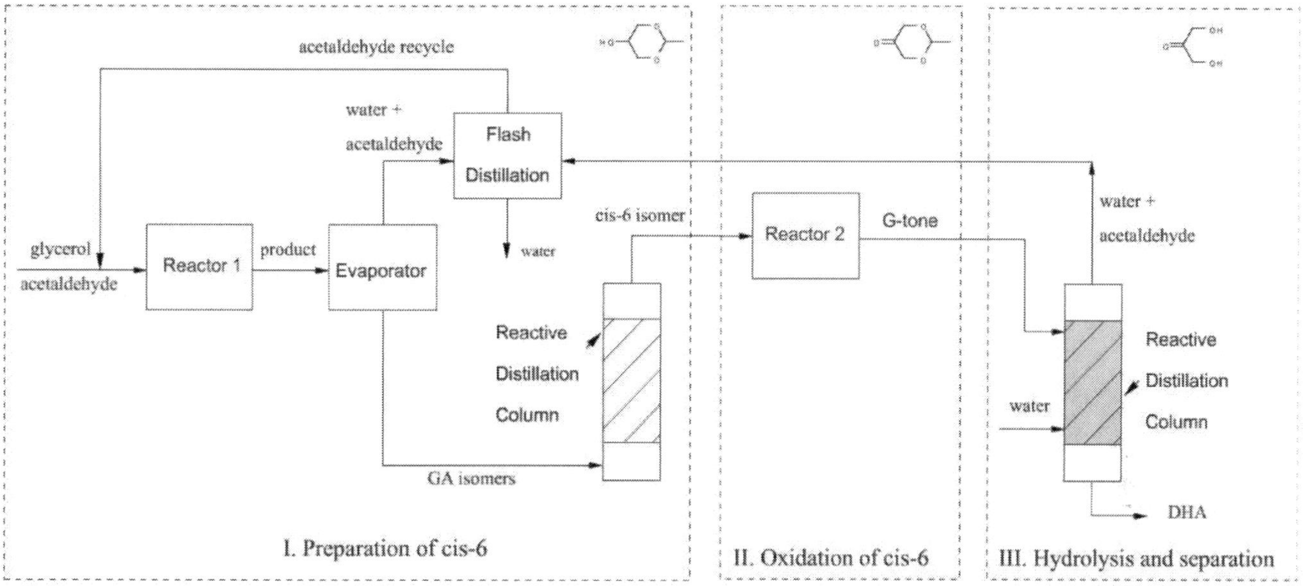

Figure 3. A novel process to produce DHA from glycerol.

In the second stage of this design, cis-6 GA isomer is introduced to the oxidation Reactor 2 to form 2-methyl-1,3-dioxan-5-one (G-tone). A preferred operation for this stage is to use ruthenium tetroxide-related oxidation.

In the last stage, G-tone and an excess amount of water are reacted, perhaps in a reactive-distillation column, to hydrolytically decompose G-tone to the desired product DHA and acetaldehyde, which can be recycled or reformed into DEE. The possible application of reactive distillation along with a large amount of water will overcome any chemical equilibrium limitations of this reaction, such that a high yield of DHA can be achieved from this process.

Summary

A novel process to make 1,3-dihydroxyacetone (DHA) from glycerol is presented in this paper. The process involves three major steps: 1) formation of glycerol acetal (GA) and reactive distillation to get the 6-membered GA isomer 5-hydroxyl-2-methyl-1,3-dioxane (cis-6); 2) oxidation of cis-6 GA isomer to the corresponding ketone product 2-methyl-1,3-dioxan-5-one (G-tone); 3) conversion of G-tone to DHA via hydrolysis. Experimental results show that in each step, over 90% yield of the desired product can be achieved. Thus, the new method holds a good potential for commercial success.

References

1. Mishra, R, Jain, S. and Kumar, A. (2008). Microbial production of dihydroxyacetone. Biotechnology Advances. 26, 293-303.
2. Garcia, R, Besson, M. and Gallezot, P. (1995). Chemoselective catalytic oxidation of glycerol with air on platinum metals. Applied Catalysis A: General. 127, 165-176.
3. Hu, W, Knight, D, Lowry, B. and Varma, A. (2010). Selective Oxidation of Glycerol to Dihydroxyacetone over Pt-Bi/C Catalyst: Optimization of Catalyst and Reaction Conditions. Industrial & Engineering Chemistry Research. 2010, 49, 10876–10882.
4. Kimura, H. (1993). Selective oxidation of glycerol on a platinum-bismuth catalyst by using a fixed bed reactor. Applied Catalysis A: General. 105, 147-158.
5. Da Silva Ferreira, A, Barbe, J. and Bertrand, A. (2002). Heterocyclic acetals f from glycerol and acetaldehyde in port wines: Evolution with aging. Journal of Agricultural and Food Chemistry. 50, 2560-2564.
6. Câmara, J, Marques, J, Alves, A. and Da Silva Ferreira, A. (2003). Heterocyclic acetals in Madeira wines. Analytical and Bioanalytical Chemistry. 375, 1221-1224.
7. Deutsch, J, Martin, A. and Lieske, H. (2007). Investigations on heterogeneously catalysed condensations of glycerol to cyclic acetals. Journal of Catalysis. 245, 428-435.
8. Aksnes, G, Albriktsen, P. and Juvvik, P. (1965). Studies of cyclic acetal and ketal isomers of glycerol. Acta Chemica Scandinavica. 19, 920-930.
9. Tojo, G. and Fernandez, M. (2006). Oxidation of alcohols to aldehydes and ketones. NY: Springer.

(641d) Catalytic Oxidative Dehydration of Butanol Isomers: 1-Butanol, 2-Butanol and Isobutanol

Ivan C. Lee, US Army Research Laboratory, Adelphi, MD

The catalytic production of economically important four-carbon olefins was achieved through catalytic oxidative dehydration of 1-butanol, 2-butanol, and isobutanol using a millisecond contact time reactor. Both alumina foam and rhodium-alumina foam catalysts convert these four-carbon alcohols into four-carbon olefins, with contact time equal to 25 ms over a wide range of equivalence ratios (f) from 0.176 to 6.338 (or C/O_{air} ratio from 0.0659 to 2.11) . The cracking of the carbon backbone was found to be minimal. A mixture of butene isomers was obtained due to the isomerization of the carbocation intermediates. The degree of this isomerization was determined by analyzing the ratios of butenes/(total olefins), 1-butene/2-butene, cis-2-butene/trans-2-butene, and isobutene/(linear butenes). In addition, a reaction mechanism of the catalytic oxidative dehydration of butanol in short contact times was proposed. In this mechanism, the C-O bond of the majority of butanol was broken to form olefinic species on the alumina surface. These surface olefinic species would undergo successive oxidation to form CO and/or CO_2, depending on the availability of oxygen.

See more of this Session: Chemical and Catalytic Conversions and Processes for Renewable Feedstocks
See more of this Group/Topical: Sustainable Engineering Forum

(641e) Effects of Reaction Conditions On the Acid-Catalyzed Hydrolysis of Miscanthus Dissolved In An Ionic Liquid

Sean J. Dee and Alexis T. Bell, Chemical and Biomolecular Engineering, University of California - Berkeley, Berkeley, CA

Miscanthus, a perennial grass, is an attractive biofuel feedstock compared with corn and poplar due to its high energy yield per hectare and low agricultural inputs such as water and soil nutrients. Recent studies have shown the three primary components of miscanthus, cellulose, hemicellulose, and lignin, can be dissolved in ionic liquids (ILs) at levels approaching 20 wt%. An interesting question is then whether the cellulosic and hemicellulosic components of the dissolved miscanthus can be hydrolyzed to sugars in high yield. While acid-catalyzed hydrolysis of cellulose and hemicellulose has been studied extensively in ILs, the conversion of biomass is not as well understood. Recent literature has claimed that effective chemical conversions of biomass in ILs requires either thermochemical delignification which can destroy both hemicellulose and cellulose, or multistage processing that is inefficient and makes IL recycling energy intensive. This study investigated the acid-catalyzed hydrolysis of miscanthus dissolved in the ionic liquid 1-ethyl-3-methylimidazolium chloride ([Emim][Cl]). The effects of varying temperature, acid concentration, miscanthus concentration, and the gradual addition of water were undertaken. The similarities and differences between the rate of hydrolysis of microcrystalline cellulose (Avicel) and amorphous hemicellulose (Xylan from birchwood) compared to the cellulosic and hemicellulosic components of miscanthus were determined. The influence of miscanthus delignification on the hydrolysis rate was also explored. The present work has demonstrated that 90% biopolymer conversion from miscanthus could be achieved, with 93% selectivity to saccharine products in a single stage hydrolysis without pretreatment.

See more of this Session: Chemical and Catalytic Conversions and Processes for Renewable Feedstocks
See more of this Group/Topical: Sustainable Engineering Forum

(641f) Secondary Thermochemical Xylo-Oligomer Hydrolysis of High-Solids Dilute-Acid Pretreatment Slurries

Erik M. Kuhn[1], Joseph Shekiro[1], Nick J. Nagle[2] and Richard T. Elander[1], (1)National Renewable Energy Laboratory, Golden, CO, (2)National Bioenergy Center, National Renewable Energy Lab, Golden, CO

Dilute-acid pretreatment is a critical step in the biochemical conversion of biomass to fuels and chemicals, releasing hemicellulose as oligomeric and monomeric xylose and rendering the cellulose fraction readily digestible by enzymes. It has been difficult to achieve very high yields of monomeric xylose during pretreatment, since acid hydrolysis also converts xylan to various oligomers and non-fermentable sugar degradation products. An alternative strategy for the production of high yields of monomeric xylose is to perform secondary thermochemical hydrolysis on the dilute-acid pretreated material. This secondary hydrolysis, typically performed at lower temperatures than the primary dilute-acid pretreatment step, liberates additional monomeric xylose while minimizing the yield of sugar degradation products. This work presents the results of laboratory and pilot scale primary and secondary thermochemical hydrolysis experiments over a wide range of conditions. Our results suggest that reduced severity conditions for primary dilute-acid pretreatment provide the highest overall yields of monomeric xylose after secondary thermochemical hydrolysis for the conditions tested. The results of our LC-MS investigations suggest that higher-molecular weight xylo-oligomers are converted to xylose during secondary thermochemical hydrolysis, but that a small fraction of the xylan in corn stover is converted to stable but non-fermentable compounds, such as xylo-oligomers, during both primary and secondary hydrolysis.

See more of this Session: Chemical and Catalytic Conversions and Processes for Renewable Feedstocks
See more of this Group/Topical: Sustainable Engineering Forum

(650a) Physical Properties and Rheological Characteristics of Deasphalted Oil Produced From a Solvent Deasphalting Process

Jung Moo Lee, Ji Won Hwang, Sang Chul Shin, Hyun Wook Jung and Ki Bong Lee, Department of Chemical & Biological Engineering, Korea University, Seoul, South Korea

Because of conventional oil being-exhausted and its increasing price, the technologies to use unconventional oil and low value crude residues, which have not been fully exploited, are receiving more and more interest. Unconventional oil contains a lot of asphaltene which hinders utilizability of unconventional oil. As the asphaltene has high viscosity and contains large amounts of heavy metals, it decreases the fluidity of oil and causes catalyst poisoning. Furthermore, the process of desulfurization and denitrification are necessary to use unconventional oil because the asphaltene has a high sulfur and nitrogen content. As the unconventional oil has these drawbacks, an extraction process called solvent deasphalting is commonly used to acquire deasphalted oil(DAO) through the removal of asphaltene. The DAO is generally used for lube base oil, and besides, it is converted into transportation fuel and chemical raw materials by additional refinement. In this study, the physical properties and rheological characteristics of DAO were measured for more efficient utilization of DAO. Pentane and hexane were added into DAO to investigate the effect of the solvents on the viscosity of DAO solution. As the content of pentane and hexane was increased, the viscosity of DAO was decreased, resulting in the improved fluidity of DAO. The degree of decreasing rate in viscosity was declined with increasing amount of solvent. The viscosity of DAO was characterized using a rheometer under various temperature conditions to investigate the effect of temperature on the fluidity. The fluidity of DAO was rapidly increased with increasing temperature and the DAO showed the behavior of a Newtonian fluid above 70 ^{O}C, which is called the softening point. Additionally, the fluid behavior of DAO in a pipe was simulated using computer softwares, GAMBIT and FLUENT. The velocity profile of DAO in a pipe was rapidly changed around the softening point.

See more of this Session: Heavy Oil and Flow Assurance I
See more of this Group/Topical: Energy and Transport Processes

(650b) Asphaltene Instability Measured by Small-Angle Neutron Scattering

Michael P. Hoepfner, Chemical Engineering, The University of Michigan, Ann Arbor, MI, Claudio Vilas Boas Favero, Chemical Engineering, Universidade Estadual de Maring, Maring, Brazil and H. Scott Fogler, Chemical Engineering, University of Michigan, Ann Arbor, MI

Asphaltenes are defined as the petroleum fraction that precipitates with the addition of a normal alkane; however, it was only recently discovered that there is a strong time-dependency for precipitation in oil/n-alkane mixtures [1]. It was found that at low precipitant concentrations, asphaltenes would eventually precipitate if one simply waited, sometimes up to several months. Asphaltenes were also recently found to deposit before the precipitation could be detected by optical microscopy, further revealing asphaltene instability [2]. Small-angle neutron scattering (SANS) was selected as a technique that could detect subtle changes in the asphaltene size and shape once a precipitant is added. Optical microscopy requires asphaltene aggregates to grow to ~0.5 μm before the precipitation can be detected. With SANS, changes in size or shape of a few nanometers can easily be measured.

Experiments were performed on the CG-2 SANS instrument at Oak Ridge National Laboratory. Oil samples were diluted with heptane at various concentrations, and were allowed to mix for times ranging from less than 1 day and up to 6 months prior to performing the scattering experiments. The goal was to investigate the asphaltene stability over a wide range of precipitant concentrations and to monitor the aggregation process over time. If asphaltenes are found to be stable at low enough precipitant concentrations, the scattering results will indicate that there is no change in the morphology of an asphaltene nanoaggregate and that any change in the scattering would be due to dilution and scattering contrast effects.

It was found that asphaltenes grew in size with the small addition of precipitant, as low as 10 vol. % heptane, for an oil with an instantaneous onset condition of 40 vol. % heptane. The asphaltenes continued to grow over time. Attempting to detect this instability with optical microscopy may not be possible due to the long-term precipitation/aggregation kinetics [1]. Experiments were also performed on a second oil and a model oil consisting of 3 wt. % asphaltenes dispersed in d8-toluene. Additionally, the asphaltenes remaining in solution were much smaller than those that precipitated, suggesting that the largest of the asphaltene nanoaggregates are the most unstable and precipitate first.

These findings question whether asphaltenes are ever truly stable and a small fraction of asphaltenes may precipitate with any slight perturbation. Future models to understand asphaltene stability need to consider this effect and the assumption of equilibrium may not be valid for a wide range of conditions.

[1] T. Maqbool, A. T. Balgoa, and H. S. Fogler, "Revisiting Asphaltene Precipitation from Crude Oils: A Case of Neglected Kinetic Effects," *Energy & Fuels*, vol. 23, no. 7, pp. 3681-3686, Jul. 2009.

[2] M. Hoepfner and H. S. Fogler, "Effects of Preciptant Concentration on Asphaltene Deposition," presented at the *10th Annual International Conference on Petroluem Phase Behavior and Fouling*, Rio de Janiero, Brazil, 2009.

See more of this Session: Heavy Oil and Flow Assurance I
See more of this Group/Topical: Energy and Transport Processes

(650c) Modeling of Asphaltene Deposition In a Vertical Tubing

Dmitry Eskin, John Ratulowski, Kamran Akbarzadeh and Simon Andersen, DBR Technology Center, Schlumberger, Edmonton, AB, Canada

Asphaltene depositionphenomena are investigated both theoretically and experimentally. A Couette device, where the inner cylinder rotates andparticles deposit on the outer wall, is used for laboratory studies of thedeposition process. The modeling approach, recently proposed by the authors(Eskin et al.[1]), is improved. Thedeposition model is composed of the sub-models describing asphalteneagglomerate growth and their transport to the wall. The agglomerate growthmodel is simplified without reducing its accuracy and the more accuratemodel of particle transport to the wall is employed. According to thedeposition model, tuned by the experimental data, only sub-micron asphaltene particles can deposit on the pipe wall and onlya very small fraction of particles touching the wall sticks to itssurface. Those phenomena are explained based on an analogy between thedeposition and water in oil emulsion stabilization by asphaltenes,and on an analysis of interaction of asphaltenemolecules. The deposition model performance is illustrated by modeling ofoil production, accompanied with asphaltenedeposition, from a cylindrical reservoir through vertical tubing. Thecalculations (see figure below), performed for a reservoir depleting over time,demonstrate an excellent visual agreement with the filed datareported in literature[2].

References

1. Eskin D., Ratulowski J., Akbarzadeh K., Pan S. Modeling Asphaltene Deposition in Turbulent Pipeline Flows. Canadian Journal of Chemical Engineering. 2011(in Press).

2.Haskett CE and Tartera M. Apractical solution of the problem to the asphaltenedeposits - Hassi MessaoudField, Algeria. Journal of Petroleum Technology. 1965;April: 387-391.

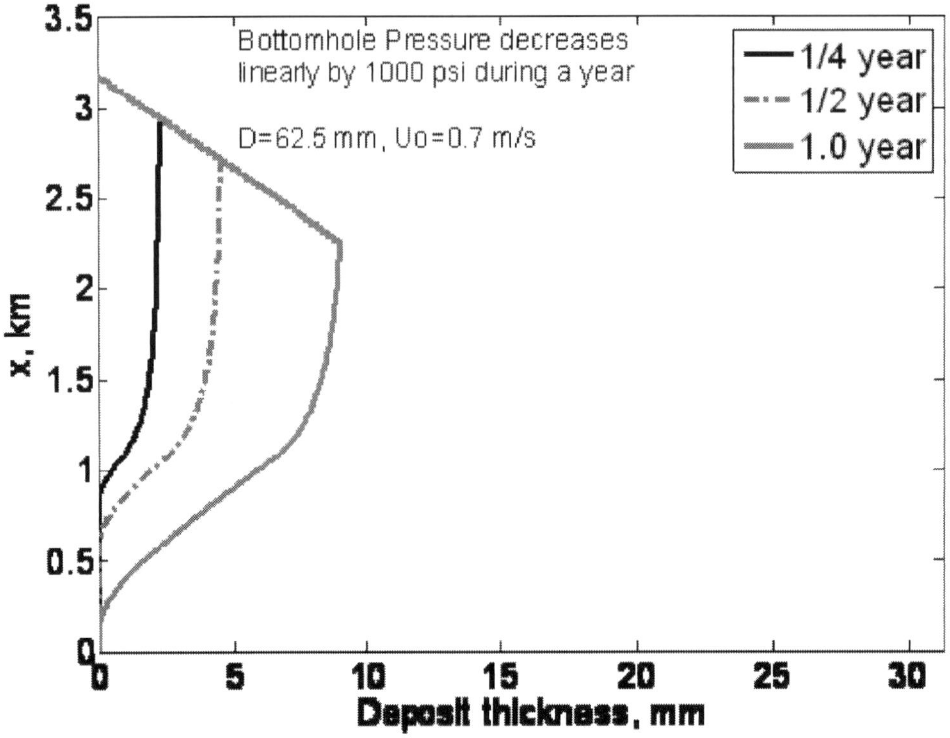

See more of this Session: Heavy Oil and Flow Assurance I
See more of this Group/Topical: Energy and Transport Processes

Phase behavior of asphaltenes + organic diluents

Kasra Nikooyeh and John M. Shaw
Department of Chemical and Materials Engineering
University of Alberta
Corresponding Author: John M. Shaw (jmshaw@ualberta.ca)

Asphaltenes, defined operationally on the basis of ASTM standards that include mixing and macroscopic filtration steps, are typically asserted to be soluble in toluene [1]. A similar definition is used to classify organic diluents as solvents and non-solvent for asphaltenes. In the analysis of phase separation mechanisms of asphaltenes from hydrocarbon media, the assumption of solubility in the original medium is a key one. From a solution thermodynamics standpoint, dissolution of a solid into a liquid includes the enthalpy of transition to liquid for the solute plus the enthalpy associated with homogeneous mixing at a molecular level [2]. The present work challenges the application of the concept of solubility to mixtures of asphaltenes with organic diluents and suggests an alternative explanation for the behavior of asphaltenes in organic liquids based on thermodynamic measurements.

Solution calorimetry and differential scanning calorimetry (DSC) were performed on mixtures of asphaltenes with a variety of diluents ranging from n-alkanes to bicyclic aromatics. Athabasca n-pentane asphaltenes were mixed with diluents at compositions ranging from 0.1 to 10 wt%. In each case 25 mg of asphaltenes were added to 25 ml of a mixture with an underlying composition. Solution calorimetry measurements performed using a TA Instruments TAM Precision solution calorimeter, in the range 15 to 80 °C provided information on phase change with composition at fixed temperature. DSC measurements with a TG-DSC 111 (Setaram, France) calorimeter, provided information on phase change with temperature at fixed composition and on the reversibility of such phase behavior.

These solution calorimetry measurements reflect the net enthalpy change resulting from addition of asphaltenes to diluents. The value of enthalpy of solution is a weak function of temperature and composition if dissolution is complete, i.e.: the solute undergoes a solid-liquid phase change. This behavior was demonstrated for the benchmark case pyrene (solute) + toluene (diluent). The enthalpies measured for asphaltenes in all diluents used in this study, both integral [ΔH^{sol}] and differential, [$d\Delta H^{sol}/dx$] measurements, however, change significantly as temperature and composition are varied.

The enthalpies trend to higher values with increasing temperature and to lower values at increasing asphaltene concentration at fixed temperature on a per unit mass of asphaltene basis as shown in Figure 1. These results are in agreement with the hypothesis that only a fraction of asphaltenes undergo a solid-liquid phase change upon mixing with diluents and that fraction is a function of diluent properties, global composition and temperature.

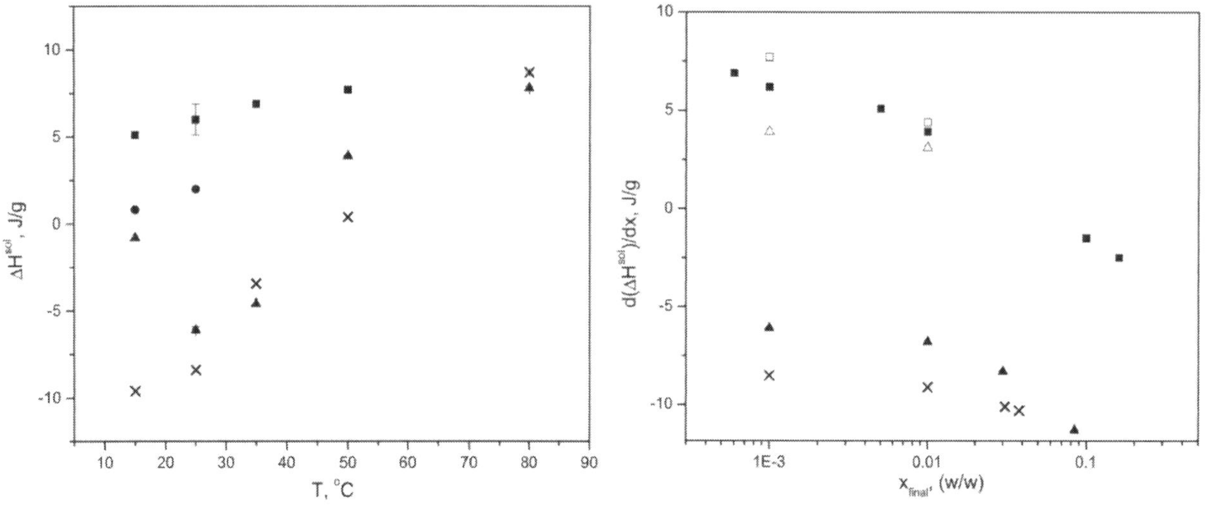

Figure 1 a) Enthalpy of dispersion of Athabasca C_5 Asphaltenes at 0.1 wt % in toluene (■), tetrahydrofuran (●), 1-methylnaphthalene (▲) and quinoline (×). b) Enthalpy of dispersion of Athabasca C_5 Asphaltenes in toluene at 25 °C(■) and 50 °C (□) and 1-mthylnaphtalene at 25 °C(▲) and 50 °C (Δ) quinoline (×) as a function of concentration

Differential scanning calorimetry measurements for asphaltene + toluene and 1-methylnaphthalene mixtures allow for determination of the enthalpy of solution at higher concentrations. The results with toluene, an example of which is plotted in figure 2, show that mixing asphaltenes with toluene at high concentrations (5 wt% and above) is largely exothermic, a result consistent with the solution calorimetry data.

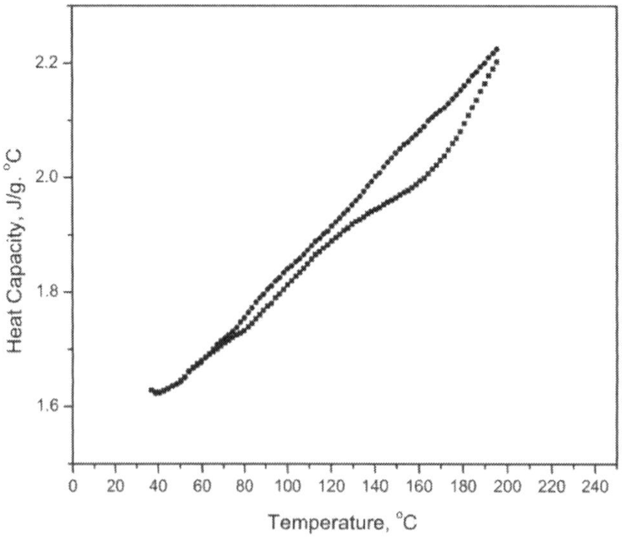

Figure 2. Apparent heat capacity of C_7 asphaltenes added to toluene, derived from differential scanning calorimetry measurements. The heating rate was set to 5 °C/min and the composition range was 5 to 20 wt%. First run (■), second run, (●)

The apparent heat capacities for asphaltenes derived from the DSC thermograms for toluene + asphaltene mixtures also illustrate that the impact of mixing on phase behavior is irreversible, in the temperature interval ~120 to ~200 °C as shown in Figure 2 where the values differ significantly between repeated runs with the same sample.

Over all, these results are inconsistent with the application of dissolution thermodynamics concepts, as classically defined, to asphaltene + diluent mixtures. The ranges of values, and their trends with temperature and composition are consistent with overlapping exothermic and endothermic processes occurring simultaneously that involve only a fraction of the asphaltenes.

References

1. ASTM-D6560, *Standard Test Method for Determination of Asphaltenes (Heptane Insolubles) in Crude Petroleum and Petroleum Products*. 2000, ASTM International: West Conshohocken, PA.
2. Zhang, Y., et al., *Estimating the Interaction Energy of Asphaltene Aggregates with Aromatic Solvents*. Energy Fuels, 2005. **19**(3): p. 1023-1028.

(650e) Study of Asphaltene Precipitation From Brazilian Petroleum

Mara T. Fukunaga[1], Eduardo T. Iamazaki[1], Felipe M.R. Cardoso[2], Watson Loh[1] and **Paulo T.V. Rosa**[1], (1)Institute of Chemistry, State University of Campinas, Campinas, Brazil, (2)State University of Campinas, Campinas, Brazil

Petroleum is a complex fluid containing several classes of compounds dispersed or solubilized in a liquid phase. During the petroleum production and processing, changes in pressure, temperature, and composition can decrease the stability of the dispersed systems, causing precipitation of a solid phase, composed mainly by asphaltenes. Asphaltenes are recognized as a mixture of species of high molecular weight, with a reasonable polarity, and different functional classes[1]. The aggregative nature of asphaltenes is related to the presence of heteroatoms in its structure that results in a superficial activity for this compounds[2,3]. The precipitation and deposition of asphatenes is a serious problem that can affects the production and refining process. Some methods are used to prevent deposit or its removal, but the use of precipitation inhibitors and/or dispersants is a feasible solution[4]. This work presents a kinetic study of asphaltene precipitation in two different types of Brazilian crude oil (P1 and P2). N-heptane was used as precipitant and three different types of precipitation inhibitor or dispersant (I1, I2 and I3) were tested. The efficiency of the additive was accessed by the sedimentation curves of the precipitated asphaltenes followed using a Hewlett Packard spectrophotometer, model 8453. Samples were prepared by weighing 0.1 g of oil, followed by the addition of different concentrations of additive, and approximately 10 g of n-heptane. After homogenizing the solution, 3 mL was placed into a quartz cuvette and the absorbance was monitored at 850 nm for a period of 3600 seconds. All tested additives presented some efficiency on the precipitation of asphaltenes but I2 needed lower concentration to stabilize the system.

[1] Loh, W., Mohamed, R.S., Ramos, A.C., "*Aggregation of asphaltenes obtained from a brazilian crude oil in aromatic solvents*", Petroleum Science and Technology, 17, 147-163, 1999.

[2] Mohamed, R.S., Loh, W., Ramos, A.C., "*Aggregation behavior of two asphaltenic fractions in aromatic solvents*", Energy & Fuels, 13, 2, 323-327, 1999.

[3] Mansoori, G.A., "*Modeling of asphaltene and other heavy organics deposition*", J. Pet, Sci. Eng., 17, 101-111, 1997.

[4] Stephenson, W.K., "*Producing asphaltenic crude oils: problems and solutions*", Petroleum Eng. International, 24-31, 1990.

See more of this Session: Heavy Oil and Flow Assurance I

(650f) Characterization of Asphaltenes In Stock-Tank Oil and Prediction of Its Phase Behavior At Reservoir Conditions

Francisco M. Vargas[1], Senthilmurugan Balasubramanian[2], Sameer Punnapala[1], Alfredo Halique[1], Sama Al Hallaq[1], Salama Al Marzouqi[1], Sara Al Hashmi[1] and Ismail Alkhatib[1], (1)Chemical Engineering, The Petroleum Institute, Abu Dhabi, United Arab Emirates, (2)Petroleum Engineering, The Petroleum Institute, Abu Dhabi, United Arab Emirates

Asphaltene deposition is one of the most widely found flow assurance problems in upstream and downstream operations. It is well known that asphaltene precipitation is a necessary but not a sufficient condition for deposition. Although substantial amount of work has been done on this area, the mechanisms by which asphaltene precipitate and deposit are still being actively investigated.

Current methods used to determine the stability of asphaltenes in crude oil systems are based on near-infrared spectrometry and high pressure microscopy. Although these methods are widely used and accepted in the industry, they require time and can be expensive. This is the case particularly when a comprehensive study requiring different gas injections and temperatures needs to be performed.

In this work we propose as an alternative method the determination of the asphaltene behavior in stock-tank oil at room conditions by determining the onset of precipitation upon addition of different n-alkanes, which are known to be asphaltene precipitants. Determination of onset of asphaltene precipitation is done by using NIR spectrometry at room conditions. These data can then be used to tune asphaltene parameters of a model. This model will then be used to extrapolate the behavior at reservoir conditions including the associated gas. Comparison with data at high pressure and temperature is also presented and discussed.

Because the precipitation and flocculation of asphaltenes is a process greatly affected by kinetic factors, systematic experimentation was done to determine the effect of time on determining the asphaltene precipitation onset. Including this information, a detailed experimentation procedure was developed that matches the experiments done at high pressure and temperature. Different crude oils were studied to validate the proposed methodology.

Finally, the spectrometric methodology was also applied to the study of the effect of asphaltene inhibitors. Advantages of the proposed methodology over current experimental methods will be highlighted in this presentation. Also, important insight into the mechanism of action of asphaltene inhibitors will be presented and discussed.

(690a) Development of New Redox Flow Battery Systems

Liyu Li[1], Wei Wang[2], Soowhan Kim[2], Vijayakumar Murugesan[2], Zimin Nie[2], Baowei Chen[2], Qingtao Luo[2], Feng Chen[2], Jun Liu[2], Jianzhi Hu[2] and Gary Yang[2], (1)Hydrocarbon Processing Group, Pacific Northwest National Laboratory, Richland, WA, (2)Pacific Northwest National Laboratory, Richland, WA

Concerns over the environmental consequences of burning fossil fuels and their remaining reserves have led to an increasing use of renewable energy generated from sources such as solar and wind. However, the intermittent, varied nature of the renewable resources makes it difficult to integrate these valuable energies into the electrical grids. A solution to this problem is to employ large-scale electrical energy storage (EES), which is widely considered as an effective approach to improve the reliability, power quality, and economy of these renewable energies. Among the most promising large-scale EES technologies are the redox flow batteries (RFBs). RFBs are special electrochemical systems that can repeatedly convert and store multi-megawatt-hours (MWhs) of electricity energy to chemical energy and convert the chemical energy back to electricity energy when needed. The advantages of RFBs include short response time, low self discharge, long life-time, and independent tunable power and storage capacity, which make it possible to independently scale up the electricity storage capacity and power generation capacity. With the support of US Department of Energy's Office of Electricity, PNNL has carried out extensive efforts in the development of advanced redox flow batteries. Substantial progress has been achieved in the past two years, including two new redox chemistries (V-Fe system and all vanadium mixed acid system), which will be reported in this presentation

See more of this Session: Energy Storage and Electrochemical Processes I
See more of this Group/Topical: International Congress on Energy 2011

Study on preparation and characterization of the composite bipolar plate for Vanadium Redox Flow Battery

XIE Xiao-Feng[1], WANG Wen-Pin[2] , WANG Shu-Bo[1], WANG Jin-Hai[1],

LV Ya-Fei[2], MATHUR V. K[3]

1 Institute of Nuclear and New Energy Technology, Tsinghua University, Beijing 100084, China

2 College of Material Science And Engineering, Beijing University of Chemical Technology, Beijing 100029,China

3.Department of Chemical Engineering, University of New Hampshire, NH 03824, USA

Bipolar plate, as one of the main components of all-vanadium redox flow battery (VRB), can collect the current generated by electrochemical reaction, and can separate positive and negative electrolyte. The electrolyte is a sulfuric acid solution containing vanadium ions (V(IV/V,II/III)/ V(V)). Among these ions, the V(II) has the most strong oxidation ability. Therefore, the bipolar plate not only should have good conductivity, but also have high resistance to chemical and electrochemical corrosion.

Bipolar plate is a carbon composite material, which is a mixture of carbon materials and a polymer. The polymers used are polyethylene, nylon, polypropylene (PP), rubber modified polypropylene, polytetrafluoroethylene and polyvinylidene fluoride, etc. Polymer is helpful to form the conductive network, and improve the corrosion resistance of bipolar plate.

Maleic anhydride grafted polypropylene(g-PP) is a non-polar molecules in the main chain with a strong polar side groups maleic anhydride. It can increase the bonding and the compatibility between polar materials and non-polar materials for enhancing the dispersion between the conductive filler and the polymer matrix, and to improve conductive property and corrosion resistance of the bipolar plate.

In this work, composite bipolar plates, by polypropylene (PP), polypropylene-maleic anhydride graft copolymer (g-PP) as polymer matrix, and flake graphite(GC) as conductive filler, were prepared using a solution intercalation method. Four probes method, potentiodynamic experiments, potentiostatic and cyclic voltammetry(CV) were used to investigate the conductivity, corrosion resistance and electrochemical performance of bipolar plates. The results show that g-PP made the graphite uniformly dispersed in the polymer, which significantly improved compatibility between the conductive filler and the polymer matrix. The conductivity of the composite bipolar plate was about 77.8 $S \cdot cm^{-1}$. When the voltage was 1.2 V, the corrosion current was about $5.74 \times 10^{-2} A \cdot cm^{-2}$. The prepared bipolar plate could meet the requirements of VRB.

EXPERIMENTALS

The conductivity of samples (5 cm×1.5 cm) was measured by a four probe resistance tester. Other tests carried out were: Cyclic voltammetry: -1.5~2.0V, scanning speed

10mv·s-1; potentiodynamic polarization curves: -1.5~2.0V, scanning speed 5v·s-1; potentiostatic test (1.2 V, scan time 3600 s). Electrochemical measurements used three-electrode cell with the sample as the working electrode. A Pt electrode as the counter electrode and a saturated calomel electrode as reference electrode (SCE) were used. A homemade VRB electrolyte was used for the experiments.

A certain amount of GC, PP, g-PP and xylene were placed in three-necked bottle, stirred, heated to reflow temperature and kept under reflux for a period of time. The product was precipitated by a solvent, leached, dried, and hot-pressed to form the composite conductive plate.

RESULTS AND DISCUSSION

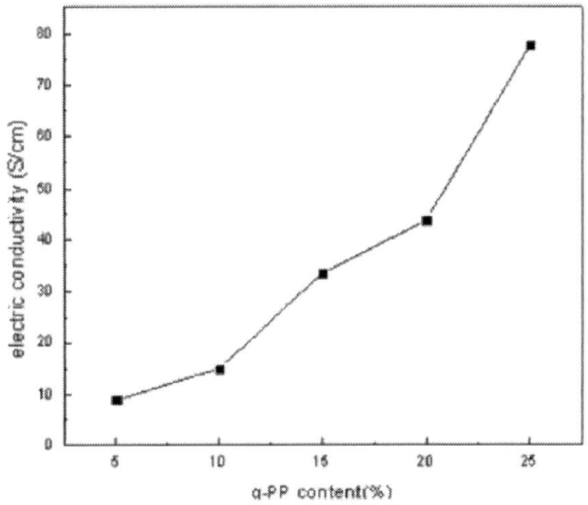

Fig.1 Effect of g-PP contents on conductivity

GC is the most familiar of the natural graphite materials. It has good conductivity. For a GC content of 70wt%, bipolar plates were prepared by using the g-PP content 10wt%, 15wt%, 20wt% or 25wt%. The effect of g-PP contents on conductivity of bipolar plates are shown in Fig. 1. It can be seen that with an increase in g-PP content, the conductivity of bipolar plate increases steeply. It means that g-PP addition contributes to the uniform distribution of the GC in the matrix, resulting in the conductive network structure formation and thus conductive property improvements. Polyacrylonitrile graphite felt (PAN-based) as electrode materials is widely used in VRB. Its conductivity commonly is about 50 $S \cdot cm^{-1}$. It is preferred to have the g-PP content greater than or equal to 20 wt% for collecting current.

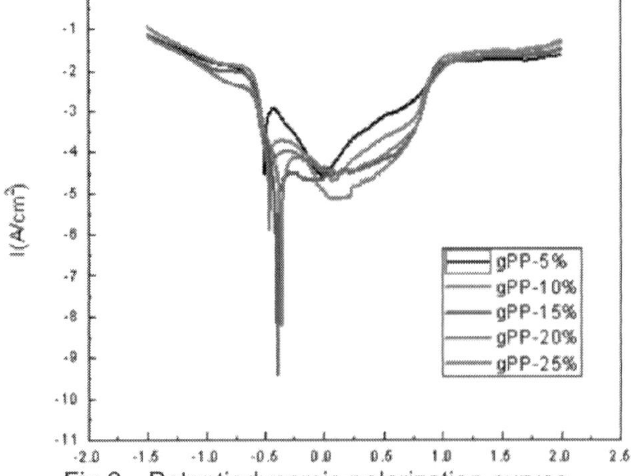

Fig.2 Potentiodynamic polarization curves

Fig.2 shows the potentiodynamic polarization curves of bipolar plates. It can be seen that as the voltage increases, anode passivation of bipolar plate occurs. That means the corrosion of bipolar plate is mainly due to oxygen corrosion. As the g-PP content increases, the corrosion current of bipolar plate decreases first and then increases. When the g-PP content of bipolar plate is 20 wt%, the corrosion current is the lowest.

The effect of g-PP contents on corrosion current are showed in Fig.3. It can be seen that with the increase in g-PP content, the electrochemical corrosion current of bipolar plate decreases rapidly, and then runs through the transition state into the steady-state. When g-PP contents is about 5%, the corrosion current density is 1.8×10^{-2} A/cm^2. With the g-PP content increasing, the corrosion current density reduces to 9.1×10^{-3} A/cm^2. Moreover, once corrosion occurs, it will cause intergranular corrosion or pitting corrosion, and then the corrosion rate further increases. When the g-PP content is 20wt%, the corrosion current of bipolar plate is close to the minimum, which is approximately 8.9×10^{-3} A·cm^{-2}.

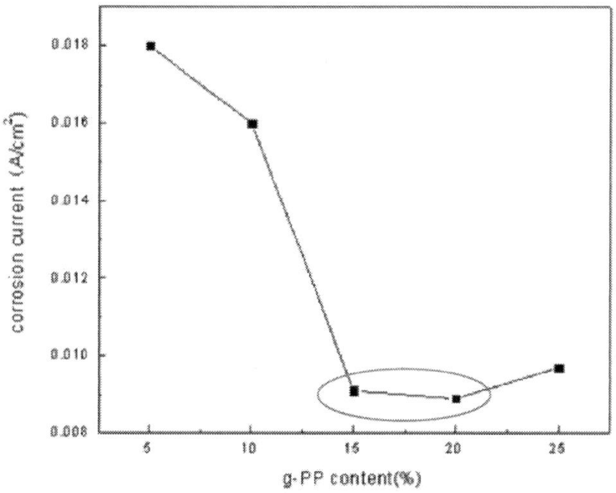

Fig.3 Effect on g-PP contents on corrosion current

Fig.4 shows cyclic voltammetry of bipolar plates under various g-PP contents. With the g-PP content increasing, more V(II) and V(III), V(IV) and V(V) transformation peaks were observed with reversibility. When g-PP is about 10%, it has the high activity. If it was used as bipolar plate in battery, the bipolar plate was easily corroded. For oxygen evolution peak of bipolar plate at anode, the higher the oxygen evolution potential, the more stable the battery is. When the g-PP content is 20wt%, there is low activity and high oxygen evolution potential of the bipolar plate providing the best the corrosion resistance of the bipolar plate.

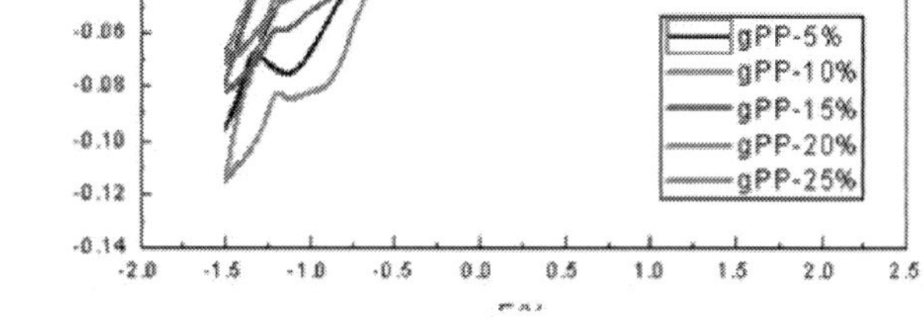

Fig.4 Cyclic voltammetry of bipolar plates

With the increase of g-PP content, the dispersion of GC in the polymer matrix is more uniform, the holes in the bipolar plate surface are reduced, and vanadium ions have difficulty in attaching to the surface of bipolar plate. As the g-PP content increases to 25wt%, the functional group COO^- content of the bipolar plate surface increases. COO^- can contribute to vanadium ion adsorption from solution to the surface. With the reaction of vanadium ion adsorption and desorption, the charge transfer should obviously change. It makes the electrochemical reaction of the vanadium ions on the surface of the bipolar plate to easily happen. These results are in agreement with the previous investigation.

CONCLUSIONS

Analysis of potentiodynamic, potentiostatic experiments and cyclic voltammetry shows when the GC content is constant, within a certain range, conductivity of bipolar plate is proportional to the g-PP content. The g-PP is added to make the dispersion between GC and the polymer matrix and improve the compatibility between the conductive filler and the polymer matrix significantly. Addition of g-PP leads to improving the corrosion resistance of bipolar plate. Conductivity and corrosion resistance of composite bipolar plate can meet the requirements of VRB.

ACKNOWLEDGEMENTS

The work was funded by Tsinghua University Initiative Scientific Research Program (201010800623), State Key Laboratory for Automotive Safety and Energy(KF10132), State Key Lab of Power Systems(SKLD10KZ03).

REFERENCES

[1] WANG Wenpin , WANG Jinhai, WANG Shubo, XIE Xiaofeng , LV Yafei. Properties of Conductive Bipolar Plate for All-Vanadium Redox Flow Battery [J]. The Chemical Industry and Engineering Society of China, 2011, 62(SI): 203-207

[2] **http://www.asbury.com/Natural-Flake-Graphite.html**

[3] Xia Liping, Lin Changwu, Luo Dongmei, Zhang Aimin. Study on properties of fluoro rubber composite current collector for vanadium redox battery [J]. Power Sources, 2010, 134(7): 672-675

[4] Chen maobin, Zhang Shengtao, Li Xiaobing, Liu Lian, Liu Xiaojiang, Meng Fanming. Preparation and performance of polytetrafluoroethylene conducting plastic on vanadium redox flow battery collector [J]. Polymer Materials Science And Engineering, 2009, 25(6): 121-124

[5] Chen Maobin, Lan Wei, Zhang Shengtao, Cui Yanhua, Meng Fanming, Liu Xiaojiang. Performance of Conductive Polyethylene Composite [J]. Applied Chemistry, 2009, 26(5): 619-621

Investigation on electrode reaction mechanism for vanadium redox flow battery

XIE Xiaofeng[1], LI Xiaoshan[1], YANG Chun[1], WANG Jinhai[1], WANG Shubo[1], Shang Yuming[1], MATHUR V. K[2]

1. Institute ofNuclear andNew EnergyTechnology, Tsinghua University, Beijing100084, China

2.Department of Chemical Engineering, University of New Hampshire, NH 03824, USA

Abstract:

The effect of glycerine addition to the anodic electrolyte, on the anodic reaction, was investigated. The results show that 1% glycerine added to the electrolyte had positive effect on the performance of electrolyte reaction on the surface electrode. The solution resistance, capacitance value and polarization resistance also improved. The reaction of VO_2^+ to VO^{2+} was helped as a result of lower polarization resistance.

Key words: Vanadium Redox Flow Battery ; Glycerine ; Electrochemical impedance spectroscopy; Electrolyte, Cyclic voltammetry; Potentiodynamic

1. Intruduction

Vanadium redox flow battery (VRB) employs vanadium ions of different valences as liquid electrolytes. It is independent of storage capacity, output power, normal atmospheric temperature operation, fast response, and long operating life. The VRB system can widely be used for load leveling, distributing power, standby power supply, and municipal transport. Cyclic voltammetry and Electrochemical impedance spectroscopy (EIS) methods are useful in characterizing the polarization resistance corrosion rate and electrochemical mechanism.

When the positive electrolyte vanadium concentration is up to 2M or the temperature is elevated above 310K, the fully charged electrolyte has poor stability. The precipitate of V(V) would block the electrode and prevent the electrolyte from flowing through the pipe. In order to keep the electrolyte from precipitating, and improve the energy storage density of electrolyte used in VRB, one method is to increase the concentration of vanadium ions. Many workers have chosen inorganic and organic additives to be added to the electrolyte for solving this problem. Some additives added to electrolyte can broaden the electrochemical window and increase electrochemical activity. The dissolved V(V) in concentrated sulfuric acid solution becomes more stable and effective at high concentration of vanadium. Glycerine has been used to enhance the conductivity and stability of anodic electrolyte for improving the performance of VRB. In this work, the effect of glycerine addition, to the anodic electrolyte, on the mechanism has been studied.

2 EXPERIMENTS

2.1 Preparation of anodic electrolyte

$VOSO_4 \cdot 3H_2O$ was dissolved in the deionized water, and H_2SO_4 was added to form $VOSO_4$+3 M solution. The solution was added to anodic electrolytic cell, and H2SO4 of the same concentration was added to cathodic electrolytic cell. The electrolytic cell was separated by an ion exchange membrane (Nafion 117, DuPont), and the lead electrodes were used for anode and cathode electrolysis. The prepared electrolyte was 1.8 mol/L, with the active area of $16cm^2$ (4cm × 4cm) and electrolytic current density of 20 ~ $200mA/cm^2$.

2.2 Electrochemical tests

A three-electrode electrochemical cell was used for electrochemical tests. The graphite electrode (with 5mm diameter) was a working electrode. The reference electrode was SCE, and a platinum mesh was used as auxiliary electrode. Different concentration (v/v) of glycerine was added in the anodic electrolyte.

The electrode characterization included the determination of the polarization resistance and electrochemical mechanism. A Zanher IM6ex instrument was used to conduct the potentiodynamic, cyclic voltammetry and EIS measurements. Potentiodynamic tests were used to measure the corrosion resistance. The initial potentials were -0.15 V, and the final potentials were 2.0 V, the scan rate was 1 mVs-1. Cyclic voltammetry was used to measure the electrochemical activities. The scan range was -0.8 ~ 1.6 V. The EIS measurements were performed at different potential and the perturbation amplitude was 5 mV. The applied frequencies ranged from 100K to 10^{-1} Hz.

3. RESULTS AND DISCUSSION

A potentiodynamic test is commonly employed to measure a material's resistance to corrosion. Potentiodynamic tests were conducted to estimate the corrosion resistances. The "corrosion resistances" here was considered as the electrolyte's ionization after the glycerine was added. Fig. 1 presents the potentiodynamic polarization scans at different glycerine contents. From the Tafel curves in Fig. 1, the related polarization kinetic parameters were analyzed by potentiodynamic polarization model in Zanher IM6ex electrochemical workstation. From the potentiodynamic polarization results and using Eq. (1), polarization resistance, Rp was estimated. Table 1 presents the polarization parameters of anodic electrolyte. The polarization resistances for 0.1% and 10% glycerine were 44744 Ω·cm2 and 13027 Ω·cm2, respectively.

$$R_P = \frac{\beta_a \beta_c}{2.3 i_{corr}(\beta_a + \beta_c)} \quad (1)$$

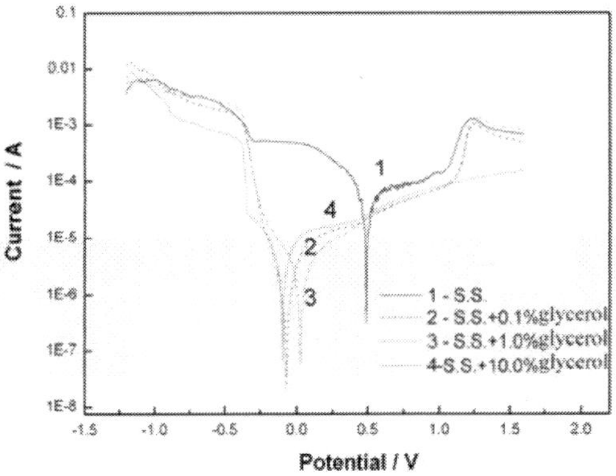

Fig. 1 Potentiodynamic polarization scans of anodic electrolyte

Table 1 Polarization parameters of anodic electrolyte

	βa/mV	-βc/mV	Rp/Ω	Icorr/μA
Non glycerin	86.2	79.9	1654	10.9
0.1% glycerin	64.1	62.3	44744	0.307
1.0% glycerin	54.0	52.9	28617	0.406
10% glycerin	57.1	60.7	[illegible]	[illegible]

Cyclic voltammetry or CV is a type of potentiodynamic electrochemical measurement. Cyclic voltammetry was used to study the electrochemical properties of the glycerine in solution. The results are shown in Fig. 2. When glycerine was added to the anodic electrolyte, the quantity and status of vanadium is changing between the electrolyte and electrode. With the glycerine content increased from 0.1% to 10%, the oxidation peak current decreased. The hydroxy groups on glycerine are bound with the vanadium ions, resulting in the decrease of surface electrode charge, transferring more vanadium ions to the bulk solution.

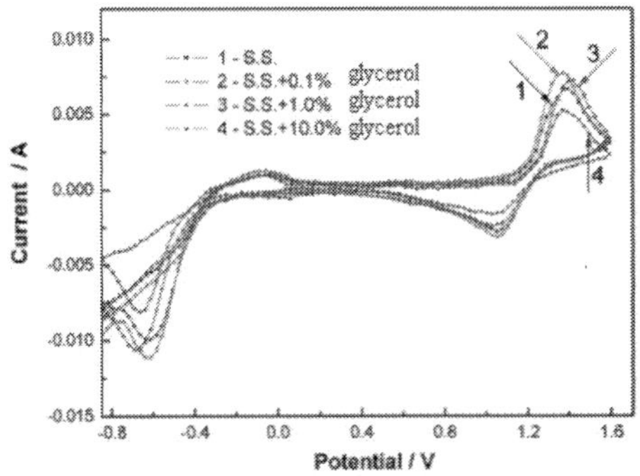

Fig. 2 Cyclic voltammetry scans of anodic electrolyte

EIS reveals information about the reaction mechanism of an electrochemical process: different reaction steps will dominate at certain frequencies, and the frequency response shown by EIS can help identify the rate limiting step. The EIS was employed to investigate the electrode reaction mechanism in anodic electrolyte containing glycerine as additive. The spectrums were simulated and analyzed through the equivalent circuit of $R_{\Omega}(C_d(R_pW))$. The reaction mechanism of glycerine added to electrolyte is shown in Fig 3.

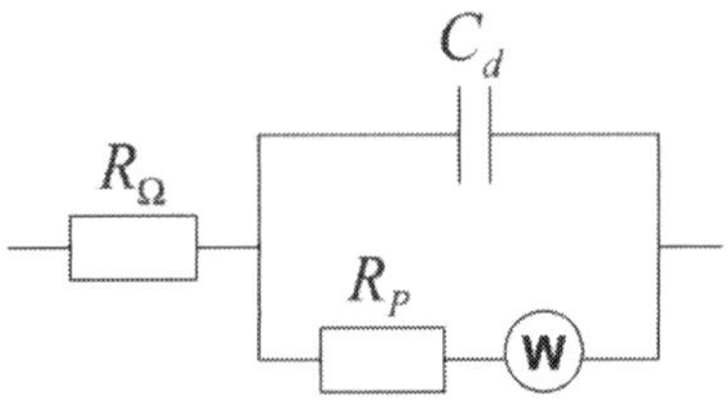

Fig 3. Equivalent circuit using RΩ(Cd(RpW))

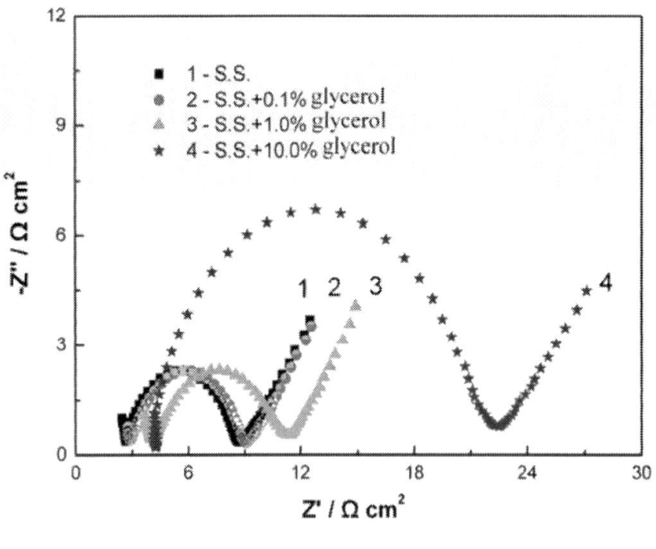

Fig.4 Impedance spectra for anodic electrolytes

The results of EIS are shown in Fig 4. From the impedance curve, with the glycerine concentration increasing, the impedance seems to increase. It means there is an optimal amount for glycerine that can be added to the electrolyte. Therefore, we should conduct the simulation of equivalent circuit to explore the mechanism of reaction with different concentration of glycerine added in. The results show the solution resistance increases by the addition of glycerine. It is difficult for glycerine to ionize in water. When the non-conducting glycerine is added to the electrolyte, the solution resistance should increase. With the increase in glycerine concentration, the resistance of electrolyte increases from 2.795Ω·cm^2 to 4.16 Ω·cm^2. In fact, the molecular structure of glycerine has three hydroxyls,

Table 2 Parameters of the equivalent circuit

	RΩ/(Ω·cm^2)	Cd/(S·cm^2)	Rp/(Ω·cm^2)	W/(S·cm^2)
Non glycerin	2.572	7.735	6.106	0.2478
0.1% glycerin	2.795	8.809	6.25	0.2676
1.0% glycerin	3.855	4.04	7.413	0.245

which are prone to bind with vanadium ions. With the glycerine content increasing from 0% to 10%, the polarization resistance increased. It means the glycerine added to electrolyte is unfavorable for reaction VO_2^+ to VO^{2+}. When the concentration exceeds 1%, the unnecessary glycerine might occupy the reaction site which can catalyze the redox reaction of vanadium ions. Therefore, when more glycerine is added to electrolyte, the reaction impedance increases. It means less glycerine is helpful for VO_2^+ to react to VO^{2+}. The results are in agreement with potentiodynamic test and Cyclic voltammetry. It indicates that glycerine contents when increased beyond certain limits will decrease the reaction rate.

4 . CONCLUSIONS

The results show that glycerine added to anodic electrolyte greatly improved oxidation peak current, reaching a maximum when glycerine was about 1%(v/v). Meanwhile, with the increase in glycerine content, the solution resistance (R_Ω), the double electric layer capacitor (C_d) and the polarization resistance (R_p) reached their minimum. However the concentration polarization reached its maximum when glycerine added was about 1%(v/v).

The addition of glycerine increased solution resistance (R_Ω) which is disadvantageous to the vanadium ion movement in electrolyte solution. Moreover, the glycerine could greatly decrease the corrosion of electrolyte materials. With the increase in additive, the anti-corrosion tends to decrease. When the addivties was up to 0.1%(v/v), the anti-"corrosive ability" was the best.

ACKNOWLEDGEMENTS

The work was funded by Tsinghua University Initiative Scientific Research Program (201010800623), State Key Laboratory for Automotive Safety and Energy(KF10132).

REFERENCES

1) Bard, Allen J.; Larry R. Faulkner (2000-12-18). Electrochemical Methods: Fundamentals and Applications (2 ed.). Wiley. ISBN 0471043729.

2) http://en.wikipedia.org/wiki/Electrochemical_impedance_spectroscopy

(690d) Active Species Concentration Effects On Performance for a Non-Aqueous All-Vanadium Redox Flow Battery

Aaron A. Shinkle, Alice E. S. Sleightholme, Levi T. Thompson and Charles W. Monroe, Department of Chemical Engineering, University of Michigan, Ann Arbor, MI

Redox flow batteries(RFBs) are being developed for large-scale energy storage, uninterruptablegrid-independent power supplies and load-leveling systems for solar or windpower. Aqueous chemistries are used for all current commercial RFBs, therefore, cell potentials are limited to thestability range of water (01.23 V vs. SHE). Since a battery's cellpotential determines its energy and power density, aqueous chemistries limitthe possibilities for the performance of RFBs.

Several non-aqueous systems have beenreported recently, including systems based on ruthenium, uranium, chromium,manganese, and vanadium coordination complexes [1-5]. In the case ofvanadium acetylacetonate, the use of a non-aqueous electrolyte results in a stablecell potential window of 2.2 V [5], more than 50% wider than the aqueoussingle-metal vanadium chemistry (1.26 V [6]). However, very few resultsfrom battery charge/discharge experiments have been shown for this non-aqueousvanadium system.

This talk will examine thecharge/discharge profile for the vanadium acetylacetonate ($V(acac)_3$)RFB system in acetonitrile, with tetraethylammonium tetrafluoroborate ($TEABF_4$)support. The effects of active-species concentration on batteryperformance were examined. Linear sweep voltammetry will be used todetermine the maximum charge/discharge rate of the battery cell as a functionof electrolyte composition. Figure 1 shows the first five cycles of acharge/discharge result for 0.4 M $V(acac)_3$,0.5 M $TEABF_4$ in acetonitrile. Coulombic, energy, and powerefficiency for charge/discharge experiments were examined as a function ofconcentration to illustrate the effects of active species crossover. Efficiency and energy density were balanced to determine an optimal activespecies concentration range for operation of the battery. Lastly, thenon-aqueous system was compared to the optimized aqueous all-vanadium system toidentify areas of future work.

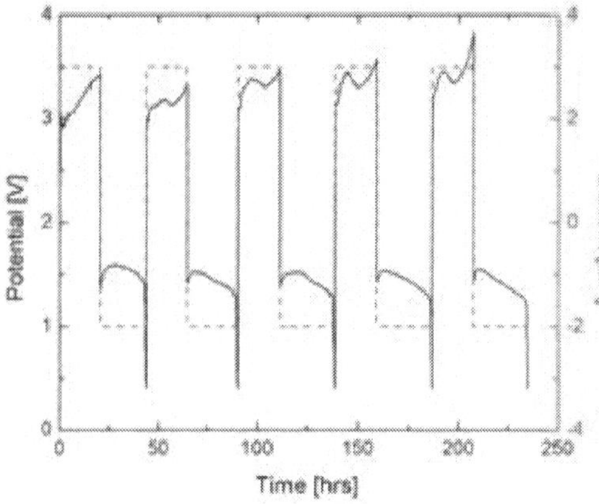

Figure1: First five cycles of a charge/discharge experiment for an all-vanadiumnon-aqueous redox flow battery chemistry. The system was charged at 3 mA until 50% theoretical stateof charge and discharged at 2 mA until 0.4 V. The system shown is 0.4 M vanadium acetylacetonate supported by 0.5 Mtetraethylammonium tetrafluoroborate in acetonitrile solvent. An H-cellconfiguration was used with graphite plate electrodes and a NeoseptaAHA anion exchange membrane.

REFERENCES

1. Matsuda Y, Takasu Y, Morita M, Tanaka K, Okada M, Matsumura-Inoue T(1985) Denki Kagaku 53:632

2. YamamuraT, Shiokawa Y, Yamana H,Moriyama H (2002) Electrochim Acta48:43

3. LiuQ, Shinkle AA, Li Y, Monroe CW, Thompson LT, Sleightholme AES (2010) Electrochem Comm 12:1634

4. SleightholmeAES, Shinkle AA, Liu Q, Li Y, Monroe CW, Thompson LT (2011) J Power Sources196:5742-5745

5. LiuQ, Sleightholme AES, Shinkle AA, Li Y, Thompson LT (2009) ElectrochemComm 11:2312

6. Poncede Leon C, Frias-Ferrer A, Gonzalez-Garcia J, Szanto DA, Walsh FC (2006) J Power Sources 160:716-732

See more of this Session: Energy Storage and Electrochemical Processes I
See more of this Group/Topical: International Congress on Energy 2011

(690e) Energy Storage and Generation From Thermopower Waves: Covalently Functionalized Thermal and Electrical Conduits

Joel T. Abrahamson[1], Changsik Song[2], Jennifer H. Hu[1], Jared M. Forman[1], Sayalee G. Mahajan[1], Nitish Nair[1], Wonjoon Choi[3], Eun-Ji Lee[2] and Michael S. Strano[1], (1)Department of Chemical Engineering, Massachusetts Institute of Technology, Cambridge, MA, (2)Department of Chemistry, Sungkyunkwan University, Suwon, South Korea, (3)Mechanical Engineering, Massachusetts Institute of Technology, Cambridge, MA

We have demonstrated through simulation and experiment that the nonlinear coupling between an exothermic chemical reaction and a nanowire or nanotube with large axial heat conduction accelerates the thermal reaction wave along the nano-conduit [1-3]. The thermal conduit rapidly transports energy to unreacted fuel regions, and the reaction wave induces a concomitant thermopower wave of high power density, producing electrical current in the same direction. At up to 7 W/g, this can be substantially larger than the power density offered by current micro-scale power sources (e.g. fuel cells, batteries) and even about seven times greater than that of commercial Li-ion batteries. MEMS devices and wireless sensor networks would benefit from such high power density sources to enable functions such as communications and acceleration hampered by present power sources.

Recently, we have covalently attached energetic molecules of mono-, di-, and trinitrobenzenes to single-walled carbon nanotubes via diazonium chemistry. Although covalent functionalization introduces defects that can scatter electrons and phonons, thermopower waves can still rapidly propagate and produce electricity on these decorated nanotubes. Differential scanning calorimetry confirms that the energetic molecules release heat, but thermopower waves require additional fuel adsorbed on the nanotubes to propagate.

[1] W. Choi, S. Hong, J. T. Abrahamson, *et al. Nature Materials*, 9, (2010), 423.

[2] W. Choi, J. T. Abrahamson, J. M. Strano, M. S. Strano. *Materials Today,* 13, (2010), 22.

[3] J. T. Abrahamson, *et al. ACS Nano* 5, (2011), 367.

See more of this Session: Energy Storage and Electrochemical Processes I
See more of this Group/Topical: International Congress on Energy 2011

Size and Structure of Asphaltene Aggregates in the Bubble Pressure Region

Jesús Leonardo Amundaraín Hurtado*, Martin Chodakowski, Bingwen Long**, J. M. Shaw***

University of Alberta, Edmonton, Alberta, Canada

* Now working at Schlumberger, Russia

** Visiting professor, Beijing University of Chemical Technology, presenter

*** **Corresponding Author:** E-mail: jmshaw@ualberta.ca.

Introduction

Asphaltene precipitation is typically most pronounced adjacent to the liquid-vapour to liquid phase boundary of reservoir fluids. Knowledge of the size and shape of asphaltene aggregates along such boundaries is essential to the reliably of models for the phase behavior of asphaltene rich fluids [1,2]. Small-angle X-ray scattering has been used extensively to study the aggregate structure of asphaltenes in solutions at the nanometer length scale [3,4]. In this work, SAXS measurements were employed to study the size and nano-structure of Athabasca asphaltene aggregates in 1-methyl naphthalene (aromatic) and *n*-dodecane (*n*-alkane) over broad concentration, temperature, and pressure ranges. The key feature of the measurements is that they were carried out along the vapor-liquid saturation curves at fixed global composition. The structural information extracted from the SAXS emissions is presented and then interpreted and discussed from the perspective of the phase behavior of neat asphaltenes, and asphaltenes + diluent mixtures.

Experimental Section

Asphaltenes from Athabasca vacuum residue (AVR) were prepared based on a modified ASTM D3279 protocol where n-heptane is substituted for *n*-pentane [5]. Asphaltene mixtures of different concentrations were prepared gravimetrically with an analytical balance (resolution 0.00001 g). The SAXS experiments were performed at the BESSRC-CAT 12ID-C experimental station at the Advanced Photon Source (APS) facility in the Argonne National Laboratory. The scattering experiments were performed with an X-ray energy of 18 keV, for which the beamline optics had been adjusted. The CCD images were collected at a sample-to-detector distance of 3.7 m, corresponding to a usable scattering wave vectors q range from 0.005 to 0.2 $Å^{-1}$. The experiments were carried out from 50 to 310 °C at 20°C intervals with the asphaltene volume range from 0.6 to 9.7% in the solution, which corresponds to the vapour pressure up to 348 and 620 kPa for 1-methylnaphthalene and *n*-dodecane, respectively. Multiple emissions measurements were obtained at each experimental condition. Special designed measuring tube, magnetic mixing and multi-sample holder are applied to ensure the measurements were carried out along the saturation curves of the solutions at fixed global compositions. The emission data, after being subjected to calibration, normalization, averaging and solvent background signal subtraction, was converted to an absolute scale for subsequent analysis.

Quantitative Analysis of the Structure Parameters

The SAXS emission curves suggest that asphaltenes have a two-level aggregate structure co-existing in the solutions. Combinations of standard plots and non-linear fitting methods were employed to characterize the two-level structures. The size of the larger, secondary, structures could not be estimated because a Guinier plateau was not observed at low q.

Kratky and Porod plots [6], Figures 1 and 2 respectively, show that the smaller, primary, asphaltene aggregates have a rod–like structure and they tend to form three-dimensional networks in 1-methylnaphalene at low temperatures. With increasing temperature, the networks gradually decompose and at 310°C the networks are absent. Asphaltene + *n*-dodecane mixtures appear to behave similarly. However, at high concentration, strong inter particle interactions make the asphaltene structures become more compact and they exhibit a globular conformation.

a) b)

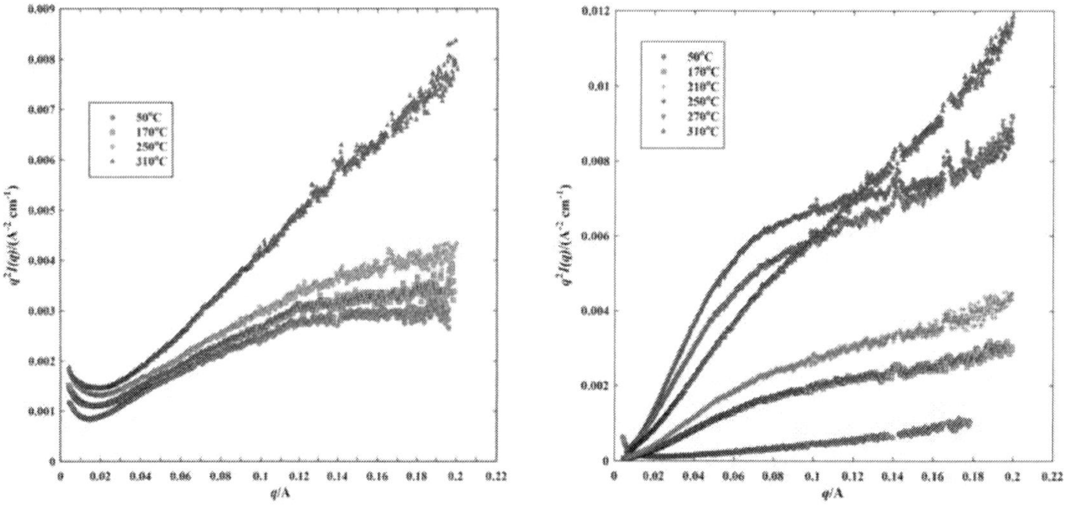

Figure 1. Kratky plots of SAXS data for 2% asphaltene in diluents. a) 1-Metheylnaphaltene; b) *n*-dodecane.

a) b)

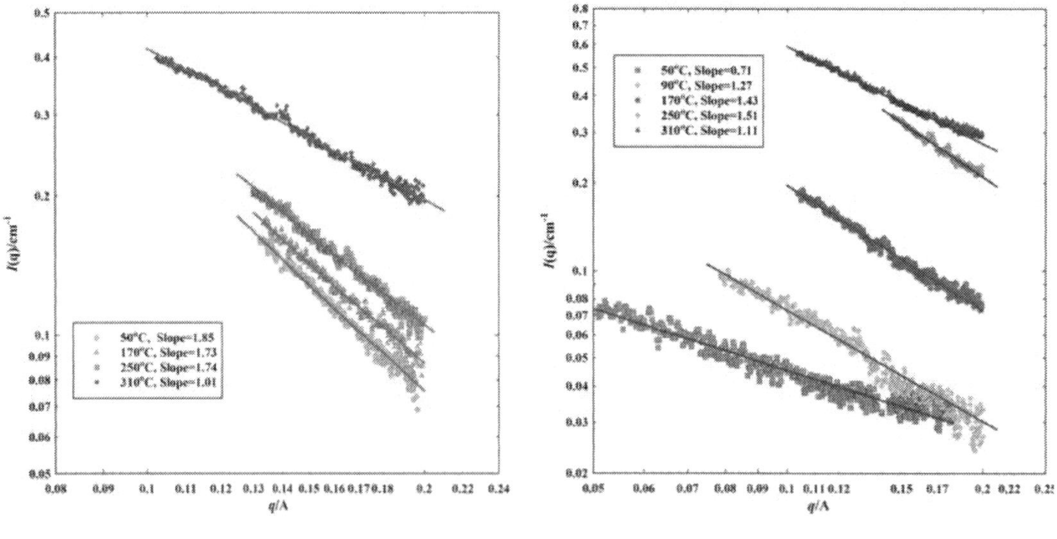

Figure 2. Porod plots of SAXS data of 2% asphaltene in dilutes. a) 1-Metheylnaphaltene; b) *n*-dodecane.

A two-level Beaucage unified model [7,8] was used to model the emission data. This five-parameter model reproduces the experimental scattering curves with an overall average relative deviation between the calculated and experimental scattering intensity is 1.9% for all the 93858 data points. A typical fitting result of the asphaltene mixture SAXS data is shown in Figure 3.

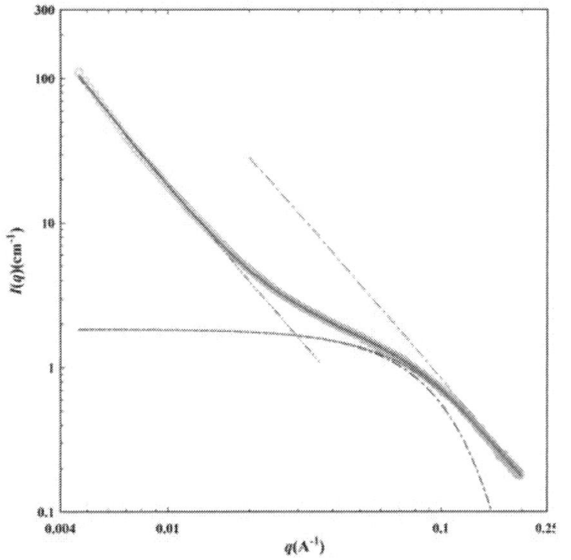

Figure 3. Fitting SAXS data with the Beaucage unified model (8.2% AVR asphaltene in *n*-dodecane at 90 C). (Open circle): experimental scattering intensity; (red solid line): two-level unified model fitting; (brown dashed line): level 2 Porod's law fitting; (pink dashed line): level 1 unified model fitting; (blue dashed line): level 1 Guinier fitting; (green dashed line): level 1 Power-law fitting;

From the Beaucage model, the radius of gyration R_g values for the primary aggregates are in the range of 10 to 35 Å and the values appear to be invariant with respect to temperature and concentration at low concentrations (< 5% wt asphaltenes). The size of primary asphaltene aggregates in 1-methylnaphalene appears to be somewhat larger than in *n*-dodecane. The primary asphaltene aggregates begin to break up starting at 250°C and by 310°C the rod network is absent for 1-methylnaphthalene, while the network appears to persist in *n*-dodecane.

Results from the standard plots and the Beaucage model fits both suggest that the basic unit of primary asphaltene aggregates is rod-like structure (P=1), which is also confirmed by the molecular weight ~ *Rg* relation. These aggregates appear to form a loose three dimensional network connected by rod-rod interactions leading to scattering coefficients P_1 around 2 at low temperature. Concentration seems to have little effect on these structures. The scattering coefficients of secondary asphaltene aggregates clearly show mass fractal behavior with a dense structure (P_2, 2.3~3.4). These large aggregates show greater thermal stability in 1-methylnaphalene than in *n*-dodecane.

In summary, the thermal and concentration variations of asphaltene properties in the two diluents are shown to exhibit complex and contrasting relationships. The SAXS data analysis shows that solvent has the strongest influence on the size and structure of asphaltene aggregates, followed by concentration. Temperature has little effect on structure prior to thermal conversion reactions that take place at high temperature.

References

1. Maham, Y., M. G. Chodakowski, et al. (2005). "Asphaltene phase behavior: prediction at a crossroads." Fluid Phase Equilibria **227**(2): 177-182.
2. Diaz, O. C., J. Modaresghazani, et al. (2011). "Modeling the phase behavior of heavy oil and solvent mixtures." Fluid Phase Equilibria **304**(1-2): 74-85.
3. Tanaka, R., E. Sato, et al. (2004). "Characterization of asphaltene aggregates using X-ray diffraction and small-angle X-ray scattering." Energy and Fuels **18**(4): 1118-1125.
4. Eyssautier, J. l., P. Levitz, et al. (2011). "Insight into Asphaltene Nanoaggregate Structure Inferred by Small Angle Neutron and X-ray Scattering." The Journal of Physical Chemistry B **115**(21): 6827-6837.
5. Hurtado, J. L. A. (2010). Characterization of Athabasca Asphaltenes Separated Physically and. Chemically Using Small-Angle X-Ray Scattering. Master, Univeristy of Alberta.
6. Glatter, O. and O. Kratky (1982). Small-Angle X-ray Scattering. New York, Academic Press Inc.
7. Beaucage, G. (1995). "Approximations Leading to a Unified Exponential / Power-Law Approach to Small-Angle Scattering." Journal of Applied Crystallography **28**(6): 717-728.
8. Beaucage, G. (1996). "Small-Angle Scattering from Polymeric Mass Fractals of Arbitrary Mass-Fractal Dimension." Journal of Applied Crystallography **29**(2): 134-146.

(694c) Destabilization and Aggregation Kinetics of Asphaltenes

Nasim Haji Akbari Balou, Chemical Engineering, The University of Michigan, Ann Arbor, MI and H. Scott Fogler, Chemical Engineering, University of Michigan, Ann Arbor, MI

Destabilization and Aggregation Kinetics of Asphaltenes

Nasim Haji Akbari Balou & H. Scott Fogler

Asphaltenes comprise the heaviest and most polar components of crude oil and their precipitation and deposition can have serious economical impacts in the oil industry. For years it was believed that asphaltene mixtures reach thermodynamic equilibrium shortly after changing the operational conditions. However, recent work in our group demonstrated that kinetics plays an important role in the precipitation of asphaltenes from crude oil-precipitant mixtures- especially at low precipitant concentrations [1]. It is expected that the kinetics of precipitation depend on the properties of the crude oil; the goal of this research is to investigate this relationship. Due to the complexity of crude oil, model mixtures are used as alternatives for this purpose. Model mixtures are prepared by extracting asphaltenes from crude oil and dissolving them in various solvents with varying concentrations. The precipitation rates are then estimated by adding different quantities of heptane as precipitant and measuring the time needed for the particles to reach to a size of ~ 0.5 μm via optical microscopy. We show that model mixtures represent crude oils reasonably well in terms of precipitation kinetics and an exponential relationship exists between detection times and precipitant concentrations which is similar to what was previously reported by Maqbool et al [1] for crude oil-precipitant mixtures. We also investigate the effect of asphaltene concentration and the medium in which they are stabilized on the precipitation kinetics. The knowledge from this study can lead to a better understanding of the destabilization and growth processes which can in turn give rise to new predictive models to foresee precipitation kinetics under different operational conditions.

[1] T.Maqbool, A.T. Balgoa, and H. S. Fogler, " Revisiting Asphaltene Precipitation from Crude Oils: A Case of Neglected Kinetic Effects," *Energy & Fuels*, vol. 23, no. 7 , pp. 3681-3686, Jul. 2009.

See more of this Session: Heavy Oil and Flow Assurance II
See more of this Group/Topical: Energy and Transport Processes

THERMAL RESPONSE DURING RHEOLOGICAL TRANSITIONS IN CYCLOPENTANE HYDRATE-FORMING EMULSIONS

Genti Zylyftari, Prasad U Karanjkar, Jae W Lee, Jeffrey F Morris
City College of City University of New York, New York, NY

Introduction

Light hydrocarbon molecule stabilized hydrogen-bonded water cages result in a stable crystalline compound called clathrate hydrate. Hydrates are thermodynamically stable at low temperatures and high pressures, conditions predominantly found in subsea pipelines [1-3]. Clathrate hydrate formation within pipelines is a major problem in the oil and gas industry [1,2]. Hydrate induced plug formation causes product flow interruption, severe equipment damage and even personnel fatalities.

The goal of flow assurance is to prevent blockage of flowlines in an efficient and cost effective manner. Mitigation of plug formation requires a mechanical understanding of the process involved in hydrate formation. A mechanism of shell fracture in the hydrate system controlled by the stress imposed in the system is reported.

Results and Discussion

The viscosity and temperature evolution of the 40% v/v aqueous fraction hydrate and ice forming systems are shown in figure 1. The emulsions are quenched from room temperature to -7°C. At shear rate of $\dot{\gamma}$ = 100 s^{-1}, the viscosity increase of the hydrate forming system is sharper and a significant heat generation rate is observed, implicitly deduced from the plotted temperature behavior, when compared to the ice forming system in figure 1. Neglecting temperature dependent corrections, the latent heat of ice is 333 kJ/kg [4] and the heat of formation of hydrate is 348.9 kJ/kg water converted. The heat of formation of hydrate on a converted water basis is computed from the heat of cyclopentane hydrate dissociation, 283.9 kJ/(kg hydrate) [5], and the ratio 1:17 of cyclopentane to water molecules forming the hydrate [6]. The difference between ice and cyclopentane hydrate heat of formation is 4.8%, which is not large enough to explain the difference in temperature behavior, assuming the rate of formation is similar. Therefore, the conclusion to be drawn from figure 1 is that at same temperature and shear rate, hydrate formation rate is significantly higher than that of ice. During hydrate formation, the experimental apparatus cooler heat removal capacity is not large enough to maintain an isothermal condition in the Couette geometry during phase transition from emulsion to hydrate; by contrast, for ice formation, quasi-isothermal conditions are achievable for shear rate of $\dot{\gamma}$ = 1 s^{-1} and 100 s^{-1}.

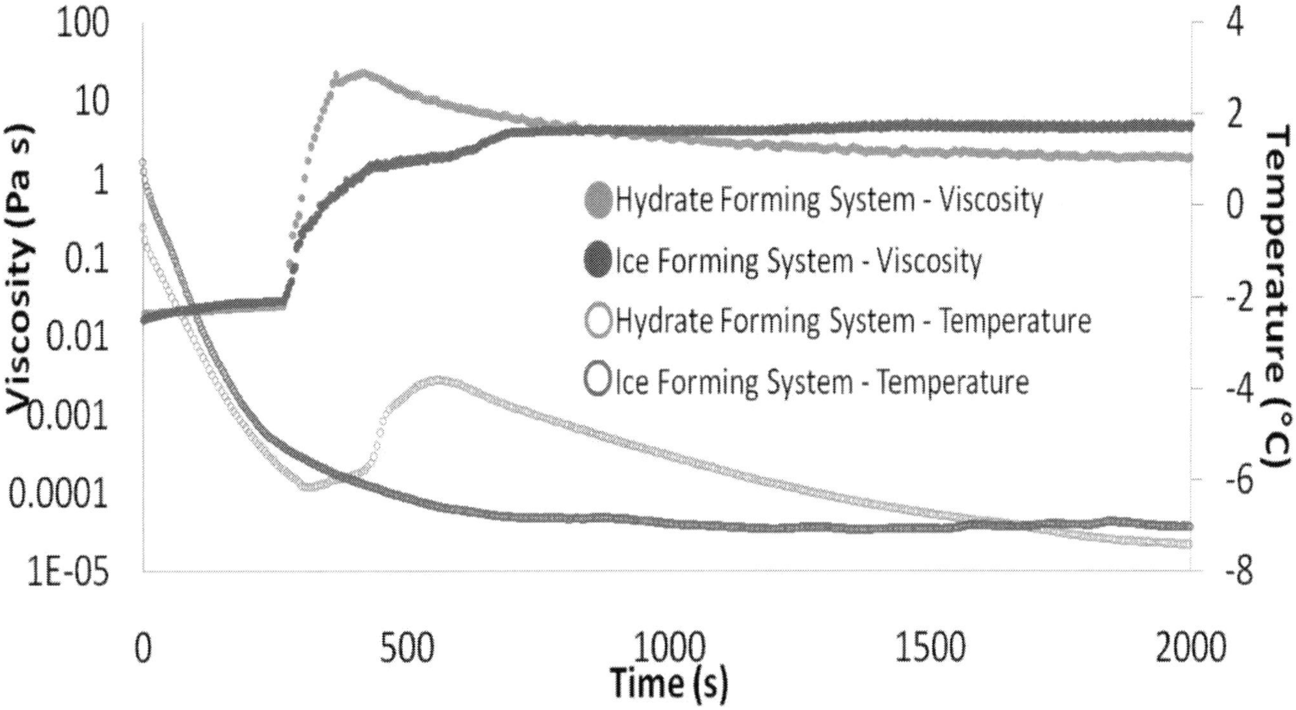

Figure 1. Viscosity and temperature plot with respect to time of the 40% v/v aqueous fraction hydrate (green) and ice (red) forming systems under shear rate of $\dot{\gamma} = 100\ s^{-1}$, from room temperature to -7°C.

The thermal responses in the hydrate and ice forming systems were investigated for different stresses, $\tau = 100$ and 300 Pa, applied during the transition of emulsion to a solidified state. At a controlled shear stress value of $\tau = 100$ Pa, no significant heat generation rate is observed for the hydrate, shown in figure 2, and ice forming systems. When the stress imposed on systems is 300 Pa, the hydrate forming system shows a significant heat generation, shown by the temperature response in figure 3. Therefore a three-fold increase in the stress is able to fracture the hydrate shells quicker and generate more interfacial area in a short period of time. Furthermore, the stress evolution in the ice system during phase transition is not high enough to reach the set value as observed in figure 4. In the ice forming system, the shear rate reaches considerably higher values than those imposed on the hydrate system without being able to attain the desired stress values. The inability of the stress in the ice forming system to reach the same value of the stress quickly achieved in the hydrate system implies that the solidification processes of the two systems must be intrinsically different. The observation in stress development at constant stress experiments is in agreement with the viscosity evolution of the hydrate and ice forming systems in figure 1. The viscosity evolution of the hydrate forming system is rapid and overshoots the final viscosity. In comparison with hydrate behavior, the ice forming system slowly and smoothly reaches its final viscosity value. Apparently, hydrate formation is significantly impacted by stress the system experiences during the crystallization period.

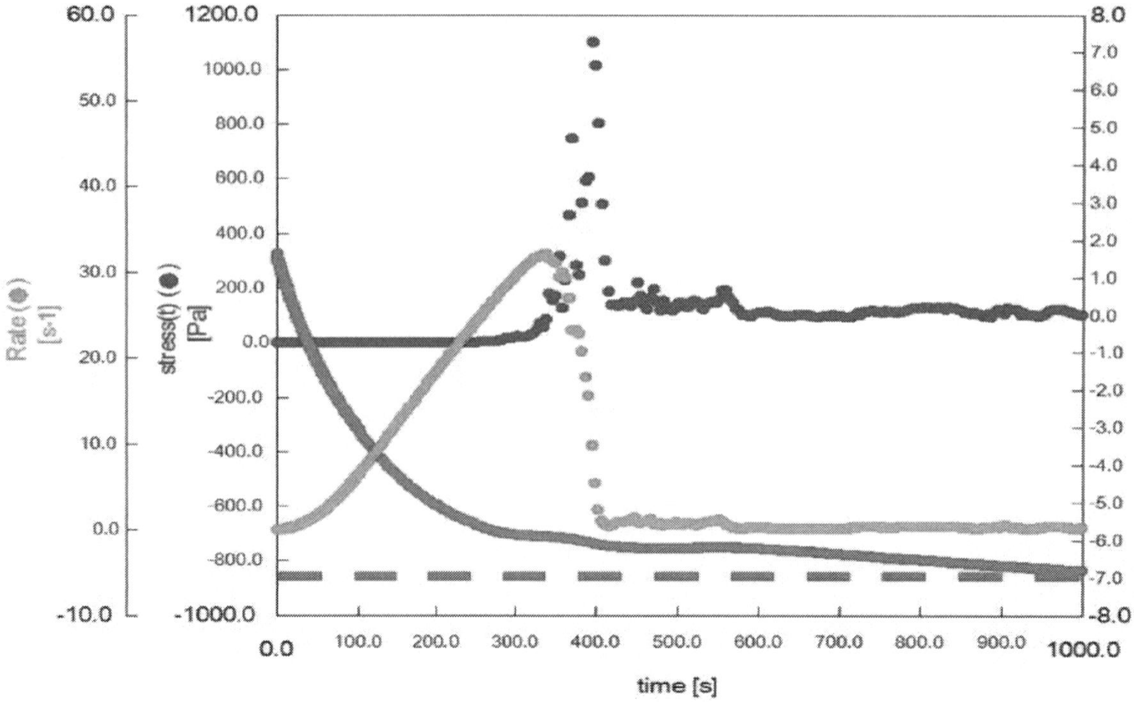

Figure 2. Temperature, stress and shear rate with respect to time of the 40% aqueous fraction hydrate forming system under shear stress of $\tau = 100$ Pa quenched from room temperature to -7°C.

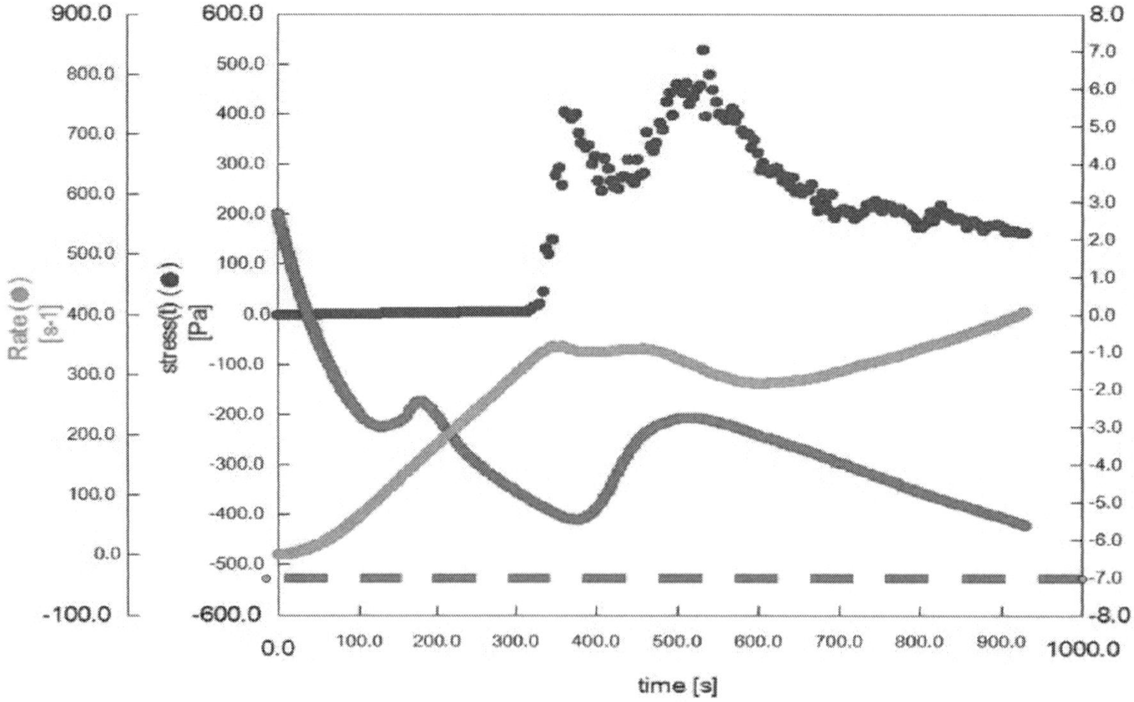

Figure 3. Temperature, stress and shear rate with respect to time of the 40% aqueous fraction hydrate forming system under shear stress of $\tau = 300$ Pa quenched from room temperature to -7°C.

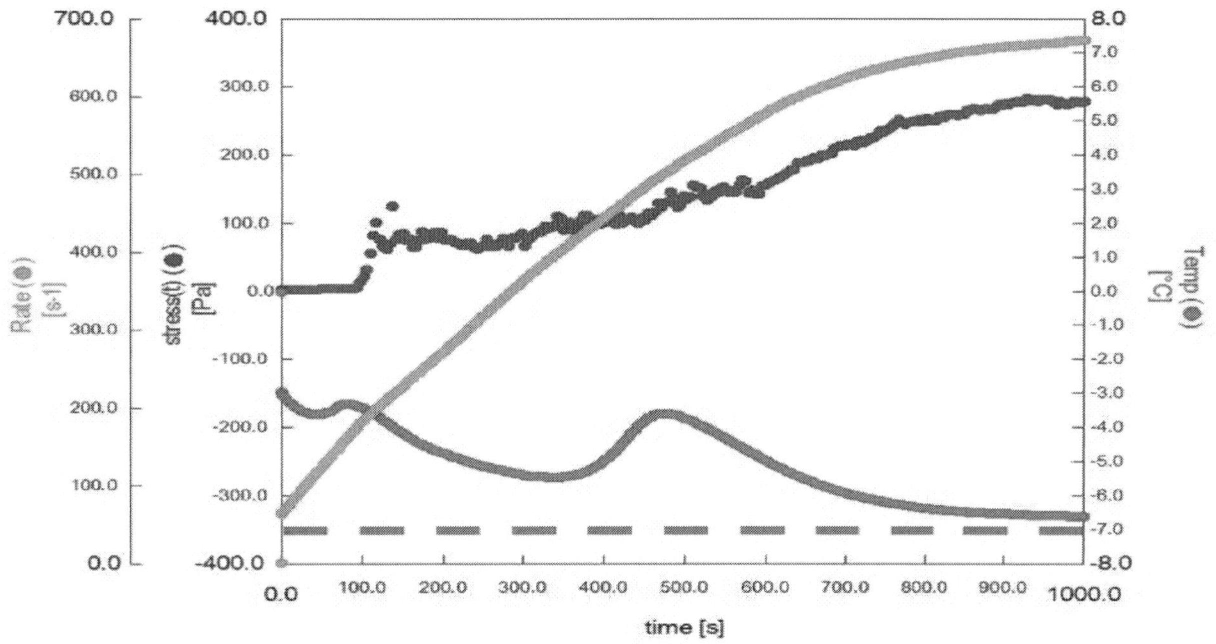

Figure 4. Temperature, stress and shear rate with respect to time of the 40% aqueous fraction ice forming system under shear stress of $\tau = 300$ Pa quenched from room temperature to -7°C.

Conclusion

Hydrate formation rate is dependent on the availability of hydrate former and water to be in close proximity. In an emulsion system, hydrate formation occurs initially at the interface region between the aqueous and hydrate former containing oil phase. After crystal nucleation occurs, a relatively rapid hydrate coverage of the whole droplet interface is attained. Thus, a thin hydrate polycrystalline shell forms a barrier between the oil and the aqueous phase. The hydrate encapsulation of water inhibits further evolution of hydrate due to mass diffusion limitations across the shell. Fracturing of the formed hydrate shell by flow shear stress apparently allows hydrate growth to proceed further. When the hydrate shell is cracked, water trapped inside can flow through the opening to form a new interface in contact with the hydrate former. A regeneration mechanism of the interfacial area aided by the stress imposed fracturing of the hydrate shell has been proposed.

Hydrate formation rate is an indication of the regeneration rate of the interfacial area. Since interfacial area generation rate is controlled by stress applied in the system to fracture the shells, hydrate formation rate is shown to be dependent on the stress applied to the complex fluid. The method used in analyzing the hydrate formation rate is examination of the exothermic heat generation rate. The hydrate formation rate, as implied by the temperature behavior, increases with higher stresses. Ice formation rate is less dependent on the stress imposed in the system. Hydrate forming system stress evolution is abrupt, while ice forming system is characterized by a slow evolution. The difference in formation rate and stress evolution of the hydrate and ice forming systems can plausibly be ascribed to the difference in crystallization processes. Hydrate is an interface controlled process, whereas ice formation is a bulk process, not dependent on the availability of oil-aqueous phase interfacial area.

References

1. Sloan, E., D., *A changing hydrate paradigm from apprehension to avoidance to risk management*. Fluid Phase Equilibria, 228-229, 67-74, (2005).
2. Sloan, E., D., Koh, C., A., Sum, A., K., McMullen, N., D., Shoup, G., Ballard, A., L., Palermo, T., and Chevron, USA, Inc. *Natural Gas Hydrates in Flow Assurance*. Gulf Professional Publishing, Elsevier Inc. ISBN 978-1-85617-945-4, (2011).
3. Chatti, I., Delahaye, A., Fournaison, L., and Petitet, J., *Benefits and drawbacks of clathrate hydrates: a review of their areas of interest*. Energy Conversion and Management, 46, 1333-1343, (2005).
4. Fournaison, L., Delahaye, A., and Chatti, I., *CO_2 Hydrates in refrigeration processes*. Ind. Eng. Chem. Res., 43, 6521-6526, (2004).
5. Nakajima, M., Ohmura, R., and Mori, Y., H., *Clathrate hydrate formation from cyclopentane-in-water emulsions*. Ind. Eng. Chem. Res., 47, 8933-8939, (2008).
6. Zhang, Y., Debenedetti, P., G., Prudhomme, R., K., and Pethica, B., A., *Differential scanning calorimetry studies of clathrate hydrate formation*. J. Phys. Chem. B, 108, 16717-16722, (2004).

(694e) Modeling of Reservoir Connectivity and Tar-Mat Using Gravity-Induced Asphaltene Compositional Grading

Sai R. Panuganti[1], Francisco M. Vargas[2], Anjushri S. Kurup[1] and Walter G. Chapman[1], (1)Chemical and Biomolecular Engineering, Rice University, Houston, TX, (2)Chemical Engineering, The Petroluem Institute, Abu Dhabi, United Arab Emirates

Reservoir compartmentalization is one of the major issues in both onshore and offshore reservoirs. High capital costs are involved especially in deepwater exploration and production, making it essential to assess prior to production the extent of compartmentalization within a reservoir. Seismic data and discontinuous variation of fluid properties (e.g. chemistry, density, and viscosity) are commonly used in an attempt to assess the level of compartmentalization. Within a single oil column, fluid properties vary along the depth of a reservoir due to compositional grading. Considerable fluid flow is required to yield such compositional gradients, suggesting connectivity. In particular, reservoir connectivity can be best understood based on continuous characteristics of asphaltene compositional gradient, because to equilibrate asphaltene, the heaviest and least mobile crude oil component, necessitates substantial permeability.

Compositional grading studies available in the literature account for gas-to-oil ratio and composition variations, but do not analyze compositional grading related to asphaltenes. In this work, we present an algorithm based, in part, on Whitson's analysis [1] and use the PC-SAFT equation of state to address asphaltene compositional grading due to the attainment of thermodynamic equilibrium in a uniform gravitational field. The results are compared with field data to evaluate the compartmentalization of a reservoir. Based on the solution thermodynamics, an approximate analytical solution for calculating the asphaltene gradient is also presented. The analytical solution will be helpful for sensitivity analysis and approximate estimate of the asphaltene compositional gradient.

Asphaltene compositional gradient leads to significant variations in crude oil viscosity [2], and extreme cases of asphaltene compositional grading can lead to tar-mat formation [3]. The PC-SAFT asphaltene compositional grading introduced in this study is extended to model the possibility of a tar-mat formation due to gravitational segregation of asphaltene.

References

1. Whitson C H, Trondheim U, Belery P. Centennial Petroleum Symposium, 1994. Tulsa, USA. SPE 28000:443-459.

2. Hirschberg A. Journal of Petroleum Technology, 1988. 40(1):89-94.

3. Carpentiera B, Arab H, Pluchery E, Chautru J M. Journal of Petroleum Science and Engineering, 2007. 58(3-4):472-490.

See more of this Session: Heavy Oil and Flow Assurance II
See more of this Group/Topical: Energy and Transport Processes

(694f) Model Ranking Methodology for Accurate Pick-up Velocity Predictions to Initiate the Motion of Sand Particles At Low Sand Particle Concentration In Horizontal Single Phase Fluid Flow

Frits Byron Soepyan[1], Selen Cremaschi[1], Cem Sarica[2] and Hariprasad J. Subramani[3], (1)Department of Chemical Engineering, The University of Tulsa, Tulsa, OK, (2)McDougall School of Petroleum Engineering, The University of Tulsa, Tulsa, OK, (3)Flow Assurance, Chevron Energy Technology Company, Houston, TX

The incipient motion of sand particles from the bottom of pipelines has been a topic of interest due to the desire to prevent the accumulation of sand particles in pipes. In order to initiate the movement of the sand particles, the magnitude of the fluid velocity must be equal to at least that of the pick-up velocity of the sand particle. There are multiple sand transport models that predict the pick-up velocity, but for the same operating conditions, the velocity estimations of these models might range over four orders of magnitude. Therefore, given a set of operating conditions, it is necessary to determine the models that best predict the pick-up velocity for those conditions.

This need is addressed with the introduction of a novel model optimization and ranking methodology that combines data clustering approaches with statistical analysis. The data clustering component collects the representative data points that lie within the range of the operating conditions, and the models are optimized using those data points to ensure that they perform optimally within the range of the operating conditions. Then, those models are ranked to determine the ones that make the most accurate velocity predictions within the range of the operating conditions. Lastly, the uncertainty of the velocity predictions of the models is quantified to determine the upper bound of the velocity predictions, thus providing more conservative results. The methodology is tested using field operating conditions, and the results yield consistent pick-up velocity predictions when compared to the observed ones.

See more of this Session: Heavy Oil and Flow Assurance II
See more of this Group/Topical: Energy and Transport Processes

(714a) The Role of Cellulose Accessibility On Enzymatic Saccharification of Lignocelluloses

Junyong Zhu, USDA Forest Products Laboratory, Madison, WI, Zhiguang Zhu, Biological Systems Engineering, Virginia Polytechnic Institute and State University (Virginia Tech), Blacksburg, VA, Y.-H. Percival Zhang, Biological Systems Engineering Department, Virginia Polytechnic Institute and State University, Blacksburg, VA, Yonghao Ni, 3Limerick Pulp and paper Center, University of New Brunswick, New Brunswick, Canada, ZiBin He, Limerick Pulp and Paper Centre, Virginia Polytechnic Institute and State University (Virginia Tech), Blacksburg, VA and Xiaolin Luo, South China University of Technology, Guangzhou, China

So much research has been dedicated to understand the effects of various factors on enzymatic hydrolysis of lignocelluloses. It has been understood that removal of hemicelluloses is very important to improve lignocellulosic substrate digestibility. It has also been identified that improving enzyme accessibility to cellulose is more important than delignification to enzymatic saccharification. This study is attempted to develop a comprehensive understanding of various factors on enzymatic hydrolysis of lignocelluloses by taking a progressive approach. In the first phase, a set of lignocellulosic substrates were produced by drying to different degrees from the same never dried sample. Therefore, these dried substrates have exactly the same chemical composition and structure. However, drying induced fiber hornification produces irreversible change of cell wall structure when rewetting, i.e., fibrils are tightly packed together through hydrogen bonding to produce permanent fiber pore shrinkage and loss of pores in cell wall. As a result, the enzyme accessibilities of these samples were significantly reduced comparing with the never- dried sample and varied with the degree of drying. This set of samples is an excellent candidate to verify the effect of enzyme accessibility on enzymatic hydrolysis. In the second phase, two sets of samples, one produced from fractionation of the same sample, and another from disk milling of the same chemically pretreated biomass to different degrees, were used to determine the effect of substrate size/specific surface or degree of size reduction on enzymatic hydrolysis. Finally, different pretreatment methods were used to explain the effect of chemical composition on enzymatic hydrolysis. Both enzymatic hydrolyses and measurements of cellulase accessiblity to cellulose (CAC) were carried out to quantify the contributions of various factors to enzymatic hydrolysis of lignocelluloses.

See more of this Session: Recalcitrance of Woody Biomass
See more of this Group/Topical: International Congress on Energy 2011

(714b) Structural Studies On Recrystallized Ionic-Liquid (IL) Treated Lignocellulosic Biomass

Indira Samayam[1], B. Leif Hanson[2], Paul Langan[3], Christopher J. Barr[1] and **Constance Schall**[1], (1) Chemical and Environmental Engineering, University of Toledo, Toledo, OH, (2)Instrumentation Center, The University of Toledo, Toledo, OH, (3)Biosciences Division - Los Alamos National Laboratory, Los Alamos, NM

Terrestrial lignocellulosic biomass is composed of three major components; cellulose, a linear polymer of glucose (a hexose sugar); hemicellulose, a complex branched polymer of xylose (a pentose sugar) and other sugar derivatives; and lignin, a polyphenyl propanoid macromolecular assembly that is covalently cross-linked to hemicellulose. These components can serve as a source of carbon based feedstock for fuel and chemical production in much the same way that crude oil serves as the carbon feedstock in petrochemical refineries.

The deconstruction of lignocellulosic biomass into simple sugars constitutes a core barrier for producing value added products from the sugar platform. The cellulosic component has a crystalline and oriented fibrillar structure in biomass, whereas the hemicellulosic and lignin components have less ordered structures that encrust the cellulose fibers. The crystalline structure of native cellulose fibers, called cellulose I, can be a major impediment to its hydrolysis to monomeric sugars. Pretreatment can disorder the crystalline cellulosic component and result in structural changes in the cellulose fibers. These structural changes in cellulose fiber can be achieved by modifying the pretreatment conditions. Depending upon the type of pretreatment, the conversion of cellulose fibers to other crystal forms, such as cellulose II, cellulose III_{II} and cellulose III_{I}, or amorphous forms, can greatly improve their susceptibility to hydrolysis. To investigate the effects of varying the parameters of IL pretreatment on the structure of their cellulosic components, X-ray powder diffraction (XRD) and X-ray fiber diffraction from biomass samples was used with unoriented and partially (fiber) orientated cellulosic components . A comparative study on enzymatic hydrolysis and sugar analysis related to the observed changes in cellulose structure to biomass digestibility was also performed. The improvement in hydrolysis is hypothesized to be primarily due to structural changes in the cellulose constituent.

See more of this Session: Recalcitrance of Woody Biomass
See more of this Group/Topical: International Congress on Energy 2011

(714c) Kinetics and Mechanism of Cellulase Hydrolysis of Nanocrystalline Cellulose

Xiao Zhang, Chemical Engineering and Bioengineering, Washington State University, Richland, WA and Xiaohui Ju, Washington State University, Richland, WA

Kinetics and Mechanism of Cellulase Hydrolysis of Nanocrystalline Cellulose

Xiaohui Ju, Xiao Zhang

School of Chemical Engineering and Bioengineering, Centre for Bioproducts and Bioenergy, Washington State University,

Abstract

Decrystallization of cellulose is a major bottleneck in enzymatic hydrolysis of lignocellulosic biomass. The kinetic and mechanism of enzyme degradation of nanocrystalline cellulose prepared from bleached kraft pulp were investigated. The efficiency of enzymatic degradation of nanocrystalline cellulose by three commercial enzymes was determined. All these cellulase preparations demonstrated the ability to degrade nanocrystalline cellulose. The mechanism of enzyme decrystallization of cellulose was investigated by XRD and NMR. This presentation will discuss some of the new findings in cellulose crystal structural changes during enzymatic hydrolysis.

See more of this Session: Recalcitrance of Woody Biomass
See more of this Group/Topical: International Congress on Energy 2011

(714d) Incremental Disassembling of Woody Biomass and Production of Biochemicals

Shijie Liu, Paper and Bioprocess Engineering, SUNY College of Environmental Science and Forestry, Syracuse, NY

Woody biomass is a reliable renewable source of raw materials for energy and chemicals. Woody biomass is structurally composed of cellulose, lignin, hemicelluloses and other minor components. The compositions vary from species to species. The complexity / heterogeneity of woody biomass presents challenges to conversion technologies, which is perceived as recalcitrance. Pretreatment and detoxification have been two unfavorable but widely discussed processes todate. At SUNY ESF, a strategy of incremental disassembling of woody biomass is adopted in which hemicelluloses and extractives are separated first through hot-water extraction, cellulose and lignin are next. Hot-water extraction uses water as solvent without any other added chemicals. Acetyl groups in the woody biomass dissolve into the liquor and serves as catalyst. In each step of the separation, high value applications are examined. For example, hemicelluloses are hydrolyzed and fractionated via membrane for xylo-oligomers, xylose, acetic acid, methanol, and aromatics, etc. The residual solid mass containing mostly cellulose and lignin has wide applications: fiber board, wood fuel pellet, pulp and paper, and sugars / aromatics. Incremental disassembling is emphasized for value preservation of various components. A clever integration of the processes makes the conventional pretreatment and detoxification processes obsolete. The sugar stream produced at SUNY ESF has been successfully fermented to ethanol, butanol, lactic acid, and PHA. There is no deliberate detoxification that generates waste byproducts.

See more of this Session: Recalcitrance of Woody Biomass
See more of this Group/Topical: International Congress on Energy 2011

(714e) Effect of Steam Explosion On Degradability and Accessiblity of Loblolly Pine In Bioethanol Applications

Yuzhi Kang, Sang Beom Kim, Prabuddha Bansal, Dr Matthew Realff and Andreas S. Bommarius, School of Chemical and Biomolecular Engineering, Georgia Institute of Technology, Atlanta, GA

Abstract

Among the various biomass pretreatment methods, steam explosion is one of the most widely used due to its cost-effective disruption of the plant structure and hence enhanced enzymatic accessibility. The operational parameters including temperature, residence time and acid concentration have pronounced effect on crystallinity (CrI) of the feedstock loblolly pine. CrI was determined from X-ray diffraction pattern with the method of least squares and it was found to decrease as pretreatment severity increase.

Three different substrates, Avicel, phosphoric acid swollen cellulose (PASC), and SO_2 steam exploded Loblolly pine (SELP), were subjected to enzymatic hydrolysis and subsequent adsorption study. Both SELP samples (2.9-419-5 and 5-417-5) were found more readily degraded by the CelluclastÒ enzyme system than Avicel. Despite the complexity of both substrates and enzyme mixtures, the SELP adsorption isotherm can be fitted to a Langmuir isotherm with $R^2 > 0.99$. The saturation point was reached at ~6.24 mg/ml total enzyme loading. Adsorption data was correlated with initial rates and a linear relationship between initial rate and adsorption amount was confirmed.

See more of this Session: Recalcitrance of Woody Biomass
See more of this Group/Topical: International Congress on Energy 2011

CHANGE OF STRUCTURE OF CORNCOB WITH DILUTE ACID AND LIME PRETREATMENT AND A POTENTIAL PROCESS FOR XYLOSE CO-PRODUCTION OF CELLULOSIC ETHANOL

Ji-hong Li[1], Shi-Zhong Li[1], Zhi-pei Yan[1] , Qing Li[2], Liang-cai Peng[3]

[1]Institute of New Energy Technology, Tsinghua University, Beijing, China

[2]Colledge of Science, Huazhong Agricultural University , Wuhan , China

[3]Colledge of Plant Science & Technology, Huazhong Agricultural University , Wuhan , China

Abstract

Degree of polymerization (DP) of cellulose is considered an important factor to effect on the conversion of Lignocellulose. The higher DP of cellulose, the more difficult conversion of lignocellulose is. In this study, corncobs were pretreated with dilute acid process for production of xylose, and then the residues were pretreated with lime. The enzymatic hydrolysis and the structures involved the composition, DP and crystallinity index of these two types of feedstock were determined. Comparison of structures resulted from these two pretreatment revealed that DP was not the crucial factor to determined efficiency of enzymatic hydrolysis. Lime pretreatment could improve the conversion of cellulose although the residues with lime pretreatment had higher DP and percentage of crystallized cellulose. Moreover, the results also confirmed lignin was not the key factors for enzymatic hydrolysis of lignocellulose because the residues with dilute acid pretreatment had the similar percentages of lignin to that with lime pretreatment. The crucial effect of lime pretreatment was to remove the inhibitor for cellulase and increase the percentages of cellulose. With obtained xylose, the residues with further lime pretreatment had expected efficient enzymatic hydrolysis of cellulose. So this process could be a potential process for xylose co-production of cellulosic ethanol to decrease the cost of cellulosic ethanol.

(714g) Green and High Efficient Enzyme Hydrolysis Process for Sweet Sorghum Bagasse

Xinshu Zhuang, Wen Wang, Qiang Yu, Wei Qi, Qiong Wang, Xuesong Tan, Yu Zhang and Zhenhong Yuan, Guangzhou Institute of Energy Conversion,CAS, Guangzhou, China

In order to improve enzymatic hydrolysis efficiency, several approaches including liquid hot water (LHW) pretreatment, adding surfactant, magnetic stirring and fed-batch hydrolysis were investigated, in which Sweet sorghum bagasse (SSB) was chosen as the feedstock. The experiment results showed that with the LHW pretreatment, the SSB conversion ratio reached 35.91% , nearly 75% xylan and 33% lignin were removed. Using the LHW-pretreated SSB, glycan conversion ratio raised from 16.6% to 37.9% by adding surfactant Tween80 with the optimum dosage of 0.175ml/g DM. The virulence test showed that Tween80 nearly had no influence to the growing of ethanol fermentation yeast-*Saccharomyces cerevisiae*. Comparing stirring and shaking mode for the enzyme hydrolysis, the glycan conversion rate was 57.3% in stirring mode, higher than 44.4% obtained from shaking mode. The image of Scanning Electron Microscope (SEM) showed that LHW pretreatment could destroy the tight structure of SSB and the comminution of pretreated material could smaller the particles to enhance the hydrolysis process. Finally the LHW-pretreated SSB, Tween80 and stirring mode were applied to fed-batch process, the concentration of glucose, cellobiose and xylose reached 88.8g/l, 30.7g/l and 18.9g/l respectively, and the glycan conversion ratio was 89.3% under the optimum condition. These research works lay the foundation for the construction of a green and high efficient sugar platform to facilitate the cellulosic ethanol development.

These research work are financially supported by the National Key Technology R&D Program (2011BAD22B01), National High Technology Research and Development Program of China (2010AA101606) and the Knowledge Innovation Project of the Chinese Academy of Sciences (KSCX1-YW-11-A3, KSCX2-YW-G-063)

[1]Corresponding author. yuanzh@ms.giec.ac.cn (Z. Yuan)

See more of this Session: Recalcitrance of Woody Biomass
See more of this Group/Topical: International Congress on Energy 2011

(718a) Lifetime and Efficiency Studies of Zinc/ Bromine Electrochemical Energy Storage In Large Scale Renewable Energy Applications

Alex Bistrika[1], Eunice Naswali[2], Chianna Alexander[2], Hai-Yue Han[2], David Naviaux[2], Annette von Jouanne[2], Ted Brekken[2] and Alexandre Yokochi[1], (1)School of Chemical, Biological and Environmental Engineering, Oregon State University, Corvallis, OR, (2)School of Electrical Engineering and Computer Science, Oregon State University, Corvallis, OR

As the world's power needs grow, the demand for power from renewable resources, such as wind or solar, is increasing. One major drawback associated with these renewable resources is that the power output is dependent on environmental factors, such as cloud cover and wind speeds. This allows the possibilities of either power output exceeding or falling short of forecast levels which may lead to grid instabilities. Therefore, Large Scale Energy Storage (LSES) systems are critical to store excess power when the output exceeds demand in order to supplement output power when it falls short of demand.[1] Some promising LSES contenders are: 1) Zinc bromide flow cells (FC), 2) Vanadium redox FC, 3) Sodium sulfur batteries, and 4) Lead acid batteries.[2] Of these available technologies the zinc bromide technology was selected for further investigation because of previously reported long cycle-life (CL) and high efficiencies.[3]

In this work, the cell efficiency dependence on temperature of a bench scale Zinc Bromide battery manufactured in our lab was investigated to confirm the theoretical efficiency. Efficiency cycling was voltage limited on the discharge cycle at a 1.5 / V cell cut-off potential. Standard method of measuring columbic efficiency was implemented. At room temperature, ~80% efficiency was achieved when cycling the system between 20 - 100% State of Charge (SOC), which is in good agreement with published values.[4] The cell efficiency was demonstrated at over 90% at 50°C and cycling between 20 - 100% SOC.

In addition, given that deep discharge is perceived as highly damaging to electrochemical energy storage, the battery was cycled through the Zero-Point (ZP) 10,000 times. The cell was continuously cycled between 0 - 5% (SOC). Battery performance was evaluated from transfer rate and exchange current density, determined using Tafel scans. No degradation of the electrodes was found, and in fact a slight increase in the exchange current density was observed. We interpret this exchange current increase as a result expected for electrode conditioning of the polished carbon electrode substrates used in our test cells.[5]

These along with further more recent results will be presented.

1. Dell, R. M.; Rand, D. A. J. *Energy storage: a key technology for global energy sustainability. J. Power Sources* **2001,** *100*, 2-17.

2. (a) Schoenung, S. M. *Characteristics and Technologies for Long- vs. Short-Term Energy Storage: A Study by the DOE Energy Storage Systems Program*. Sandia National Laboratories SAND2001-0765 **2001** p 59; (b) Schoenung, S. M.; Eyer, J. *Benefit/Cost Framework for Evaluating Modular Energy Storage*. Sandia National Labratories SAND2008-0978 **2008** p 40.

3. Manla, E.; Nasiri, A.; Hughes, M. *Modeling of zinc energy storage system for integration with renewable energy. Industrial Electronics* **2009**, IECON '09 35th Annual Conference of IEEE, 3-5, p 3987-3992.

4. Lex, P.; Jonshagen, B. *The zinc/bromine battery system for utility and remote area applications. Power Engineering Journal* **1999**, *13*, 142-148.

5. Kautek, W.; Conradi, A.; Fabjan, C.; Bauer, G. *In situ FTIR spectroscopy of the Zn-Br battery bromine storage complex at glassy carbon electrodes. Electrchimica Acta* **2001**, *47*, 815-823.

See more of this Session: Sustainable Electricity: Generation, Transmission, and Storage
See more of this Group/Topical: Sustainable Engineering Forum

(718b) Parametric Sensitivity of Reverse Electrodialysis for Electricity Production Using Salinity Gradient

Rahul Patil and Glenn Lipscomb, Department of Chemical and Environmental Engineering, University of Toledo, Toledo, OH

Global population growth has led to increased energy demand and stress on energy supplies. Concerns over the stability, sustainability, and environmental impact of global energy consumption have generated renewed interest in alternative energy sources. The ideal source would be locally distributed and renewable with minimal economic and environmental impact on our current lifestyle.

The sun is the source of virtually all energy. Solar energy can be utilized in the form of fossil fuels, wind energy, photovoltaic or thermal solar, and hydroelectric. Growing interest exists in using the salinity difference between fresh water and sea water to produce power from mixing the two streams. Membrane processes such as Reverse Electrodialysis (RED) and Pressure Retarded Osmosis (PRO) can be used to produce electricity [1, 2]. Literature reports the effects of membrane type, membrane and gasket thickness, electrode systems and flow rates on power production [3].

In the current study, we demonstrate the optimization of power output using RED. The effects of operating parameters such as current density, concentration of process streams and the concentration and composition of electrode rinse are investigated. We also examined the relationship between electrons generated (current) and ion transfer between the solutions. The efficiency of RED can be determined from the ratio of current to ion transfer. We examined possible sources of energy loss and inefficiency. Shunt currents and membrane area occupied by spacer mesh are believed to be significant contributors. The origin and reduction of these factors to increase the efficiency is discussed as well as the effect of other cell design variables.

References:

1. Post, J. et. al., *Salinity-gradient power: Evaluation of pressure-retarded osmosis and reverse electrodialysis*, Journal of Membrane Science 288 (2007) 218–230.
2. Dlugolecki, P. et. al., *Practical Potential of Reverse Electrodialysis As Process for Sustainable Energy Generation*, Environmental Science and Technology 43 (2009) 6888–6894.
3. Veerman, J. et. al., *Reverse electrodialysis: evaluation of suitable electrode systems,* Journal of Applied Electrochemistry 40 (2010) 1461–1474.

See more of this Session: Sustainable Electricity: Generation, Transmission, and Storage
See more of this Group/Topical: Sustainable Engineering Forum

(718c) Energy Storage with Molten Metal Oxide

Sounak Roy[1], Ashay D. Javadekar[1], Abhimanyu Jayakumar[2], Douglas Buttrey[1], John Vohs[2] and Raymond J. Gorte[2], (1)Chemical Engineering, University of Delaware, Newark, DE, (2)Chemical and Biomolecular Engineering, University of Pennsylvania, Philadelphia, PA

Energy storage will play a critical role for electrical power generation as we move towards using renewable resources, since many renewable energy sources are intermittent in nature. Reversible fuel cells have been proposed as a possible way to store energy; however, energy losses associated with the conversion between H_2O and H_2 and the volumes required for storing H_2 are serious impediments to this technology. To solve these problems, we propose to use the electrochemical oxidation of Sb and reduction of Sb_2O_3 in a solid oxide fuel cell (SOFC) as a means of energy storage. Both Sb (m.p. 903K) and Sb_2O_3 (m.p. 929K) are molten at typical SOFC operating temperatures (e.g. 973K). Furthermore, initial data for Sb electrode overpotentials indicates that these losses are very low. Although the Nernst potential for Sb-Sb_2O_3 equilibrium is relatively low, this is inconsequential for the energy-storage application. Finally, the volume associated with 1MW-h of energy is only 45 liters, based on cell performance that has already been achieved. The main goals of work in this area are to understand how cell performance is affected by Sb-Sb_2O_3 conversion. Mixing the metal and oxide phases is found to be a critical factor in determining cell performance.

See more of this Session: Sustainable Electricity: Generation, Transmission, and Storage
See more of this Group/Topical: Sustainable Engineering Forum

(718d) Evaluation of Organic Rankine Cycle Geothermal Power Plant and Considerations

Kirtipal Barse[1], Michael D. Mann[1], William D. Gosnold[2] and Hossein Salehfar[3], (1)Chemical Engineering, University of North Dakota, Grand Forks, ND, (2)Geology, University of North Dakota, Grand Forks, ND, (3)Electrical Engineering, University of North Dakota, Grand Forks, ND

The objective of this work is to evaluate the feasibility of an Organic Rankine Cycle (ORC) power plant to generate electricity from low flow rates of a geothermal resource (~875 gpm) at low temperatures (T~100°C). A model of ORC was developed using HYSYS process simulator. DOE's Cost of Renewable Energy Spreadsheet Tool (CREST) model was used for economic modeling of the power plant. First law efficiency of the power plant was studied using energy analysis. Exergy analysis was used to study the second law efficiency of the power plant. A comparison was made between shell and tube heat exchangers and plate and frame heat exchangers for performance and cost evaluation. Plate and frame heat exchangers have advantages such as higher overall heat transfer co-efficient and smaller footprint compared to shell and tube heat exchangers. The quality of water used with respect to geothermal development was investigated, including dissolved solids, toxic minerals, additives, scale formation, corrosion and non-condensable gas. This paper will summarize the results of this study.

See more of this Session: Sustainable Electricity: Generation, Transmission, and Storage
See more of this Group/Topical: Sustainable Engineering Forum

(737a) Development and Testing of An Asymmetric Supercapacitor Based On a Carbon Foam Electrode Impregnated with Nickel Hydroxide

Wen Nee Yeo[1], Tony N. Rogers[1] and Bahne C. Cornilsen[2], (1)Department of Chemical Engineering, Michigan Technological University, Houghton, MI, (2)Department of Chemistry, Michigan Technological University, Houghton, MI

Supercapacitors have attracted significant attention in recent years due to their potential to power hybrid vehicles, portable audio equipment, and memory back-up devices. The trend in supercapacitors is to combine a double-layer electrode with a pseudo-capacitance electrode in an asymmetric capacitor configuration.[1] The double-layer electrode is usually an activated carbon (AC) since it has high surface area, good conductivity, and relatively low cost.[2, 3] The pseudo-capacitance electrode usually consists of RuO_2, MnO_2, $Ni(OH)_2$, etc., with the most commonly studied material being $Ni(OH)_2$.[4-7]

In this work, a graphitic carbon foam was impregnated with nickel hydroxide active mass using electrochemical deposition to make a nickel-carbon foam electrode. It served as the pseudo-capacitance electrode in an asymmetric supercapacitor. Carbon foam was used as a conductive holder of the active mass because of its electrical conductivity, pore size, and pore connectivity. The objective was to reduce the cost and weight of the supercapacitor while maintaining or increasing its capacitance and gravimetric energy storage density. The specific capacitance of the nickel-carbon foam electrode at different current densities was measured in constant-current flooded cell cycling tests. The measured specific capacitance of 2641 F/g at 5 mA/cm^2 was higher than a reported value of 2080 F/g for an Al-substituted α-$Ni(OH)_2$ electrode.[6] The nickel-carbon foam electrode was then combined with an AC at different ratios of the negative and positive electrode masses to study asymmetric cell performance. The mass ratio for optimum cell discharge capacity was found to be ~3.7.

References

[1] A.I. Belyakov, 3rd European Symposium on Supercapacitors and Applications (ESSCAP) Roma, Italy, 2008.

[2] K. Babel, K. Jurewicz, *Journal of Physics and Chemistry of Solids* 65 (2004) 275-280.

[3] P.-L. Taberna, Geoffroy Chevallier, P. Simon, D. Plee, T. Aubert, *Materials Research Bulletin* 41 (2006) 478-484.

[4] Jong Hyeok Park, Sungwook Kim, O Ok Park, J.M. Ko, *Applied Physics A* 82 (2006) 593-597.

[5] D.-D. Zhao, S.-J. Bao, W.-J. Zhou, H.-L. Li, *Electrochemistry Communications* 9 (2007) 869-874.

[6] J.-W. Lang, L.-B. Kong, M. Liu, Y.-C. Luo, L. Kang, *Journal of Solid State Electrochemistry* 14 (2010) 1533-1539.

[7] Q. Huang, X. Wang, J. Li, C. Dai, S. Gamboa, P.J. Sebastian, *Journal of Power Sources* 164 (2007).

See more of this Session: Energy Storage and Electrochemical Processes II
See more of this Group/Topical: International Congress on Energy 2011

(737b) Effects of Annealing On $Zr_8Ni_{19}X_2$ (X = Ni, Mg, Al, Sc, V, Mn, Co, Sn, La, Hf) for Ni/MH Battery Operation

Jean Nei[1], Kwo Young[2], Simon K. Y. Ng[1] and Steven O. Salley[1], (1)Chemical Engineering and Materials Science, Wayne State University, Detroit, MI, (2)Energy Conversion Devices Inc./Ovonic Battery Company, Rochester Hills, MI

Zr, Ti-based AB_2 metal hydride (MH) alloys are considered to be one of the most promising candidates to replace the misch metal based AB_5 alloys presently used as the negative electrode in nickel/metal hydride (Ni/MH) batteries. Since most of the studies on AB_2 MH alloys focus on multi-element and highly disordered families, the understanding of basic Ti-Ni and Zr-Ni intermetallic alloys is crucial for fully utilizing the synergetic effects between the main AB_2 Laves phases and the non-Laves secondary phases.

For the best balance between large storage capacity (lower hydride heat of formation) and high discharge capability (higher hydride heat of formation), a stoichiometry of A:B close to 1:2 is desirable. However, there is no AB_2 intermetallic alloy for either Zr-Ni or Ti-Ni binary systems. The Zr-Ni phase diagram shows several intermetallic alloys, and the closest intermetallic compounds to AB_2 are Zr_7Ni_{10} and Zr_8Ni_{21}. Zr_7Ni_{10} has slightly stronger metal-hydrogen bond strength and solidifies congruently from liquid. Zr was replaced partially by Ti to reduce the metal-hydrogen bond strength. Other substituting elements were also studied in order to improve the electrochemical properties. Zr_8Ni_{21} was chosen as the focus of this report. Zr_8Ni_{21} does not solidify directly from the liquid as in the case of Zr_7Ni_{10}. Instead, the Zr_2Ni_7 phase is first solidified from the liquid with a Zr-to-Ni ratio of 8:21 and later reacts with the remaining liquid to form Zr_8Ni_{21} alloy peritectically. If the condition of equilibrium is not reached during cooling, one or more secondary phases will precipitate out. Therefore, an annealing process was adopted for this study.

The effects of annealing at 960 °C for 8 hours on the phase structures of a series of partially substituted Zr_8Ni_{21} alloys were studied. The substituting elements included Mg, Al, Sc, V, Mn, Co, Sn, La, and Hf. In general, the abundance of dominant phase increased and the number of smaller secondary phases diminished after annealing. Similar to the choice of C14/C15 main phase in AB_2 alloys, the number of outer-shell electrons plays an important role in determining the main phase of the

A_8B_{21} samples. As number of outer shell electrons of the substituting element increases, the main phase evolves from tetragonal Zr_7Ni_{10}, to orthogonal Zr_7Ni_{10}, Zr_2Ni_7 and then Zr_8Ni_{21}.

The annealing effects on the gas phase and electrochemical hydrogen storage properties were studied. It's observed that as the Zr-content in the main phase of the alloy increases (orthogonal Zr_7Ni_{10} > tetragonal Zr_7Ni_{10} > Zr_8Ni_{21} > Zr_2Ni_7), the gas phase storage capacity increases. After annealing, due to the increase in the abundance of main phase, the capacities of alloys with higher capacities before annealing increase while others decrease. The sample with Hf-supplement shows the highest electrochemical discharge capacity of 200 mAh/g. After annealing, all samples with the same main phase as the as-cast ones show degradation in electrochemical capacity due to the reduction in number and abundance of secondary phases. All supplements show improvement in surface exchange current from the base alloy. Except for La and Hf supplements, annealing reduces the exchange current density. The bulk hydrogen diffusion constant decreases for most of the supplements except V and Sn. All supplements, excluding Sn, show improvement in bulk diffusion after annealing. The gas phase full capacity has strong correlation to the full electrochemical discharge capacity. Among all samples, as-cast La-supplement shows the best overall electrochemical properties.

See more of this Session: Energy Storage and Electrochemical Processes II
See more of this Group/Topical: International Congress on Energy 2011

Optimal Charge Rates for a Lithium Ion Cell

Saeed Khaleghi Rahimian, Sean Rayman, Ralph E. White

Department of Chemical Engineering, University of South Carolina, Columbia, SC, USA

Extended Abstract

The optimum charge rate for a lithium ion cell at each cycle is determined to maximize the useful life of the cell that is defined as the number of cycles before the end of discharge voltage (EODV) drops below 3.0 V or the cell discharge capacity becomes less than 20% of the original discharge capacity. First, we showed that by applying a dynamic optimization routine the number of cycles can be increased by approximately 29.4 % with respect to the case with one optimal charge current. The dynamic optimization results indicated that the optimum charge rates are the minimum currents at which the constraints for the useful life are satisfied. This is due to the minimum charge rate resulting in minimum side reaction rate and capacity fade. Then, a new method has been developed to find the optimal charge without using optimization algorithms. The cell capacity and the EODV are both a monotonic ascending function of the charge current; as a result the optimal charge current can be obtained by some iterations of the charge –discharge processes for each cycle. The new approach improves the useful cell life by approximately 41.6 % with respect to using only one optimal charge current in a few minutes while the previous optimization algorithm takes at least one day. Thus, the new technique enables us to find the best charge profile in online application in reasonable computation time by using any rigorous model for the cell simulation.

Introduction

The capacity of the Li-ion batteries decreases with the number of charge-discharge cycles due to undesirable side reactions. Since these batteries are used for tens to thousands cycles, depending on application, there is a demand to seek the best operating conditions to improve the cycling performance to minimize the capacity fade, and thereby increasing cell life. Only a few studies have attempted to optimize the charge rates for battery longevity and energy efficiency using optimization algorithms. S. Bashash et. al. applied multiobjective genetic algorithm to obtain the charge pattern of a plug-in hybrid electric vehicle to simultaneously minimize the total cost of fuel and electricity and the battery life degradation [1]. In previous work [2] we showed that the lithium ion cell life can be improved by ~30% with respect to the case with single optimal charge rate used by optimization of charge currents, as a function of cycle number during cycling. Choosing twenty rates, the optimization needs to evaluate the objective function (number of cycles) more than 80000 times and if the objective function evaluation takes only 1 sec, at the best condition, it will take about one day to find the optimal profile. In this work, a new approach is proposed to find the best charge rate at each cycle that improves the useful life 10% more than the dynamic optimization in much less computation time.

Lithium-ion cell Simulation

A single particle (SP) physics based model, which includes capacity fade, was applied to simulate the cell performance under low-earth-orbit (LEO) cycling conditions [2]. The main assumptions associated with the SP model are as follows:

- The concentration of the electrolyte and the potential in the solution phase are constant and uniform for all time across the cell sandwich (cathode, separator, anode).

- Positive and negative electrode potentials depend on time only.
- The model considers the capacity fade by a continuous occurrence of a side reaction (reduction of ethylene carbonate) only during the charge near the surface of the negative electrode. Moreover, the state of the charge (SOC) of the positive electrode is diminished during cycling.
- The concentration of Li ions in the spherical particle is estimated using a two term polynomial approximation.

A volume average technique is applied to predict the diffusion of the lithium ions for the cathode ($LiCoO_2$ with no Ni) and the carbon anode [4]:

$$\frac{dx_{p,avg}}{dt} = \frac{-3J_p}{FR_p c_{p,\max}} \qquad x_{p,surf} - x_{p,avg} = \frac{-J_p R_p}{5FD_{s,p} c_{p,\max}} \qquad J_p = \frac{I_{app}}{S_p}$$

$$\frac{dx_{n,avg}}{dt} = \frac{-3J_n}{FR_n c_{n,\max}} \qquad x_{n,surf} - x_{n,avg} = \frac{-J_n R_n}{5FD_{s,n} c_{n,\max}} \qquad J_n = \frac{-I_{app}}{S_n} - J_s$$

where $x_{i,avg}$ is the ratio of the solid average concentration to the maximum solid concentration for each electrode ($c_{i,max}$) , $x_{i,surf}$ is the ratio of the solid surface concentration to the maximum solid. J_s, the side reaction rate expression, is calculated by using Tafel kinetics:

$$\begin{cases} J_s = -i_{0f} \exp\left(\frac{-\alpha_{c,f} F}{RT} \eta_s\right) [charge] \\ J_s = 0 \; [discharge] \end{cases}$$

$$\eta_s = \phi_n - U_{ref,f} - \frac{I_{app}}{S_n} R_{film}$$

R_{film} is defined as:

$$R_{film} = R_{SEI} + \frac{\delta_{film}}{k_f}$$

and the rate at which the film thickness increases is calculated by:

$$\frac{d\delta_{film}}{dt} = \frac{-J_s M_f}{\rho_f F}$$

The SOC of the positive electrode updated at the end of each charging process as follows [4]:

$$\theta_{p,N} = \theta_{p,N-1} - \theta_{1,N-1}$$

where θ_1 is the loss of SOC obtained by dividing the capacity loss to the maximum capacity of the cell (subscript N-1 indicates the previous cycle):

$$\theta_1 = \frac{Q_S}{Q_{max}}$$

$$Q_s = -\int_0^{t=chsrge\ time} (J_s S_n) dt$$

$$Q_{max} = \int_0^{t=chsrge\ time} I_{app} dt \qquad (first\ cycle)$$

LEO cycling [3], which contains the following steps, was applied for the cell simulation after the cell is completely charged initially:

1) Constant current discharge (0.6857 C-rate (40 % DOD)) for 35 min discharge time. Unless the voltage drops below 3.0 V or the cell capacity reaches the 20% of the total discharge capacity, go to step 2.
2) Constant current charging (up to 1 C rate) for 61 min charge time. If the voltage reaches the EOCV (4.05 V), go to step 3, if not go to step 1.
3) Constant voltage (4.05 V) charging for the remaining charge time, go to step 1.

Dynamic Optimization Results

To compare the results of the dynamic optimization and the new method, the best profile obtained by the optimization using single charge rate and twenty charge rates are presented in this work as well as [2]. Figures 1 and 2 show the variation of the cell capacity and the end of

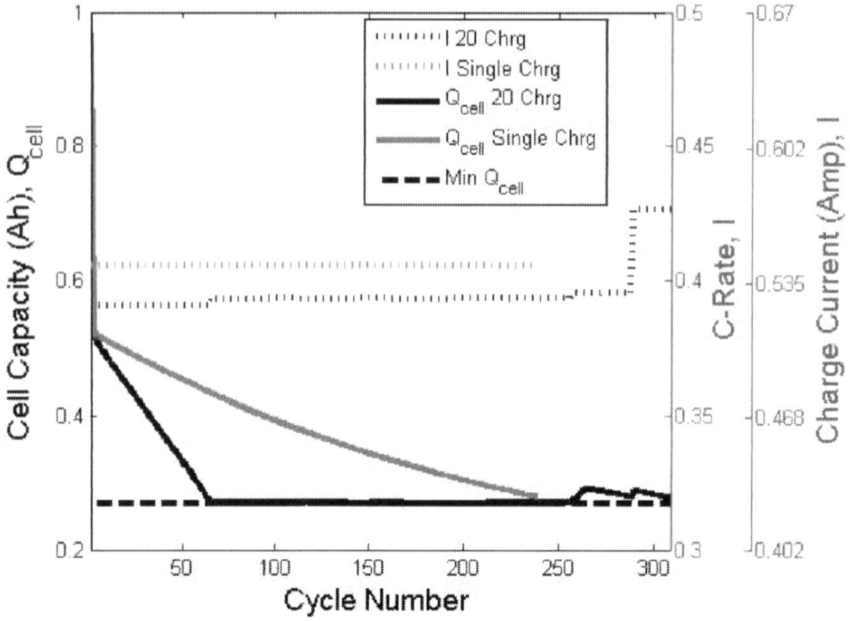

Figure 1: Remaining cell capacity at the end of discharge versus cycle number for single and 20 charge currents [2]

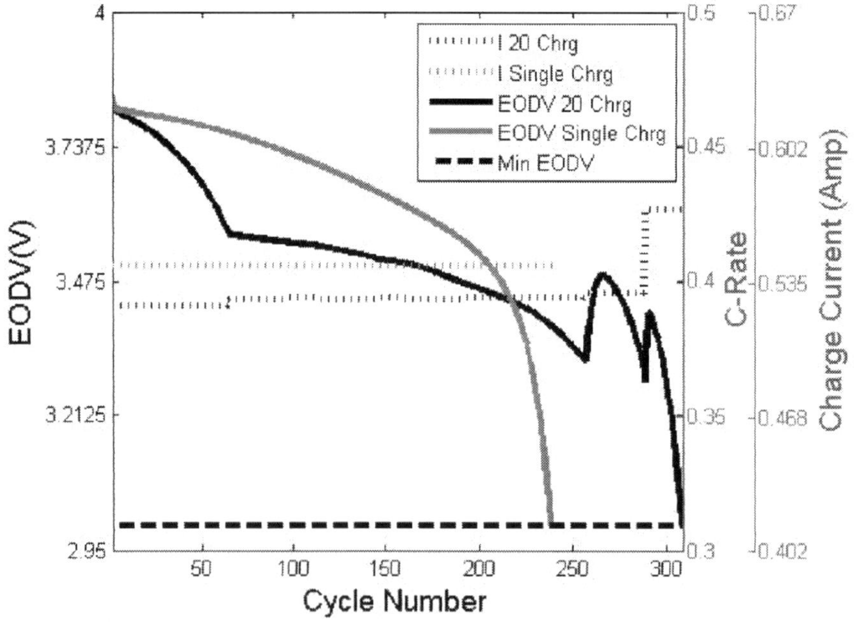

Figure 2: EODV versus cycle number for single charge and 20 charge currents [2]

voltage (EODV) respectively as a function of cycle number for single and twenty charge rates. Using twenty charge currents increases the number of cycles to 309, 29.4 % increase with respect to one optimum current.

New Approach

The optimization results indicate that the best charge current is the minimum rate at which, the cell capacity constraint and the EODV constraint are not violated. As a result we tried to develop a new method to find the optimal rate at each cycle as shown in Figure 3. By using this algorithm, any other constraints (e.g. end of charge voltage (EOCV) at 4.0 V) can be included without any difficulties.

Results and Discussion

The optimal charge rate and the cell capacity as a function of the cycle number obtained by the new approach that improves the number of cycles to 337, 41.6 % increase with respect to one optimal charge rate, shown in Figure 4. Variation of the EODV and the EOCV with the cycle number is presented in Figure 5. Adding the constraint that keeps the EOCV greater than 4.0 V (*constraint in Figure 3), decreases the number of cycles to 317. This is due to increase in charge rate in some cycles to not allow the EOCV dropping below the constraint. The computation time for the new approach is 228 sec on a Dell Precision T7500, with 2 Quad Core 2.53 GHz Zenon Processors CPUs and 12.285 GB of RAM.

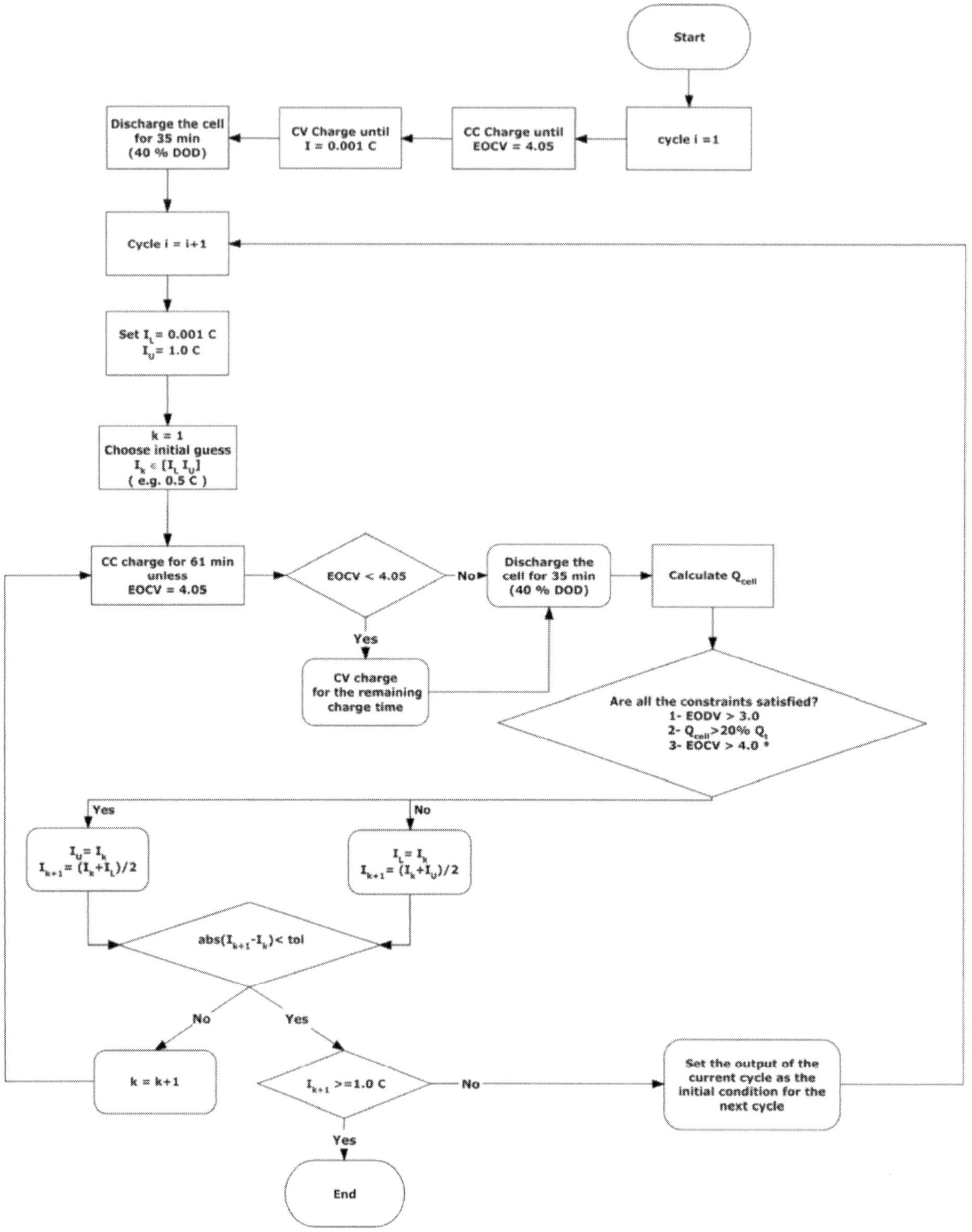

Figure 3: The new approach flowchart

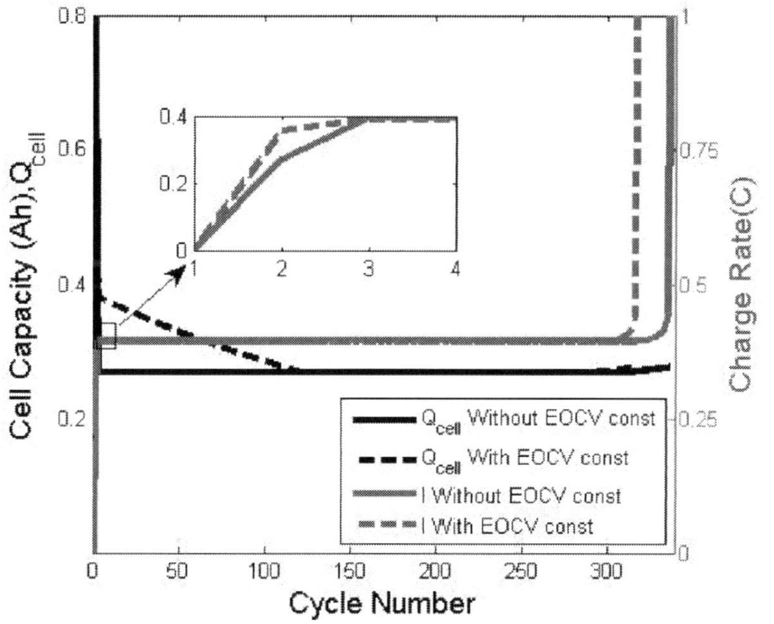

Figure 4: Remaining cell capacity at the end of discharge and Charge rate versus cycle number obtained by new approach with and without EOCV constraint

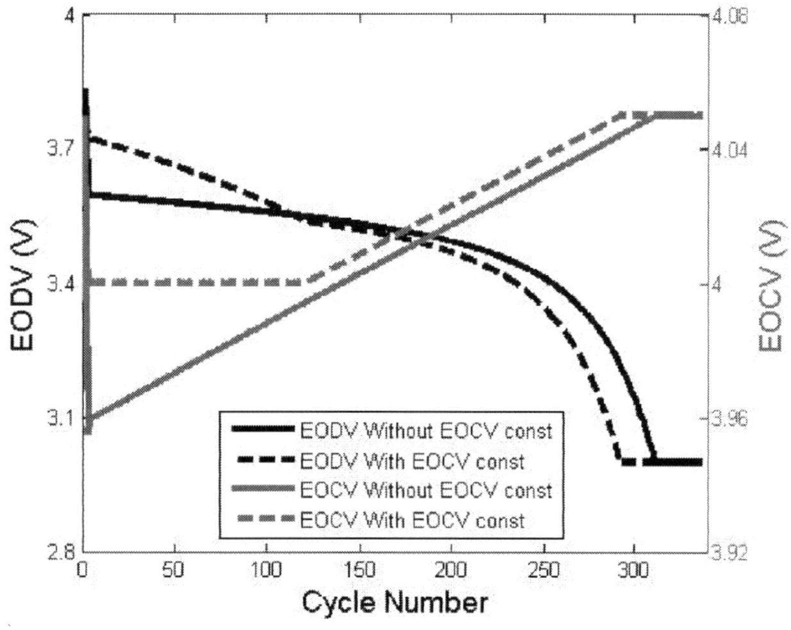

Figure 5: EODV (black) and EOCV (gray) versus cycle number obtained by new approach with and without EOCV constraint

Conclusion

A new approach was developed to provide the optimal charge rates at each cycle to maximize the useful life of a Lithium-ion cell. The computation time is greatly reduced by this approach with respect to the optimization method. For online application, the new method only requires the prediction of the next cycle by using prior cycling data while the entire cell life needs to be predicted for the optimization. Moreover, the new approach allows one to use a more rigorous model (i.e. P2D) to find the best charge rates without prohibitively increasing the computational time.

References

1. S. Bashash, S. J. Moura, J. C. Forman, H. K. Fathy, *J. Power Sources*, **196**, 541 (2011).
2. S. Khaleghi Rahimian, S.C. Rayman, and R.E. White, Maximizing the life of a lithium-ion cell by optimization of charging rates, *J. Electrochem. Soc.* **152** (12), A1302–A1308 (2010).
3. J. Lee, Y. K. Anguchamy, B. N. Popov, J. Power Sources 162 (2006) 1395.
4. P. Ramadass, B. Haran, P. M. Gomadam, R. E. White, B. N. Popov, J. Electrochem. Soc. 151(2) (2004) A196.

Basic model of secondary zinc air batteries to quantify environmental impacts

Daniel Schröder[a], Ulrike Krewer[a,b]

[a]Portable Energy Systems, Chair for Process Systems Engineering, Otto-von-Guericke University Magdeburg, Universitätsplatz 2, 39106 Magdeburg, Germany
[b]Otto Hahn Group/Portable Energy Systems, Max Planck Institute for Dynamics of Complex Technical Systems, Magdeburg, Sandtorstrasse 1, 39106 Magdeburg, Germany

---

Zinc air batteries (ZABs) offer a high theoretical specific energy density (1353 $\mathrm{Wh/kg}$[1]) and therefore appeal for application in portable and mobile devices. The availability of bi-functional air electrode catalysts, such as non precious perovskite structured metal, makes ZABs electrically rechargeable and even more attractive.
Figure 1 illustrates the reaction and transport processes occurring in a ZAB during discharge mode. In general, the battery consists of the zinc electrode, the air electrode and the separator. Equations (1)–(3) state the occurring reactions in caustic electrolyte.

$$\mathrm{Zn} + 4\,\mathrm{OH}^- \overset{r_1}{\rightleftharpoons} \mathrm{Zn(OH)_4^{2-}} + 2\,\mathrm{e}^- \tag{1}$$

$$\mathrm{Zn(OH)_4^{2-}} \overset{r_2}{\rightleftharpoons} \mathrm{ZnO} + 2\,\mathrm{OH}^- + \mathrm{H_2O} \tag{2}$$

$$\mathrm{H_2O} + \frac{1}{2}\mathrm{O_2} + 2\,\mathrm{e}^- \overset{r_3}{\rightleftharpoons} 2\,\mathrm{OH}^- \tag{3}$$

[1]Linden, E., 2004. Handbook of Batteries, 3rd ed., McGraw-Hill, p.1.13

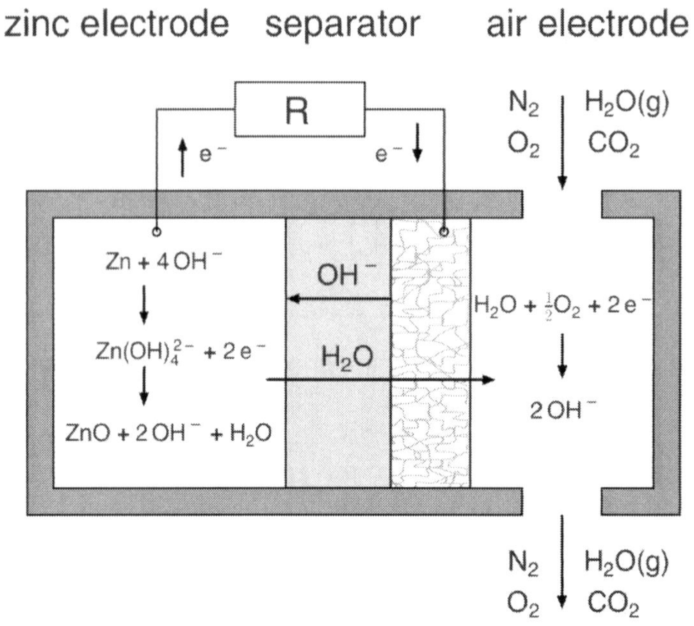

Figure 1: Schematic view of a zinc air battery showing the occurring reaction and transport processes during discharge. The charge process is then vice versa.

It is essential that OH^- and H_2O are transported through the separator during ZAB operation. As well, the ZAB is open to the environment which implies that uptake or loss of gaseous water, O_2 and even CO_2, might influence the operation behavior of the ZAB. For example, carbonates can be formed due to the reaction of the caustic electrolyte with CO_2.
Research on ZABs is almost exclusively experimental and focuses on the characterization of materials for zinc electrode, air electrode or separator, as well as on investigating the performance of entire ZABs.
It is suggested that ZABs are affected by the surrounding humidity; this holds especially for secondary batteries due to the longer operation time[2]. Models can help to analyze the influence of humidity on ZAB performance and get insight into the water balance of a ZAB.
Relatively few models exist to describe ZABs. They are mainly based on the macroscopic approach for porous battery electrodes by Newman and Tobias[3]. These models apply a set of coupled discretized non-linear dif-

[2]Haas, O., Müller, S. and Wiesener, K., 1996. Chem. Ing. Tech., 68(5), pp.524-542.
[3]Newman, J. and Tobias, C.W., 1962. J. Electrochem. Soc., 109, pp.1183-1191.

ferential algebraic equations (DAEs) to describe the transport and reaction phenomena occurring in the ZAB. They are used to analyze and optimize the ZABs design dimensions such as separator thickness and zinc electrode thickness. So far the existing models are not used to investigate the environmental impact on ZAB operation.

Our aim is to investigate the impact of the environmental gaseous water amount, O_2 amount and CO_2 amount on ZAB operation. For this, a flexible and expandable model is designed. In a first step, we introduce an isothermal basic model to describe the reaction and transport processes in secondary ZABs.

Our approach is as follows: The zinc electrode and the air electrode are described as an ideal continuously stirred tank reactor (CSTR), respectively. Each CSTR electrode consists of a fixed metal phase and a liquid phase containing the electrolyte and the dissolved and ionic species. The metal phase at the zinc electrode consists of Zn and ZnO. The metal phase at the air electrode is of inert catalyst structure. Both metal phases do not change their structure with time, however the zinc electrode metal phase ideally replaces its elements Zn and ZnO. The liquid volume in each CSTR electrode is varying with time and can be increased for example by uptake of gaseous water into the air electrode.

The here considered basic model consists of five molar balances for the ionic and solid species in the zinc electrode, three molar balances for the ionic and dissolved species at the air electrode, two charge balances for each electrode and two differential equations for the varying liquid electrolyte volume in the electrodes. These equations are a system of ordinary differential equations (ODEs). In our work, we solve the resulting ODE system with appropriate initial conditions using Matlab®.

The respective electrode over-potentials at the zinc electrode and the air electrode, originating from charge balances, and the ohmic drop through the separator, depending on the conductivity of the electrolyte, enable to calculate the total battery voltage.

The electrochemical reactions (1) and (3), are described by a Butler-Volmer approach. The chemical reaction (2) is applied as precipitation/saturation mechanism in the liquid electrolyte KOH.

Diffusion of OH^- and H_2O and migration of OH^- via the separator are included into the molar balances as exchange terms between the two electrodes. Both electrodes are apart by a certain distance, here the separator thickness in the range of μm. Due to the presence of a counter-ion (K^+) in the liquid electrolyte, electro-neutrality holds. Migration of $Zn(OH)_4^{2-}$ through the separator is omitted.

At this point we investigate three modifications of the basic model:

- Modification (a) is a reference, where gaseous water is neither entering nor leaving the ZAB. Here the O_2 amount in the air electrode is set constant and does not change with time.
- Modification (b) accounts for varying O_2 amount in the air electrode, whereas uptake or loss of water is neglected.
- Modification (c) allows gaseous water to enter or leave while the ZAB is operated with constant O_2 amount in the air electrode.

From transient simulations with galvanostatic charge and discharge cycles we obtain three results for the respective modification:

- Modification (a) shows that OH^- and H_2O transport is sufficient to maintain battery operation.
- Modification (b) shows how cycling may affect the reaction rates at the air electrode and the related ZAB performance
- Modification (c) shows how cycling may affect the water content and the related ZAB performance.

Further investigations target to include the carbonation at the air electrode.

(737e) Novel O2/H2O-Selective Membrane for Metal/Air Batteries

Wei Liu, Pacific Northwest National Laboratory, Richland, WA

This presentation will describe new membrane materials and structures that can be used to encapsulate the cathode of a metal/air battery for its operation under ambient air conditions. The membrane barrier is needed to allow transport of oxygen from air into the cathode for the electrochemical reaction and to block permeation of water vapor that can degrade the battery. In this work, two kinds of membranes are found effective for selective permeation of oxygen over water vapor: immobilized oil membrane, supported dense polymer thin-film membrane. The immobilized oil membrane is prepared by loading silicone oil into the pores of a thin (~50μm), porous support sheet of suitable pore sizes and surface property for retaining of the oil by capillary force. The supported thin-film membrane was prepared by coating a dense layer of PTFE (polytetrafluoroethylene) material (<10mm thick) on a thin porous metal sheet support (~50μm). Both membranes enable stable operation of a lithium/air battery in ambient air (~20% RH) in a much longer time than the conventional porous diffusion membrane. With attachment of a quality membrane on the cathode, the battery operated in ambient air even shows better performance than the same battery housed in a dry box. Systematic membrane preparation, characterization, and battery testing results will be presented. The membrane has large potential for other metal/air battery systems.

See more of this Session: Energy Storage and Electrochemical Processes II
See more of this Group/Topical: International Congress on Energy 2011

AUTHOR INDEX

AUTHOR INDEX

AUTHOR INDEX

AUTHOR INDEX

AUTHOR INDEX

AUTHOR INDEX

AUTHOR INDEX